LIGHT ON DARK MATTER

ASTROPHYSICS AND SPACE SCIENCE LIBRARY

A SERIES OF BOOKS ON THE RECENT DEVELOPMENTS
OF SPACE SCIENCE AND OF GENERAL GEOPHYSICS AND ASTROPHYSICS
PUBLISHED IN CONNECTION WITH THE JOURNAL
SPACE SCIENCE REVIEWS

VOLUME 124

PROCEEDINGS

LIGHT ON DARK MATTER

PROCEEDINGS OF THE FIRST IRAS CONFERENCE,
HELD IN NOORDWIJK,
THE NETHERLANDS, 10-14 JUNE 1985

Edited by

F. P. ISRAEL

Sterrewacht, Leiden, The Netherlands

D. REIDEL PUBLISHING COMPANY

A MEMBER OF THE KLUWER  ACADEMIC PUBLISHERS GROUP

DORDRECHT / BOSTON / LANCASTER / TOKYO

Library of Congress Cataloging in Publication Data

IRAS Conference (1st : 1985 : Noordwijk, Netherlands)
    Light on dark matter.

    (Astrophysics and space science library: v. 124)
    Includes indexes.
       1. Infrared astronomy–Observations–Congresses. 2. Infrared sources
–Observations–Congresses. 3. Infrared Astronomical Satellite–Congresses.
4. Astrometry–Congresses. 5. Cosmic dust–Observations–Congresses. 6.
Galaxies–Observations–Congresses. I. Israel, F. P. II. Title. III.
Series.
QB470.A1I73   1986   522'.68   86-6554
ISBN-13: 978-94-010-8577-9   e-ISBN-13: 978-94-009-4672-9
DOI: 10.1007/978-94-009-4672-9

Published by D. Reidel Publishing Company,
P.O. Box 17, 3300 AA Dordrecht, Holland.

Sold and distributed in the U.S.A. and Canada
by Kluwer Academic Publishers,
101 Philip Drive, Assinippi Park, Norwell, MA 02061, U.S.A.

In all other countries, sold and distributed
by Kluwer Academic Publishers Group,
P.O. Box 322, 3300 AH Dordrecht, Holland.

TABLE OF CONTENTS

(IR=Invited Review; W=Workshop Paper)

Editor's Preface and Workshop Preface   xv

Organization   xvii

List of Participants   xix

SECTION 1. **THE IRAS SURVEY**

A Statistical Analysis and Overview of   T. Chester   3
    the IRAS Point Source Catalog (IR)

Semi-Automated Identification of IRAS   A. Savage   23
    Point Sources Using UKST Plates   R.G. Clowes
    and the Cosmos Measuring Machine   M. Kalafi
                                            S.K. Leggett
                                            H.T. MacGillivray
                                            R.D. Wolstencroft

Submm Continuum Observations of Sources   R. Chini   29
    from the IRAS Point-Source   E. Kreysa
    Catalogue   E. Krügel
                                            P.G. Mezger
                                            H.-P. Gemünd

A Statistical Analysis of the LRS   F.M. Olnon   31
    Catalog (IR)

The Zodiacal Background in the IRAS   M.G. Hauser   39
    Data (IR)   J.R. Houck

Heliocentric Dependences of Zodiacal   R. Dumont   45
    Emission, Temperature and Albedo   A.C. Levasseur-Regourd

Spectral Decomposition of IRAS Maps   R. Braun   47
                                            R.G. Strom
                                            H. v.d. Laan
                                            H. Greidanus

Observations of Infrared Cirrus (IR)   T.N. Gautier   49

Dust at the North Galactic Pole   J. Knude   55

SECTION 2. **STARS AND STELLAR PHENOMENA**

IRAS Observations of Cool Excess around   F.C. Gillett   61
    Main Sequence Stars (IR)

The Flux Distribution of Vega for   S.K. Leggett   71
    10 μm < λ < 100 μm, and the
    Calibration of IRAS at 12 μm and
    25 μm

A Search for Infrared Excesses in   S.F. Odenwald   75
    G-Type Stars

IRAS Intrinsic Colours of Hot Stars   J. Coté   77
                                            L.B.F.M. Waters

The Infrared Excess from Stellar Winds   H.J.G.L.M. Lamers   79

The Disc Structure and Mass Loss Rates   L.B.F.M. Waters   83
    of Be Stars

IRAS Observations of Wolf-Rayet Stars   K.A. van der Hucht   87
                                            T.A. Jurriens
                                            F.M. Olnon
                                            P.S. Thé
                                            P.R. Wesselius
                                            P.M. Williams

Dust Formation in Wolf-Rayet Stellar   K.A. van der Hucht   90
    Winds (W)   P.M. Williams
                                            P.S. Thé

Observations of Young (Orion-Type)   H.J. Walker   91
    Stars with IRAS   P.L. Marsden

Mass Loss by Cool Stars (IR)   B. Zuckerman   93

Models of IRAS Observations of   M. Rowan-Robinson   101
    Circumstellar Shells   A. Lock
                                            D.W. Walker
                                            S. Harris

Luminosities of OH/IR Stars   J.H. Burger   103
                                            J. Herman

OH/IR Catalogue and Correlation with   R. Breukers   105
    the IRAS Data Base   W. van der Veen
                                              P. te Lintel
                                              M. Wiertz
                                              H. Habing

AGB Stars with High Mass Loss Rates in   W.E.C.J. van der Veen   107
    the Bulge of Our Galaxy

The Circumstellar Envelope of VX   J.M. Chapman   109
    Sagittarii   R.J. Cohen

The Internal Radius of CS Shells around   R. Papoular   111
    Cool, Oxygen-rich Stars (W)   B. Pegourie

IRAS Observations of Carbon Stars   F.J. Willems   113

From Miras to Planetary Nebulae: a   P.J. Bedijn   119
    Model of Mass Loss (IR)

Ground-based and IRAS Observations of   S. Kwok   127
    Proto-planetary Nebulae   B.J. Hrivnak
                                              E.F. Milone
                                              R.T. Boreiko

Spectra of Some IRAS Sources   J.W. Menzies   129
                                              P.A. Whitelock
                                              I.M. Coulson

IRAS Measurements of Planetary Nebulae   S.R. Pottasch   131
    (IR)

IR Observations of An Extended   A. Leene   143
    Planetary Nebula: NGC 7293 - the   S.R. Pottasch
    Helix Nebula

Novae Detected in the IRAS Point   H.L. Dinerstein   145
    Source Catalog   E.L. Robinson

IRAS Observations of Classical Novae   C.M. Callus   149
                                              J.S. Albinson
                                              A. Evans
                                              M.F. Bode

Collisional Heating of Dust in the 1985   J.S. Albinson   151
    Outburst of RS Ophiuchi   C.M. Callus
                                              A. Evans

Infrared Observations of Tycho Using   P.L. Marsden   153
    IRAS

Shock-heated Dust in Young Supernova   R. Braun   155
    Remnants   R.G. Strom
                                            H. v.d. Laan
                                            H. Greidanus

SECTION 3. **DUST GRAINS AND THEIR PROPERTIES**

Grains, What Do We Know? (IR)   H.C. van de Hulst   161

IRAS Cirrus Observations and the Nature   J.S. Mathis   171
    of Dust (IR)

Dust in Diffuse Clouds: One Stage in a   J.M. Greenberg   177
    Cycle

Infrared Extinction in Molecular   R. Hofmann   189
    Clouds: the Form of the Curve in   D.S. Davis
    Orion   H.P. Larson

UV Extinction as a Key to Grain Optical   G. Chlewicki   191
    Properties in the IR and UV   J.M. Greenberg

The Wavelength of Maximum Polarization   D.C.B. Whittet   197
    in the Chamaeleon Dark Cloud (W)   J.H. Hough
                                            J.A. Bailey
                                            M.F. Rouse
                                            T.M. Kirrane

Three Principal Heating Sources of Dust   P. Cox   201
    in the Galactic Disk   E. Krügel
                                            P.G. Mezger

Mid-IR Emission of the Interstellar   F. Boulanger   203
    Medium (W)   M. Pérault
                                            J.L. Puget

Optical Luminescence from Reflection   G. Olofsson   209
    Nebulae? (W)

Infrared Spectra and Dust Temperature   F.X. Désert   213
    Fluctuations (W)

Non-equilibrium Emission from Small   K. Sellgren   217
    Particles (W)

Evidence for a 12 Micron Water-ice   M. de Muizon   221
    Absorption Band in the IRAS LRS   L.B. d'Hendecourt
    Spectra of Protostars and Late Type   C. Perrier
    Stars

S_2 Formation in Interstellar Dust; a   R.J.A. Grim   225
    Diagnostic of the Maximum   J.M. Greenberg
    Aggregation Temperature for a Comet   L.J. van IJzendoorn
    (W)

Formation of Organic Molecules on   W. Schutte   229
    Interstellar Dust Particles (W)   J.M. Greenberg

Polycyclic Aromatic Hydrocarbons and   G.P. van der Zwet   233
    the Diffuse Interstellar Bands (W)   L.J. Allamandola

Identification of Polycyclic Aromatic   A. Léger   237
    Hydrocarbons (W)   L. d'Hendecourt

Silicate Absorption Strength; Polarized   D.K. Aitken   241
    Emission and Absorption by Aligned
    Grains (W)

Optical Properties of Simulated   J.R. Stephens   245
    Astrophysical Grains and Their   T.D. Kunkle
    Dynamics in the Near-earth   I.B. Strong
    Environment

Ultraviolet Photoprocessing and   L.B. d'Hendecourt   247
    Infrared Spectroscopy of Laboratory
    Simulated Grain Mantles (W)

Radiation Effects on Grain Materials   G. Strazulla   253
    (W)

Reflection Nebulae, Non-equilibrium   K. Sellgren   261
    Thermal Emission, and IRAS   L.J. Allamandola
                                            J.D. Bregman
                                            M.W. Werner
                                            D.H. Wooden

SECTION 4. INTERSTELLAR MEDIUM AND STAR FORMATION

Theories of Star Formation Confronted   B.G. Elmegreen   265
    by IRAS Data (IR)

IMF in Starburst Regions   H. Zinnecker   277

Point Sources in the Orion Complex (IR)   C.A. Beichman   279

The IR Emission of the Orion-Monoceros   F. Boulanger   293
    Molecular Clouds   R.J. Maddalena
                                            P. Thaddeus

Young Stars and High Density   G. Sandell   295
Condensations in the Horsehead Region   B. Reipurth
                                           C. Menten
                                           M. Walmsley
                                           H. Ungerechts

Mapping of the Coronae Austrinae Star   A. Evans   297
    Forming Region   J.S. Albinson
                                           M.F. Bode
                                           D.C.B. Whittet

Analysis of Point Sources in the   R.E. Jennings   299
    Ophiuchus and Perseus Clouds and   W. Cudlip
    CPC Observations of NGC 1333   C.J. Hirst
                                           D.H.M. Cameron

Star Forming Loops in the IRAS Sky   P.R. Schwartz   301
    Images

Models for IRAS Observations of   J. Crawford   303
    Galactic HII Regions   M. Rowan-Robinson

Far-infrared (100-200 µm) Photometry of   R.R. Daniel   305
    HII Regions with a 1m Balloon Borne   S.K. Ghosh
    Telescope   K.V.K. Iyengar
                                           T.N. Rengarajan
                                           S.N. Tandon
                                           R.P. Verma

Young Stars and Dense Cores in Nearby   P.C. Myers   307
    Dark Clouds (IR)

Water Masers Coincident with IRAS   J.G.A. Wouterloot   313
    Sources   C.M. Walmsley

IR CCD Imaging of L1551-IRS 5: Direct   A. Moneti   315
    Observations of Its Circumstellar   J.L. Pipher
    Shell   W.J. Forrest
                                           C.E. Woodward

A Model for Bipolar Sources in   M.D. Smith   319
    Molecular Clouds

Comparison of CO and IR Emission of   F. Boulanger   321
    IRAS Unidentified Sources   F. Casoli
                                           F. Combes
                                           Ch. Dupraz
                                           M. Gerin

IRAS Observations of Symbiotic Objects   P.A. Whitelock   323

A Large Scale OH Survey in Orion and   P.M.M. Jenniskens   325
    Monoceros   H.J. Habing
                                                 J.G.A. Wouterloot
                                                 P. te Linter-Hekkert
                                                 A. Blauw

SECTION 5.   **GALACTIC BULGE AND GALACTIC STRUCTURE**

The Galactic Distribution of Stellar   H.J. Habing   329
    Sources Found by IRAS (IR)

Variables, the Galactic Bulge and IRAS   M.W. Feast   339
    (IR)
Ground Based Observations of Nuclear   J.A. Frogel   349
    Bulge Stars (IR)

The Galactic Morphology of the   W.B. Burton   357
    Interstellar Dust Detected by IRAS   E.R. Deul
    (IR)   H.J. Walker
                                                 A.A.W. Jongeneelen

The Association of Clouds in the Carina   H.J. Walker   373
    Arm with IRAS Spline Maps   E.R. Deul
                                                 H.M. Butner
                                                 W.B. Burton

Dark Cloud Statistics   J.V. Feitzinger   375
                                                 J.A. Stüwe

High Latitude Molecular Clouds: L. Magnani   377
    Completeness of the Survey and   L. Blitz
    Implications for Molecular Surveys

Dust in High Velocity Clouds   B.P. Wakker   379

SECTION 6.   **GALAXIES**

IRAS Observations of the Magellanic   F.P. Israel   383
    Clouds   P.B. Schwering

Collisionally Heated Dust in LMC   J.R. Graham   397
    Supernova Remnants   W.P.S. Meikle
                                                 A. Evans
                                                 M.F. Bode
                                                 J.S. Albinson

Groundbased Infrared Observations of   J. Koornneef   399
    Magellanic Cloud HII Regions   F.P. Israel

Dust in M31 - Observations in the IR   R.A.M. Walterbos   401
                                              P.B.W. Schwering

Statistical Properties of IRAS Galaxies   G. Helou   405
    (IR)

Infrared Radiation from Normal Galaxies   E.E. Becklin   415
    (IR)

Models for IRAS Galaxies   M. Rowan-Robinson   421
                                              J. Crawford

The Identification of IRAS Galaxies   R.D. Wolstencroft   425
    (IR)   R.G. Clowes
                                              M. Kalafi
                                              S.K. Leggett
                                              H.T. MacGillivray
                                              A. Savage.

Mid Infrared Spectroscopy of IRAS   D.K. Aitken   435
    Bright Galaxies   P.F. Roche
                                              C. Smith

The Relationship between Blue and Fir   K.V.K. Iyengar   437
    Luminosities of Spiral Galaxies   T.N. Rengarajan
                                              R.P. Verma

Preliminary Results of an HI Survey of   J.Th. Armstrong   439
    a Sample of IRAS Galaxies   A. Wootten

Far-infrared Properties of Multiple   K.J. Fricke   441
    Nucleus Galaxies   W. Kollatschny

Young Supernovae in the Starburst   S.W. Unger   443
    Galaxy M82   A. Pedlar
                                              D.J. Axon
                                              P.N. Wilkinson
                                              P.N. Appleton

The Amazing Tail of NGC 2146   H.C.M. Caspers   445
                                              W.W. Shane

Starbursts in the Nuclei of Interacting   R.D. Joseph   447
    and Merging Galaxies (IR)

Starbursts in Non-interacting Galaxies   T.G. Hawarden   455
    (IR)   J.H. Fairclough
                                              R.D. Joseph
                                              S.K. Leggett
                                              C.M. Mountain

New Ground-based Studies of Two Active   C.G. Wynn-Williams   463
    IRAS Galaxies   E.E. Becklin
                                             D.L. DePoy
                                             J.N. Heasly
                                             G.J. Hill
                                             J.W. MacKenty
                                             C.A. Beichman

Spectroscopy of Active and Starburst   P.F. Roche   467
    Galaxies between 8-13 μm (W)

IRAS Observations of Active Galaxies -   G. Miley   471
    a Review (IR)   R. de Grijp

Seyfert Galaxies in the IRAS Survey and   I.S. Glass   487
    JHKL Photometry

The Most Luminous Galaxies   R.P. Norris   489
                                             P.F. Roche
                                             D.A. Allen

Submillimetre to Infrared Observations   L.M.J. Brown   491
    of Active Galaxies

Active Galactic Nuclei in the IRAS PSC   M. Dennefeld   493
                                             M.P. Veron-Cetty

SECTION 7. **COSMOLOGY**

Cosmological Results from IRAS (IR)   M. Rowan-Robinson   499

Evidence from IRAS Data for Large-scale   C.A. Collins   507
    Anisotropy in the Hubble Flow   R.D. Joseph
                                             N.A. Robertson

SUMMING-UP

Astronomy after IRAS (IR)   M.S. Longair   511

INDICES

Subject Index   525

Object Index   531

Author Index   537

EDITOR'S PREFACE

'Light on Dark Matter', held from 10-14 June 1985 in the Dutch
seaside resort of Noordwijk, was the first international conference
devoted to the results of the all-sky survey by the US-Dutch-UK
Infra-Red Astronomical Satellite (IRAS). As such, it was a hommage to
the scientists, engineers and technicians who conceived, built and
operated this extremely successful satellite. That this was generally
felt to be the case, was proven by the large number of participants
(over 200 from seventeen different nations), the lively discussions,
and the great variety of topics presented during the meeting. All
this not withstanding a typical Dutch summer: gale-force winds, heavy
cloud cover, and meter-high surf crashing onto a beach on which only
the hardy ventured. Most participants contented themselves by
watching the lonely seagulls patrolling the North Sea coastline
through the panoramic windows of the conference center.
    Parallel to the IRAS Conference, a Workshop on Infrared
Properties of Interstellar Grains was organized by J.M. Greenberg of
the Leiden Laboratory Astrophysics Group: a busy shuttling of
participants between the Workshop room and the Main Conference Hall
showed that many found it hard to choose.
    A large number of people were involved in making the Conference
a success: in the first place the scientific organizers with their
valuable advice and the conference speakers, among which I would like
to mention Dr. J.H. Van der Waals, who opened the Conference on
behalf of the Royal Dutch Academy of Sciences (KNAW), and Dr. F.J.
Low who delivered a magnificent Grand Conference Lecture on the
history of astronomical infrared research open to the public.
    Then there were the many participants putting up and explaining
their posters, the members of the local organizing committee
(especially Wanda van Grieken (Leiden) en Theo Jurriëns (Groningen)
who did a magnificent job of handling respectively the voluminous
administration and the complex finances) and the members of the local
assistance committee, many of the latter being students who did not
hesitate to sacrifice their spare time in order to run an almost
continuous minibus service between Noordwijk and Leiden. Thanks are
also due to the hotel Huis ten Duin personnel, who for the first time
since the renewal of their resort hosted an international conference
and to the staff of the Restaurant Allemansgeest who came up with a
superb conference diner. The conference poster was generously
supplied by Fokker Aeospace Industries, and I would like to thank in
particular Drs. R. van Duinen and B. Baud for their help and advice.
    Editing a conference proceedings is not easy! It always takes up
more time than even a pessimist would think. I would like to thank

all authors for their contributions, especially the few who actually
sent in their papers on time. At the conference, IRAS data were
liberally interpreted. Unfortunately, several authors also liberally
interpreted the guidelines for camera-ready copy. For this reason my
special thanks go to Anke van Vuuren and Marcel Bey for efficiently
dealing with a massive (re)typing job. Thanks also go to the Leiden
Astronomy Department and the Groningen Laboratory for Space Research
for supplying the necessary manpower and support.

        Finally, the Organizing Committee gratefully acknowledges
financial support by the Royal Dutch Academy of Sciences (KNAW), the
Leiden Kerkhoven-Bosscha Fund (LKBF), the Netherlands Agency for
Aerospace Programs (NIVR) and the Dutch Foundation for Space Research
(SRON).

They can be assured that the money spent has been paid back with
considerable scientific interest.

                            Frank P. Israël
                            Leiden, December 1985

                        WORKSHOP PREFACE

        The frequent "standing room only" attendant of the Interstellar
Dust Workshop was our measure of its success. We owe a debt of
gratitude to the speakers who inspired very lively discussions by
presenting clear and provocative lectures. Even though the topics
covered were quite varied, they maintained a clear connection with
many of the important aspects of the IRAS symposium. I should like
particularly to express my thanks to Liesbeth van der Poel for her
invaluable assistance in the organization of the meeting and to my
wife Naomi who added a tone of pleasant informality at our house by
arranging and giving a dinner for many of the symposium and workshop
participants.

                            J.M. Greenberg
                            Leiden, December 1985

ORGANIZATION

Conference Scientific Organizing Committee

H.J. Habing, Sterrewacht Leiden (chairman)
P. Clegg, Queen Mary College London
F.G. Gillett, KPNO Tucson
J.M. Greenberg, Astrophysics Lab. Leiden
J. Gunn, IAS Princeton
F.P. Israel, Sterrewacht Leiden
T. de Jong, Sterrenkundig Instituut Anton Pannekoek Amsterdam
J. Lequeux, Observatoire de Marseille
M.S. Longair, Royal Observatory Edinburgh
F.J. Low, Steward Observatory Tucson
G. Neugebauer, Caltech Pasadena

Local Organizing Committee

F.P. Israel, Sterrewacht Leiden (chairman)
W. van Grieken-Rückert, Sterrewacht Leiden (secretary)
M.H.K. de Grijp, Sterrewacht Leiden
P. Wesselius, Space Research Laboratory Groningen (treasurer)
P. te Lintel Hekkert, Sterrewacht Leiden
Th. Jurriens, Space Research Laboratory Groningen

Local Assistance Committee

E. Engelsman
H. Greidanus
H.J. van Langevelde
H.J. Latour
B. Liem
Nicole de Nies
R. Peletier
F. Steeman
W. Steemers
W.E.C.J. van der Veen

Workshop Organizers

J.M. Greenberg, Laboratory Astrophysics Leiden
E.H.J. van der Poel, Laboratory Astrophysics Leiden (secretary)

LIST OF PARTICIPANTS

J. Abolins, Ruth. Appleton Labs., Chilton, UK
S. Aiello, Universita di Firenze, Firenze, Italy
D.K. Aitken, Melbourne University, Victoria, Australia
J.S. Albinson, University of Keele, Keele, UK
D.A. Allen, Anglo-Australian observatory, Epping, Australia
J.T. Amstrong, NRAO, Charlottesville, USA
F. Baas, Lab. Astrophysics, Leiden, The Netherlands
J-P. Baluteau, Observatoire de Meudon, Meudon, France
B. Baud, Fokker, Schiphol-Oost, The Netherlands
E.E. Becklin, University of Hawaii, Honolulu, U.S.A.
P. Bedijn, Inst. für Theoretische Astroph. der Univ. Heidelberg, F.R.G.
C.A. Beichman, IPAC, ,Pasadena, U.S.A.
D.A. Beintema, Department of Space Research, Groningen, The Netherlands
J.A.M. Bleeker, Department of Space Research, Utrecht, The Netherlands
L. Blitz, University of Maryland, U.S.A.
J.B.G.M. Bloemen, Sterrewacht Leiden, The Netherlands
W. Boland, ASTRON, Den Haag, The Netherlands
A. Bonetti, Universita di Firenze, Firenze, Italy
P.B. Bosma, Vrije Universiteit, Amsterdam, The Netherlands
F. Boulanger, Goddard Institute for Space Studies, New York, U.S.A.
J. Brand, Sterrewacht Leiden, The Netherlands
R. Braun, Sterrewacht Leiden, The Netherlands
R. Breukers, Sterrewacht Leiden, The Netherlands
K. Brink, Universiteit van Amsterdam, The Netherlands
L. Brown, School of Physics and Astronomy, Preston, UK
G. Burbidge, University of California San Diego, La Jolla, U.S.A.
W.B. Burton, Sterrewacht Leiden, The Netherlands
R. Buser, University of Basel, Binningen, Switzerland
H.M. Butner, Sterrewacht Leiden, The Netherlands
C. Callus, University of Keele, UK
R. Cameron, Mt. Stromlo Observatory, Australia
Casoli, Observatoire de Meudon, France
Castets, Groupe d'Astrophysique, St. Martin d'Heres, France
E. Caux, Centre d'étude Spatiale, Toulouse, France
C. Cesarsky, Serv. d'Astroph. C.E.N. Saclay, Gif sur Yvette Cedex, France
A. Chalabaev, ESO, Santiago, Chile
J. Chapman, Jodrell Bank, Cheshire, UK
R. Chini, Max-Planck Institut für Radioastronomie, Bonn, F.R.G.
G.P. Chlewicky, Lab. Astrophysics, Leiden, The Netherlands
J. Clavel, ESA Satellite Tracking Station, Madrid, Spain
R.S. Cohen, Institute for Space Studies, New York, U.S.A.
J. Coté, Space Research Lab., Utrecht, The Netherlands

P. Cox, Mac Planck Inst. für Radioastronomie, Bonn, F.R.G.
J. Crawford, Queen Mary College, London, UK
L. Deharveng, Observatoire de Marseille, France
M. Dennefeld, Institut d'Astrophysique, Paris, France
F.X. Désert, Lab. de Physique de l'Ecole Nor. Sup., Paris, France
E. Deul, Sterrewacht Leiden, The Netherlands
H.L. Dinerstein, University of Texas at Austin, U.S.A.
S. Drapatz, Max-Planck Inst. für Extraterr. Phys., Garching bei München,
        F.R.G.
R. Dumont, Observatoire, Floriac, France
G.M.R. Duvert, Sterrewacht Leiden, The Netherlands
B. Elmegreen, IBM Thomas J. Watson Research Center, Yorktown Heights, U.S.A.
E. Engelsman, Sterrewacht Leiden, The Netherlands
N. Epchtein, Observatoire de Meudon, France
R. Ewald, Physikalisches Institut der Universität zu Köln, Köln, F.R.G.
J. Fairclough, Rutherford/Appleton Lab., Chilton, UK
M. Feast, South African Astron. Observatory, Cape Town, South Africa
J.V. Feitzinger, Ruhr Universität Bochum, F.R.G.
M. Fischer, Lab. Astrophysics, Leiden, The Netherlands
P. Fontanelli, Observatoire de Meudon, France
A. Francheschini, Inst. de Astronomia, Padua, Italy
K.J. Fricke, Universitäts-Sternwarte, Göttingen, F.R.G.
U. Frisk, ESTEC, Noordwijk, The Netherlands
J. Frogel, Kitt Peak National Observatory, Tucson, U.S.A.
T.N. Gautier, IPAC, Pasadena, U.S.A.
D. Gezari, NASA Goddard Space Flight Center, Greenbelt, U.S.A.
F.C. Gillett, Kitt Peak National Observatory, Tucson, U.S.A.
I.S. Glass, South African Astron. Observatory, Cape Town, South Africa
W.M. Goss, Kapteyn Laboratorium, Groningen, The Netherlands
Th. de Graauw, Space Research Laboratory, Groningen, The Netherlands
J.M. Greenberg, Lab. Astrophysics, Leiden, The Netherlands
H. Greidanus, Sterrewacht Leiden, The Netherlands
M.H.K. de Grijp, Sterrewacht Leiden, The Netherlands
R. Grim, Lab. Astrophysics, Leiden, The Netherlands
M.S. de Groot, Lab. Astrophysics, Leiden, The Netherlands
B. Gustaffson, Stockholm Observatory, Saltsjobaden, Sweden
H.J. Habing, Sterrewacht Leiden, The Netherlands
L. Haikala, University of Helsinki, Finland
A. Harris, ESA Satellite Tracking Station, Madrid, Spain
D. Hartmann, Sterrewacht Leiden, The Netherlands
T.G. Hawarden, Royal Observatory, Edinburgh, UK
C. Heiles, University of California, Berkeley, U.S.A.
G. Helou, J.P.L. Caltech, Pasadena, U.S.A.
L.B. d'Hendecourt, Université de Paris, France
J. Herman, ESTEC, Noordwijk, The Netherlands
W. Hermsen, Space Research Laboratory, Leiden, The Netherlands
R. Hofmann, M.P.I. für Phys. und Astroph., Garching bei München, F.R.G.
J.R. Houck, Cornell University, Ithaca, U.S.A.
K.A. van der Hucht, Space Research Laboratory, Utrecht, The Netherlands
H.C. van de Hulst, Sterrewacht Leiden, Leiden, The Netherlands
R.M. Humphreys, University of Minnesota, Minneapolis, U.S.A.

L. K. Hunt, Osservatorio Astrofisico de Arcetri, Firenze, Italy
F.P. Israel, Sterrewacht Leiden, The Netherlands
R.E. Jennings, University College London, London, UK
P. Jenniskens, Sterrewacht Leiden, The Netherlands
A.P. Jones
A. de Jong, Space Research Laboratory, Groningen, The Netherlands
T. de Jong, Universiteit van Amsterdam, The Netherlands
A.A.W. Jongeneelen, Sterrewacht Leiden, The Netherlands
R.D. Joseph, Imperial College, London, UK
T. Jurriens, Space Research Laboratory, Groningen, The Netherlands
T. Kamperman, Space Research Laboratory, Utrecht, The Netherlands
M.F. Kessler, ESTEC, Noordwijk, The Netherlands
U. Klaas, Max-Planck Institut für Astronomie, Heidelberg, F.R.G.
B. Kneissel, Ruhr-Universität Bochum, F.R.G.
J. Knude, University Observatory, Kopenhagen, Denmark
W. Kollatschny, Universitäts-Sternwarte, Göttingen, F.R.G.
J. Koornneef, ESA/Space Tel. Science Institute, Baltimore, U.S.A.
A. Krabbe, Max-Planck Institut für Astronomie, Heidelberg, F.R.G.
W. Krätschmer, Max-Planck Institut für Kernphysik, Heidelberg, F.R.G.
J. Krelowski, Institute of Astronomy, Torun, Poland
C.K. Kumar, Howard University, Washington, U.S.A.
S. Kwok, University of Calgary, Canada
H. van der Laan, Sterrewacht Leiden, The Netherlands
J.P.J. Lafon, Observatoire de Meudon, France
P-O. Lagage, Service d'Astrophysique C.E.N. Saclay, Gif sur Yvette, France
Ph. Lamy, Lab. d'Astronomie Spatiale, Marseille, France
H.J. Langevelde, Sterrewacht Leiden, The Netherlands
H.P. Larson, Max-Planck Institut für Physik, Garching bei München, F.R.G.
R.J. Laureijs, Kapteyn Laboratorium, Groningen, The Netherlands
T.J. Lee, Royal Observatory, Edinburgh, UK
A. Leene, Kapteyn Laboratorium, Groningen, The Netherlands
A. Leger, Université de Paris, Paris, France
S.K. Leggett, Royal Observatory, Edinburgh, UK
P.J. Lena, Observatoire de Meudon, France
P. Lenzini, Osservatorio di Arcetri, Firenze, Italy
B. Liem, Sterrewacht Leiden, Leiden, The Netherlands
Th.J. van der Linden, Universiteit van Amsterdam, The Netherlands
P. te Lintel, Sterrewacht Leiden, The Netherlands
M.S. Longair, Royal Observatory, Edinburgh, UK
D. Lorenzetti, C.N.R., Frascati, Italy
F. Low, Univ. of Arizona, Tucson, U.S.A.
J. Lub, Sterrewacht Leiden, The Netherlands
L. Magnani, University of Maryland, College Park, U.S.A.
G. Magni, Instituto Astrofisica Spaziale, Roma, Italy
P.L. Marsden, University of Leeds, UK
P. Martin, University of Toronto, Canada
J.S. Mathis, University of Wisconsin, Madison, U.S.A.
K. Matilla, University of Helsinki, Finland
P.G. Mezger, Max-Planck Institut für Radioastronomie, Bonn, F.R.G.
A. Monetti, Osservatorio Astrofisico Arcetri, Firenze, Italy
J. Monin, University of Grenoble, St. Martin d'Heres, France

A. Moorwood, ESO, Garching bei München, F.R.G.
M. Mountain, Royal Observatory, Edinburgh, UK
M. de Muizon, Sterrewacht Leiden, The Netherlands
P.C. Myers, Center for Astrophysics, Cambridge, U.S.A.
Nan Sheng Shao, Lab. Astrophysics, Leiden, The Netherlands
S. Odenwald, Naval Res. Lab., Washington, U.S.A.
F.M. Olnon, Radiosterrenwacht Dwingeloo, The Netherlands
G. Olofsson, Stockholm Observatory, Saltsjobaden, Sweden
O. Omont, Astrophysique CERMO, Saint-Martin d'Heres Cedex, France
J.H. Oort, Sterrewacht Leiden, The Netherlands
R. Papoular, Serv. d'Astroph. C.E.N. Saclay, Gif sur Yvette Cedex, France
R. Peletier, Sterrewacht Leiden, The Netherlands
Peppel, Max-Planck Inst. für Extraterr. Phys., Garching bei München, F.R.G.
M. Perault, Observatoire de Meudon, France
R.S. Le Poole, Sterrewacht Leiden, The Netherlands
S. Pottasch, Kapteyn Laboratorium, Groningen, The Netherlands
A. Preite-Martinez, Instituto Astrofisica Spaziale, Frascati, Italy
J.-L. Puget, Observatoire de Meudon, France
M. Raharto, Sterrewacht Leiden, The Netherlands
E. Raimond, Radiosterrenwacht Dwingeloo, The Netherlands
W.M. Reich, Max Planck Institut für Radioastronomie, Bonn, F.R.G.
B. Reipurth, University Observatory, Kopenhagen, Denmark
P. Richards, Rutherford/Appleton Lab. Chilton, UK
S.T. Ridgway, Kitt Peak National Observatory, Tucson, U.S.A.
J.K. Ridley, UK
P.F. Roche, Anglo-Australian Observatory, Epping, Australia
C. Rogers, University of Toronto, Canada
H. Röttgering, Sterrewacht Leiden, The Netherlands
M. Rowan-Robinson, Queen Mary College, London, UK
M. Salvati, Instituto di Astrofisica Spaziale, Frascati, Italy
G. Sandell, University of Helsinki, Finland
L. Sanz Fernandez de Cordoba, ESA Satellite Tracking Stations, Madrid, Spain
A. Savage, Royal Observatory, Edinburgh, U.K.
J. Schmid-Burgk, Max-Planck Institut für Radioastronomie, Bonn, F.R.G.
W. Schutte, Lab. Astrophysics, Leiden, The Netherlands
P.R. Schwartz, Naval Research Laboratory, Washington, U.S.A.
P. Schwering, Sterrewacht Leiden, The Netherlands
K. Sellgren, University of Hawaii, Honolulu, U.S.A.
G. Serra, C.E.S.R., Toulouse, France
W.W. Shane, Sterrenkundig Instituut, Nijmegen, The Netherlands
F. Sloff, Sterrewacht Leiden, The Netherlands
R. Shubert, California State University, Fullerton, U.S.A.
C. Smith, Melbourne University, Victoria, Australia
M.D. Smith, University of Leicester, England
C. Snyder, J.P.L., Pasadena, U.S.A.
H. van de Stadt, Sterrewacht "Sonneborgh", Utrecht, The Netherlands
F. Steeman, Sterrewacht Leiden, The Netherlands
W. Steemers, Sterrewacht Leiden, The Netherlands
J.R. Stephens, Los Alamos National Laboratory, Los Alamos, U.S.A.
G. Strazzulla, Osservatorio Astrofisico di Cantania, Cantania, Italy
W.M. Tobin, University of Oxford, UK

S.W. Unger, Jodrell Bank, Cheshire, UK
V. Ungerer, Space Research Laboratory, Groningen, The Netherlands
M.S. Vardya, Tata Inst. of Fundamental Research, Bombay, India
I. Vauglin, Observatoire de Lyon, St. Genis Laval, France
W.E.C.J. van der Veen, Sterrewacht Leiden, The Netherlands
K. Vedi, Queen Mary College, London, U.K.
R.P. Verma, Tata Inst. of Fundamental Research, Bombay, India
F. Viallefond, Observatoire de Meudon, Meudon, France
J. de Vries, ESTEC, Noordwijk, The Netherlands
C.P. de Vries, Sterrewacht Leiden, The Netherlands
J.M. Waalwijk, Philips Persdienst, Eindhoven, The Netherlands
H.J. Walker, Sterrewacht Leiden, The Netherlands
R.A.M. Walterbos, Sterrewacht Leiden, The Netherlands
L.B.F.M. Waters, Space Research Laboratory, Utrecht, The Netherlands
N. van Weeren, Sterrewacht Leiden, The Netherlands
P. Weissman, JPL, Pasadena, U.S.A.
P. van der Werf, Kapteyn Laboratorium, Groningen, The Netherlands
P. Wesselius, Space Research Laboratory, Groningen, The Netherlands
P.A. Whitelock, South African Astron. Observatory, Cape Town, South Africa
D.C.B. Whittet, Preston Polytechnic, Preston, U.K.
F. Willems, Universiteit van Amsterdam, The Netherlands
S.P. Willner, Center for Astrophysics, Cambridge, U.S.A.
A. Winnberg, Onsala Space Observatory, Onsala, Sweden
H. van Woerden, Kapteyn Laboratorium, Groningen, The Netherlands
R.D. Wolstencroft, Royal Observatory, Edinburgh, U.K.
J. Wouterloot, Max-Planck Institut für Radioastronomie, Bonn, F.R.G.
G. Wynn-Williams, University of Hawaii, Honolulu, U.S.A.
H. Zinnecker, Royal Observatory, Edinburgh, U.K.
B.M. Zuckerman, University of California, Los Angeles, U.S.A.
G. van de Zwet, Lab. Astrophysics, Leiden, The Netherlands

SECTION 1.

THE IRAS SURVEY

A STATISTICAL ANALYSIS AND OVERVIEW OF THE IRAS POINT SOURCE CATALOG

T. Chester
IPAC, 100-22
California Institute of Technology
1201 E. California Street
Pasadena, California 91109
U.S.A.

Abstract. The IRAS Point Source Catalog can be segregated into stars, "IR galactic components", galaxies, cirrus, and "disk and bulge stars" categories. For each source type, the types and distances of the sources are discussed, and the sky distribution, log N/log S, and identification rate are given.

1.   INTRODUCTION

The IRAS Point Source Catalog contains 245,839 sources, making it one of the the largest collections of astronomical sources with accurate positions and measured fluxes. This paper attempts to describe the statistical properties of the major source categories that constitute the bulk of the catalog. I also present some of the general characteristics of the catalog.

The general content of the catalog can be readily seen in the color equal area plot of the catalog shown in Figure 1. In this figure, 12 μm emission has been coded blue, 60 μm emission has been coded green, and 100 μm emission has been coded red. Note the tremendous concentration of sources to a thin galactic plane in the inner quadrants of the Galaxy (± 90° longitude and ± 10° latitude). This implies that the catalog contains a significant proportion of sources that can be seen through most of the Galaxy. In fact, half of all the catalog sources are in the 12% of the sky that is within 7° of the galactic plane, and two thirds of all sources are in the half of the sky within ± 90° longitude of the galactic center. If the scale height of the sources is less than one kpc, then taking 7° as an angle to one scale height implies a distance greater than 8 kpc. Only 15% of the sources are found in the half of the sky above 30° latitude.

Roughly 55% of IRAS sources are stars, seen as blue objects in Figure 1, and are strongly concentrated to the plane. The next largest category are the yellow objects, which are galaxies and the "stuff that galaxies are made of" in the infrared, such as HII regions, molecular clouds, plantetary nebulae, and star formation

3

F. P. Israel (ed.), Light on Dark Matter, 3–22.

Fig. 1. Equal area all-sky plot of the IRAS Point Source Catalog in galactic coordinates. 12 μm emission is coded blue, 60μm emission green, and 100 μm emission red. Black areas were not surveyed.

regions. These objects comprise 25% of the catalog, with galaxies 10% of the catalog and Galactic components 15% of the catalog. At high galactic latitudes, these objects are almost entirely galaxies, and are uniformly distributed. At lower galactic latitudes, these objects are usually IR components of the Galaxy, and are fairly well concentrated to the plane. "Infrared cirrus", that wispy filamentory component of the local interstellar medium, contributes almost 15% of the catalog sources. Cirrus is red in Figure 1, and is spread almost uniformly between galactic latitudes of ± 50° with only a slight concentration to the plane. Finally, disk and bulge stars (non-blue stellar sources) constitute 10% of the catalog. Due to the tight concentration to the galactic plane and bulge, these objects cannot be seen in Figure 1. (Overlap between categories and rounding makes the total 105%. See Table 1 for a more accurate breakdown.)

Table 1

SOURCE TYPES AND NUMBER IN THE CATALOG

Type	Approximate Number	Percentage
Stars		
Photospheres	18,000	7
Circumstellar Dust Shells	57,000	23
Faint Stars	59,000	24
Total	134,000	55
IR Galactic Components (HII Regions, Plantetary Nebulae, Molecular Clouds, star formation regions, etc.)	35,000	14
Cirrus sources	33,000	13
Disk and Bulge Stars	24,000	10
Galaxies	22,000	9
Total	245,839	100

2.   GENERAL CHARACTERISTICS OF THE POINT SOURCE CATALOG

The catalog covers 96% of the sky, with half of the unsurveyed sky concentrated into the two gaps readily seen in Figure 1. The remainder of the unsurveyed sky is scattered throughout the sky, with some concentration in the southern sky. The coverage of the surveyed sky was fairly uniform - 72% of the sky received three or more

separate coverages, whereas 24% of the sky received only two separate
coverages.

Much more important than any non-uniformity of sky coverage was
the non-uniformity imposed by the infrared sky itself. The sky was
often confused to our instrument within 5° of the galactic plane at
12-60 μm, and in patches up to 50° galactic latitude at 100 μm.
Sources had to meet higher standards to get in the catalog in these
regions, complicating detailed statistical analysis.

Outside confused regions, the completeness and reliability of
the catalog is high. The completeness is essentially unity above 0.4,
0.5, and 0.6 Jy at 12, 25, and 60 μm, respectively, and falls sharply
below those values. (All flux densities quoted in this paper are not
corrected for color. See Explanatory Supplement 1985.) The
completeness at 100 μm is probably unity above 1.6 Jy away from
cirrus regions, but the threshold is higher over most of the sky. The
reliability is outstanding, well over .998 in every flux range, as
long as one is aware that IRAS only required that point sources be
one-dimensional point sources. Thus cirrus ridges often show up as
strings of point sources in the catalog. A point source to IRAS is
one less than roughly 0.5, 0.5, 1.0, 2.0' in size (in-scan) at 12,
25, 60, and 100 μm, respectively.

The amount of spectral information in the catalog is given in
Table 2. Data are displayed separately for all sources with any high
or moderate quality flux in a band and for all sources with any high
or moderate quality flux above the thresholds of 0.8, 1.0, 1.2, and
2.0 Jy at 12, 25, 60, and 100 μm, respectively. The latter criterion
removes the dual problems of overestimation of fluxes near threshold
and the larger uncertainties of such fluxes. (Although even higher
thresholds should probably be used within a few degrees of the
galactic plane.)

Table 2

SPECTRAL INFORMATION IN THE CATALOG

No. of Adjacent bands	No. of Sources in Catalog	%	No. At 2 x Flux Threshold	%
1	126,486	52	39,448	16
2	95,904	39	30,181	12
3	17,106	7	6,622	3
4	6,343	3	3,584	1

The positional accuracy of the IRAS sources can be approximated
by a Gaussian about two orthogonal axes. The One Sigma uncertainties
range from 2-4" on one axis and 8-16" on the other. However, the

distribution on the larger axis is not quite Gaussian. Most sources
have a somewhat smaller uncertainty, but there is a small tail of a
few sources with a significantly larger uncertainty which pumps up
the overall sigma.

The photometric accuracy is of the order of 5-15%, depending on
source flux and band.

3.   SOURCE CLASSES

Figure 2 shows and infrared color- color plot for all IRAS sources
above galactic latitude 50° that have 12, 25, and 60μm flux densities
above two times the flux thresholds in each band. Three source
classes are clearly defined:

1. PHOTOSPHERES, stellar sources with $f_\nu(12) > 3 f_\nu(25)$ $(m_{12}-m_{25}$
   $< .37)$ which have naked photospheres (little circumstellar
   dust) with blackbody temperatures above 1500 K. Most of these
   sources hve effective temperatures greater than 2500 K, which
   gives them $f_\nu(12) \sim 4 f_\nu(25)$ for all temperatures due to the
   insensitivity of the IRAS bands to temperatures above 2000 K.

2. Circumstellar Dust Shells (CDS), i.e., stellar sources with
   circumstellar dust shells, which
   produce $f_\nu(25) < f_\nu(12) < 3 f_\nu(25)$ over most of the sky $(.37$
   $< m_{12} - m_{25} < 1.4)$.

3. GALAXIES, with $f_\nu(25) < f_\nu(60)$ $(m_{25} - m_{60} > 1.88)$.

Figure 3 shows the same color-color plot for the entire sky, which
adds the following source classes.

4. IR GALACTIC COMPONENTS, which now dominate the galaxy region
   of the color-color diagram, and are apparently responsible
   for the colors seen in external regions. Most of these
   objects are tightly restricted to the galactic plane.

5. DISK AND BULGE STARS, which continue the CDS region of the
   color-color diagram below the $f_\nu(12) = f_\nu(25)$ line. These
   stars constitute the bulk of the "other" class mentioned in
   the Explanatory Supplement, and in fact are defined here in
   the same way, as all sources which do not fit in any of the
   other categories. While there is no apparent distinct region
   in the color-color diagram delimiting this class, its sudden
   appearance below 10° galactic latitude and apparently unique
   restriction to the bluge and disk argue for a separate class.

Note also that the ratio of numbers of CDS to photospheres has
changed dramatically at lower latitudes. Finally, a sixth source
class exists which does not appear in these plots.

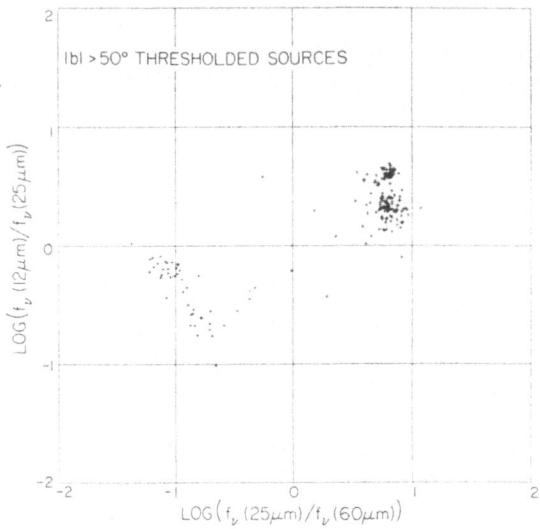

Fig. 2. Infrared color-colot plot for all 12-25-60 μm sources with
        fluxes above 0.8, 1.0, and 1.2 Jy, respectively above
        galactic latitude 50°

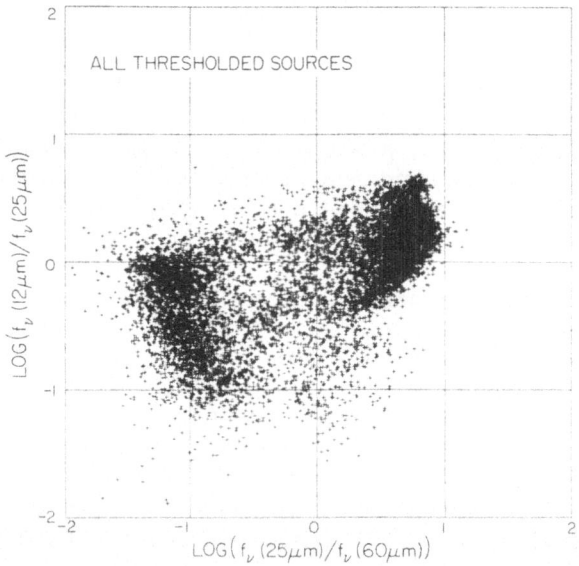

Fig. 3. Infrared color-color plot for all 12-25-60 μm sources with
        fluxes above 0.8, 1.0, and 1.2 Jy, respectively, for all
        galactic latitudes.

6. CIRRUS, which are defined as 100 μm only objects. These objects are sometimes true condensations in the wispy cirrus filaments, but sometimes are only "one dimensional point sources" produced by IRAS scans across the axis of a cirrus filament.

These objects are discussed more fully below. Table 3 presents typical density ranges for each component, averaged over galactic longitude.

Table 3

DENSITY RANGES FOR EACH SOURCE CLASS AVERAGED OVER GALACTIC LONGITUDE

CLASS	DENSITY PER SQUARE DEGREE	
	MIN	MAX[*]
Stars		
Photospheres	0.1	1.1
CDS	0.05	12
Faint Stars	0.3	5
Total	0.5	18
IR Galactic Components	0	11
Cirrus Sources	0	2
Disk and Bulge Stars	0	9
Galaxies	0.6	0.6

[*] Maximum is set by high source density processing in some cases.

Finally, Figure 4 presents the association rate with other catalogs for each component.

4. Stellar Sources

Stellar sources are defined here as having a 12 μm detected flux density (high of moderate quality) with $f_\nu(12) > f_\nu(25)$. The disk and bulge population, with $f_\nu(12) < f_\nu(25)$ are also probably stellar, but are discussed separately below. Stellar sources can be divided into three categories, based upon the status and value of their 25 μm flux. PHOTOSPHERES have a detected 25 μm flux, with $f_\nu(12) > 3 f_\nu(25)$ ($m_{12} - m_{25} < .37$). Almost all such sources probably have $f_\nu(12) \leqq 4 f_\nu(25)$, but their observed density falls off at a ratio of

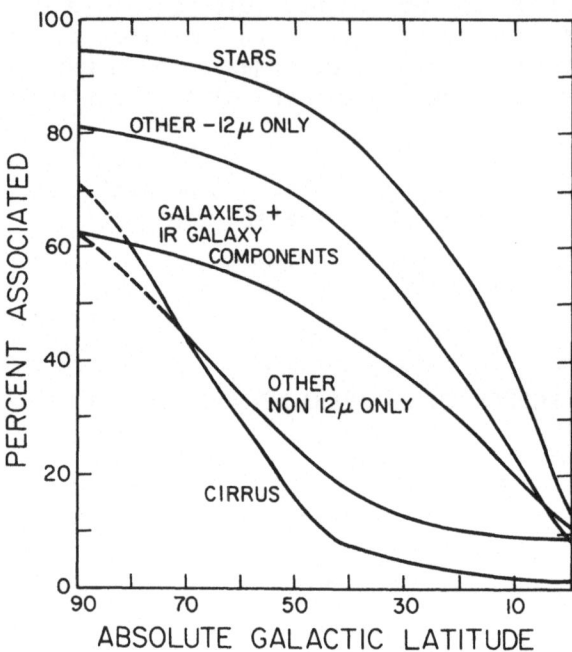

Fig. 4. Association rate, as given in the IRAS catalog, versus
        galactic latitude for the various source classes. The 12μm
        only sources have been separated from the OTHER catagory
        (disk and bulge stars) in the figure.

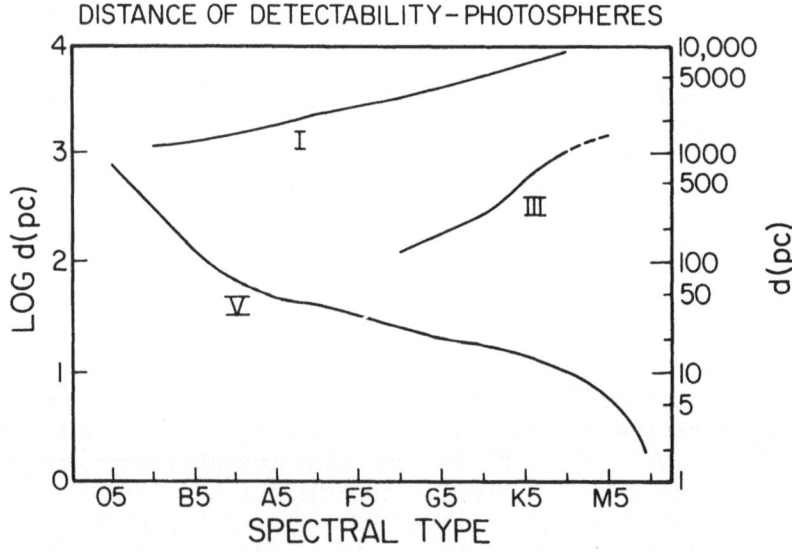

Fig. 5. Limiting distance at which stars with purely photospheric
        emission would have been present in the catalog.

3, which includes the IRAS photometric errors. CDS, or circumstellar dust shells, have a detected 25μm flux, with $3\ f_\nu(25) > f_\nu(12) > f_\nu(25)$ $(.37 < m_{12} - m_{25} (1.4)$ At greater than 5° from the galactic plane, almost no stellar sources have $f_\nu(12) < f_\nu(25)$, which accounts for that otherwise arbitrary cutoff. Very few sources are observed by IRAS with $f_\nu(12) \simeq 3\ f_\nu(25)$, which corresponds to a color temperature of 1100 K. This is caused primarily by the fact that the color points $f_\nu(12) = 4\ f_\nu(25)$ and $f_\nu(12) = 2\ f_\nu(25)$ are accumulation points - i.e., all stars with $T_{eff} > 2000$ have $f_\nu(12) = 4\ f_\nu(25)$ and all stars with thick dust shells tend to have color temperatures of 300-600 K. Also, it does not take much optical depth to quickly move a star from photospheric ratios to one with typical CDS ratios. There is also a selection effect in that IRAS preferentially picks up the more luminous CDS stars with thicker shells, depopulating the intermediate color region. Finally, those objects with no 25 μm detected flux but with $f_\nu(12) > f_\nu(25)$ (upper limit) are assigned to the class FAINT STARS.

a) PHOTOSPHERES

Figure 5 presents the distances to which IRAS would have detected a star with pure photospheric emission (no CDS). The assumptions used were $m_{12} < 5$, $M_{12} - M_V$ given by the blackbody function, and M_V taken from Allen (1972). These inputs probably cause no more than one magnitude inaccuracy in the results, or 50% in distance. One can use these distances along with the densities of different spectral types given in Allen (1972) to estimate that about 1000-5000 main sequence dwarfs should be in the catalog, and only 100-500 bright enough to have a 25 μm detected flux to be in the photosphere category here. The overwhelming majority of the IRAS stars are thus late spectral type M except for the fainter stars at high galactic latitudes. In those areas we have almost completely run out of M stars as we see significantly beyond their scale heights (see Figure 6).

If the scale heights of late-type giants are 250 pc (Bahcall and Soneira 1980), then at 1.5 Jy, the completeness limits of the photospheres class, that class contains objects more distant than a scale height for spectral types later than K5. Thus one expects the photospheric class to concentrate somewhat towards the galactic plane. This expectation is borne out by Figure 7, which shows the spatial distribution of photospheres.

The log N/log s plots also substantiate that most of the photospheres are confined to a disk of height smaller than the limiting distance to which we can see them. Figure 8 shows the log N/log S plot for two ranges of galactic latitude - above 50% and from 15° to 30°. Both have a slope only slightly greater than 1.0, confirming that we are running out of these sources perpendicular to the plane. Note that above 50°, the log N/log S curve is depressed below 4 Jy. The slope does not differ significantly from 1.1 until 2° galactic latitude, below which the slope is 1.3. This is consistent with a scale height of 250 pc and a limiting distance of 7 kpc, implying that the class is dominated by giants in the plane.

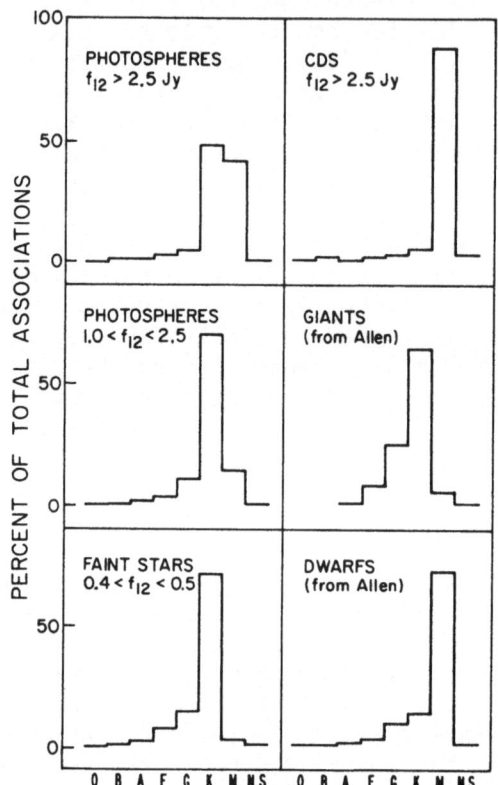

Fig. 6. Histogram of associations with SAO stars versus spectral
class for the three classes of stars defined in the text, for
galactic latitudes above 50°. Note that although the SAO
catalog theoretically begins to be incomplete for M5 stars
below 1 Jy, probably few such stars are actually missed
because that flux corresponds to distances well beyond the
scale height of M giants.

b) Circumstellar Dust Shells (CDS)

Figure 9 shows the distances to which IRAS would have detected a star
that is surrounded by a thick dust shell that absorbs all the stellar
radiation and reradiates it at 500 K. This should give a rough upper
limit to the distance to which any stellar source could be detected.
Note that all supergiants in our galaxy would have been detected if
they have these thick shells around them. The bulk of the CDS stars
are probably late-type giants, which we detect to distances of order
the distance to the galactic center. Figure 6 shows the distribution
of SAO spectral types associated with CDS stars at high galactic

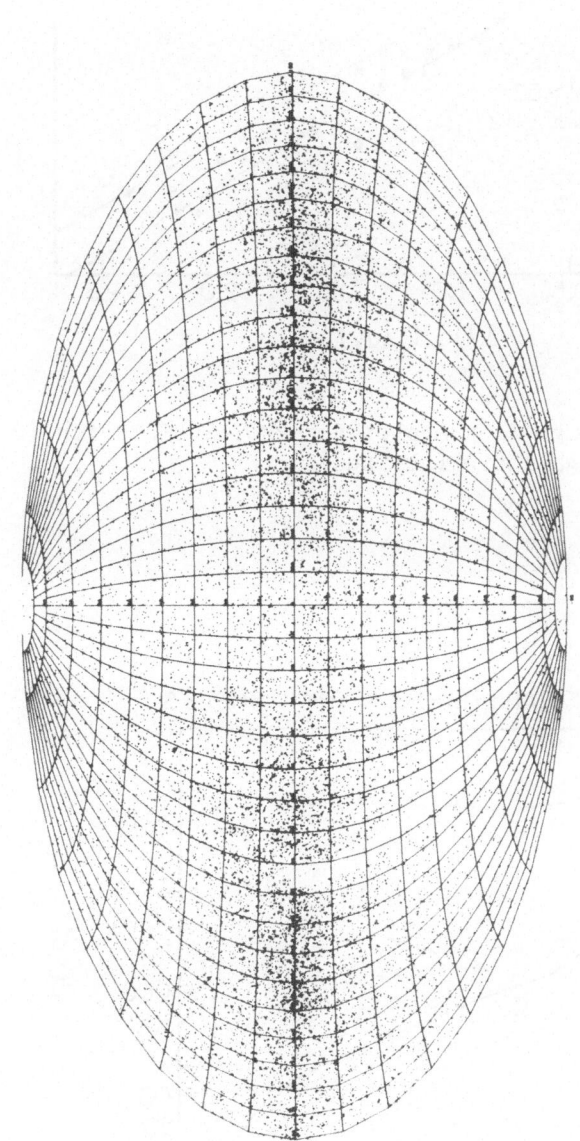

Fig. 7. Equal area all-sky plot of PHOTOSPHERES; Note that the apparent density decrease in the galactic plane near the galactic center is probably due to higher thresholds imposed in that area.

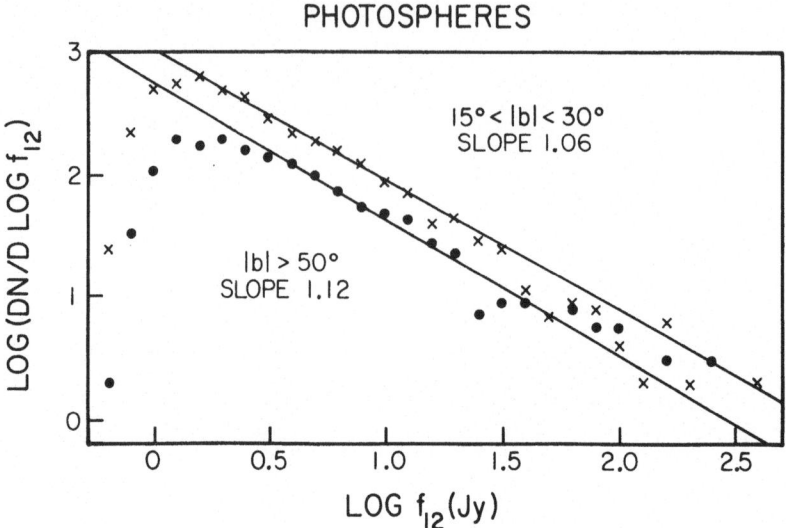

Fig. 8. Differential log N/log S plot for PHOTOSPHERES for 12 μm for
two ranges of galactic latitude.

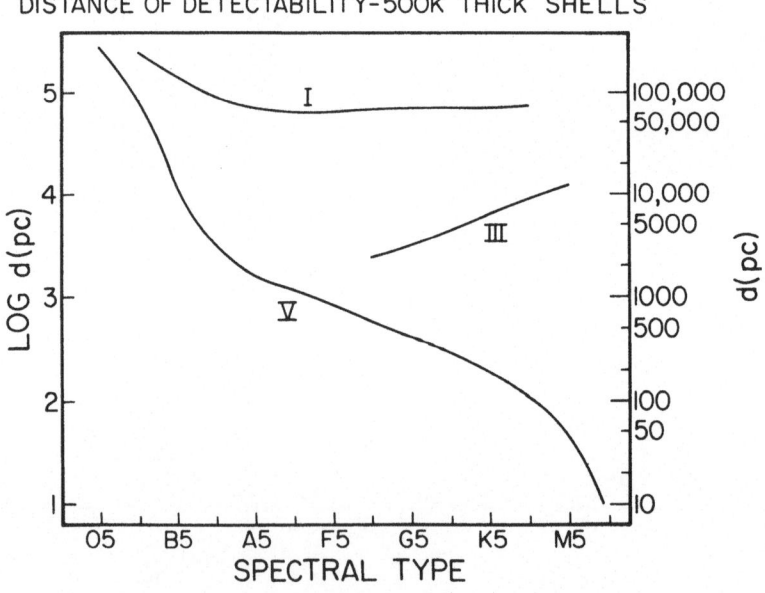

Fig. 9. Limiting distance at which stars with thick 500 K dust shells
would have been present in the catalog.

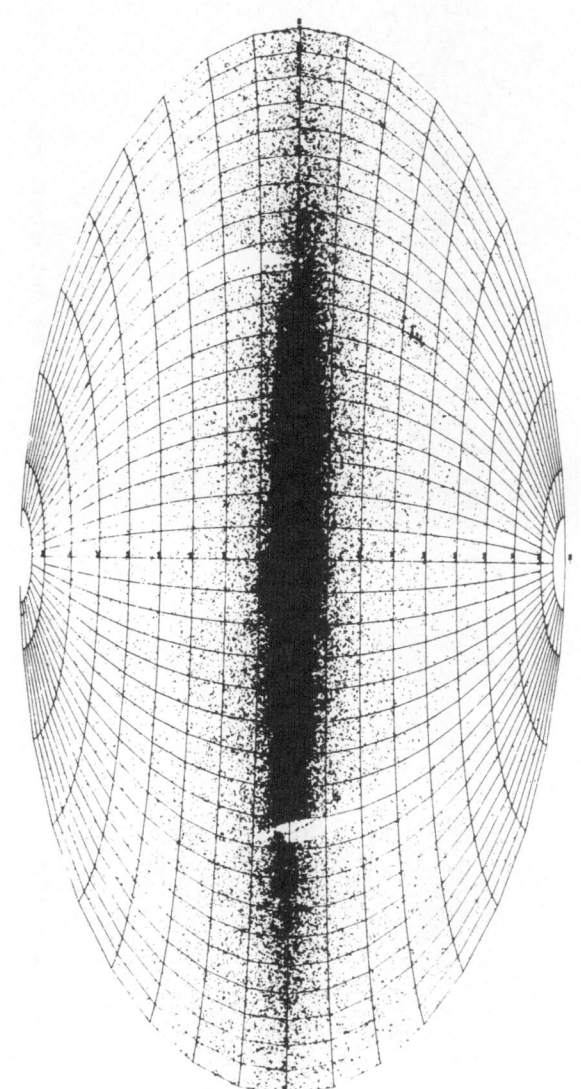

Fig. 10. Equal area all-sky plot of stars with circumstellar dust shels.

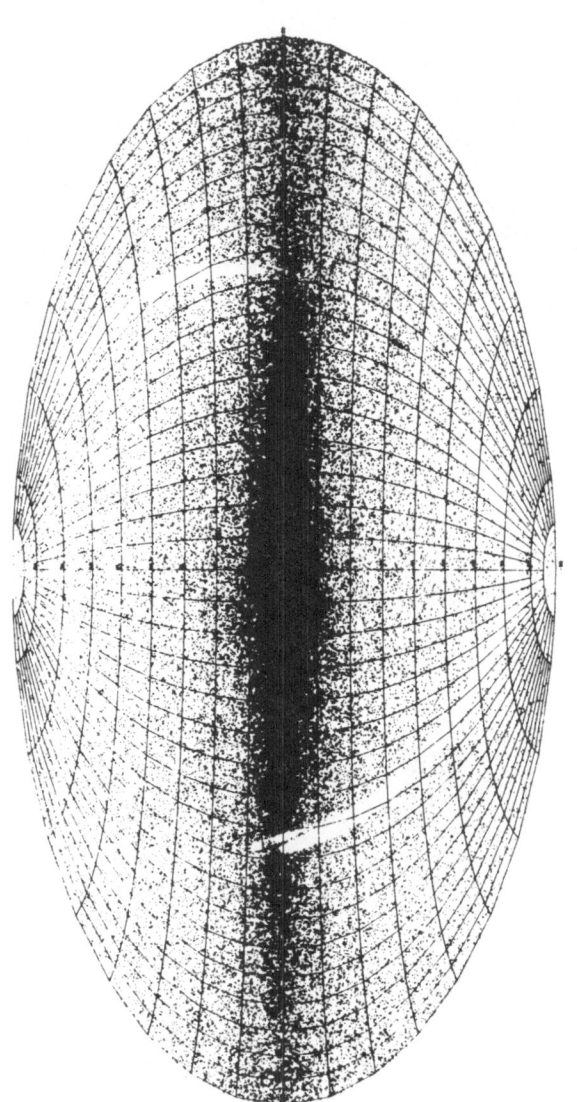

Fig. 11. Equal area all-sky plot of faint stars (those detected only at 12 μm.

latitudes, which confirm the expectation that late-type giants dominate this class.

Since we see these stars at great distances, it is not surprising that they show a tremendous concentration towards the galactic plane. Figure 10 shows the sky distribution of CDS stars. The visibility of the galactic bulge in this figure confirms that distances of at least 10 kpc are reached. The log N/log S plots for CDS stars are anemic until about 30° galactic latitude. They do not reach a slope of 1.0 until 10°, and the slope never exceeds 1.2.

c) FAINT STARS

The sky distribution of faint stars (those detected only at 12 μm) is shown in Figure 11. Above 10°, the faint stars are mostly photospheres, and below 10°, mostly CDS stars.

5. GALAXIES AND IR GALACTIC COMPONENTS

Galaxies and IR Galactic Components are discussed together since it is difficult to separate them by their IRAS colors. They are defined here as having a 60 μm detected flux with $f_\nu(25) < f_\nu(60)$.

Rowan-Robinson et al (1985) have studied the sky distribution of galaxies with b > 60°. Known clusters of galaxies are quite apparent in their data, which allow one to estimate the distances that IRAS samples. The Coma cluster, at about 100 Mpc, is quite evident, implying that a large number of galaxies can be seen at 100 Mpc. The most distant IRAS galaxy is at a redshift of 0.37, or 2000 (H/50) Mpc (Lonsdale et al 1985).

At least two effects other than distance determine wether a galaxy is a strong enough infrared emitter to be detectable by IRAS. First, it must have dust to absorb starlight and reradiate in the infrared. Thus dusty spiral galaxies are preferentially detected (Soifer et al 1984). Second, something must be heating the dust and the usual excuse is ongoing star formation, perhaps caused by recent gravitational interaction with one or more galaxies (Lonsdale et al 1984).

Figure 12 shows the sky distribution of galaxies and IR galactic components, and Figure 13 shows a histogram of the number of such objects versus galactic latitude. Note that down to about 30° , galaxies heavily dominate, and even down to 10° galaxies are more numerous than IR galactic components. The total number of galaxies can be estimated from Figure 12. The fitted line corresponds to a galaxy density of 0.67 per square deg, which gives 27,000 over the whole sky. However, 1000 galaxies must be subtracted due to the lower density in the south, and 2000-5000 must be subtracted due to higher thresholds in the galactic plane. (The galactic plane begins to raise our flux threshold at 10° deg where it is 0.6 Jy at 60 μm, to 0.8 Jy form 5°-10° and to 1.0 Jy at 0°-5°.) Thus the total number of galaxies are probably 21,000 - 24,000. This is supported by taking the association rate of 55% in figure 4 form 50°-90°, the 11,400

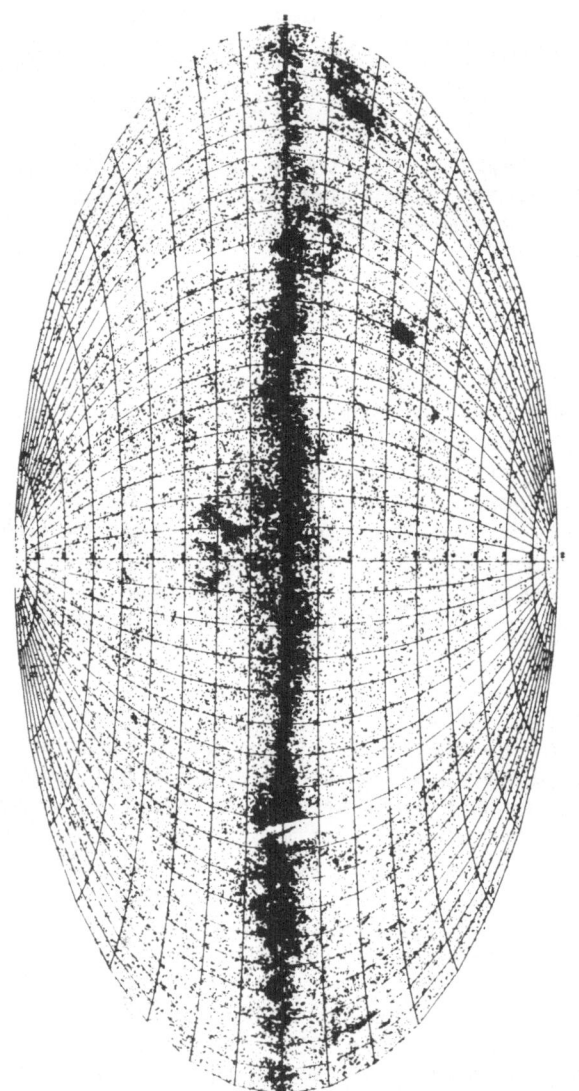

Fig. 12. Equal area all-sky plot of galaxies and IR galactic components.

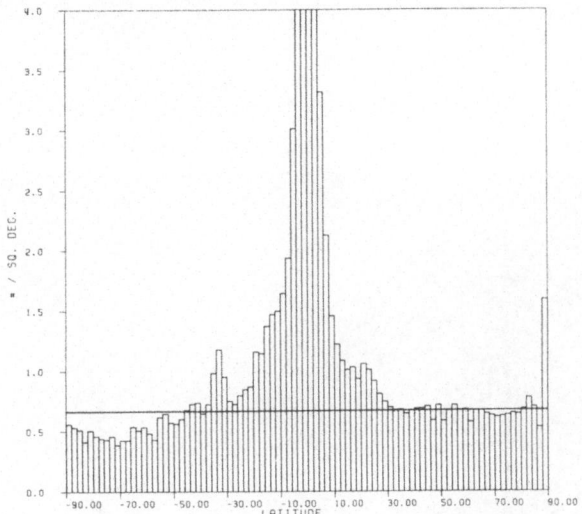

Fig. 13. Histogram of density of galaxies and IR galactic components
versus galactic latitude. The bump at -30° to -36° is caused
by the LMC.

associated galaxies in the catalog, and an optical incompletenes
modelled by a galactic latitude cutoff of 10°-15°, to give 21,000 -
26,000. As expected, the log N/Log S curve is consistent with a slope
of 1.5 down to about 5°. IR galactic components will not be discussed
further here, due to their great diversity.

6.   CIRRUS

Cirrus sources are defined here as sources with a detected flux only
at 100 μm. Except for a small number of galaxies, most of these
sources represent a cool phase of the local interstellar medium.
These sources are present in the catalog because only a one-
dimensional source detector was used. Thus when wisps of cirrus were
oriented perpendicular to the scan direction, a string of IRAS "point
sources" was produced.
     Figure 14 shows the cirrus sources present in the point source
catalog. Note that cirrus roughly follows Gould's Belt, with an
angular scale height of perhaps 15°-20°. If the scale height of the
ISM is 100 pc, then a rough estimate of the distance of these sources
is 300-400 pc. The log N/log S curves indicate the peculiar nature of
these sources - the slope is higher than 3.0 at all latitudes above
10°.

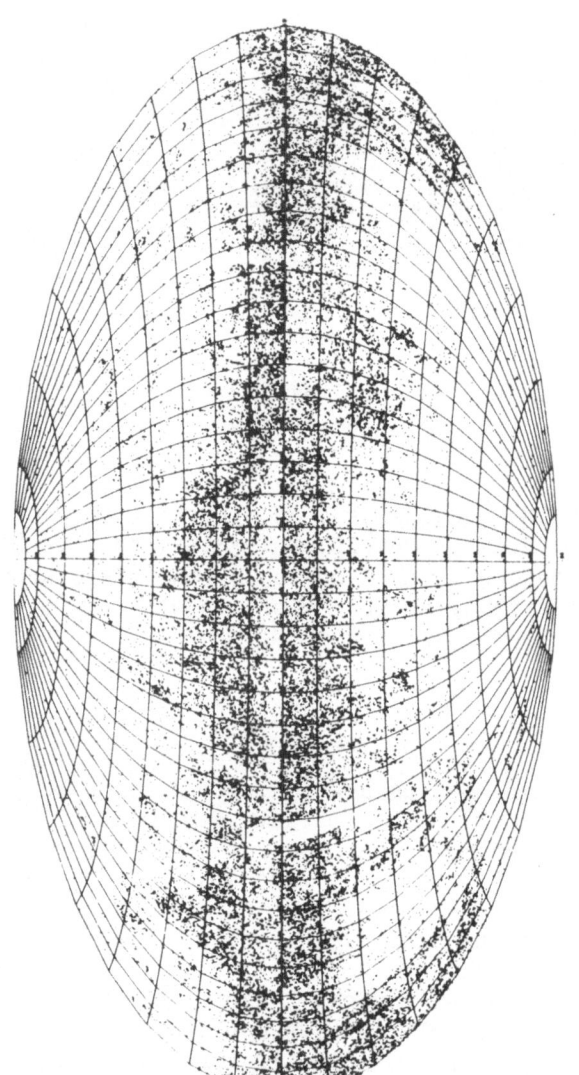

Fig. 14. Equal area all-sky plot of cirrus sources.

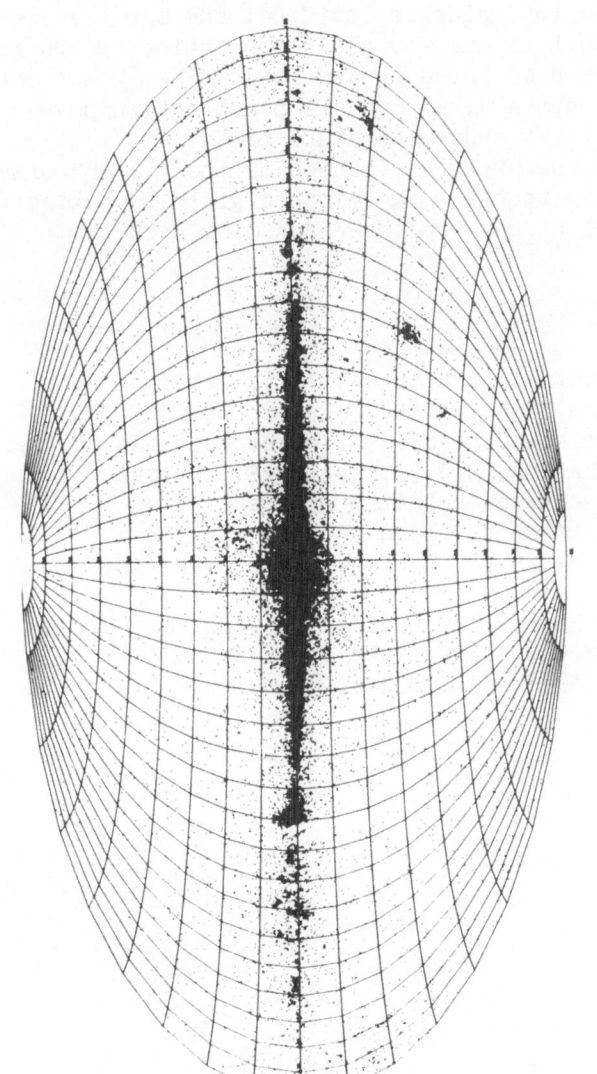

Fig. 15. Equal area all-sky plot of disk and bulge stars.

7.   DISK AND BULGE STARS

The sources which do not fit into any of the above categories were
put into the "other" class in the Explanatory Supplement. It has
since become clear that the "other" class has two significant
constituents. First, above 10° galactic latitude, the sources are
virtually all stars which happened to have poor 25μm upper limit
fluxes. Second, below 10° galactic latitude, the sources are
substantially identical to the sources which define the bulge
population, which are also found in the inner disk. These sources
dominate the overall numbers, so for all practical purposes, the
"other" class is the disk and bulge stars class.
     Figure 15 shows the sky distribution of the "other" class. Note
the strong resemblance to the bulge picture given by Habing (these
proceedings); see the discussion there for more details.

References

Allen, C.W., Astrophysical Quantities, 1976, Athlone Press.
Bahcall, J.N., Soneira, R.M., 1980, Ap.J. Supp., 44, 73.
IRAS Explanatory Supplement, 1985, edited by Beichman, C.A.,
     Neugebauer, G., Habing, H.J., Clegg, P.E., Chester, T.J., The
     United States Government Printing Office.
Habing, H., these proceedings, page 239.
Lonsdale, C.J., Persson, S.E., Matthews, K., 1984, Ap. J., 287, 95.
Lonsdale, C.P., et al, 1986, in preparation.
Rowan-Robinson, M., Chester, T., Soifer, T., Walker, D., and
     Fairclough, J., M.N.R.A.S., 1986, in press.
Soifer, B.T., et al. 1984, Ap. J. Letters, 278, L71.

SEMI-AUTOMATED IDENTIFICATION OF IRAS POINT SOURCES USING UKST PLATES AND THE COSMOS MEASURING MACHINE

A. Savage, R.G. Clowes, M. Kalafi, S.K. Leggett, H.T. MacGillivray and R.D. Wolstencroft
Royal Observatory
Blackford Hill
EDINBURGH EH9 3HJ
Scotland

ABSTRACT. We have initiated a large-scale programme to identify IRAS point sources, and now report on our initial study covering three hundred square degrees, centred on the South Galactic Pole (SGP) and containing 300 IRAS point sources. At ROE we have the ideal facilities to undertake such a large programme, viz: the rapid scanning capabilities of the COSMOS measuring machine to exploit the depth and resolution of the United Kingdom Schmidt Telescope (UKST) J survey plates. This automated procedure is more rapid with higher accuracy than visual identification procedures and will ultimately prove to be more objective.

1.   PROCEDURE

Accurate positions ($\leq$ 1 arcsec), isophotal magnitudes and other image parameters for objects on the UKST J survey plates are obtained by scanning the central 29 square degrees of each 40 square degree plate with the automated plate measuring machine COSMOS (see the review of MacGillivray and Stobie 1985). The use of COSMOS and the large area Schmidt plates leads to an enormous reduction in the manual effort involved by removing the need for individual position measurements and magnitude estimates. This is a great simplification of identification techniques and helps to eliminate human biases and errors.

Sophisticated software is used to output all positional coincidences in the COSMOS data within a search circle of radius one arcmin centred on each IRPS position (Beichman et al. 1985). The error distribution from the positional coincidences with bright stars is used to confirm the overall error ellipsoids and position angles in this small region. Identifications are made on the basis of positional coincidence alone for those brighter candidates where the chance coincidence rates are low. Morphological types have been assigned to galaxies after visual inspection following the classification scheme of de Vaucouleurs (1959). To avoid biases, such as those in early radio identification programmes (Savage et al. 1982), recourse to any additional criteria is being kept to a minimum - except occasional use of the infra-red spectrum to distinguish between stellar or galaxy identifications.

F. P. Israel (ed.), Light on Dark Matter, 23–28.
© *1986 by D. Reidel Publishing Company.*

Two SERC I survey plates (7150Å-9000Å) have also been inspected. Many
of the stellar identifications show enhanced images on I when compared
with J (3900Å-5400Å). Many of the bright galaxy identifications show
enhanced nuclear images on I when compared with images of galaxies
which have the same apparent magnitude on J.

Table I gives the surface densities of various types of objects as a
function of magnitude. The Table has been compiled from the following
sources: Lauberts (1982), Cohen et al. (1977), Griffiths et al.
(1982) and Savage et al. (1982). For 95% completeness of
identifications Jauncey et al. (1982) quote a two sigma search area of
$4\pi\sigma_\alpha\sigma_\sigma$ and a chance coincidence rate of $4\pi\sigma_\alpha\sigma_\sigma$ multiplied by the
density of objects.

TABLE I. Density of optical objects as a function of magnitude

Objects	Magnitude	Density per square arcsec	Chance Coincidence $\sigma_A=\pm15"$ $\sigma_B=\pm3"$
Bright galaxies	$B_J < 14.0$	0.00×10^{-4}	0.00006
Galaxies	$M_J \leqq 18.0$	0.05×10^{-4}	0.003
Galaxies	$21.0 > M_r > 18.0$	2.58×10^{-4}	0.15
All galaxies to $M_r \sim 21.5$ POSS limit		2.63×10^{-4}	0.15
All galaxies to $B_J \sim 22.5$ SERC limit		3.3×10^{-4}	0.19
All stars to $M_b \sim 21.5$ POSS limit		1.3×10^{-4}	0.08
All stars to $B_J \sim 22.5$ SERC limit		6.0×10^{-4}	0.34
All stars & galaxies to $B_J \sim 22.5$		9.3×10^{-4}	0.53

2.   RESULTS

Figures 1 and 2 illustrate the typical galaxy identification content
of this survey. One of the problems associated with large error
ellipsoids and high densities of faint objects is that at fainter
magnitudes one or two faint objects are likely to fall in the IRAS
error ellipse by chance and an unambiguous identification cannot be
made. 00016-3056 (21 mag. galaxy 30 arcsec off) and 23515-2917 (19
mag. galaxy 7 arcsec off) are two such examples; both are too faint
to have images on the I plates. If these objects are claimed as the
identification it should be remembered that they have an extremely
high probability of being chance coincidences. Source 01367-3010
illustrates another type of problem encountered. Both galaxies fall
outside the 2σ formal error box, the brighter face-on galaxy has a
lower chance coincidence rate, is definately a spiral and the IRAS
fluxes support it as the identification. However the edge-on fainter
galaxy is nearer. With finding charts readily available for
inspection 01472-2756 is obviously an interesting candidate interacting galaxy.

In all some 5% of this IRAS sample are problematical identifications and in such cases details of all the nearby optical candidates will be given. About 3% are genuine empty fields whilst the remainder of the identifications divide equally between late-type (K,M) bright stars (B_J < 16.0) and galaxies. Comparison with the percentage of morphological types of field galaxies of Kirshner et al. (1979) show that in this IRAS selected sample spirals predominate. Further details on the statistics of the firm galaxy identifications are given in the review of Wolstencroft et al. (these proceedings) and full details of this survey will be published in Monthly Notices of the Royal Astronomical Society (in preparation).

3. REFERENCES

Beichman, C.A., Neugebauer, G., Habing, H.J., Clegg, P.E. & Chester, T.J., 1985. Explanatory Supplement to IRPS.

Cohen, A.M., Porcas, R.W., Browne, I.W.A., Daintree, E.J. & Walsh, D., 1977. Mem.R.astr.Soc. 84, 1.

Griffiths, R.E., Murray, S.S., Giacconi, R., Bechtold, J., Murdin, P., Smith, M., MacGillivray, H.T., Ward, M., Danziger, J., Lub, J., Peterson, B.A., Wright, A.E., Batty, M.J., Jauncey, D.L. & Malin, D.F., 1983. Ap.J. 269, 375.

Jauncey, D.L., Batty, M.J., Gulkis, S. & Savage, A., 1982. A.J. 87, 763.

Kirshner, R.P., Oemler, A. & Schechter, P.L., 1979. A.J., 84, 951.

Lauberts, A., 1982. The ESO/Uppsala Survey of the ESO(B) Atlas.

MacGillivray, H.T. & Stobie, R.S., 1985. Vistas in Astronomy, 27, 433.

Savage, A., Bolton, J.G. & Wall, J.V., 1982. Mon.Not.R.astr.Soc., 200, 1135.

Savage, A., Jauncey, D.L., Batty, M.J., Gulkis, S., Morabito, D.D., & Preston, R.A., 1983. Astronomy with Schmidt-type telescopes, IAU Coll. 78, 481.

de Vaucouleurs, G., 1959. Handbook of Physics, Astrophysics IV: Stellar Systems, 8, 275.

Wolstencroft, R.D., Clowes, R.G., Kalafi, M., Leggett, S.K., MacGillivray, H.T. & Savage, A., these proceedings, page 425.

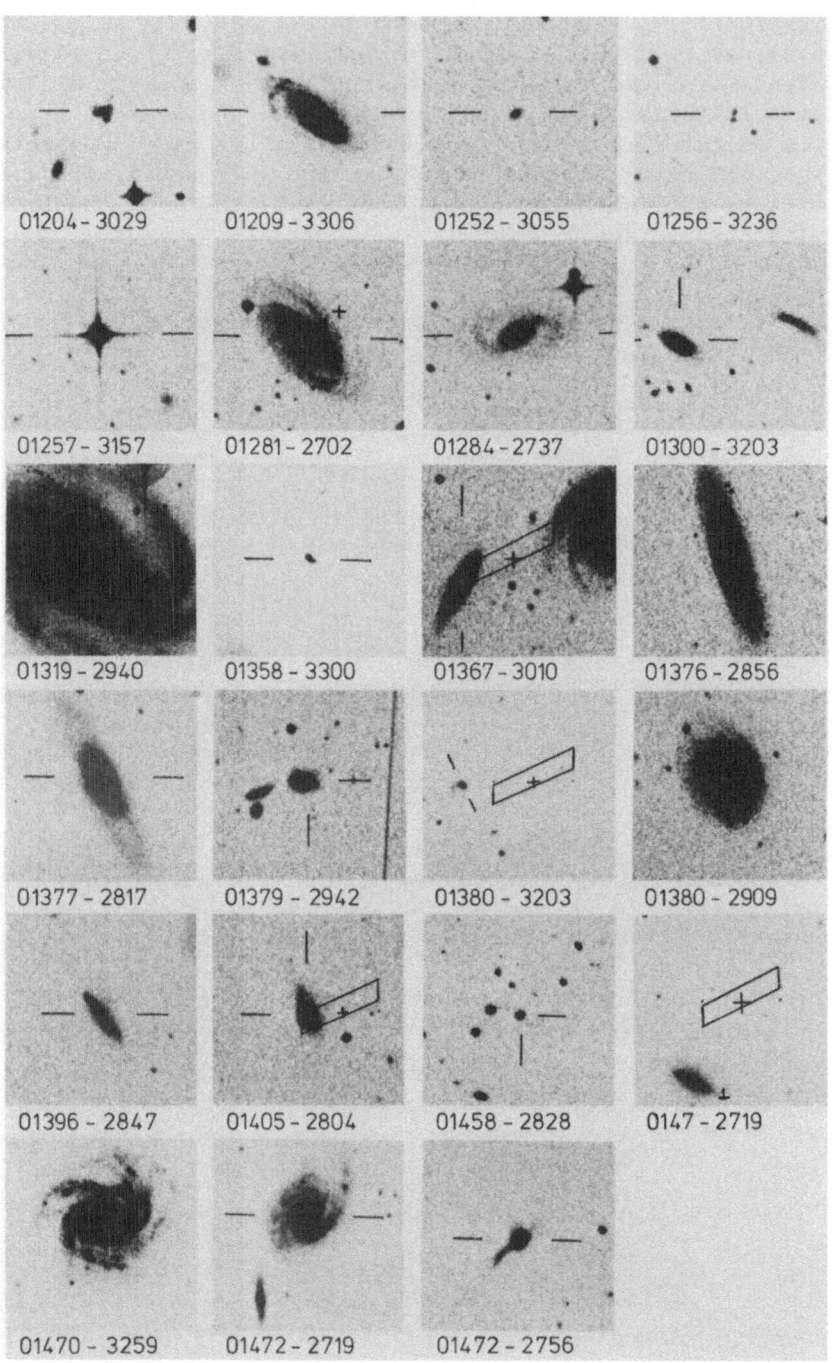

01204 – 3029   01209 – 3306   01252 – 3055   01256 – 3236

01257 – 3157   01281 – 2702   01284 – 2737   01300 – 3203

01319 – 2940   01358 – 3300   01367 – 3010   01376 – 2856

01377 – 2817   01379 – 2942   01380 – 3203   01380 – 2909

01396 – 2847   01405 – 2804   01458 – 2828   0147 – 2719

01470 – 3259   01472 – 2719   01472 – 2756

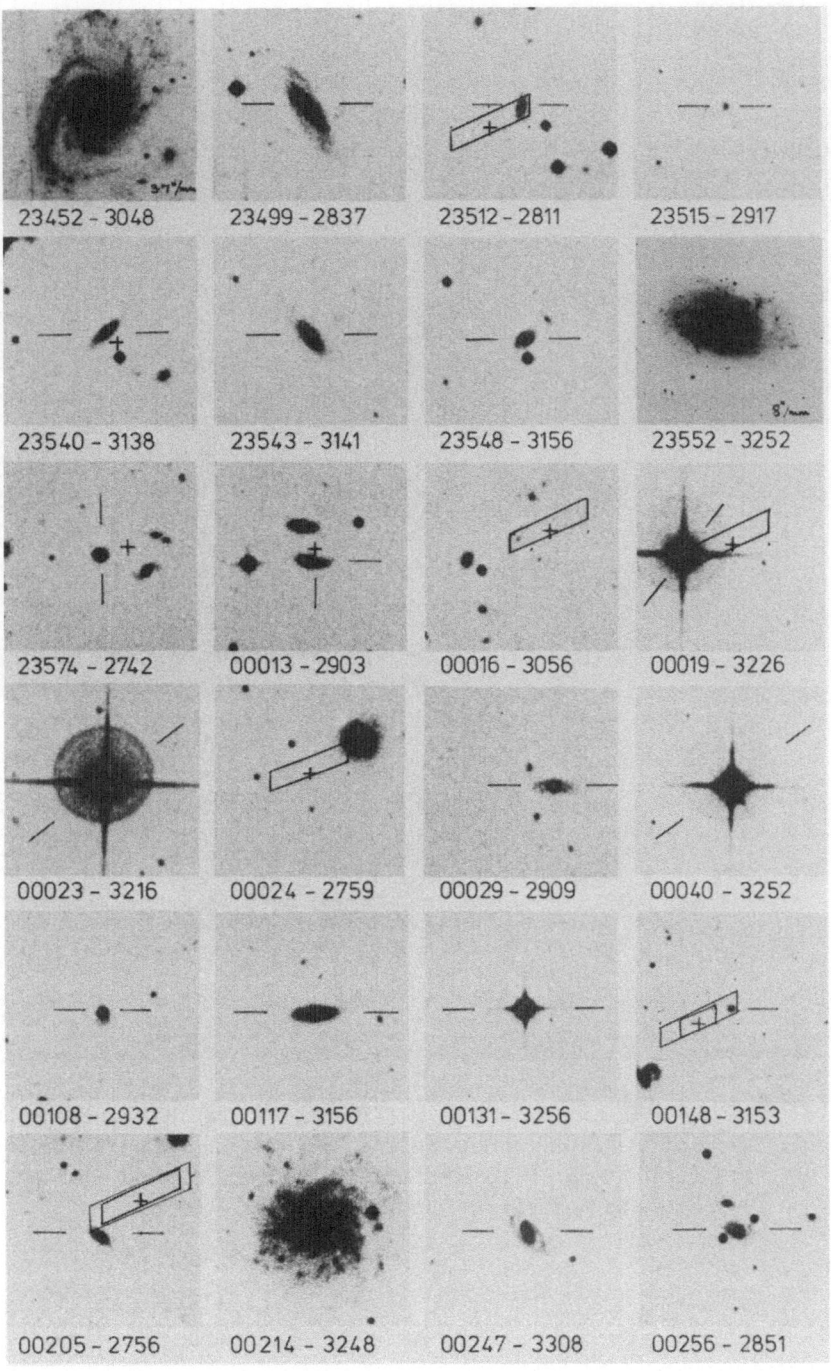

23452 – 3048 23499 – 2837 23512 – 2811 23515 – 2917

23540 – 3138 23543 – 3141 23548 – 3156 23552 – 3252

23574 – 2742 00013 – 2903 00016 – 3056 00019 – 3226

00023 – 3216 00024 – 2759 00029 – 2909 00040 – 3252

00108 – 2932 00117 – 3156 00131 – 3256 00148 – 3153

00205 – 2756 00214 – 3248 00247 – 3308 00256 – 2851

IRAS Field 51 $\alpha = 4^h 00^m$, $\delta = +30°$, HCON-3, 12 μm. Field in Taurus
                  and Perseus, showing NGC 1579 (top left), NGC 1499
                  (top center) IC 348 and NGC 1333 (right) and the
                  Pleiades (M45) at bottom right.

SUBMM CONTINUUM OBSERVATIONS OF SOURCES FROM THE I R A S POINT-SOURCE CATALOGUE

R. Chini, E. Kreysa, E. Krügel, P. G. Mezger and H.-P. Gemünd
Max-Planck-Institut für Radioastronomie
Auf dem Hügel 69, 5300 Bonn 1, F.R.G.

ABSTRACT. More than 120 galactic and extragalactic objects, taken from the IRAS pointsource catalogue have been observed at 350 and 1300 μm. Supplementary near IR data and extensive model fits over this large spectral range allow the determination of the total luminosity, dust temperatures, the wavelength dependence of dust opacity, the total amount of extinction towards embedded stellar objects and the density distribution of dust around central heating sources within HII regions and dust clouds.

1. INTRODUCTION

A large fraction of objects associated with interstellar dust shows increasing energy distributions in the four IRAS bands. To derive from these spectra physical properties like dust temperatures and luminosity, observations at longer wavelengths are essential. This work reports on continuum observations at 350 and 1300 μm of pointlike FIR sources detected by IRAS. We concentrated our investigations on sources with $S_{12\mu} < S_{25\mu} < S_{60\mu} < S_{100\mu}$ (flux quality flag 2 or 3) observable from the northern hemisphere (Dec $\geq -25^\circ$).

2. OBSERVATIONS

The observations have been carried out at the IRTF and the 88" telescope of the University of Hawaii on Mauna Kea. We used a ^{3}He cooled Ge composite bolometer, developed at the MPIfR, Bonn. It is operating at 0.27 K and is background photon noise limited (NEP = 5 10^{-16} W Hz^{-1} ^{2}). The wavelengths bands are defined by resonant mesh filters, optimized for the atmospheric windows. The beam size at 350 μm was 30" and at 1300 μm about 90" and 130".

3. RESULTS

3.1 HII Regions
Most of the strong sources from the IRAS pointsource catalogue with 100 μm flux densities >1000 Jy are compact HII regions. So far we observed about 60 objects of this class at (sub)mm wavelengths. From the

F. P. Israel (ed.), Light on Dark Matter, 29–30.

luminosities of the integrated spectra we derive spectral types generally earlier than 08 with a maximum around 06. The FIR part of the spectrum is dominated by emission from dust of about 20-35 K. When fitting
the observations by a radiative transfer calculation for a spherical
dust distribution around a central star, some common features can be
found with all sources:
i) The HII regions are very dust depleted.
ii) Only a small amount of dust is located near the star.
iii) At larger distances the density increases to a thick dust shell.
iv) The average ratio of total to ionizing luminosity is about 10.
v) For a selection of extensively studied HII regions we derived the
    wavelength dependence of dust opacity m to be close to 2.

3.2 Dark Clouds and Globules

In total we observed about 30 FIR sources (50 Jy $\leq S_{100} \leq$ 500 Jy) embedded
within dust clouds and globules. At $\lambda > 100$ µm the emission from these objects is dominated by dust of about 20 K and densities of $n_H \sim 10^5-10^6 cm^{-3}$.
As a typical object of this category the spectrum of the large globule
L 810 is shown in Fig. 1. L 810 is a spherical, isolated dark cloud
of about 5' in   diameter at a distance of

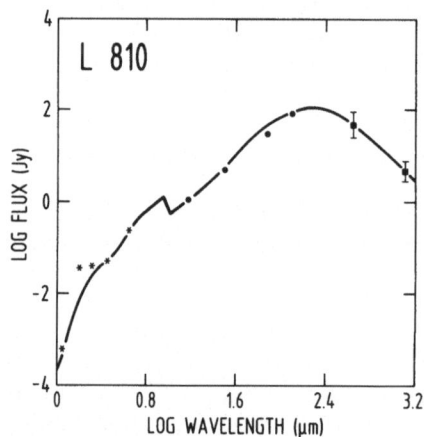

1.5 kpc. The total luminosity of 350 $L_\odot$
corresponds to a spectral type of B7 if
the star has already reached the main sequence. The solid line shows a fit resulting from a radiative transfer calculation. The observations shortward of 20µm
require a warm dust shell ($n_H \sim 10^6 cm^{-3}$,
200 K $< T_d <$ 600 K) near the central heating
source ($r < 1.2 \ 10^{15}$ cm) whereas the emission at longer wavelengths originates from
cool dust ($n_H \sim 10^5 cm^{-3}$, T_d 20 K) at distances of about 7 10^{17} cm. The total extinction of this density distribution is
15.6 mag towards the embedded star.

Fig. 1: Spectrum of L 810
* Th. Neckel (1985, priv.
comm.); • IRAS, ▪ this work;
solid line: model fit de-
scribed in the text.

3.3 Galaxies

We observed 35 identified galaxies from the IRAS pointsource catalogue
with $S_{100} \geq 50$ Jy and several unidentified objects with galaxy-like spectral shape. The flux density at 1300 µm for most of the galaxies is below 1 Jy. This is consistent with a rather cool dust emisstion ($T_d <$ 20 K)
and a wavelength dependence of dust opacity of m = 2 for $\lambda > 100$ µm.

For sources detected at 350 and 1300 µm, the emission longward of 100 µm
can be interpreted by dust of 18 K (m = 2), while the observations at
shorter wavelengths indicate the presence of dust of several hundred
degrees.

A STATISTICAL ANALYSIS OF THE LRS CATALOG

F. M. Olnon
Netherlands Foundation for Radio Astronomy
Postbus 2
7990 AA Dwingeloo
The Netherlands

ABSTRACT. The IRAS catalog of low-resolution spectra contains 8-22 μm spectra of 5425 relatively bright point sources distributed over the whole sky. The various types of spectra (LRS classes) can readily be associated with well-known groups of objects. About 96% of the sources are stars, most of them with oxygen- or carbon-rich envelopes, five sources are galaxies, and the others are compact HII regions, planetary nebulae, and 'peculiar stars'.

1. INTRODUCTION

The IRAS instrumentation included a low-resolution spectrometer (LRS) which measured the 8-22 μm spectra of the brighter point sources. The instrument, the observations, and the data processing are described in the IRAS Explanatory Supplement (1985). In Section 2 we summarize how the LRS catalog was made, and in Section 3 we analyze its contents.

2. PRODUCTION OF THE LRS CATALOG

2.1. The Instrument

The LRS was basically an objective prism spectrograph, oriented in such a way that the dispersion was aligned with the scan direction. There-fore, it could only measure useful spectra for relatively isolated point sources.

Two overlapping wavelength bands were scanned simultaneously, one ranging from 7.7 to 13.4 μm and the other from 11.0 to 22.6 μm. In both bands the resolution increased from 20 at the short-wavelength to 60 at the long-wavelength side. The integration time per resolution element was only 60 msec, which resulted in noise levels equivalent to 1.5 Jy at the shorter and 3 Jy at the longer wavelengths.

The aperture mask in the focal plane measured 6 arcmin in-scan by 15 arcmin cross-scan. This wide field of view ensured a sky coverage identical to that of the IRAS survey: 96% of the sky was observed at

31

F. P. Israel (ed.), Light on Dark Matter, 31–38.
© *1986 by D. Reidel Publishing Company.*

least twice, and 72% even three times. To reduce spatial confusion, the
aperture width was covered by three detectors at the short wavelengths
and two at the long wavelengths. The detectors were sampled continually,
and the data were received on the ground together with the data from the
survey array.

2.2. Data Processing

Whenever the data from the survey detectors indicated that a point-like
source brighter than about 2 Jy at 12 or 25 μm had crossed the LRS aperture, the corresponding sections of the LRS data stream were extracted.
At the end of the mission, the database contained 170,000 uncorrected
spectra linked to 50,000 survey sources, and a number of calibration
tables. However, most of the spectra were rather useless due to the very
liberal thresholding.
      All the spectra were interpolated to a standard wavelength grid,
various gain corrections were applied, and a number of quality checks
were performed. For each source, the best spectrum was selected, and if
there was at least one other spectrum of the source that correlated well
with this reference, an average spectrum was made. The spectra were then
classified, as will be described in the next subsection. All this was
done automatically.
      Visual inspection led to the rejection of about one hundred ob-
viously bad spectra (mostly confused or slightly extended sources). The
remaining 5425 spectra form the LRS catalog.
      The overall shape of the spectra is accurate to 2-4%, and the flux-
densities agree with the broad-band 12 μm fluxes within 10-15%. The LRS
catalog is certainly less complete than the point-source catalog due to
the additional selection criteria applied. Even at high galactic latitudes, it contains only 90% of the IRAS point sources brighter than 28
Jy at 12 μm (Hacking et al., 1985). Some very bright red stars, like
Mira and R Leo, are conspicuously absent.

2.3. Classification

Each LRS spectrum is characterized by a two-number code (class), which
is included in the point-source catalog. The first number (main class)
describes the overall character of the spectrum, and the second number
(subclass) gives some more detailed information. The LRS classes are
briefly described in Table I.
      The classification is based on spectrometer data only. Spectral
indices, band strengths and line fluxes were derived from the fluxes in
a number of well-chosen narrow bands, typically two resolution elements
wide.
      The spectral index, β, is defined by $F(\lambda) \propto \lambda^{\beta}$, where $F(\lambda)$ is the
flux per unit wavelength interval, and λ the wavelength. 'Blue' spectra
are those with $\beta < -1$ between 14 and 22 μm, and the others are called
'red'. Band strengths are defined as $B = \ln(F/Fc)$, where F is the fluxdensity at 9.8 or 11.4 μm (the central wavelengths of the broad silicate
and SiC bands), and Fc the interpolated continuum fluxdensity at the
same wavelength.

TABLE I. LRS Spectral Classes

Main class		Type of	Subclass
'Blue'	'Red'	spectrum	
	0	weak	nearest main class
1	5	featureless	spectral index
2	6	10 μm em. band	band strength
3	7	10 μm abs. band	band strength
4		11 μm em. band	band strength
	8	11.3 μm 'line'	excitation level
	9	only ionic lines	excitation level

Spectra with significant emission lines were put in main class 8 or 9, depending on the presence or absence of the 11.3 μm 'line'. (Note, that we only speak of a line, to indicate the difference with the broad SiC band at about the same wavelength.) The subclass (0 to 6) indicates the excitation level, as determined by the strongest ionic line in the spectrum. HII regions are found in classes 81 and 91 (12.8 μm [NeII]), and planetary nebulae in class 95 (15.5 μm [NeIV]).

Spectra with significant bands and $B(10) > B(11)$ were put in main class 2 or 6, those with $B(10) < 0$ in main class 3 or 7, and the others in main class 4. The subclass (0 to 9) is equal to $20×B$ for emission and $-10×B$ for absorption bands. Stars with strong silicate emission bands were classified as 29, and stars with moderately strong SiC bands as 43.

Spectra without features stronger than 5×sigma or 10% of the continuum level, were put in main class 1 or 5. The subclass is equal to twice the absolute value of $β(8-13)$ and $β(14-22)$, respectively. E.g., Rayleigh-Jeans type spectra ($β = -4$) were classified as 18. When the spectrum was so noisy, that even a medium-strong band would not have been recognized, it was classified as 01 or 05.

This simple scheme led to correct classifications for more than 98% of the LRS spectra, and most errors occur there, where we would expect them. Weak 'blue' spectra disappear in the noise at the longest wavelengths, which sometimes led to a 'red' classification. For spectra with emission bands just above the 10% level, the choice between main classes 2 and 4 is quite arbitrarily. Spectra with self-absorption in the 10 μm band look very much like class 4 spectra, and only when the 18 μm band of silicate is visible, we would know how to choose (the classification program did not).

3. STATISTICAL ANALYSIS

3.1. Fluxdensity Distributions

In Figure 1 we show some $\log(N)-\log(S)$ curves for the LRS sources. The curve for the total sample indicates, that the catalog is 'complete' above a fluxdensity of 20 Jy at 12 μm (but, see Section 2.2) and contains only a small fraction of the sources below 10 Jy. The slope is -1

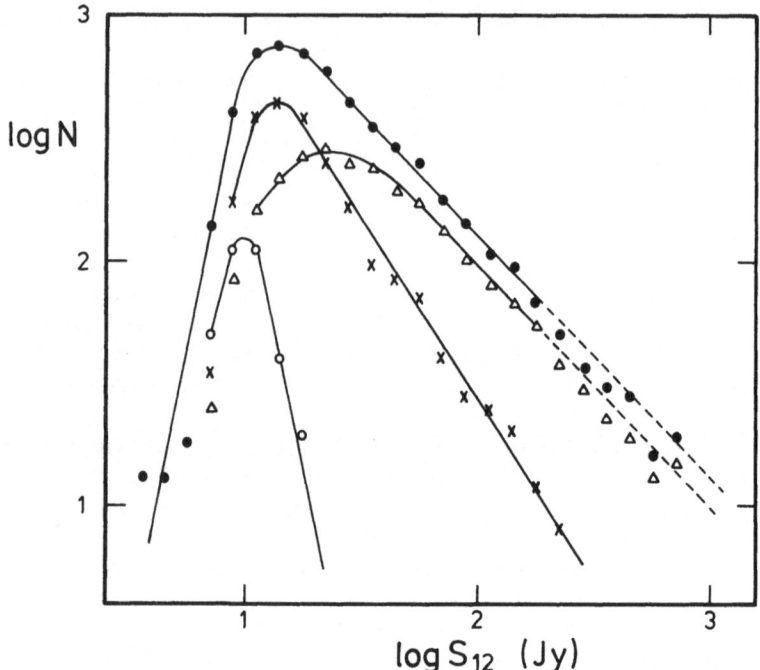

Figure 1. Differential log(N)-log(S) curves for LRS sources. N is the
number of sources in 0.1 wide bins of log(S12), where S12 is the 12 μm
survey fluxdensity in Jy. Filled circles: all classes, open circles:
class 0, crosses: class 1, and triangles: classes 2, 3 and 4.

over most of the range and tends to -1.5 at the highest fluxdensities,
as we expect for a population of galactic sources.

The curves for the three subgroups are strongly influenced by the
properties of the classification scheme. Above 40 Jy, most of the spectra show broad bands and fall in the main classes 2 (46%), 4 (20%), and
3 (5%). At lower fluxdensity levels, weaker bands are no longer recognized and more and more spectra are classified in the featureless main
class 1.

The curve for class 1 has a constant slope of -1.5 down to the
completeness limit at 20 Jy, indicating a very local population. But,
the migration effect is not visible. As we will see later, there are two
distinct groups of class 1 spectra: Rayleigh-Jeans type (classes 17-19)
and migrated band-type spectra (14-16). The curve for the first group
has a slope of -1.5 above 40 Jy and then becomes less steep, just where
the number of sources in the second group rapidly increases.

3.2. Colour-Colour Diagrams

Figure 2 shows the LRS colour-colour diagram for the brightest 'blue'
sources. By definition, all these sources lie below the -1 line, but
some come close to this limit (e.g., class 29). Each main class occupies
a well-defined area, although some of them partly overlap. There are

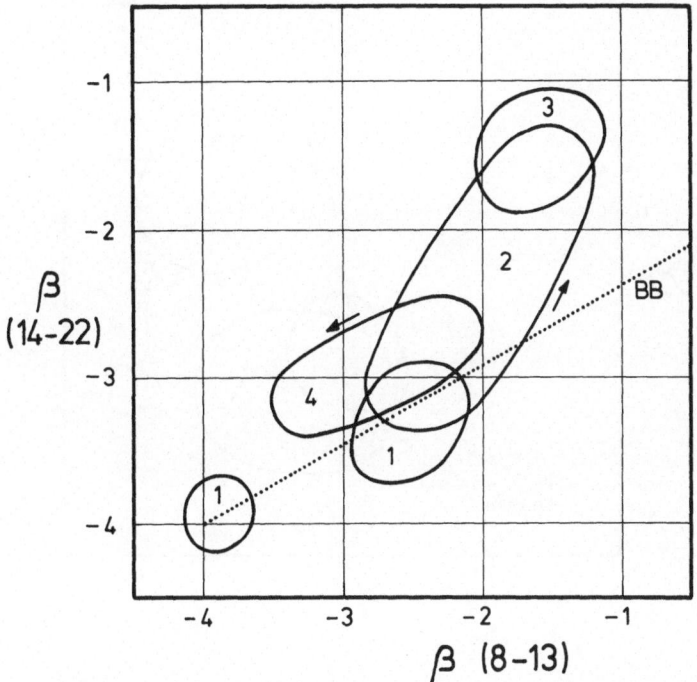

Figure 2. Colour-colour diagram for the 'blue' LRS sources. β(8-13) and
β(14-22) are spectral indices determined from the spectra. The areas
occupied by the various classes are labeled with the corresponding main-
class codes. Arrows indicate the direction of increasing band strength
(subclass) for classes 2 and 4. The dotted line is the locus of blackbody radiators, from high temperatures (lower left) down to 300 K.

clearly two groups of class 1 spectra: one in the lower left-hand corner
of the diagram (class 18, hot blackbodies) and the other in the lower
centre (class 15). The second group extends the class 2 region at the
lower subclasses, suggesting that most bright class 15 sources are stars
with 10 μm band strengths below 10%.
At lower fluxdensity levels, the 15 region grows at the cost of the
class 2 and 4 regions, as we would expect. We also find a sharp upper
boundary, which means that some spectra must have crossed into the 'red'
regime (see Section 2.3).
    Figure 3 shows a colour-colour diagram for the brightest 'red'
sources, but this time we used the broad-band survey colours. Main classes 7, 8 and 9 occupy about the same regions, but there are clearly two
groups. The upper one contains mainly HII regions and the lower one
(overlapping with class 6) consists of 'peculiar stars', classified
rather vaguely as low-excitation and proto-planetaries, emission-line
stars, and PMS stars.
    At lower fluxdensities, a third group of class 9 sources becomes
visible: the planetary nebulae (dashed outline). We also find more and
more class 6 sources (mostly 69) in the region occupied by class 2. They
are faint stars of type 29 that got classified as 'red'.

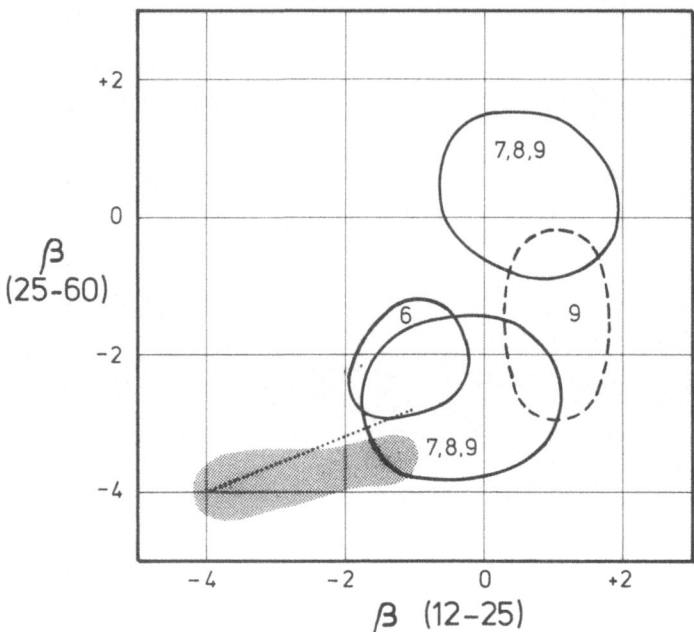

Figure 3. Colour-colour diagram for the 'red' LRS sources. β(12-25) and
β(25-60) are spectral indices derived from the broad-band survey fluxes.
The areas occupied by the various 'red' classes are labeled, and the
shaded area contains the 'blue' sources.

3.3. Associations

     The positions of all IRAS point sources have been compared with
other catalogs (see IRAS Explanatory Supplement). For each LRS source,
we first looked for a stellar, then for an extragalactic, nebular, or
infrared association. These prime associations appear to be proper identifications. Table II summarizes the results, where we excluded the five
galaxies that were found (four of them in class 8).
     The 'blue' classes 01 and 1 to 4, which contain 94% of the LRS
sources, are clearly composed of stars. The 'red' classes 7 to 9 are
populated by HII regions, planetary nebulae and 'peculiar stars' (mostly
emission-line stars). Classes 05, 5 and 6 contain a similar mixture, but
are dominated by stars that were 'accidentally' classified as 'red'. The
number of stars in the LRS catalog is so large, that a 2% accident rate
is sufficient.

3.4. Stellar Properties

Table III shows the spectral-type distribution for the optically classified 'blue' LRS sources, only 35% of the whole sample. This distribution is certainly biased in favour of the earlier spectral types, as was
shown for the subsample of bright stars (Hacking et al., 1985).

TABLE II. Association statistics of LRS sources

LRS Class	Total number	Associations (% of total)			
		Stars	Nebulae	Infrared	None
17-19	905	96	-	1	3
01	331	67	1	7	25
10-16	1333	51	2	13	34
2	1730	38	2	8	53
4	538	36	3	13	49
05	32	44	9	9	38
3	229	17	5	8	70
6	78	18	5	8	69
5	63	33	16	11	40
7	67	12	28	12	48
8	65	13	42	3	36
9	49	2	86	-	12
all	5420	50	4	8	38

TABLE III. Properties of LRS stars

LRS class	Total number	Spectral types (% of total)						Variability	
		<K	K	M0-M4	>M4	N/C	S	ID	IRAS
17-19	831	7	25	54	8	4	2	15	1
01	168	6	27	42	7	15	3	17	8
14-16	296	4	8	29	41	12	6	77	11
21-23	107	7	3	25	45	10	10	81	14
24-26	70	16	3	29	49	3	-	86	24
27-29	80	1	2	28	61	4	4	95	35
3	24	17	21	40	6	12	4	48	32
4	167	1	1	1	3	89	5	91	32

Classes 17-19 and 0 consist mainly of stars earlier than M5, the classes 14-16 and 2 are dominated by oxygen-rich stars later than M4 (Hacking et al. find even 90%), and class 4 is populated by carbon-rich stars. Due to the classification problem for weak bands (Section 2.3), some carbon stars are found in the classes 21-23. S-type stars have weak emission bands, if any: they are found in both groups of class 1 and in the lower subclasses of 2 and 4. Class 3 contains a high percentage of stars with early spectral types. The 10 μm absorption in these cases is probably caused by foreground extinction.

The last two columns of Table III contain two variability indices. The column headed ID gives the percentage of identified stars included in Kukarkins variable star catalog, and the last column gives the percentage of sources for which the IRAS data indicate variability above the 90% confidence level (see IRAS Explanatory Supplement). The last number is a severe underestimate.

Both variability indices show the same behaviour: stars in classes 17-19 and 01 are essentially not variable, those in the other classes are essentially all variable. The only exception is class 3: the IRAS index indicates that most stars are variable (like in classes 27-29), whereas only half the identified stars are variable.

The optically identified class 3 sources, only 20% of the total, are clearly not representative for the whole class. Most of the other sources are probably late-M type stars with very thick envelopes, which made them even invisible for earlier infrared surveys. Some of them have been identified with OH/IR stars, and the others are at least good candidates for OH searches.

4. SUMMARY

1) The LRS catalog contains average spectra of 5425 sources, with 12 µm fluxdensities above 10 Jy, and appears to be complete above 20 Jy. The LRS class, together with the broad-band survey colours, is a good indicator for the source type.

2) About 96% of the LRS sources are stars. Those without circumstellar envelopes (not variable) are found in classes 17 to 19, and those with envelopes (variable) in main classes 2 and 3 (oxygen-rich), class 4 (carbon-rich) and classes 14 to 16 (either oxygen- or carbon-rich).

3) The remaining 4% consists of about equal numbers of planetary nebulae (classes 92 to 96) and compact HII regions (the redder sources in classes 7, 8 and 91), and a smaller number of 'peculiar stars', like proto-planetaries and emission-line stars (the remaining sources in classes 6 to 9).

4) The LRS catalog was made by a machine that was tuned to give the best overall performance. It is, therefore, very well suited for statistical studies (e.g., mass loss of late-type stars), but the interpretation of single (averaged) spectra is hazardous. In that case, one should go back to the original spectra.

4. REFERENCES

Hacking, P., Neugebauer, G., Emerson, J.P., Beichman, C.A., Chester, T.J., Gillett, F.C., Habing, H.J., Helou, G., Houck, J.R., Olnon, F.M., Rowan-Robinson, M., Soifer, B.T., Walker, D. 1985, P.A.S.P., in press.

IRAS Catalogs and Atlases, Explanatory Supplement 1985, US Government Printing Office.

THE ZODIACAL BACKGROUND IN THE IRAS[*] DATA

M.G. Hauser
Lab. for Extraterrestrial Physics
Goddard Space Flight Center
Greenbeld, Maryland 20771

J.R. Houck
Center for Radiophys. & Space Res.
Cornell University
Ithaca, New York 14853

ABSTRACT. The IRAS sky survey yielded a very extensive picture of
bright diffuse thermal emission from interplanetary dust grains at
wavelengths from 12 to 100 micrometers. These data provide insights
into the character, spatial distribution, and perhaps origin of these
particles which both confirm and expand upon those gained from
previous zodiacal light studies. The zodiacal emission is a major
component of the large scale infrared brightness of the sky at all
IRAS wavelengths. It must therefore be carefully characterized both
to facilitate study of large-angular-scale sources in the Galaxy and
beyond using IRAS data, and to permit optimum design of future space
infrared instrument.

1. OBSERVATION SUMMARY

Thermal emission from the interplanetary dust, though often described
as a 'background', provides a bright foreground through which all
space infrared instruments must observe the cosmos. Study of this
emission and the scattering of sunlight by the same particles
(zodiacal light) promises to enhance substantially our understanding
of the nature, physical condition, distribution, and origin of the
interplanetary dust. The IRAS survey provided extensive data
particularly well-suited to study of the zodiacal emission, and
preliminary analyses of these data have already provided some
surprises. Some of these results have been reported in the literature
(Hauser et al. 1984a; Low et al. 1984; Neugebauer et al. 1984; Hauser
et al. 1984b); others have been reported at various meetings (Gautier
et al. 1984; Hauser and Gautier 1984; Good et al. 1984; White et al.
1985; Rickard et al. 1985). We shall have review the implications of
these initial studies and the status of attempts to characterize the
zodiacal emission itself.

* The Infrared Astronomical Satellite (IRAS) was developed and
operated by the Netherlands Agency for Aerospace Programs (NIVR), the
United States National Aeronautics and Space Administration (NASA),

F. P. Israel (ed.), Light on Dark Matter, 39–44.

and the United Kingdom Science and Engineering Research Council
(SERC).

The nearly 800 IRAS survey scans were each obtained at a fixed
viewing angle relative to the Sun (elongation), with angle about the
satellite-Sun line (inclination) increasing at a constant rate in
each scan. Elongation angles were restricted to the 80 to 100 degree
range during the first half year, which provided redundant coverage
of the full celestial sphere except for a small gap around the great
circle at ecliptic longitudes 160 and 340 degrees. During the final
four months of the mission, about 70% of the sky was reobserved with
elongation angles ranging from 60 to 120 degrees. These data provide
the most complete geometric viewing of the interplanetary dust cloud
presently available. Two data products released by the IRAS Project
in November 1984 are particularly valuable for study of the zodiacal
emission (IRAS Catalogs and Atlases Explanatory Supplement 1984): the
zodiacal observation history file (time-ordered data at 1/2-degree
sampling) and the low resolution allsky maps (Aitoff Projections in
galactic coordinates at 1/2-degree sampling).

Broadly peaked emission near the ecliptic plane clearly
dominates the large angular scale sky brightness at 12 and 25
micrometers in individual scans and in the all-sky images; similar
structure is evident at 60 and 100 micrometers, but emission
associated with the galactic plane becomes relatively more prominent
at these wavelengths. Preliminary analysis of data from a few
selected scans showed that zodiacal emission accounts for essentially
all of the largest scale infrared sky brightness except at 100
micrometers, where a small galactic component varying as cosec(b) can
be identified (Hauser et al. 1984a, Fig. 1). Other noteworthy
features in the data are the high brightness, seasonal variation of
properties such as polar brightness and apparent position of the peak
brightness, and the entirely unexpected structures in the profiles
near the ecliptic plane attributed to interplanetary dust bands (Low
et al. 1984, Fig. 2; Neugebauer et al. 1984, Fig. 3). We first review
briefly some of the characteristics of these features and their
implications for the dust cloud.

2. THE INTERPLANETARY DUST CLOUD

2.1. Grain Properties

The peak brightness in the ecliptic plane at elongation 90 degrees is
46, 78, 24, and 14 MJy/sr at 12, 25, 60, and 100 micrometers
respectively. (NOTE: these values in the final data products for a
scan on 24 June 83 are 10-30% lower than the preliminary values
reported for the same scan by Hauser et al. 1984a due to final
calibration corrections.) The optical zodiacal light brightness in
the same direction is fainter than this by about a factor of 11
(Levasseur-Regourd and Dumont 1980), implying black particles with
apparent albedo ~ 0.09.

The spectrum of the zodiacal emission both at the ecliptic plane and poles, shown to be consistent, within calibration uncertainties, with optically thin emission from particles with emissivity spectral index 0 (Hauser et al. 1984a), implies large grains of radius at least 15-20 micrometers. Briotta (1976) reported evidence of a silicate emission feature in the zodiacal emission at 10 micrometers, which, for particles of this size, would imply grain temperatures somewhat below 300 K at 1 AU (Roser and Staude 1978). The polar blackbody temperature of 275±57 K reported by Hauser et al. (1984a), though not conclusive, is consistent with silicate materials. The low albedo could be caused by surface blackening in the interplanetary environment.

2.2. Geometry of the Interplanetary Dust Cloud

The variation of zodiacal emission brightness (ignoring for the moment brightness) with ecliptic latitude and elongation angle is generally consistent with the picture of a Sun-centered, azimuthally symmetric, flattened dust distribution found in previous zodiacal light studies (Hauser et al. 1984a). However, the evidence suggests deviation from strict azimuthal symmetry, that is, the symmetry 'plane' is not really plane. This conclusion, previously suggested by Misconi (1980) and supported by the IRAS data (Hauser and Gautier 1984), is based on the apparently significantly different properties (inclination and ecliptic longitude of ascending node) obtained for this plane when observations are made in different parts of the cloud.

Previous optical zodiacal light studies from the ground and Helios spacecraft, which primarily observed the region interior to 1 AU, obtained an inclination and ascending node for the symmetry plane of about 3 deg. and 87 deg. respectively (Misconi and Weinberg 1978; Leinert et al. 1980). The seasonal variation Ecliptic Poles in the IRAS data (see section VI.B.3.b and Figs. VI.B.¼.¼ and VI.B.1.2 of the IRAS Catalogs and Atlases Explanatory Supplement 1984), which probes the region at 1 AU, has been analyzed by Hauser and Gautier (1984) and in more detail by Rickard et al. (1985). The longitude of the ascending node consistently lies in the 74-78 degree range. These analyses give a symmetry plane inclination of 2-2.5 deg. if one assumes a North-South dust density scale height at 1 AU of 0.48 AU (Leinert et al. 1978). A preliminary analysis of the seasonal variation of the ecliptic latitudes of the apparent symmetry plane in the IRAS data, which probes the region exterior to 1 AU, yields an ascending node of about 55 deg. and an inclination of 1.5 deg. (Hauser and Gautier 1984). If one takes the uncertainty estimates in these various determinations at face value, there is a systematic variation of symmetry surface parameters as one moves outward from the Sun, i.e., the symmetry surface is not a plane. In any event, the IRAS data confirms that the interplanetary dust cloud is not symmetric with respect to the ecliptic plane, implying that the zodiacal emission foreground in space infrared observations is dependent upon time of year as well as observation angles relative in the Sun.

2.3. Zodiacal Dust Bands and Dust origin

The bands of enhanced emission near the ecliptic plane are subtle but
perhaps the most significant features of the interplanetary dust
cloud discovered in the IRAS data. A reason for high interest in
these bands is their implication of an asteroidal origin of at least
part of the cloud. Appearing in individual scans as three small
brightness enhancements, roughly 5% of the total, at the ecliptic
plane and roughly 10 degrees on either side of it (when viewed at 90
degrees elongation), these features are found to be present at all
ecliptic longitudes. Low et al (1984) determined a color temperature
of 165-200 K, implying a heliocentric distance for grey particles of
2.7±0.5 AU and leading to the suggestion that they are produced by
asteroidal collision products in the main belt. Subsequent IRAS data
obtained at elongation angles ranging from 60 to 120 degrees showed
variation in apparent separation of the outer band consistent with
circular, heliocentric orbits at 2.5 AU inclined at 8.1 deg. with
respect to the ecliptic plane (Gautier et al. 1984).
         Dermott et al. (1984a; 1984b) noted that the three most
prominent Hirayama families of asteroids have semi-major axes and
inclination quite close to those needed to account for these emission
bands, the Koronis and Themis families contributing to the outer
bands, and the Eos family the band at the ecliptic plane. These
investigators also noted that the surface area required to produce
the observed optical depth in the bands ($\sim 10^{-8}$) is consistent with
the particle size spectrum expected in these families. On the
assumption that only gravitational forces are significant, they
derived detailed orbital parameters (inclinations and eccentricities)
which lead to predictions of latitudinal variations in positions and
longitudinal variation in brightness of the bands. If these
predictions are confirmed by further analyses of the band data, a
positive link between asteroids and at least some of the
interplanetary dust particles will have been established.

3. STATUS OF ZODIACAL EMISSION MODELING

Models of the zodiacal emission are desired for at least three
purposes: to characterize the dominant infrared foreground emission
so that future space infrared investigations and instruments can be
optimally designed; to provide a convenient form of the zodiacal
emission data for use in studying properties of the interplanetary
dust cloud; and to permit removal of the bright foreground from IRAS
and future instrument data to facilitate study of galactic and
extragalactic infrared sources. Each of the objectives imposes
different quality criteria on such models; because of the dominance
of the zodiacal intensity over other large-scale sources in much of
the sky, the third objective may be the most difficult to satisfy.
         Preliminary modeling with the IRAS data has taken two forms:
empirical characterization of the sky brightness, and attempts to
determine parameters in physical dust cloud models which can be used
to reproduce the observed brightness. The first approach has been

pursued by Good et al. (1984), who fit a four-parameter curve to the
band-averaged brightness distribution in each scan in each IRAS band.
The scan data were analyzed in polar coordinate form, with intensity
and inclination angle as the radial and angular coordinates
respectively. An ellipse was chosen as a fitting function, and the
four parameters were the semi-major and semi-minor axes, a tilt angle
between the major axis and the ecliptic plane, and an offset of the
center of the ellipse in the minor axis direction. Each fit was done
in an iterative fashion to the lower envelope of the data, assuming
that positive excursions generally represented galactic sources. This
scheme yielded reasonable fits for most scans at the shorter
wavelengths, but often failed to converge at 100 micrometers. An
attempt was made to construct a 'global' model by fitting each
parameter (at each wavelength) as a function of elongation angle and
time. As expected, the parameters were tightly correlated with
elongation angle and showed seasonal variations. However, small
linear time-dependent trends in the parameters also indicate that
residual calibration errors remain: complete photometric closure of
the data has not yet been achieved.

     The model of Good et al. is a reasonable first approximation to
the zodiacal emission brightness, with typical residuals at the
several percent level. Its most obvious shortcomings are lack of
definitive discrimination between large-scale solar B picture of the
100 micrometer emission, and some systematic deviations from the
shape of the intensity profiles, especially near the ecliptic plane.
Low-resolution all-sky images have been made from the residuals from
these fits to get a first look at the extra-solar system sky (White
et al. 1985). While the zodiacal emission bands, not modeled in these
fits, are a prominent feature in these images, the galactic plane and
regions of galactic cirrus are also clearly visible.

     Attempts to construct physical models of the emission of the
main dust cloud are still at a very early stage. The efforts to
contribute to this process by provinding some of the needed geometric
properties of the cloud. Rickard et al. (1985) have examined simple
models using Sun-centered, azimuthally symmetric dust density
distributions with power law radial variation and exponential
vertical variation with scale height proportional to radius. The
grain emissivity was assumed to vary with wavelength as a power law.
Parameters were adjusted to try to represent the scans of 24 June 83
(where the galactic and ecliptic planes were maximally separated).
While the profiles of the three shortest wavelength bands could be
reasonably reproduced with density distribution parameters similar to
those found in optical consistently underestimated. This may
represent an inadequacy of the simple model, a misinterpretation of
the amount of extra-solar system emission at 100 micrometers, or,
perhaps, a remaining zero point error in the calibration. Each of the
possibilities needs further investigation.

4. CONCLUSION

The IRAS survey produced a detailed picture of the zodiacal emission,

which dominates the large-scale sky brightness over much of the
infrared spectrum. The faint zodiacal emission bands discovered by
IRAS may show that at least part of the interplantary dust is
asteroidal in origin. Initial attempts to characterize the zodiacal
emission both empirically and with physical models provide reasonably
accurate representations of the sky brightness at 12, 25, and 60
micrometers. However, much additional work will be needed to
construct accurate maps of the extra-solar system sky, especially at
100 micrometers.

NOTE added 'in proof': A 1/2-degree position error (in the scan
direction) was recently found in the zodiacal observation history
file and all-sky maps. Corrected cloud geometry parameters are
expected to differ only modestly from those reported here.

ACKNOWLEDGEMENTS

The results reviewed here are compiled from the work of many
colleagues, both within and outside of the IRAS project, for whose
collaboration and helpful discussions we are grateful. This work was
supported by NASA under the IRAS Extended Mission program.

REFERENCES

Briotta, D.A. 1976 Ph.D. Thesis, Cornell University.
Dermott, S.F., Nicholson, P.D., Burns, J.A., Houck, J.R.
        1984a, Nature, 312, 505;
        1984b, 'An Analysis of IRAS' Solar System Dust Bands',
        published in Proc. of IAU Colloquium 85, "Properties and
        Interactions of Interplanetary Dust".
Gautier, T.N., Hauser, M.G., Low, F.J. 1984, Bull. Am. Astr. Soc.,
        16, 442.
Good, J.C., Gautier, T.N., Hauser, M.G. 1984, Bull. Am. Astr. Soc.,
        16, 921.
Hauser, M.G. et al. 1984a, Ap. J. (Letters), 278, L15.
Hauser, M.G. et al. 1984b, "IRAS Observations of the Interplanetary
        Dust Emission",   published in Proc. of IAU Colloquium 85,
        "Properties and Interactions of Interplanetary Dust".
Leinert, C., Hanner, M., Fitz, E. 1978, Astr. Ap., 63, 183.
Leinert, C., Hanner, M., Richter, I., Fitz, E. 1980, Astr. Ap., 82,
        328.
Levasseur-Regourd, A.C. Dumont,R. 1980, Astr. Ap., 84, 277.
Low, F.J. et al. 1984, Ap. J. (Letters), 278, L19.
Misconi, N.Y. Weinberg, J.L. 1978, Science, 200, 1984.
Misconi, N.Y. 1980, "Solid Particles in the Solar System", I.
        Halliday B.A. McIntosh (eds.), 49-53.
Neugebauer, G. et al. 1984, Science, 224, 14.
Rickard, L.J., Dwek, E., White, R.A., Hauser, M.G. 1985, Bull. Am.
        Astr. Soc., 17, 591.
Roser, S., Staude, H.J. 1978, Astr. Ap., 67, 381.
White, R.A., Good, J.C., Rickard, L.J., Wieland, J., Hauser, M.G.
        1985, Bull. Am. Astr. Soc., 17, 591.

HELIOCENTRIC DEPENDENCES OF ZODIACAL EMISSION, TEMPERATURE AND ALBEDO

R. Dumont
Observatoire de Bordeaux
B.P. 21
F - 33270 Floirac
France

A.C. Levasseur-Regourd
Service d'Aéronomie du C.N.R.S.
B.P. 3
F - 91370 Verrières-Le-Buisson
France

ABSTRACT. Values of the temperature, of the global emissivity, and (by comparison with zodiacal light optical observations) of the albedo of interplanetary dust are retrieved from IRAS data (Hauser et al., 1984) in two localized regions of the line-of-sight, near 0.98 and 1.40 a.u. heliocentric distance. The model-dependence is reduced, compared to the values averaged on the line-of-sight. The temperature probably does not exceed 250 K at the distance of the Earth. The albedo is found to decrease from ∿ 0.08 at the distance of the Earth to ∿ 0.06 at the distance of Mars.

1. APPLICATION TO IRAS OF THE LOCALIZING METHOD "OF THE SECANT"

The (line-of-sight integrated) zodiacal emission in the ecliptic plane, as measured in mid-February 1983 by IRAS, had the following brightnesses, I_ν (MJy ster^{-1}), at the following solar elongations, ε :

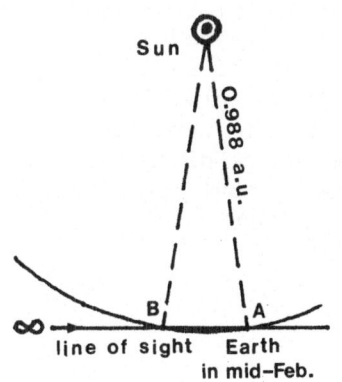

Wavelength (µm) ⟍ Elongation (°)	81.5	98.4
12	59	43
25	105	82

Since these elongations are symmetric w.r.t. 90° within 0.05° error, then the differences

$$59 - 43 = 16$$
$$105 - 82 = 23$$

represent the contributions to the integrated emissions by a chord of angular aperture 16.9° which crosses the Earth's orbit.

All the points between B and A have the same heliocentric distance r = 0.98 a.u. within 1 % error. The ratio 16/23 = 0.696 being that of the brightnesses at 12 and at 25 µm of a black- or grey-body at 253 K, this has to be the temperature of the zodiacal dust located near 0.98 a.u. and assumed to radiate like a grey-body.

F. P. Israel (ed.), Light on Dark Matter, 45–46.

In a more elaborate treatment we have taken into account :
  i) the precision and the accuracy claimed for the data,
  ii) the inverting possibilities developed in the optical case by us (1983, 1984, 1985a and 1985b).

Our "method of the secant" allowed to retrieve the local contributions at two "nodes of lesser uncertainty" located one near 1.5 a.u. heliocentric distance, the other near the subsolar point of the line-of-sight. In the thermal case the locations are slightly different : $r = 0.98$ and 1.40 a.u. in the present case. Then we have access to two localized values : of the temperature T ; of the monochromatic emissivities $\mathcal{E}_\nu$ at 12 and 25 μm, therefore of the global emissivity $\mathcal{G} = \int_0^\infty \mathcal{E}_\nu \, d\nu$; and (by comparison with the values of the directional scattering coefficient $\mathcal{S}$ available in our abovementioned optical papers) of the albedo $\mathcal{A} = \mathcal{S}c/(\mathcal{G} + \mathcal{S}c)$, where c denotes the solar irradiance at the relevant distance. This albedo is defined for $\theta = 90°$ scattering angle (see Hanner, 1980).

	$r = 0.98$ a.u.	$r = 1.40$ a.u.	Difference(***)
T (K)	253 ± 43	216 ± 45	37 ± 10
$\mathcal{G}$ (*)	$4.62 \, ^{+0.40}_{-0.10}$	$1.86 \, ^{+0.45}_{-0.23}$	2.76 ± 0.20
$\mathcal{A}$	$0.080 \, ^{+0.010}_{-0.015}$	$0.060 \, ^{+0.018}_{-0.021}$	0.020 ± 0.014

(*) In 10^{17} w.(a.u.)$^{-3}$ ster^{-1}.
(***) Less sensitive to calibration errors, as the observers emphasize.

2. DISCUSSION AND CONCLUSION

Compared to those averaged along the line-of-sight, our results are in rough agreement on the temperatures (Hauser et al., 1984) and in good agreement on the albedo (Hauser et al., 1985). Our localizing process leads us to a heliocentric decrease of the albedo, already suggested by us (1984, 1985b). This corroborates the heliocentric change of polarization observed by the Helios probes (Leinert et al., 1981) and formally derived by the inversion approach (Dumont, 1983), to definitely invalidate the homogeneity of the zodiacal cloud. The observed change of brightness $R^{-2.3}$ interpreted up to now as a $r^{-1.3}$ change of density, could be partly due to the fall of albedo, while the fall of density, proportional to r^{-1}, would then correspond to a better theoretical likelihood.

REFERENCES

Dumont R. 1983, *Planetary Space Sci.* **31**, 1381
Dumont R., Levasseur-Regourd A.C. 1985a, *Planetary Space Sci.* **33**, 1
Dumont R., Levasseur-Regourd A.C. 1985b, *IAU Coll.* **85**.
Hanner M.S. 1980, *Icarus* **43**, 373
Hauser M.G. et al. 1984, *Astrophys. J.* **278**, L-15
Hauser M.G., Gautier T.N., Good J., Low F.J. 1985, *IAU Coll.* **85**.
Leinert C., Richter I., Pitz E., Planck B. 1981, *Astron. Astrophys.* **103**, 177
Levasseur-Regourd A.C., Dumont R. 1984, *C.R. Acad. Sci. Paris* **300 II**, 109

SPECTRAL DECOMPOSITION OF IRAS MAPS.

R. Braun, R.G. Strom, H. v.d. Laan, H. Greidanus
Sterrewacht
Postbus 9513
2300 RA Leiden
Netherlands

The large-scale emission present in IRAS maps is usually dominated by zodiacal light, with a spectrum that resembles dust at ~ 190 K, and emission from cold interstellar dust at ~ 20 K. The confusion due to this background tends to obscure other, weaker emission, such as e.g. the emission from the shock-heated dust (T ~ 60 K) we have been looking for in supernova remnants (Braun et al 1986).

We were able to remove much of this background however, using the following procedure.

Assuming IRAS in-band flux ratios for emission from the three components zodiacal light, shock-heated dust and cold dust, the intensity of each component could be determined in every mappoint separately, by a least squares fit to the four IRAS fluxes. In subsequent iterations, assumed flux ratios were tuned so as to minimize the fit-residuals and minimize the spatial correlation between components found. Using the two-dimensional information present in the maps in this way, actual in-band flux ratios could be determined from initial estimates.

The results of this technique are illustrated by the maps of the Cygnus Loop taken from Braun (1985) (Fig. 1 and 2). Only bands 2 and 4 are shown here; in bands 1 and 3 similar improvement is attained.

References.

Braun, R. 1985, Ph.D. Thesis
Braun, R., Strom, R.G., v.d.Laan, H., Greidanus, H. 1986, in these
    proceedings, page 155.

F. P. Israel (ed.), Light on Dark Matter, 47–48.
© 1986 by D. Reidel Publishing Company.

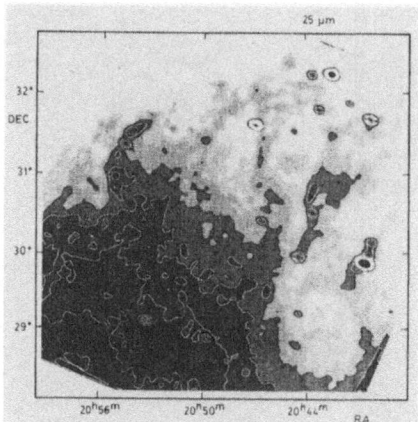 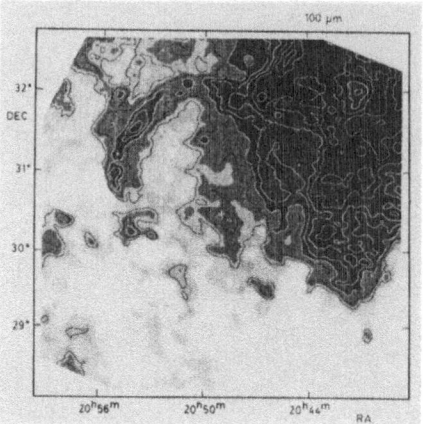

Figure 1. IRAS images in the 25 and 100 micron band of the Cygnus Loop. Contours increase logarithmically by a factor of 1.2. The 25 μ image (lowest contour 1.3 MJy/sr) is dominated by zodiacal light emission (SE corner), while the 100 μ image (lowest contour 10 MJy/sr) is badly contaminated with emission from cold dust (NW corner).

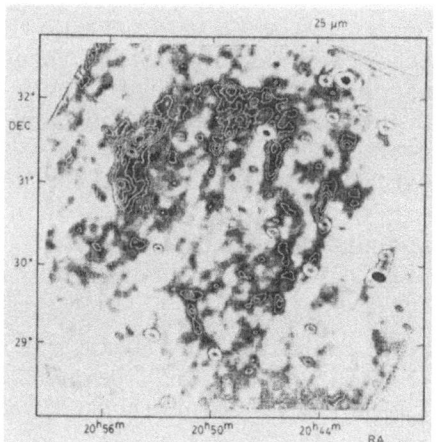 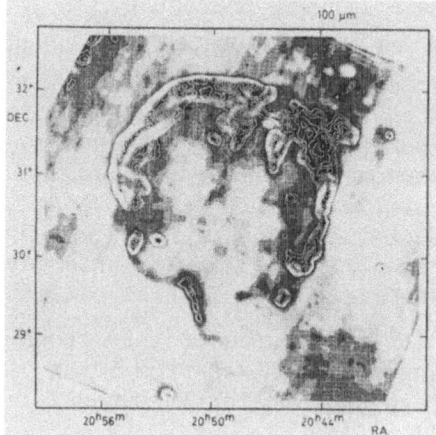

Figure 2. The maps of figure 1 after subtraction of the cold dust component and a second order surface fit to the zodiacal light component. In the 25 μ image the contours increase logarithmically by a factor of 1.3 from the lowest contour of 0.32 MJy/sr; in the 100 μ image the factor is 1.1 and the lowest contour is 2.82 MJy/sr.

Observations of Infrared Cirrus

T.N. Gautier
IPAC, California Institute of Technology
Pasadena, California

A striking feature of the infrared sky seen by the IRAS survey is the presence of thin, wispy clouds of infrared emission over nearly all the sky at high Galactic latitudes. The strong resemblance of the appearance of this interstellar material to that of cirrus clouds in the Earth's atmosphere suggests the name "infrared cirrus". Emission from this cirrus is apparently the infrared manifestation of the dust which can be seen as regions of reflection nebulosity or additional interstellar absorption in good quality photographs and which is expected to be associated with diffuse atomic hydrogen clouds detected at high latitudes. Infrared data on the material, when combined with measurements at other wavelengths, can provide an important key to understanding the composition and distribution of interstellar material. This paper reviews the observations and our current understanding of infrared cirrus, describing the morphology and observed spatial distribution of the cirrus and compares the IRAS data with observations at other wavelengths. The measured properties of the cirrus material are related to models of interstellar dust. Finally some of the unique observational opportunities presented by the cirrus are mentioned along with a review of some recent observational results.

1. MORPHOLOGY

Figure 1 shows the appearance of infrared cirrus in a field at Galactic latitude -70°. This example of cirrus has many of the common small scale cirrus structures; long, spider-like filaments, clumps and long, arcing structures composed of a series of small wisps, filaments and clumps. Small scale structure is seen down to the resolution limit of the IRAS data, 2 to 4 arc minutes. Cirrus structures like these are seen over the whole sky although they blend into the edges of large dust complexes like Orion and Ophiuchus and lose definition at low latitudes toward the central parts of the Galaxy. Larger scale structures also appear in the cirrus. Many nearly circular "holes" of lower surface brightness than their surroundings can be found ranging in diameter from 1 or 2 degrees to more than 10 degrees.

F. P. Israel (ed.), Light on Dark Matter, 49–54.
© 1986 by D. Reidel Publishing Company.

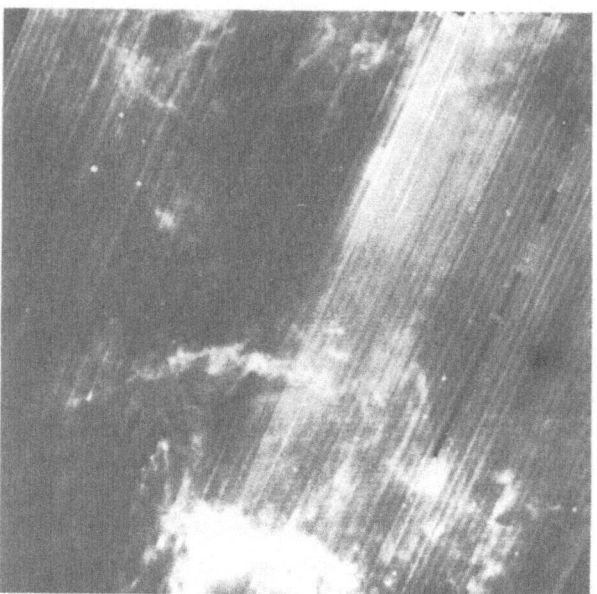

Figure 1.   100μm IRAS map
of 16.5 x 16.5   degree
field centered at α=1^h,
δ=0° in the constellation
Cetus.   Lighter areas
have   higher surface
brightness.   The bright
cloud in the center of
the lower edge is   cloud
A of Low et al. (1984).
North is up and east to
the left.

        The distribution of cirrus over the whole sky is shown in Figure 2.
Outside the burned out areas surrounding the Galactic plane and the
dusty regions of Ophiuchus, Taurus and Orion, clumps and arcs of cirrus
emission are seen to cover the sky all the way to the Galactic poles.

2. CIRRUS AT OTHER WAVELENGTHS

        Nearly exact correspondence between cirrus emission and optical
reflection nebulosity is seen wherever optical nebulosity has been
measured.   A good example is the deep optical plates made by Sandage
(1976) of the region around M81 and M82 which clearly show the same
material in reflection that IRAS shows in emission at 60 and 100μm.   De
Vries (1985) has also shown excellent correspondence in detail between
small scale structures of infrared cirrus and visible nebulosity seen on
ESO/SERC southern sky survey plates.   Details of structure smaller than
one arc minute can be seen in the nebulosity on the ESO/SERC plates.
Boulanger, Baud and van Albada (1985) have shown a clear but not perfect
correlation of small scale features in HI surface brightness with 100μm
surface brightness.

        Large scale cirrus features seen in Figure 2 show good correlation
with the distribution of atomic hydrogen.   The arc like structures in
the cirrus rising out of the Galactic Plane between l=0° and l=90° and
descending back between l=180° and l=270° correspond well with similar
features seen in the HI maps made by Heiles and Jenkins (1976).

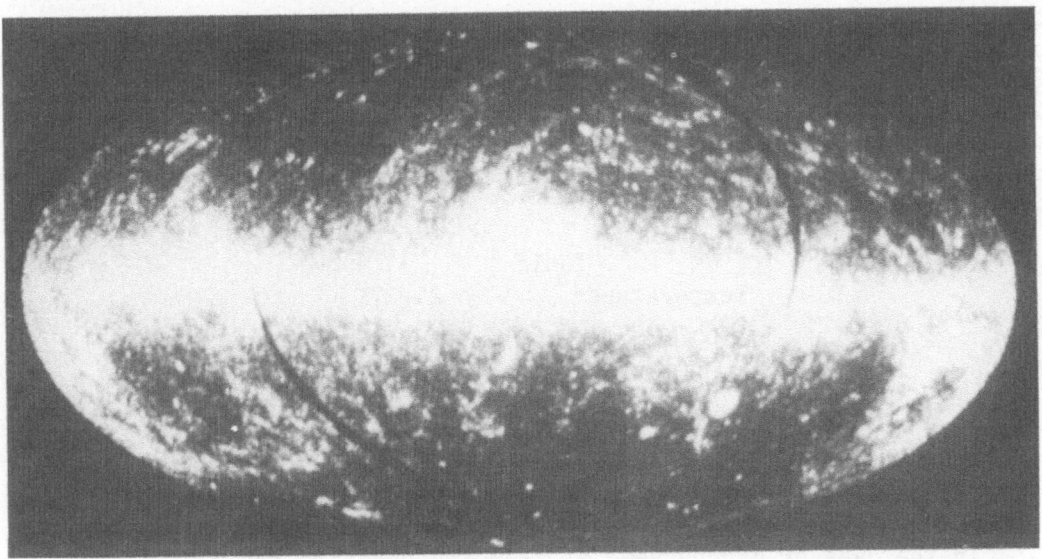

Figure 2. Full sky map of infrared cirrus at 100μm projected in
Galactic coordinates. This map was produced with a form of high
passfiltering of the time ordered IRAS survey data. l=0°, b=0°
is at the center, l=180° is on the left and right edges,
Galactic north is up, longitude increases to the left.

3. PHYSICAL PROPERTIES OF CIRRUS DUST

Cirrus is most easily seen in the IRAS data at 100μm but is also
plainly visible at 60μm. Emission from cirrus at 25μm is much fainter
than at 60 or 100μm. Only clouds with very high surface brightness at
100μm or clouds near large molecular clouds and HII regions where dust
may be strongly heated by nearby stars are visible at 25μm. At 12μm,
however, many of the same cirrus features seen at 60 and 100μm are again
visible, implying a high temperature for the material composing the
cirrus. The ratio of 12 to 100μm brightness is highly variable among
cirrus clouds, ranging from about 0.1 to less than 0.02. Clouds with
bright 12μm emission are in the minority. The spatial distribution of
bright 12μm cirrus is not yet completely understood but examples do
occur at large distances from any apparent local heating source where
the equilibrium temperature should not exceed a few tens of Kelvins even
for small grains. No cirrus clouds have been seen at 12μm that are not
present at 60 and 100μm.

Assuming that the long wavelength emission from cirrus dust is due
to a thermal continuum, temperatures and optical depths can be obtained.
Table 1 gives these properties, abstracted from the results of Low et
al. (1984), for a representative cirrus cloud similar to the bright
cloud in Figure 1. The temperature would go down by about 2K if an
emissivity law of λ^{-2} were assumed. This would bring the temperature

and emissivity model into agreement with other recent results. These
temperatures are appropriate for ordinary interstellar dust grains
heated by the general interstellar radiation field (Mezger, Mathis and
Panagia, 1982). The optical depth range given in Table 1 should not be
taken as representative of all cirrus. Some clouds are substantially
thicker than $6x10^{-5}$ at 100μm and the detection limit for cirrus on a
typical IRAS 100μm map is 10 to 20 times thinner for the same
temperature.

Table 1

Temperature*	21-27K
$\tau_{100\mu m}$	3-6 $x10^{-5}$
A_V (from HI)	.1 - .2 mag

*from the 60μm/100μm brightness ratio assuming emissivity
proportional to $\lambda^{-1.5}$

In the case of cirrus seen at 12μm the observed ratio of 12μm to
100μm emission requires a dust temperature of more than 80K and, where
25μm emission is measurable, the 12μm to 25μm ratio requires
temperatures between 400 and 1000K (Gautier and Beichman, 1985). These
same clouds show a 60 to 100μm brightness ratio appropriate to a
temperature of 21 to 27K. The best explanation available at this time
for the bright 12μm emission envokes non-equilibrium heating to ~1000K
of very small grains, ~10A radius, by the absorption of a single
ultraviolet photon. This emission mechanism was suggested several years
ago (Harwit 1975, Purcell 1976) and was recently applied to the near
infrared emission of circumstellar reflection nebulae by Sellgren and
her co-workers (Sellgren et al., 1983, Sellgren, 1984). This mechanism
allows uv photons in the general interstellar radiation field to warm
very small grains in cirrus clouds to temperatures needed for short
wavelength emission. Differences in 12μm brightness among clouds could
then be explained by variations in the relative number of very small
grains in the dust composing the cloud.

Léger and Puget (1984) have hypothesized that the small grains
which are responsible for the short wavelength emission in Sellgren's
model are large carbon molecules consisting of approximately 50 atoms.
Puget, Léger and Boulanger (1985) have computed spectra expected from
such small carbon particles in typical interstellar radiation fields.
Sample spectra from their paper are reproduced in figure 3. The
emission lines seen between 3 and 12μm in the synthetic spectra are due
to hydrocarbon bonds and are proposed by Leger and Puget to explain the
previously unidentified emission features seen in this region of the
spectrum of many Galactic and extra galactic objects. These emission
lines also fall within the pass band of the IRAS 12μm observations and
can probably explain the 12μm emission from cirrus clouds. An infrared
spectrum of an example of 12μm cirrus is needed to confirm the presence
of emission in the hydrocarbon bands.

4. CIRRUS OBSERVATIONS

Cirrus clouds at higher Galactic
latitudes present a unique opportunity
to determine properties of interstellar
dust and interstellar radiation fields.
Single, unconfused high latitude clouds
can be observed in a (probably) simple
radiation field. The combination of
optical/uv measurements of an isolated
cloud's extinction and reflected
surface brightness with infrared
measurements of emitted surface
brightness should allow an accurate
determination of properties of dust
within the cloud and the interstellar
radiation field around the cloud.

Two recent pieces of observational
work on cirrus are particularly
interesting. Wieland, Blitz and co-
workers (Wieland et al., 1985) have
combined the IRAS data with the Blitz,
Magnani and Mundy (1984) high latitude
CO survey to find that where CO is
found at high latitudes cirrus is
always present. There is good
morphological correspondence between
the CO and the cirrus. The typical
distance to a high latitude CO cloud
was established statistically as 100 pc
making the CO and presumably the
associated cirrus quite local. The
temperature of the dust as derived from
60μm to 100μm brightness ratios,
assuming a λ^{-2} emissivity, varies from
17K to 27K and clouds typically show an
infrared temperature gradient becoming
colder towards the center.

De Vries and Le Poole (1985) and
de Vries (1985) have compared the
optical and infrared properties of two
cirrus clouds. Optical information was
obtained from ESO/SERC IIIa-J southern
survey plates where the surface
brightness of the cirrus was measured
directly and the extinction through the
clouds was determined by differential
star counts. Both clouds have
temperatures of ~20K based on 60 and

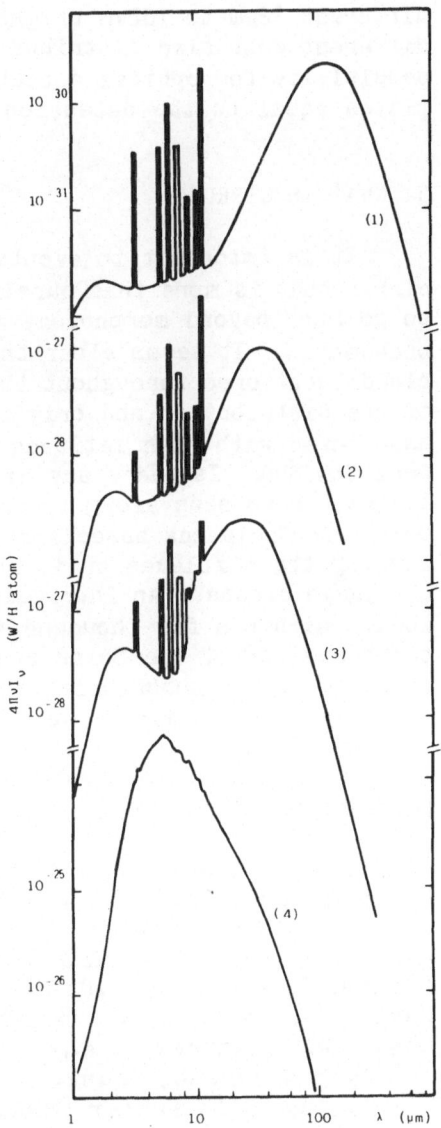

Figure 3.   Synthetic spectra
of polycyclic hydrocarbon
grains for several different
radiation environments. 1)
local interstellar radiation
field; 2) 1pc from 05 star;
3) 0.33pc from 05 star; 4)
100 stellar radii from M0
star. Reproduced from Puget,
Leger and Boulanger (1985).

100μm surface brightnesses and, while the clouds differ by a factor of 3
in 100μm optical depth, both have a ratio of visual extinction to 100μm
surface brightness of about 0.11 mag/MJy/sr. The two clouds have very
different 12μm to 100μm brightness ratios probably indicating quite
different dust size distributions. De Vries and Le Poole find the
sensitivity for optical detection of cirrus on good quality ESO/SERC
plates equal to the detection sensitivity of the IRAS 100μm data.

5. WHAT IS CIRRUS?

    It is important to eventually produce a definition of infrared
cirrus that is more than purely morphological. However, it is premature
to go much beyond morphology at this stage in our understanding of the
phenomenon. It seems clear that cirrus is made up of tenuous dust
clouds scattered throughout the interstellar medium but many questions
of the distribution and origin of the clouds remain. Cirrus clouds
associated with high latitude CO appear to be quite local, about 100pc
from the Sun. Is there any high latitude cirrus which is more distant?
Is the cirrus seen along lines of sight near large dusty complexes like
Orion and Ophiuchus associated with those complexes? If so, is it
forming the complexes or is it the result of dissolution of the edges of
the large clouds? An intriguing possibility is that some cirrus is very
local, within a few thousand AU of the Sun, and may be left over from
the formation of the Solar system. These questions and speculations
indicate that a substantial amount of research needs to be done before
"infrared cirrus" can be more than a morphological description.

References

Blitz, L., Magnani, L., Mundy, L. 1984, ApJ (Letters), 282, L9.
Boulanger, F., Baud, B., van Albada, G.D. 1985, Astr. Ap., 144, L9.
Gautier, T.N., Beichman C.A. 1985, Bull. Amer. Astr. Soc., 16, 968.
Harwit, M. 1975, ApJ, 199, 398.
Heiles, C., Jenkins, E.B. 1976, Astr. Ap., 46, 333.
Leger, A., Puget, J.L. 1984, Astr. Ap. 137, L5.
Low, F.J., Beintema, D.A., Gautier, T.N., Gillett, F.C., Beichman, C.A.,
    Neugebauer, G., Young, E., Aumann, H.H., Boggess, N.,Emerson, J.P.,
    Habing, H.J., Hauser, M.G., Houck, J.R., Rowan-Robinson, M., Soifer,
    B.T., Walker, R.G., Wesselius, P.R. 1984, ApJ (Letters), 278, L19.
Mezger, P.G., Mathis, J.S., Panagia, N. 1982, Astr. Ap., 105, 372.
Puget, J.L., Leger, A., Boulanger, F. 1985, Astr. Ap., 142, L19.
Purcell, E.M. 1976, ApJ, 206, 685.
Sandage, A. 1976, AJ, 81, 954.
Sellgren, K. 1984, ApJ, 277, 623.
Sellgren, K., Werner, M.W., Dinerstein, H.L. 1983, ApJ, 271, L13.
de Vries, C.P., 1985, preprint.
de Vries, C.P., Le Poole, R.S. 1985, Astr. Ap., 145, L7.
Wieland, et al. 1985, in preparation.

DUST AT THE NORTH GALACTIC POLE

J. Knude
Copenhagen University Observatory
Øster Voldgade 3
DK-1350 København K
Denmark

ABSTRACT.   Results on the most reddened part of the NGP are reported.
Large fractions of the pole zone b>+70° are apparently reddened by more than E(b-y) = 0.05 mag. The reddened areas may be organized in parallel string-like features. The IRAS clouds B and C are located in the NGP
zone and a reasonable agreement is obtained for the 100μ intensities
estimated from the optical absorption observations and the infrared
emission measurements. It is further indicated that the gas/dust ratio
varies across the polar area.

1.   GENERAL DISTRIBUTION

A systematic reddening survey of the NGP have been completed by observing all A5-G0 stars brighter than V~11.3 mag above b=+70°.A5-G0 stars
within this limiting magnitude are abundant nearer than 400 pc. The resulting network for the 1200 sq.deg. surveyed has more than 4*/sq.deg.
allowing detection of even quite small structures. The error of one color excess is thought to be smaller than 0.01 mag.Substantial amounts

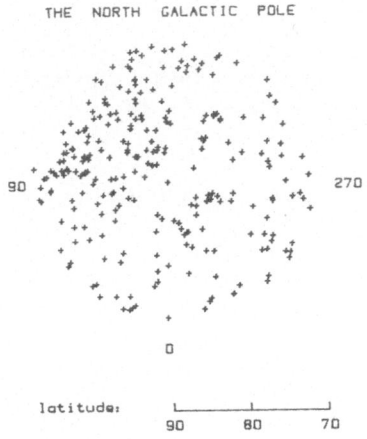

THE NORTH GALACTIC POLE

Figure 1. Lines of sight with E(b-y) > 0.05 mag.

F. P. Israel (ed.), Light on Dark Matter, 55–58.
© *1986 by D. Reidel Publishing Company.*

of reddening are present almost in the entire cap. Several directions have A_v>0.2 within 100 pc, Knude(1985). A dust free cap would have an average excess ~0 and very few excess above 5σ~0.05 mag. But as Fig.1 shows
more than 250 lines of sight are observed to have E(b-y) > 0.050 mag.
The large excesses are concentrated in the segment 1: 40°- 210° but are
also present in the remaining part of the cap. The only dust feature
reported previously is that at (l,b)~(273,84) observed by Hill,Hilditch
and Barnes (1983). The 100μ intensities above 3MJysr^{-1} at these high
latitudes will probably show a distribution similar to Figure 1.

2.   INDIVIDUAL CLOUDS

Figure 2 shows some ~25 deg^2 regions containing the IRAS clouds B and C,
Low et al. (1984). The dust column distribution in these areas are unusual by being bimodal. There is a high probability to encounter excesses
above 0.03 mag compared to the frequency of clear lines of sight.
Cloud B: <E(b-y)> = 0.018 resulting in I(100μ)~1.3 MJysr^{-1} using Draine
and Anderson (1985) with χ=1 and a(min)=0.001 μm and assuming a standard
gas/dust ratio. It may however be more appropriate to use exclusively the
profile symmetric around the bin 0.03-0.04. This results in I(100μ)=2.5
MJysr^{-1}. The gas/dust ratio may however not be normal in the cloud B direction. From Burstein and Heiles (1982)N(H)=5.2 10^{20}atoms cm^{-2}implying
N(H)/E(b-y)=1.6 10^{22}atoms cm^{-2} mag^{-1}twice the standard ratio. The same
high gas/dust ratio is found in the cloud C direction. If the Draine and
Anderson calculation is corrected for the abnormal gas/dust ratio an

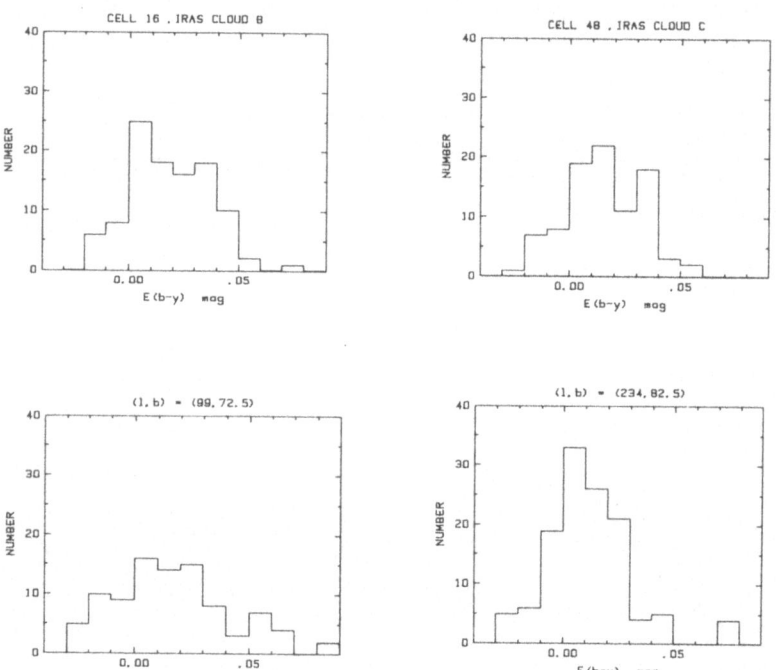

Figure 2. Dust column distributions in four ~25 deg^2 zones.

emission $I(100 \mu) = 5.0$ is expected for cloud B; 5.0 compares well to the observed 4.3 MJysr^{-1} ,Low et al.(1984). Similar considerations for cloud C result in the intensity estimates 1.8 MJysr^{-1} with the standard ratio and 3.6 MJysr^{-1} if the observed ratio for the cloud C direction is used. The IRAS intensity for cloud C is 1.3 MJysr^{-1}. A more detailed spatial comparison may refine the already good agreement. Figure 2 also shows a 25 deg^2 zone centered on $(l,b)=(99,72.5)$ where relatively many excesses above 0.05 have been found according to Figure 1. For this region the BH map shows little neutral hydrogen and the $N(H)$ maximum is given as $1.75 \ 10^{20}$ atoms cm^{-2}. Relating the average of the high excess profile for this area to the 21 cm maximum results in a gas/dust ratio $N(H)/E(b-y)= 3.1 \ 10^{21}$ atoms cm^{-2} mag^{-1} which is low by a factor of two compared to the standard ratio. This particular zone which may be typical for 'dust without gas' areas could be interesting to test for 100 μ emission. The intensity is expected to be in the range from 4 to 8 MJysr^{-1} depending on the gas/dust ratio. Figure 2 finally shows what the dust column density distribution is like in clear zones. The median column is clearly shifted toward lower excesses.

The total absorption computed from the high excess profiles in the cloud B and C directions are $A_v = 0.14$ and $A_v = 0.11$ mag respectively fairly close to 0.13 and 0.07 mag etimated from the infrared emission, Low et al. (1984).

3. REFERENCES

Burstein,D., Heiles,C.: 1982, *Astron.J* __87__, 1165
Draine,B.T., Anderson,N.: 1985, *Astrophys.J.* __292__, 494
Hill,G., Hilditch,R.W., Barnes,J.V.: 1983, *Mon.Not.R.astr.Soc.* __204__, 241
Knude,J. : 1985, *Mat.Fys.Medd.Dan.Vid.Selsk.* __41:1__,in press
Low.F.J. et al.: 1984, *Astrophys.J.(Letters)* __278__, L19

IRAS Field 49 $\alpha = 2^h00^m$, δ = +30°, HCON-3, 100 μm. Field in
                  Triangulum, showing M33 (right) and extensive
                  Galactic cirrus (left), connected to the Taurus dark
                  clouds.

SECTION 2.

STARS AND STELLAR PHENOMENA

IRAS OBSERVATIONS OF COOL EXCESS AROUND MAIN SEQUENCE STARS

F. C. Gillett
Kitt Peak National Observatory
National Optical Astronomy Observatories
P. O. Box 26732
Tucson, Arizona 85726-6732

ABSTRACT.   IRAS observations of the main sequence stars α Lyrae, α Piscis Australis, β Pictoris and ε Eridanus are reviewed and updated. All four stars exhibit cool excess emission that is attributed to thermal emission by orbiting circumstellar grains. A simple thermal model is used to determine surface areas and temperature ranges for the radiating material, and the state of this material is briefly discussed in terms of the formation and evolution of planetary systems.

1. INTRODUCTION

The discovery by IRAS of a large infrared excess associated with the bright star Vega (α Lyrae, BS7001) was the first detection of cool circumstellar material associated with a main sequence star. This excess was interpreted as thermal emission from solid grains in orbit around the star, which have grown to more than a millimeter in size as a result of the same processes that led to the formation of our planetary system (Aumann et al. 1984, paper I).

As the first direct evidence for such processes at work around another star, these observations and those of other main sequence stars with cool excess, have received considerable attention. IRAS observations of similar excesses associated with three other main sequence stars, α PsA (BS8728), β Pictoris (BS2020) and ε Eridanus (BS1084) (Gillett et al., 1984) and statistics of survey observations of stars within 20 pc (Aumann, 1984) have been presented.

The intent of this review is four-fold: 1) Review and update the IRAS results on α Lyr, α PsA, β Pic, and ε Eri. These stars are the brightest, best observed examples of the Vega phenomenon. 2) Interpret the excess in terms of a simple thermal shell model, deriving the temperature range and surface area, 3) Discuss other observations of these stars since the IRAS discoveries and 4) Briefly present ideas and issues concerning the evolutionary state of this material with respect to the formation of planetary systems.

F. P. Israel (ed.), Light on Dark Matter, 61–69.

2. OBSERVATIONS

TABLE I.

STAR	12μM	25μM	60μM	100μM
		TOTAL FLUX DENSITY (JY)		
α LYR	28.0 ± 0.8	9.2 ± 0.5	9.3 ± 0.5	7.5 ± 0.8
α PSA	11.6 ± 0.4	4.0 ± 0.2	9.8 ± 0.5	11.3 ± 1.1
β PIC	2.8 ± 0.1	10.4 ± 0.5	18.8 ± 0.9	11.2 ± 1.0
ε ERI	6.4 ± 0.2	2.0 ± 0.1	1.7 ± 0.1	1.9 ± 0.2
		PHOTOSPHERIC COMPONENT (JY)		
α LYR	28.0	6.45	1.1	0.4
α PSA	11.3	2.60	0.45	0.17
β PIC	1.16	0.27	0.05	0.02
ε ERI	6.3	1.45	0.25	0.09
		EXCESS (JY)		
α LYR	0.0 ± 0.8	2.75 ± 0.5	8.2 ± 0.5	7.1 ± 0.8
α PSA	0.3 ± 0.4	1.40 ± 0.2	9.35 ± 0.5	11.1 ± 1.1
β PIC	1.64 ± 0.1	10.1 ± 0.5	18.8 ± 0.9	11.2 ± 1.0
e ERI	0.1 ± 0.2	0.60 ± 0.1	1.45 ± 0.1	1.8 ± 0.2

The photometric measurements from pointed observations are presented in Table I. The photometry is based on the November 1984 absolute calibration (Beichman et al., 1985). The flux density uncertainties are due primarily to uncertainty in the absolute calibration. Also included in the table are the assumed stellar photospheric contributions determined by extrapolation from 3.4 μm measurements (Koornneef 1983) (Campins et al., 1985) assuming that [L]-[12] = 0.00. The photospheric contributions are extrapolated beyond 12um assuming $f_\nu \sim \nu^2$, i.e. with a hot blackbody energy distribution. The energy distributions of α Lyr, α PsA, and ε Eri are quite similar with no significant excess at 12μm, and a small but significant excess at 25μm which becomes dominant at 60 and 100μm. On the other hand, β Pic shows a large excess even at 12μm which dominates at 25μm and longer.

The pointed observations can be analyzed to provide information on the relative positions and sizes of the emitting regions (Gillett et al., 1985). The current estimates of the angular size of the 60μm emitting regions are summarized in Table II. The position difference between the centroids of the 12μm emission, assumed to define the position of the star, and the 60μm emission is less than 5" for all four stars.

3. RESULTS

In Paper I, it was argued that the excess emission associated with α Lyr was due to thermal radiation from solid grains in orbit around the star. The same arguments apply for all four stars considered here (see however, the discussion of ε Eri), and the following is based on this assertion. The basic line of reasoning is that: 1) the spectral shape of the excess emission is consistent with one or more Planck functions, while it is inconsistent with free-free continuum emission, 2) main sequence stars in general, and α Lyr in particular, show no evidence for significant current mass loss which could result in the formation of circumstellar grains, and 3) the predicted size of the dust shell, assuming blackbody particles in thermal equilibrium with radiation from the star, is consistent with the measured size of the radiating region.

Given this assumption, there are three constraints on the orbiting particle sizes that can be applied:

1) Very small grains will be removed from the vicinity of the star by radiation pressure. For spherical, black grains of diameter a in microns, radiation pressure exceeds gravitational force if $a < 1.2 \, L_s/\rho \, M_s$ where ρ is the particle density in gm/cc and L_s, M_s are the stellar luminosity and mass in solar units.

2) Over very long periods of time, the Poynting-Robertson (P-R) effect will cause particles to spiral into the star. The lifetime, t, of an orbiting particle initially R AU from the star is given by $t = 350 \, a \, \rho \, R^2/L_s$ years. Minimum grain sizes due to radiation pressure and the P-R effect are included in Table II.

3) The assumption of blackbody absorption and emission is largely a requirement that the grains be optically thick at 100μm. Likely grain constituents become increasingly transparent at the longer infrared wavelengths and grain mixtures, e.g. silicates or carbon plus ice, generally form low albedo, highly absorbing grains at optical wavelengths (Weissman, 1984). It is estimated based on optical constants of water ice (Irvine and Pollack, 1968) and silicates (Day, 1981), that grains must be substantially in excess of 100 μm diameter in order that the blackbody assumption be valid to 100μm.

The observed excesses have been fitted with a simple thermal model. In this model, the radiating cloud is assumed to be optically thin, and the individual particles are assumed to have optical emissivity $E_{vis} = 1$ and infrared emissivity $E_{ir} = 1$ (blackbody grains) or $E^{ir} = const/\lambda$. The grains are further assumed to be perfectly conducting, in thermal equilibrium with radiation from the star and distributed as $dn \sim r^{-\gamma}$ with $\gamma = 1$, 2 or 3 for $r_{min} < r < r_{max}$. R_{min} and R_{max}, or equivalently T_{max} and T_{min}, are adjusted to fit the observed excess. A spatial distribution established by the P-R effect would have $\gamma = 1$.

The thermal model results are summarized in Table II and Figure 1. The α Lyr, α PsA and ε Eri excesses can be well fitted with $\gamma = 1$ and $E^{ir} = 1$, although for α Lyr and α PsA the observed energy distribution is sufficiently narrow that other values of γ are equally acceptable. The excess around β Pic is not well reproduced for Eir = 1 and $\gamma = 1$, 2 or 3 (a typical fit for E^{ir} 1, and $\gamma = 1$ is shown in Figure 1). In addition, all $E^{ir} = 1$ models for β Pic predict shell sizes about 10x smaller than observed. Both the observed energy distribution and angular size can be reproduced for $E^{ir} = 3.7 \, m/\lambda$ and $\gamma = 1$. The temperature limits for ε Eri are not well defined with the dashed curve illustrating a high temperature cutoff at T = 500K.

In all cases, the lower model temperatures are such that water ice is stable for longer than the stellar lifetime. At temperatures above about 105K centimeter-sized bodies of ice would sublimate away in less than 10^8 years (Isobe, 1970). Thus, an inner edge with T above about

105K (i.e. for α Lyr, β Pic and ε Eri) would be the result of accretion or ejection of small non-icy grains spiralling into this regions by a massive unseen body inside the edge or the grains are sufficiently large that the P-R effect is negligible.

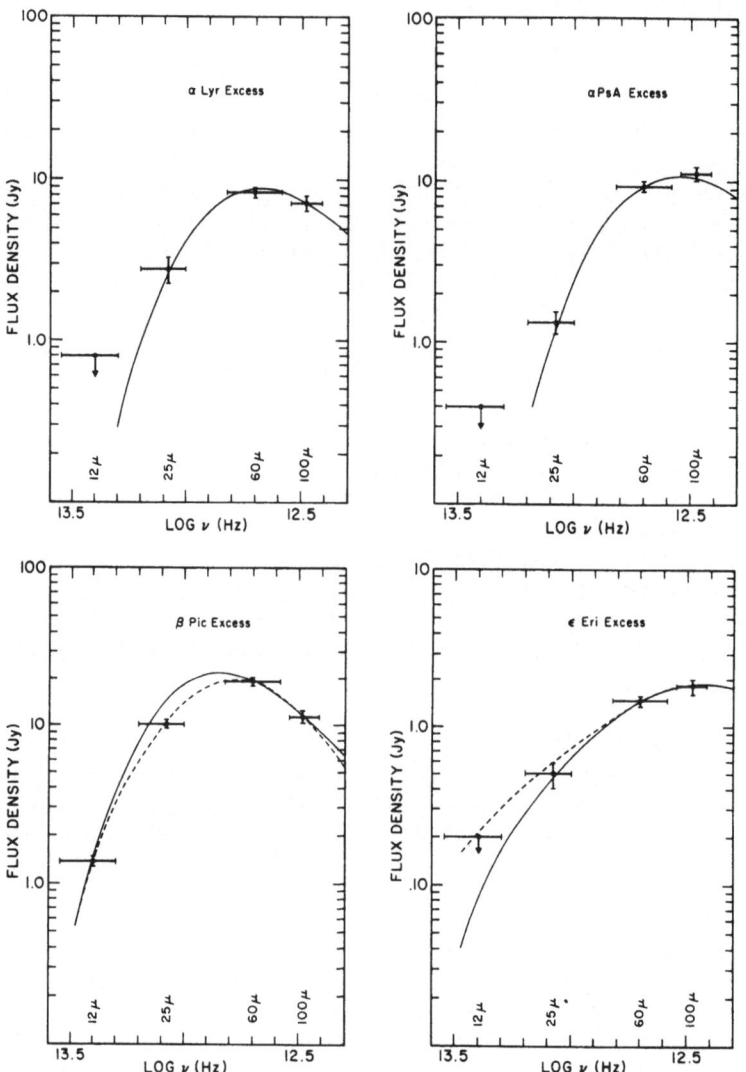

FIGURE 1 - Flux density distributions for the excess emission associated with the program stars. The solid curves for α Lyr, α PsA and ε Eri show fits for the blackbody models described in Table II. The solid curve for β Pic shows fit for blackbody model with γ = 1 and the dashed curve shows fit for the small particle model included in Table I. The dashed curve for ε Eri is for a model with T_{max} = 500K.

TABLE II

	α LYR	α PSA	β PIC	ε ERI	NOTES
STELLAR PROPERTIES					
Spectral Type	A0V	A3V	A5V	K2V	1
Distance (pc)	8.1	7.0	16.6	3.3	1
T_{eff}	9700K	8800K	8000K	5000K	1,2,3
Luminosity($L_{\odot}$)	60	13	6.5	0.37	1,2,3
Mass($M_{\odot}$)	2.5	1.9	1.7	0.9	4
MS Lifetime(yrs)	5E8	1E9	2E9	~ 2E10	5
Age(years)	~ 5E8	---	---	~ 1E9	6
Blowout Grain					
Diameter(μm)	14	4	2.3	0.24	7

	α LYR	α PSA	β PIC	ε ERI	NOTES
THERMAL MODEL SUMMARY					
Blackbody Grains: $E^{vis} = E^{ir} = 1$					
Temp Range (K)	130-60	100-45	--	250-30	
Spatial Range(AU)	36-170	28-140	--	0.75-52	
Small Grains: $E^{vis} = 1$, $E^{ir} = 3.7um/\lambda$					
Temp Range (K)	--	--	180-48	--	
Spatial Range(AU)	--	--	14-400	--	
Luminosity($L_{\odot}$)	1.6E-3	1.1E-3	1.9E-2	5.7E-5	
Grain Area(m^2)	2.5E23	5.4E23	7.8E25	5.5E22	
P-R Grain					
Diameter(μm/yr)	1100/1E8	900/2E8	130/4E8	6000/1E9	7

	α LYR	α PSA	β PIC	ε ERI	NOTES
ANGULAR SIZE SUMMARY					8
Measured FWHM at 60um					
Scan Direction					
W of N (deg)	5	29	20	20	
in-scan(")	25±3	36±3	< 14	< 17	
x-scan(")	29±5	< 13	22±6	< 11	
Model FWHM					
at 60um(")	22	19	20	5.8	

	α LYR	α PSA	β PIC	ε ERI	NOTES
MASS ESTIMATES					
Min. Dia. (um)	1000	1000	2.3	6000	9
Min. Mass ($M_{\oplus}$)	0.015	0.03	0.01	0.02	10
Asteroid Size					
Distribution($M_{\oplus}$)	450	900	7000	250	11

NOTES: 1) Hoffleit (1982); 2) Code et al.(1976); 3) Johnson (1966); 4) inferred from VandenBerg and Bridges (1984); 5) Iben (1965); 6) see text; 7) ρ = 2 gm/cc; 8) Gillett et al. (1985); 9) P-R constraint except for β Pic where blowout size is more appropriate; 10) ρ = 2 gm/cc, $M = \rho/6(a_{min}Ag)$; 11) $M = \rho/6(a_{min} a_{max})^{1/2}Ag$, ρ = 2 gm/cc.

The radiating surface area is determined by the thermal model, however the grain sizes (and thus the associated mass) are very

uncertain. The size distribution and maximum sizes of the radiating
bodies are not constrained by the IRAS observations (Paper I). A mass
estimate more realistic than the minimum mass can be obtained by
assuming a power law size distribution, e. g. dn ~ $a^{-3.5}$da deduced for
asteroids (Dohnanyi, 1969) and a_{max} = 1000 km, an order of magnitude
estimate of the maximum size of planetismals formed by gravitational
instability in a thin dust disk at 100AU (Goldreich and Ward, 1973,
Greenberg et al., 1984). Both mass estimates are shown in Table II.

4. DISCUSSION

Table II lists relevant properties of the program stars. Alpha
PsA, β Pic and ε Eri are very close to the zero age main sequence (ZAMS)
while α Lyr is substantially above the ZAMS and is likely to be near the
end of its MS lifetime of about 5E8 years (Iben, 1965).

UV spectra of α Lyr show no evidence for mass loss (Lamers et al.,
1978), with an upper limit of about 1E-10 $M_{\odot}$/yr (Lamers, private
communication).

For α Lyr, the model size is in reasonably good agreement with the
observed 60um extent, strongly supporting, as was initially shown in
Paper I, that the emission mechanism is thermal emission from large
blackbody particles. The nearly equal x-scan and in-scan extent
indicate that if the radiation material around α Lyr is in the form of a
disk, we are probably viewing it nearly face on.

Harper et al. (1984) have reported a 2 sigma measurement of α Lyr
at 193μm which is significantly below the blackbody extrapolation of the
IRAS data. This result has been interpreted in terms of emission from
grains ~ 10um in size produced by sublimation from comet-like bodies.
This interpretation is inconsistent with the blackbody model derived
from the IRAS observations which require grain sizes $\gtrsim$ 100μm. An
alternative explanation for the Harper et al. result is emission by
grains which are partially transparent at 193μm, i.e. sizes $\lesssim$ mm,
roughly consistent with the P-R lifetime constraint.

Interstellar MgII and CI absorption features in UV spectra of α PsA
have been reported (Kondo et al., 1978 and Bruhweiler and Kondo, 1982),
but a circumstellar origin of these features is now a serious
alternative (Kondo and Bruhweiler, 1985).

The blackbody model size at 60μm is about 1/2 the observed in-scan
FWHM. Given the simplified nature of the model and size measurements,
the agreement is quite good and supports the model of thermal emission
from large near-blackbody grains. The agreement could be improved by
supposing that the grain emissivity is significantly less than unity at
100um and beyond. This would simultaneously decrease the grain
temperatures inferred from the observations and increase their thermal
equilibrium temperature at a fixed distance from the star.

The substantial difference between the measured x-scan and in-scan
sizes is suggestive that the emitting material is in a disk around α PsA
which is being viewed nearly edge on.

Beta Pic has been classified as a shell star because of strong,
narrow CaII H and K absorption and narrow resonance and metastable
absorptions of FeII (Sletteback and Carpenter, 1983). Circumstellar

absorption features of CI and MgI are also present (Kondo and Bruhweiler, 1985).

Recently, an optical image of a nearly edge on circumstellar disk around β Pic was obtained by Smith and Terrile (1984). Presumably this optical disk is due to scattered starlight from the same particles responsible for the infrared excess. Qualitatively, this observation provides a very strong argument for the interpretation of the cool infrared excess around β Pic in terms of orbiting solid material. Quantitatively, 1) the angular extent of the infrared emitting region is similar to the extent of the optical disk, 2) the optical observations have been interpreted in terms of scattering by micron-sized particles distributed as $n(r) \sim 1/r^3$. The particle size is consistent with the thermal model results, however, the spatial distribution is much too steep for the thermal model, and 3) the orientation of the optical disk is only marginally consistent with the IRAS measurements.

Epsilon Eridani is a nearby K2 star, much cooler than the other stars discussed here, and frequently suggested as a nearby candidate for a planetary system. It exhibits strong chromospheric activity, indicative of relative youth, (Vaughan et al., 1981 and references therein), and an age of about 1E9 years can be estimated from its rotation rate (Vaughan et al. (1981) and Skumanich (1972)).

The scale of thermal model distances is smaller than for the other stars in this sample and is very similar to the planetary distances in our solar system. The model angular size for blackbody grains is about a factor of 2 smaller than the observed upper limits, but the difference is small enough that major deviations from blackbody grains appear to be ruled out. This is particularly surprising since radiation pressure is not effective in removing small particles from around this star.

Chromospheric emission could possibly provide an alternative explanation for the excess around ε Eri. An upper limit to the chromospheric emission, F_c, can be estimated as $F_c/F_s = \tau_c(T_c/T_s)$ where T_c and τ_c are the maximum chromospheric temperature and the optical depth above the temperature minimum, respectively. With chromosphere parameters from Kelch (1976), one obtains $F_c \cong 0.02$Jy from 12 to 100 µm, which is quite small compared to the observed excess beyond 12µm.

5. EVOLUTIONARY STATE OF MATERIAL

The temperature ranges generally indicated for the material around program stars suggest comet-like bodies, i.e., dirty snow balls, as the likely condensates from a proto-planetary nebula although this is clearly not the case for inner portions of the β Pic and ε Eri clouds. The spatial scales correspond roughly to the location of an inner Oort cloud (see e.g. Greenberg et al. 1984 and references therein). Weissman (1984) and Harper et al. (1984) have suggested that the emission around α Lyr is due to small (~ 10µm) grains generated by sublimation or collisions in a cloud of comet-like bodies. Weissman pointed out that it was improbable that planetary-sized bodies were present in the α Lyr cloud because the accretionary time scales for large bodies at this distance from the star are too long. However, Smith and Terrile (1984) have interpreted the β Pic observations in terms of a very young disk in which planetary formation is occurring now or has recently been completed. The mass estimates for dn ~ a$^{3.5}$da, (Table II), are in reasonable agreement with estimates of the mass of the Oort cloud (Bailey, 1983), except possibly for β Pic.

Obvious questions concerning the interpretation of the present data include 1) The age of ε Eri is ≥ time of formation of all planets in our solar system. What is the evolutionary state of the material around this star? 2) What are the implications of the UV circumstellar features of CI, MgI, FeII and CaII? 3) What process or processes are responsible for the inner and outer limits on the radiating surface area apparent for α Lyr, α PsA and β Pic?

Clearly, further study of these objects and others with similiar cool infrared excess will provide uniquely valuable information for furthering our understanding of these processes.

ACKNOWLEDGEMENTS: This review is largely based on a paper in preparation by F. C. Gillett, H. H. Aumann, G. Neugebauer, F. J. Low and R. Waters. The contributions of all the authors, and those of the many people responsible for the IRAS project are gratefully acknowledged.

REFERENCES
Aumann, H. H., Gillett, F. C., Beichman, C. A., de Jong, T., Houck, J. R., Low, F. J., Neugebauer, G., Walker, R. G., Wesselius, P.R. (1984), Ap. J. (Letters), **278**, L23.
Aumann, H. H., (1984), Bulletin Amer. Astron. Soc., **16**, 483.
Bailey, M. E. (1983), M.N.R.A.S., **205**, 47p.
Beichman, C. A., Neugebauer, G., Habing, H. J., Clegg, P. E. Chester, T. J. (1985), IRAS Catalogues and Atlases, Explanatory Supplement, NASA Publ. #   .
Dohnanyi, J. S. (1969), J. Geo. Res., **74**, 2531.
Goldreich, P., Ward, W.R. (1973), Ap. J., 183, 1051.
Greenberg, R., Weidenschilling, S. J., Chapman, C. R., Davis, D., R., (1984), Icarus, **59**, 87.
Harper, D. A., Lowenstein, R. F., Davidson, J.A. (1984), Ap. J., **285**, 808.
Bruhweiler, F. C., Kondo, Y. (1982), Ap. J. (Letters), 260, L91.
Campins, H., Rieke, G. H., Lebofsky, M.J. (1985), A.J., 90, 896.
Code, A. D., Davis, J., Bless, R. C., Hanbury-Brown, R. (1976), Ap. J., **203**, 417.
Day, K. L. (1981), Ap. J., **246**, 110.
Gillett, F. C., Aumann, H. H., Low, F. J., Neugebauer, G., Waters, R. (1984), Paper presented at Protostars and Protoplanets Conference, Tucson, Arizona.
Gillett, F. C., Aumann, H. H., Low, F. J., Neugebauer, G., Waters, R. Bachman, D. (1985), in preparation.
Hoffleit, D. (1982), The Bright Star Catalog Fourth Revised Edition, Yale University Observatory, New Haven, Conn.
Iben, I., Jr. (1965) Ap. J., **141**, 993.
Irvine, W. M., Pollack, T.B. (1968), Icarus, 8, 324.
Isobe, S. (1970), Publ. Astr. Soc. Japan, **22**, 429.
Johnson, H. L. (1966), Ann. Rev. Astr. Astrophys., **4**, 193.

Kelch, W. L. (1978), Ap. J., **222**, 931.

Kondo, Y., Talent, D. L., Barker, E. S., Dufour, R. J., Modisette, J. L. (1978), Ap. J. (Letters), **220**, L97.

Kondo, Y., Bruhweiler, F.C. (1985), Ap. J. (Letters), 291, L1.

Koorneef, J. (1983), Astr. and Astrophys., **128**, 84.

Lamers, H. J. G. L. M., Stallo, R., Kondo, Y. (1978), Ap. J., 223, 207.

Skumanich, A. (1972), Ap. J., **171**, 565.

Sletteback, A., Carpenter, K.C. (1983), Ap. J. Supple., 53, 869.

Smith, B. A., Terrile, R.J. (1984), Science, 226, 1421.

VandenBerg, D. A., Bridges, T.J. (1984), Ap. J., 278, 679.

Vaughan, A. H., Baliunas, S. L., Middelkoop, F., Hartmann, J. W., Mihalas, D., Noyes, R.W., Preston, G.W. (1981), Ap. J., 250, 276.

Weissman, P. R. (1984), Science, **987**, 1984.

THE FLUX DISTRIBUTION OF VEGA FOR 10 μm < λ < 100 μm, AND THE CALIBRATION OF IRAS AT 12 μm AND 25 μm.

S.K. Leggett
Astronomy Department, University of Edinburgh
Blackford Hill
Edinburgh EH9 3HJ
UK

ABSTRACT. The flux distribution of Vega for 10 μm < λ < 100 μm is investigated using the IRAS data. Using the bands 1 and 2 fluxes, relative to Vega, of 23 stars known T_e, the Vega flux is adjusted to reproduce fluxes calculated by model atmospheres. These Vega fluxes agree with those given by the pointed IRAS observations and calibration, and show Vega in excess at all bands over modelled fluxes. A comparison with Sirius implies a similar excess, however a comparison with 33 AO-type survey stars suggests that Vega is normal at band 1. There appears to be a problem either with the survey AO comparison or with the calculated infrared continuum of hot stars.

1. METHODS.

Fluxes for Vega have been derived around 12 μm (band 1) and 25 μm (band 2) by two methods. One os to use the band 1 and band 2 pointed observations for Vega together with the calibration (Rieke et al., 1984), as given in the Explanatory Supplement by Beichman et al. (1985). The other is in the application of the infrared flux method (Blackwell et al., 1980) to derive effective temperatures for 23 bright stars. Using Vega as calibration, the band 1 and band 2 fluxes are adjusted until the temperatures derived for these stars at these wavelengthd agree with those derived at (1-5) μm by Blackwell et al., 1985, and Leggett et al., 1985, who used the absolute calibration of Vega by Mountain et al., 1985. The 12 μm and 25 μm fluxes determined in these two ways agree well with each other. These values are compared to those calculated by Dreiling & Bell (1980, using the angular diameter measured by Hanbury Brown et al., 1974) and show Vega in excess over the model, as shown in Fig. 1. An extrapolation is made to 60 μm and 100 μm. Vega is then compared to stars of similar spectral types; the pointed observations of Sirius (α CMa) and 33 survey stars are used. The latter have fluxes known to better than 8% at band 1, and are single, non-variable sources. Sirius implies that Vega has an excess at bands 1, 2 and 3 in agreement with the IRAS calibration and the model, but the survey stars suggest that Vega's V-N colour is normal for an AO star (see Fig. 2). Three of the survey stars appear to have large 12 μm excesses themselves.

F. P. Israel (ed.), Light on Dark Matter, 71–74.

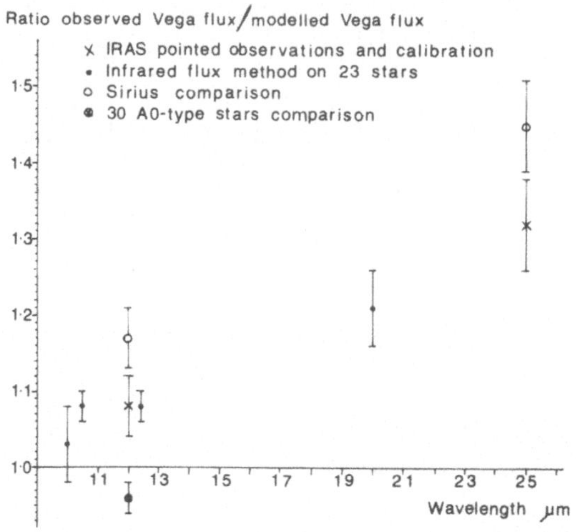

Ratio observed Vega flux/modelled Vega flux

x IRAS pointed observations and calibration
• Infrared flux method on 23 stars
o Sirius comparison
• 30 A0-type stars comparison

Figure 1: The ratio of the observed Vega fluxes to the modelled, for 10 μm < λ < 25 μm. The model by Dreiling & Bell (1980) was used together with the angular diameter measured for Vega by Hanbury Brown et al. (1974). The IRAS pointed Vega observations and calibration is given in Beichman et al. (1985).

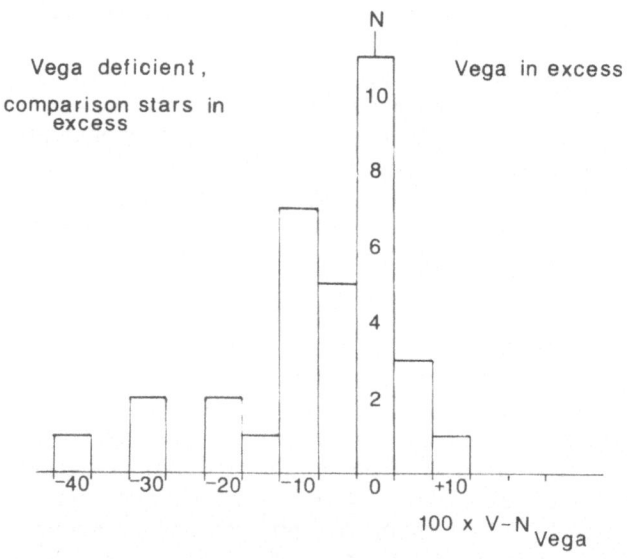

Figure 2: Histogram of the V–N value implied for Vega from a sample of 33 B8–A2 type stars. The bin size is 0.04 magnitudes.

2. CONCLUSION

There are two possible conclusions. One is that for some reason the survey comparison of 33 AO type stars is inaccurate. Then the Sirius comparison agrees with Vega being in excess as given by the IRAS calibration, the application of the infrared flux method to 23 stars, and the calculations of model atmospheres. However, there seems to be no reason why the 33 survey stars should not be a representative sample.

The other conclusion is that Vega is not anomalous in V-N, and does not have a 12 μm excess. This means that the flux distribution of Sirius (A1Vm) is very different (17%) from a normal A1V star; this difference seems larger than that might be due to its metallicity. This conclusion also means that the infrared (or at least 12 μm) continuum opacities have been miscalculated for hot stars. As the dominant opacity source for these stars is atomic hydrogen this seems unlikely. However this conclusion may be supported by the fact that the application of the infrared flux method using the Kurucz (1979) models at 10 μm and 20 μm gives a smaller Vega flux than that using the MARCS (Gustafsson et al., 1975) generated models (see Fig. 1); this would be consistent with the hot ($T_e > 6000K$) star models having too large an opacity.

Although the state of the modelling of Vega's infrared flux distribution is not clear, it is at least satisfying that the application of the infrared flux method to 23 bright stars of various spectral types gives 12 μm and 25 μm fluxes for Vega that are in close agreement with the calibration of IRAS by Rieke et al. Also the three AO stars with large band 1 fluxes look of interest for future investigation. A full, detailed, version of this paper has been submitted to Astron. & Astrophys.

REFERENCES

Beichman, C.A., Neugebauer, G., Habing, H.J., Clegg, P.E. and Chester, T.J.: 1985, Editors, Explanatory Supplement to the IRAS catalogues.

Blackwell, D.E., Booth, A.J., Leggett, S.K., Petford, A.D., Mountain, C.M. and Selby, M.J.: 1985, Mon. Not. R. astr. Soc., accepted.

Blackwell, D.E., Petford, A.D. and Shallis, M.J.: 1980, Astron. Astrophys., 82, 249.

Dreiling, L.A. and Bell, R.A.: 1980, Astrophys. J., 241, 736.

Gustafsson, B., Bell, R.A., Eriksson, K. and Nordlund, A.: 1975, Astr. Astrophys., 42, 407.

Hanbury Brown, R., Davis, J. and Allen, L.R.: 1974, Mon. Not. R. astr. Soc., 167, 407.

Kurucz, R.L.: 1979, Astrophys. J. Suppl., 40, 1.

Leggett, S.K., Mountain, C.M., Selby, M.J., Blackwell, D.E., Booth, A.J., Haddock, D.J. and Petford, A.D.: 1985a, in preparation.

Mountain, C.M., Leggett, S.K., Selby, M.J., Blackwell, D.E. and Petford, A.D.: 1985, Astron. Astrophys., accepted.

Rieke, G.H., Lebofsky, M. and Low, F.J.: 1984, preprint.

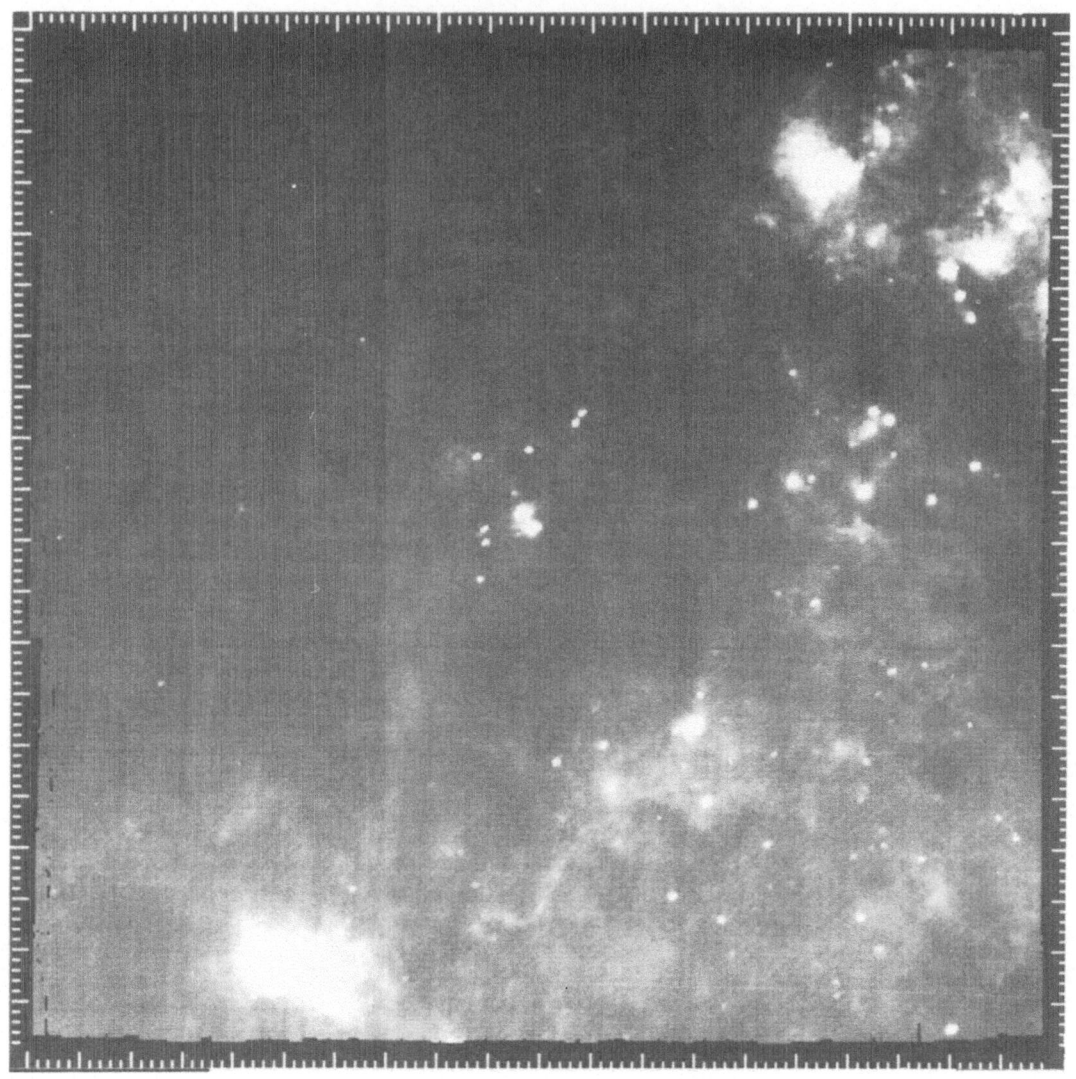

IRAS Field 53 $\alpha = 6^h00^m$, $\delta = +30°$, HCON-3, 60 µm. Field in Auriga,
                  Taurus and Gemini showing the Galaxy with NGC 1912,
                  (M38), NGC 1907, NGC 1931, IC 410 and IC 405 at top
                  right, as well as the SNR/HII region S237/IC 443 at
                  bottom left.

A SEARCH FOR INFRARED EXCESSES IN G-TYPE STARS

S. F. Odenwald
Naval Research Laboratory and
Sachs/Freeman Associates Inc.
Code 4138-0
Washington, DC   20375-5000

ABSTRACT. The IRAS Point Source Catalogue (PSC) has been searched for infrared sources associated with stars of spectral class G. Of the 1069 IRAS sources identified, 15 have enhanced emission at $\lambda > 25$ μm. This group includes 4 supergiants, 8 variable stars, and 3 stars known to have dusty circumstellar medii (RU Cen, R Sge, HD-101584). The preliminary findings from this study are described.

1. INTRODUCTION

Recent IRAS observations of stars such as Vega, Beta Pictorus and Fomalhaut have shown that even apparently normal stars have IR-excesses and that this phenomenon may be common to all stars. To address the issue of how typical our sun is as an IR source, all of the IRAS sources positionally coincident with G-type stars were examined for IR-excesses.

2. THE SURVEY

The PSC was searched for all sources containing a G-type star located within the 95% confidence ellipse of the IRAS position. The flux ratio $R=F(12$μm$)/F(25$μm$)$ was determined for each star based on the color corrected IRAS fluxes. For 1033 of the stars $R = 4.0 \pm 1.0$ consistent with normal blackbody continua. Of the remaining 36 stars, 21 were rejected due to contamination with near-by diffuse emission or for lack of HCON-3 SKYFLUX data. The remaining 15 stars (Table I) had point-like images at all IR wavelengths.

3. PRELIMINARY RESULTS

1) Half of the stars are known to be variable or are emission-line stars and three of the sources, R Sge, RU Cen and HD-101584 have been studied by Gehrz, Ney(1972), Gehrz, Woolf(1970) and Humphries and Ney(1974). The spectra between 2-10 μm are consistent with the presence of circumstellar material. R Sge and HD-101584 are known to be supergiant stars.

F. P. Israel (ed.), Light on Dark Matter, 75–76.

2) Of the 4 IR-excess stars with luminosity class information, all are supergiants. IR excess emission from supergiant stars has a frequency of at least 5% among the 74 supergiants in the sample and at least 25% among the 15 G-type stars with IR excesses.

3) The IR spectra can be well-represented by single-temperature blackbodies with T = 130 ± 60K though 75% show evidence for additional emission with T = 500 ± 200K not reproducible by the spectrum of the cold component alone. Circumstellar, isothermal shells with blackbody dust grains may be consistent with the spectrum of the cold component however the hot component requires either additional warm dust grains interior to these shells or free-free emission. The radii of these shells vary from 20 to 7000 AU.

4. SUMMARY

IR excesses are not a common feature of G-type stars (1%). When present, and for those cases where luminosity class information is available, the stars are always supergiants rather than giants or main-sequence stars. The excess emission is consistent with the existence of isothermal dust shells at large distances from these stars, however other mass distributions are possible provided that the bulk of the IR emission occurrs at large distances from the star where the dust grains have T = 130 ± 60 K. Future studies of these stars will include: Obtaining improved spectral and luminosity classes: Photometric data betwen 1-10 μm, and more sophisticated modeling of their IR continua.

Table I.   G-TYPE STARS WITH INFRARED EXCESSES

IRAS	SAO	Misc.	mv	Sp.	Ref.	Comments
05170+0535	112630		+8.6	G0		
07134+1005	96709		+8.3	G5		
09079-1942	154972	TU Pyx	+8.6v	G0		
10282-5231	238126		+8.6	G5		
11059-7721		Kuk-05099	+10.6v	Ge		HRC-249
11108-7627		Kuk-05138	+10.7v	Ge		HRC-247/T Tau
11294-5909	239145	o Cen	+5.1v	G2Ia-F7Ia	1,7	
11385-5517	239288	HD-1015184	+7.0	G0-F2eIp	2,3	Binary star?
12067-4508	223245	RU Cen	+8.7v	G0p-Fp	4	RV Taurid
15420-3408		Kuk-07226	+10.3v	Ge		HRC-248/T Tau
15532-4210	226389		+8.0	G0		
15556-2248	183986		+9.5	G5		
17109-3942	208569	HD-155603	+6.6v	G5Ia-K0Ia	7	
18384-2800	187137		+8.5	G0		
20117+1634	105871	R Sge	+8.6v	G0Ib-G8Ib	5,6	RV Taurid

1)Stift(1979) A.A. 80, 134; 2)Humphreys, Ney(1974) Ap.J. 190, 339
3)Humphreys, Ney(1974) Ap.J. 187, L75; 4)Gehrz, Ney(1972) PASP 84, 768.
5)Gehrz, Woolf(1970) Ap.J. 161, L213; 6)Gehrz(1972) Ap.J. 178, 715.
7)Michigan Spectral Line Survey

IRAS INTRINSIC COLOURS OF HOT STARS

J. Coté & L.B.F.M. Waters
Laboratory for Space Research
Beneluxlaan 21
3527 HS Utrecht
The Netherlands

ABSTRACT. The O, B and A type stars observed with IRAS at 12 μm are used to derive an intrinsic $((B-V)_o, (V-[12])_o$ - relation for "normal" stars. We find that many of the O, B and A stars have excess fluxes at 12 μm. From a comparison of the relation to models we conclude that : a) The extrapolation of the Kurucz (1979) models from the visual to 12 μm results in a 12 μm flux that is about 10 percent too low, and b) The energy distribution for hot stars from 0.4 to 12 μm cannot be described with a one-temperature blackbody.

1. SELECTION OF THE STARS.

The stars are selected from a comparison between the IRAS Point Source Catalogue (1985) and the O, B, and A stars of the Bright Star Catalogue (BSC) Hoffleit, 1982). The used visual photometry (V and (B-V)) was taken from the BSC and corrected for extinction using $A_V = 3.10\ E(B-V)$ and the intrinsic $(B-V)_o$ colours of FitzGerald (1970). We assumed that extinction at the IRAS wavelengths was negligible. The observed IRAS fluxes were converted to magnitudes, using the IRAS definition for the zero magnitudes as given in the IRAS Explanatory Supplement (1985). From this set we excluded all stars that are known to show IR excess due to e.g. circumstellar dust shells or free-free emission from stellar winds, i.e. supergiants, peculiar and emission line stars. For the same reason we excluded stars, that had a 12 and 25 μ detection, with $[12]-[25] \geqslant 0.25^m$, based on the assumption that hot stars radiate as black bodies with a $F_\lambda \propto \lambda^{-4}$ energy distribution at IRAS wavelengths.

2. THE INTRINSIC RELATION.

The following intrinsic relationship was derived:

(a) $(V-[12])_o = (4.22 \pm 0.20)(B-V)_o + (0.23 \pm 0.04)$ $\qquad (B-V)_o \leqslant -.09$

(b) $(V-[12])_o = (2.35 \pm 0.09)(B-V)_o + (0.07 \pm 0.01)$ $\qquad (B-V)_o \geqslant -.09$

F. P. Israel (ed.), Light on Dark Matter, 77–78.

We compared the colour-colour relation with the predicted relations
based on the grid of model atmospheres by Kurucz (1979) and on a black
body energy distribution. The intrinsic relation based on the Kurucz
models results in an excess for all stars, of about 10 percent. Since
the Kurucz models reproduce the UV and visual energy distributions of
hot stars with high accuracy (see e.g. Lamers et al., 1984), this
suggests an error in the Kurucz models for longer wavelengths. The
excess of about 10 percent was already reported for a limited number of
early A-type dwarfs by Rieke et al. (1985). Mountain et al. (1985) find
an excess of 5 percent at 5 μ for α Lyr relative to the Kurucz model.
The results of this study have severe consequences for the study of the
IR excess from early type stars. Especially for small excesses the use
of Kurucz models in the IR, to estimate the photospheric continuum,
introduces systematic errors, in the sense that the excesses are
overestimated. We find that the IR excess for early type stars at 12 μ
relative to the Kurucz model can be removed by extrapolation of the
models beyond 2 μ with a $F_\lambda \propto \lambda^{-4}$ energy distribution (i.e. a constant
brightness temperature as a function of wavelength). From the comparison
to the black body model we can conclude that O, B and A stars do not
behave as one-temperature black body in the wavelength range $0.4 \leqslant \lambda \leqslant$
12 μ. The origin of this deviation is the visual photometry (B-V) rather
than the (V-[12]) colour.

A detailed study of the intrinsic colours of stars will be published
elsewhere (Waters, Coté and Aumann). The authors express their gratitude
for helpful suggestions to Prof. Lamers and Dr. H.H. Aumann.
L.B.F.M. Waters acknowledges financial support from ZWO.

REFERENCES

FitzGerald, M.P., 1970, Astron. Astrophys., 4, 234
Hoffleit, D. 1982, The Bright Star Catalog (New Haven: Yale University
    Observatory)
IRAS Explanatory Supplement 1985, eds. Beichman, C.A., Neugebauer, G.,
    Habing, H.J. Clegg, P.E., Chester, T.J.
Kurucz, R.L., 1979, Astrophys. J. Suppl., 40, 1.
Lamers, H.J.G.L.M., Waters, L.B.F.M., Wesselius, P.R. 1985, Internal
    Report Space Research Utrecht I.V.R.O. 133
Mountain, C.M., Leggett, S.K., Selby, M.J., Blackwell, D.E., Petford,
    A.D. 1985 (preprint)
Rieke, G.H., Lebofsky, M.J., Low, F.J., 1985, Astron. J. 90, 896
Savage, B.D., Mathis, J.S. 1979, Ann. Rev. Astr. Ap., 17, 73

THE INFRARED EXCESS FROM STELLAR WINDS

Henny J. G. L. M. Lamers
SRON Laboratory for Space Research and
Sonnenborgh Observatory
Beneluxlaan 21
NL-3527 HS Utrecht, The Netherlands

ABSTRACT. The IR excess can be used to study the density or velocity structure in the lower parts of the winds of early type stars. The depth of the emitting region and the excess in magnitudes is given for typical O-stars, as a function of λ and $\dot{M}$. The first results of an ongoing study of the IRAS data are discussed briefly.

1. INTRODUCTION

Ultraviolet observations with the Copernicus satellite and IUE have shown that luminous early-type stars with $L \gtrsim 2 \times 10^4 \, L_\odot$ are losing mass by means of a stellar wind at a rate of $10^{-8} \lesssim \dot{M} \lesssim 10^{-4} \, M_\odot/yr$ and with velocities of the order of 1000 to 3000 km/s (e.g. see Lamers, 1981). This mass loss is probably due to radiation pressure by UV spectral lines (Castor et al., 1975). The high mass loss rate changes the evolution of the massive stars ($M \gtrsim 30 \, M_\odot$) drastically and it is responsible for, e.g., the formation of Wolf-Rayet stars, and the lack of luminous red supergiants with $L > 5 \times 10^5 \, L_\odot$. The stellar winds provide a source of energy input into the ISM which is comparable to that of supernovae. So the study of the stellar winds is important for understanding stellar evolution and the ISM.

The UV observations of line profiles enable us to study the structure of the winds at distances larger than about 1.5 R_* where $v \gtrsim 200$ km/s. However, the very interesting layers at $1 \lesssim r \lesssim 1.5 \, R_*$ where the onset of the wind occurs can be studied best by means of the IR excess. This is the subject of an ongoing study of IRAS data at the Laboratory for Space Research in Utrecht.

2. THE PRINCIPLE OF THE IR EXCESS FROM STELLAR WINDS

The free-free and bound-free absorption at long wavelengths in an ionized stellar wind is

$$\kappa_\nu = 1.98 \times 10^{-23} \, Z^2 (g+b) \, \lambda^2 \, n_i n_e \, T^{-3/2} \qquad (1)$$

F. P. Israel (ed.), Light on Dark Matter, 79–82.
© 1986 by D. Reidel Publishing Company.

where Z is the mean atomic charge, g and b are the Gaunt factors for free-free and bound-free absorption, n_i and n_e are the ion and electron density, and λ is the wavelength in cm. For a stationary spherically symmetric wind the density is given by

$$n_e(r) = (n_e/\rho) \; \dot{M}/4\pi \; r^2 \; v(r) \tag{2}$$

where $n_e/\rho = 5.2 \times 10^{23} \; g^{-1}$ for the ionized wind of Population I stars. The temperature $T(r)$ is determined by the energy balance, which, in the case of winds from hot stars, is dominated by radiative equilibrium, so $T(r) = \alpha T_{eff}$ with $\alpha \simeq 0.80$. For a given velocity structure $v(r)$ and resulting density structure $n_e(r)$ the transfer equation can be solved and the IR energy distribution can be calculated. At $\lambda \lesssim 1 \; \mu$ the winds are transparent and the energy distribution is that of the stellar photosphere. At $\lambda \gtrsim 10 \; \mu$, the opacity of the wind increases strongly with wavelength and the energy distribution shows an excess relative to the photosphere. The excess increases with wavelength in a way which depends completely on $v(r)$ and $\dot{M}$. So, by studying the IR excess as a function of λ we obtain information on the velocity structure and the mass loss rate.

An asymptotic solution of the transfer equation of radio wavelength has been described by Wright and Barlow (1975) and Panagia and Felli (1975). A general and easily applicable solution to a large variety of models has been presented in the form of curves of growth (Lamers and Waters, 1984; Waters and Lamers, 1984).

3. THE IR EXCESS OF A TYPICAL EARLY-TYPE STAR

To give an impression of the region in the wind where the free-free and bound-free radiation is emitted, we show in Table I the distance $x_{0.25}$ (in R_*) where the radial optical depth is 0.25 (x = 1 corresponds to the photosphere). We have adopted a typical model of T_{eff} = 40,000 K, $R_* = 13.5 \; R_\odot$, and T_{wind} = 30,000 K. The velocity law is assumed to vary as

$$v(r) = 0.01 \; v_\infty + 0.99 \; v_\infty \; [1 - (1/x)] \tag{3}$$

(Castor and Lamers, 1979) with v_∞ = 2000 km/s. A Gaunt factor g+b = 2 has been adopted. The excess, in magnitudes, is also given. The data are derived from Lamers and Waters (1984).

From Table I we see that an excess of 0.5 mag can be expected at 10 μ of $\dot{M} = 10^{-5}$ and at 100 μ if $\dot{M} = 10^{-6} \; M_\odot/yr$. The region of the wind between 1 and 2 R_* can be probed by IR observations between 1 and 50 μ if $\dot{M} = 10^{-5}$ and between 10 and 500 μ if $\dot{M} = 10^{-6} \; M_\odot/yr$.

The predicted IR-, mm-, and radio-fluxes at 1 $\mu \leq \lambda \leq$ 20 cm for 200 early-type stars are listed by Waters and Lamers (1985).

TABLE I. Depth of formation and excess in magnitude.

$\lambda\dot{M}$ (μm $M_\odot$/yr)	x (τ_{rad}=0.25)	Δm (mag)	$\lambda\dot{M}$ (μm $M_\odot$/yr)	x (τ_{rad}=0.25)	Δm (mag)
1×10^{-5}	1.00	0.02	1×10^{-3}	2.95	3.07
3×10^{-5}	1.02	0.10	3×10^{-3}	5.62	4.60
1×10^{-4}	1.18	0.62	1×10^{-2}	11.5	6.21
3×10^{-4}	1.68	1.70	3×10^{-2}	25.1	7.85

4. SOME RESULTS

The observed energy distribution and velocity law for two typical
stars are shown in Figure 1. These are the stars ζ Pup, a typical
O4 If star (from Lamers et al., 1984), and the hypergiant P Cygni
B1 Ia$^+$ (from Waters and Wesselius, 1985). The energy distribution
is plotted in the left panel, together with the adopted photospheric
energy distribution (dashed line) and the fit through the data for the
resulting model (solid line). The model is characterized by a mass
loss rate and a velocity law. The resulting velocity law for the two
stars is shown in the right panel. Notice that the velocity law of
ζ Pup increases rapidly with distance, in agreement with the prediction for radiation driven wind models, whereas the velocity law of
P Cygni increases very slowly. The acceleration of the wind of this
star must be due to a different mechanism. This mechanism may be related to the instability of hypergiants and to the existence of an
upper limit for the luminosity of B stars (Lamers, 1985).

ACKNOWLEDGMENTS. This paper was written when the author was a visit-
ing scientist at JILA, University of Colorado and National Bureau of
Standards in the summer of 1985. The staff of JILA is acknowledged
for their hospitality and help in the preparation of this paper.

REFERENCES

Castor, J. I., Abbott, D. C., Klein, R. I.: 1975, Astrophys. J. **195**,
    157.
Castor, J. I., Lamers, H.J.G.L.M.: 1979, Astrophys. J. Suppl. **39**,
    481.
Lamers, H.J.G.L.M.: 1981, Astrophys. J. **245**, 593.
Lamers, H.J.G.L.M.: 1985, Astron. Astrophys. (in press).
Lamers, H.J.G.L.M., Waters, L.B.F.M.: 1984, Astron. Astrophys. **136**,
    37.

Lamers, H.J.G.L.M., Waters, L.B.F.M., Wesselius, P.R.: 1984, Astron.
    Astrophys. 134, L17.
Panagia, N., Felli, M.: 1975, Astron. Astrophys. 39, 1.
Waters, L.B.F.M., Lamers, H.J.G.L.M.: 1984, Astron. Astrophys. Suppl.
    57, 327.
Waters, L.B.F.M., Lamers, H.J.G.L.M.: 1985, Astron. Astrophys. Suppl.
    (in press).
Waters, L.B.F.M., Wesselius, P.R.: 1985, Astron. Astrophys. (in
    press).
Wright, A. E., Barlow, M. J.: 1975, Mon. Not. R. Astron. Soc. 170,
    41.

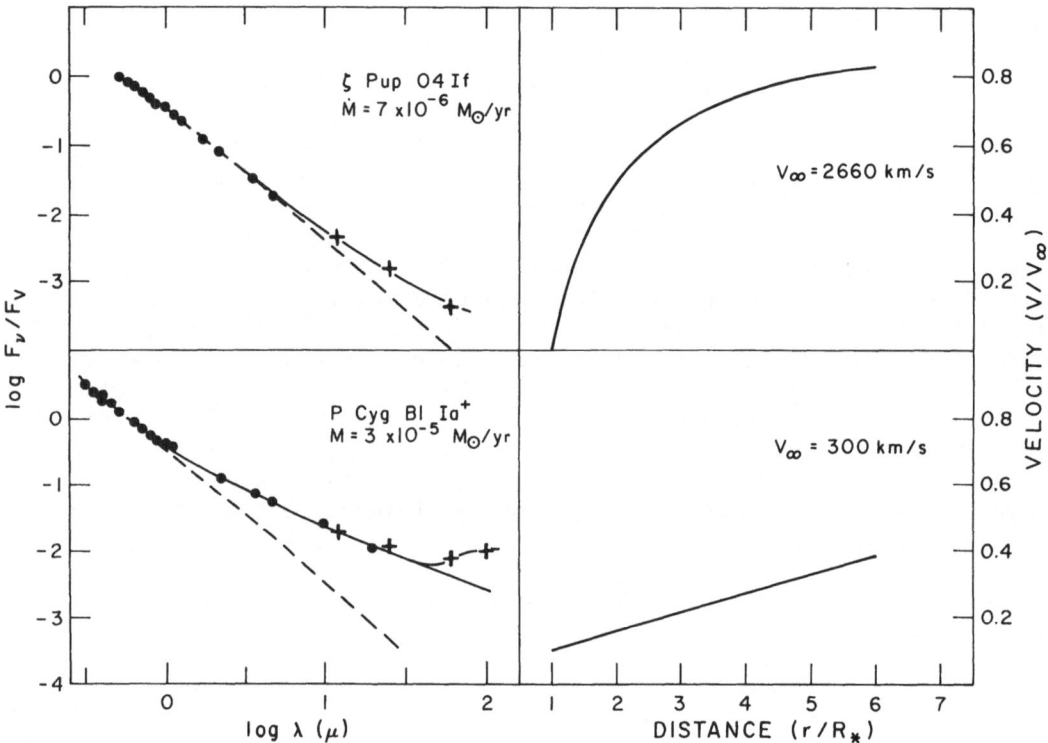

Fig. 1.   The energy distribution (left) and the velocity law (right)
          are shown for two stars: ζ Pup and P Cygni. The ground-
          based data are indicated by dots; the IRAS data by crosses.
          The dashed lines show the adopted photospheric spectrum. The
          solid line (left) shows the calculated energy distribution
          for a model with the mass loss rates and velocity laws as
          indicated. The velocity law of P Cygni is much slower than
          that of normal early-type stars. The bump in the spectrum
          of P Cygni at 60 and 100 μ is probably due to circumstellar
          dust.

THE DISC STRUCTURE AND MASS LOSS RATES OF Be STARS.

L.B.F.M. Waters
Laboratory for Space Research
Beneluxlaan 21
3527 HS Utrecht
The Netherlands

ABSTRACT. IRAS observations of four well-known Be stars, α Eri, ϕ Per, δ Cen and χ Oph are used to derive the density structure of the circumstellar discs and the mass loss rates. α Eri shows no sign of IR excess up to $\lambda = 60$ μm, but the other three have excesses in the order of a factor 10 to 70, caused by free-free radiation from the disc. The density structures of the discs suggest that matter is accelerated outwards. The mass loss rates are of the order 7 10^{-9} to 2 10^{-8} $M_\odot$/yr, i.e. a factor 100 larger than derived from UV resonance lines (Snow, 1981).

1. INTRODUCTION.

Be stars are known to show large IR excesses (Gehrz et al., 1974). These excesses are interpreted as due to free-free emission from circumstellar envelopes. The structure of the shells around Be stars is poorly known. Poeckert and Marlborough (1978) propose that, due to the rapid rotation of the stars, matter is ejected primarily in the equatorial plane, causing a disc-like wind structure. Doazan and Thomas (1982) propose that matter is ejected spherically symmetrically and after initial acceleration to high (v = 1000 km/s) velocities, is decelerated to small velocities and high densities.
The mass loss rates of Be stars are still uncertain. From an analysis of UV resonance lines Snow (1981) derives $\dot{M} \simeq 10^{-11}$–10^{-10} $M_\odot$/yr. However, the large IR excesses suggest $\dot{M} \simeq 10^{-6}$ $M_\odot$/yr if the winds were spherically symmetric. Obviously, the UV resonance lines are formed in low-density regions of the winds, whereas the IR excess originates in high-density parts.

2. THE DENSITY STRUCTURE OF THE DISCS.

The IR energy distribution of the three stars that show IR excess can be used to derive the density structure of the winds, assuming a wind

F. P. Israel (ed.), Light on Dark Matter, 83–86.

geometry. We will assume that the wind has a disc-like structure with radius R_{disc}, opening angle θ, and density structure $\rho(r) = \rho_0 (r/R_*)^{-n}$. The slope α of the far-IR energy distribution $F_\nu \propto \nu^\alpha$ is related to the parameter n as

$$\alpha = \frac{2n-3}{n-0.5}$$

We find n=2.4 for δ Cen and χ Oph. For ϕ Per, we find n=3.1 if $R_{disc} \gg R_*$ and n=2.4 if R_{disc}=6.5 R_*. So the IRAS data suggest that material is accelerated outwards (n > 2).
The mass loss rates found from the IRAS data are in the order of 7 10^{-9} to 2 10^{-8} $M_\odot/yr$, i.e. about a factor 100 larger than derived from UV resonance lines (Snow, 1981).

3. COMPARISON WITH DECELERATED WIND MODEL.

The IRAS observations are not in agreement with the decelerated wind model (Doazan and Thomas, 1982). In their model, the spherically symmetric wind consists of three regions: i/ an inner region with low densities where the wind is rapidly accelerated to high (v $\approx$ 800–1000 km/s) velocities, ii/ a region where the wind is decelerated to small velocities and high densities, and iii/ the region beyond the decelerated part. The slope of the far-IR spectrum suggests that the IR radiation does not originate from region ii, since a decelerated wind would yield n < 2. The densities in region i are too low to explain the observed IR excess (adopting the UV mass loss rates (Snow, 1981)). Therefore the far-IR flux is likely to originate from region iii, and the slope of the spectrum requires that matter is slowly accelerated outwards in that region. It is not clear what mechanism could be responsible for acceleration of the wind after it has been decelerated.

4. CONCLUSIONS.

The IRAS far-IR spectrum of 4 Be stars shows that three of these stars (ϕ Per, δ Cen and χ Oph) have large IR excesses that are in agreement with free-free radiation from circumstellar material. One star, α Eri shows no sign of excess larger than about 10 percent up to λ = 60 μm. The IRAS observations are not in agreement with the decelerated wind model as proposed by Doazan and Thomas (1982).
The mass loss rates inferred from the data are about a factor 100 larger than derived from UV resonance lines (Snow, 1981). In terms of a disc model, the UV resonance lines originate from the low-density high-velocity polar regions, whereas the IR excess originates from the high-density disc. However, the theory of radiation driven winds around rapid rotators (Abbott, 1980; Marlborough and Zamir, 1984) predicts a mass loss rate that is constant over the surface of the star. This suggests that some other mechanism than radiation pressure is responsible for the enhanced mass loss in the equatorial plane.

A detailed study of the IR energy distribution of α Eri, ϕ Per, δ Cen and χ Oph will be published elsewhere. This study was carried out with the financial support of ZWO. The author gratefully acknowledges many stimulating discussions on this topic with Prof. H. Lamers, J. Coté and K. Bjorkman.

REFERENCES.

Abbott, D.C.: 1980, Astrophys. J. 242, 1183
Doazan, V., Thomas, R.N.: 1982, in proceeding of the Third European IUE
    Conference, p. 287
Gehrz, R.D., Hackwell, J.A., Jones, T.W.: 1974, Astrophys. J. 191, 675
Marlborough, J.M., Zamir, M.: 1984, Astrophys. J. 276, 706
Poeckert, R., Marlborough, J.M.: 1978 Astrophys. J. Suppl. 38, 229
Snow, T.P.: 1981, Astrophys. J. 251, 139

IRAS Field 66 α = 19^{h}00^m, δ = +30°, HCON-$\dot{3}$, 100 μm. Galactic
                  cirrus in Lyra and Vulpecula; NGC 6870/6823 is at
                  bottom left.

IRAS OBSERVATIONS OF WOLF-RAYET STARS

K.A. van der Hucht, SRON Laboratory for Space Research Utrecht
T.A. Jurriens, SRON Laboratory for Space Research Groningen
F.M.Olnon, Netherlands Foundation for Radio Astronomy
P.S. Thé, Astronomical Institute Anton Pannekoek, Amsterdam
P.R. Wesselius, Laboratory for Space Research, Groningen
P.M. Williams, The Royal Observatory, Edinburgh

ABSTRACT. IRAS PSC, LRS, and CPC observations of Wolf-Rayet stars are used as diagnostics of hot circumstellar dust shells, cool dust in WR ring nebulae, the Ne/He abundance ratio, and the interstellar extinction. In two cases the IR energy distributions indicate WR planetary nucleus status rather than Population-I WR status.

1. INTRODUCTION

Groundbased IR photometry and spectrophotometry in the seventies had already demonstrated that WR stars can have three important characteristics in the IR: emission spectra of ionic He, C, and N recombination lines, f-f excesses caused by strong stellar winds, and thermal emission radiated by heated circumstellar dust. These data were limited to about 20 μm.

IRAS observations provide photometry at 12, 25, 60, and 100 μm, low resolution spectra (LRS) from 8 to 22 μm, and, with the Chopped Photometric Channel (CPC), 12'x9' maps at 50 and 100 μm. The observations used are from the IRAS Point Source Catalog (1985), the IRAS Spectral Atlas (1985), and IRAS-CPC observations (Wesselius et al., 1985).

Invaluable for the interpretation of IRAS data are complementary near-IR ground based observations. For this purpose, three of us (KAvdH, PST and PMW) have, over the past five years, observed WR stars at $JHKLMN_1N_2N_3Q_0$ from ESO and UKIRT.

2. THE IR SPECTRUM OF γ^2 VELORUM

The IRAS LRS spectrum of γ^2 Vel shows strong (0.9×10^{-17} W/cm^2) emission of the fine structure line [NeIII]λ15.5μm.

Shortward of 13 μm the γ^2 Vel spectrum had been observed and identified by Aitken et al. (1982), who derived an abundance ratio of N(NeII)/N(He)=0.0026. The IRAS LRS spectrum allows an evaluation of the NeIII/He and thus the Ne/He abundance ratio. We find N(Ne)/N(He)=0.009, i.e. 7.5 times the cosmic value. This reflects the evolved status of the Wolf-Rayet stars. Evolutionary calculations by

F. P. Israel (ed.), Light on Dark Matter, 87–89.
© *1986 by D. Reidel Publishing Company.*

Maeder (1983) yield for WC stars N(Ne)/N(He)=0.0066, remarkably close
to our result.

If the Ne abundance of γ^2 Vel is representative for all WC
stars, then the 38 WC stars within 3 kpc from the Sun provide a ^{22}Ne
input of 8.8×10^{-6} $M_\odot$/yr.kpc^3 for the chemical evolution of the solar
neighbourhood. Particulars of this study are given by van der Hucht
and Olnon (1985).

3. LOW IONIZATION WR STARS WITH CIRCUMSTELLAR DUST

As the IRAS observations confirm, low ionization WR stars, i.e. WN10-
11 and WC8-10 stars, have IR excesses indicative of thermal emission
by hot (~900 K) circumstellar dust. For the brightest late WC stars,
this has been known since 1974.

Speckle interferometry of WR104 (Allen et al., 1981) has shown
that its dust shell radius is of the order of 100 A.U. This means
that the grain size is of the order of 0.01 μm and that the dust is
formed within the stellar wind sphere. The dust is most probably
amorphous carbon. Because no variations have been observed in the
past 10 years, the dust must be continually replenished at a distance
of 100 A.U. from the star. As an example, the dust shell parameters
of WR104 are T(dust)=900 K, M(dust)=$10^{-6} M_\odot$, and (with d = 1.58 kpc)
L_{IR} = 3.4×10^4 $L_\odot$ ≈ 0.14 L*. The stellar winds of WR stars are among
the most hostile environments in which grains are believed to form.

Particulars of this study are given by Williams et al. (1985).

4. THE 9.7 μm ABSORPTION FEATURE

Superimposed on the smooth energy distributions caused by hot
circumstellar dust, the 9.7 μm silicate feature appears in absorption
in WR104, WR112, and WR118. These observations confirm groundbased
spectroscopy from 8 to 13 μm by Roche and Aitken (1984), who observed
the same stars together with three more WR stars. They demonstrated
that the silicate absorption features are entirely of interstellar
origin. Using the most recent groundbased photometry and the
intrinsic WR parameters given by Hidayat et al. (1985), we find for
the relation between visual extinction and 9.7 μm absorption
strength: A_V = 19.8 (±1.7) $\tau_{9.7\mu m}$.

5. TWO NEW WR CENTRAL STARS OF PLANETARY NEBULAE

The IRAS data for WR72 and WR124 reveal energy distributions, which
are significantly different from those of all other WR stars. They
are indicative of circumstellar dust with temperatures of
respectively 85 and 100 K. In that respect they resemble planetary
nebulae (Pottasch et al., 1984).

In the case of WR72 (= Sand.3), it was hinted earlier that it is
more likely the central star of a planetary nebula than a Pop.I WR
star (Barlow and Hummer, 1982). The IRAS data confirm this.

M1-67, the nebula surrounding WR124 (=209 BAC), was for many
years considered to be a planetary nebula (PK 50+3 1), till Cohen and
Barlow (1975) argued for a Pop.I WR ring nebula status. The IRAS

data, however, favour a planetary nebula status again. The fact that the central star WR124 is a single-line spectroscopic binary with a very small mass function (Moffat et al., 1982), suggests that it is a pre-cataclysmic variable, in view of current ideas on the evolution of binary central stars of planetary nebulae.

Particulars of this study are given by van der Hucht et al. (1985).

6. DUST IN WOLF-RAYET RING NEBULAE

IRAS observations of most of the known WR ring nebulae were carried out with the IRAS-CPC instrument (Wesselius et al., 1985), providing 12'x9' maps at 50 and 100 μm with 115 spatial resolution. The WR ring nebula RCW58 has 50 μm IR isophotes which coincide very well with the dark regions visible in its Hα picture. In a preliminary analysis, we find that the dust temperature in RCW58, as well as in the other with the CPC observed WR ring nebulae, are of the order of 35 K, in good agreement with size and energy balance considerations.

An elaborate version of this paper has been published by us in: Birth and Evolution of Massive Stars and Stellar Groups, Proc. of a Colloquium in honour of Adriaan Blaauw (W. Boland & H. van Woerden,eds.).

REFERENCES

Aitken, D.K., Roche, P.F. Allen, D.A.: 1982, Monthly Notices Roy. Astron. Soc. 200, 69P.
Barlow, M.J. Hummer, D.G.: 1982, in C.W.H. de Loore & A.J. Willis (eds.), Wolf-Rayet Stars, Observations, Physics, Evolution, Proc. IAU Symp. No. 99 (Dordrecht: Reidel), p. 387.
Cohen, M. Barlow, M.J.: 1975, Astrophys. Letters 16, 165.
Hidayat, B., Supelli, K.R., Admiranto, A.G., van der Hucht,K.A.: 1985, in preparation.
van der Hucht, K.A. Olnon, F.M.: 1985, Astron. Astrophys. 149, L17.
van der Hucht, K.A., Conti, P.S., Lundstrom, I., Stenholm, B.: 1981, Space Science Reviews 28, 227.
van der Hucht, K.A., Jurriens, T.A., Olnon, F.M., Thé, P.S., Wesselius, P.R., Williams, P.M.: 1985, Astron. Astrophys. 145, L13.
IRAS Point Source Catalog, 1985, Explanatory Supplement, U.S. Government Printing Office.
IRAS Spectral Atlas, 1985, Astron. Astrophys. Suppl., in press.
Maeder, A.F.J., Lamontagne, R., and Seggewiss, W.: 1982, Astron. Astrophys. 114, 135.
Pottasch, S.R., Baud, B., Beintema, D., Emerson, J., Habing, H.J., Harris, S., Houck, J., Jennings, R., Marsden, P.: 1984, Astron. Astrophys. 138, 10.
Roche, P.F. Aitken, D.K.: 1984, Monthly Notices Roy. Astron. Soc. 208, 481.
Wesselius, P.R., de Jonge, A.R., Beintema, D.A., Jurriens, T., Kester, D.K., van Weerden, J.E., de Vries, J., Perault, M.: 1985, Internal Report SRON Space Research Groningen.
Williams, P.M., van der Hucht, K.A., Thé, P.S.: 1985, in prep.

DUST FORMATION IN WOLF-RAYET STELLAR WINDS

K.A. van der Hucht, SRON Laboratory for Space Research, Utrecht
P.M. Williams, The Royal Observatory, Edinburgh
P.S. Thé, Astronomical Institute "Anton Pannekoek", Amsterdam

ABSTRACT. Wolf-Rayet stars are characterized by high mass-loss rates indicative of their transient evolutionary status. The late WC types are in addition surrounded by dust shells. Apart from episodic and variable dust formation around some late WC stars, most WC8, WC8.5, WC9, and WC10 stars show a constant termal IR excess over the history of IR astronomy.

In the past four years all 28 galactic WC stars of those types were monitored from ESO and UKIRT. The observations were supplemented with IRAS data. Typical colour temperatures are of the order of 900 K. Because of the observed smooth energy distributions, the dust consists most likely of amorphous carbon. This is not surprising with WC stellar winds having a carbon mass fraction of .34.

Typical dust shell radii are of the order of 100 A.U., i.e. in the region where the stellar winds have reached their terminal velocities ($\approx$ 1500 km/s). Because of the large mass loss rates, of the order of 3.10^{-5} $M_\odot$/yr, the high density winds carry the dust away. Since in most cases the IR characteristics are constant over many years, this implies that the dust is constantly being replenished.

In modeling the IR observations, (ρ,T) criteria are derived for the dust to form in the given carbon-rich environment. The derived carbon dust loss rates and the influence of the WC stars on the interstellar medium in this respect are compared with those of other dust producing stars.

Particulars of this study will be published in Astronomy and Astrophysics by Williams, van der Hucht, and Thé.

F. P. Israel (ed.), Light on Dark Matter, 90.
© *1986 by D. Reidel Publishing Company.*

OBSERVATIONS OF YOUNG (ORION-TYPE) STARS WITH IRAS

H. J. Walker
Sterrewacht
Leiden

P. L. Marsden
Dept. of Physics
Leeds, LS2 9JT

Forty-one young objects from Cohen's (1973, 1974) lists were found to have fluxes in more than one of the IRAS bands. Some of the sources are plotted here, using filled circles for Cohen's data and crosses for the IRAS data. Upper limits are indicated by arrows. Of the sources five were judged to have flat energy distributions, i.e. they varied by less than 0.5 decades in $\log \lambda I_\lambda$ across the range 2.2µm to 100µm. No colour correction was made to the fluxes of these objects. Fourteen of the objects were judged to show a peak in the energy distribution in the wavelength range used here. Some, such as T Tau and GW Ori, showed evidence for two distinct peaks in their energy distributions. The remaining twenty-two were classed as having energy distributions which declined steadily with increasing wavelength. In some of the energy distributions there was evidence of a feature around 10µm, but very few of these objects were bright enough to have LRS spectra available for further examination. When a colour-colour diagram was made, using the fluxes in Janskys, with $\log F_{25} - \log F_{12}$ plotted against $\log F_{60} - \log F_{25}$, there was a tendency for the sources to separate into two regions, depending on whether the energy distribution was regarded as declining or peaked/flat.

Rydgren, Strom & Strom (1976) gave an explanation for the declining energy distributions and the peaked energy distributions. For the steady decline they used a model with a star surrounded by an optically thin envelope, with a veiling factor to show how much more flux the envelope contributed to the emergent flux in comparison to the star. For the peaked distributions they used an optically thick shell. For T Tau and GW Ori there may be two independent dust shells present around the central star. The flat energy distributions may be caused by a complex mixture of dust shells.

F. P. Israel (ed.), Light on Dark Matter, 91–92.
© *1986 by D. Reidel Publishing Company.*

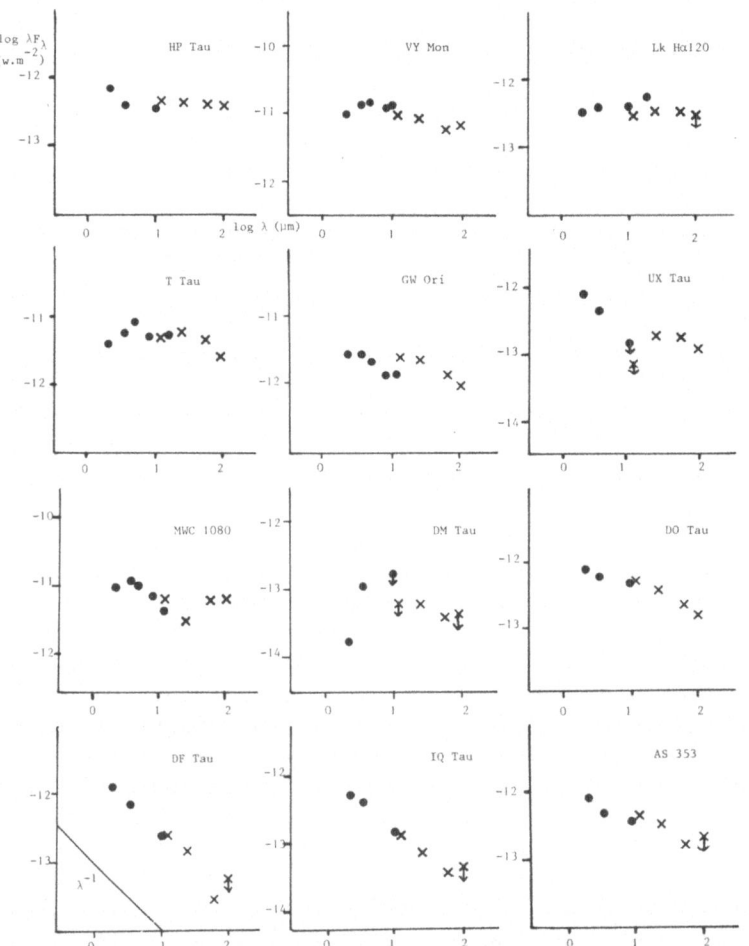

Energy distributions for some of the sources.
HP Tau, VY Mon, Lk Hα120 are classed as flat.
T Tau, GW Ori, UX Tau, MWC 1080, DM Tau are classed as peaked.
DO Tau, DF Tau, IQ Tau, AS 353 are classed as declining.
A λ^{-1} decline is also shown, on the DF Tau plot.

REFERENCES

Cohen, M., 1973. Mon. Not. R. Astr. Soc. 161, 97.
Cohen, M., 1973. Mon. Not. R. Astr. Soc. 161, 105.
Cohen, M., 1974. Mon. Not. R. Astr. Soc. 169, 257.
Rydgren, A.E., Strom, S.E., Strom, K.M., 1976. Astrophys. J. Supp 30, 307.

MASS LOSS BY COOL STARS

B. Zuckerman
Astronomy Department
UCLA
Los Angeles, CA   90024 USA

ABSTRACT. Mass loss by cool giant stars is very likely the most important mechanism for returning processed stellar material to the interstellar medium. This mass loss has profound effects on the evolution of individual intermediate mass stars and the chemical evolution of the Milky Way. The IRAS point source catalog is ideally suited for identifying those stars with large mass loss rates in dust grains. These stars, in turn, become prime targets for radio and infrared studies of outflowing gas. The total rate of return of processed material to the interstellar medium can then be obtained.

1.   INTRODUCTION

Why should one bother to study mass loss from cool giant stars instead of, for example, those faint, fuzzy objects called galaxies? Roughly one-half of all dying stars were born with main-sequence masses greater than the Chandrasekhar limit (1.4 $M_\odot$). Yet, only a few percent of these dying stars do so as supernovae. The vast majority of those that begin their lives with less than about 8 $M_\odot$ become white dwarfs with M < 1.4 $M_\odot$. Thus, during their lifetimes, some stars manage to lose up to perhaps 7 $M_\odot$. For this to be the case, mass must be lost during part of the lifetime of the star at a rate that is faster than the rate at which mass is added to an inert core. Except for very massive stars, which presumably supernovae anyway, such rapid mass loss happens only on the asymptotic giant branch (AGB) during a period of $\lesssim 10^6$ years.

AGB stars contribute the majority of all stellar material that is returned to the interstellar medium. The average main-sequence mass of a white dwarf progenitor is approximately 1.6 $M_\odot$, whereas the average mass of a white dwarf is approximately 0.6 $M_\odot$. Since the white dwarf birthrate in the Milky Way galaxy is approximately one per year, the total galactic rate of mass loss ($\dot{M}$) due to AGB stars is approximately 1 $M_\odot$ per year. The total mass lost by supernovae summed over the galaxy is probably an order of magnitude less. AGB stars and planetary nebulae very likely contribute the majority of the nitrogen and carbon that is processed inside stars and then returned to the interstellar

93

F. P. Israel (ed.), Light on Dark Matter, 93–100.

medium. Some of the preceding ideas are discussed in greater detail in chapters by Trimble and by Iben in Morris and Zuckerman (1985).

Table 1 summarizes the likely relevance of IRAS for enhancing our understanding of various aspects of mass loss on the AGB. The relevance of IRAS to loss rates in dust grains is obvious. Because $\dot{M}_{gas}$ is reasonably well correlated with $\dot{M}_{dust}$ (see below) and because IRAS contains a nearly complete sample of all AGB stars out to large distances from the Earth, the IRAS data base may be used in various ways (see below) to obtain $\dot{M}_{gas}$ for a very large sample of stars. Analysis of such a data base may help us to deduce the physical mechanism(s) that drive the mass loss. Additional radio and infrared observations of these stars should clarify our understanding of circumstellar chemistry and isotopic abundances.

2. MASS LOSS

There are at least four general IR and radio techniques for deducing mass loss rates on the AGB and these are listed in Table 2. In the third column we list the number of stars to which the indicated technique has already been or will soon be applied. We will have nothing further to say about the third technique, which has been limited in application to oxygen-rich (O/C > 1) stars but could eventually be applied to carbon-rich stars when millimeter wavelength interferometers are used to study molecules such as HCN (Jura 1983a). The second technique is based on the assumption that mass loss is driven by radiation pressure on dust grains. Here L_{IR} is the "infrared excess" defined as that fraction of the total underlying stellar luminosity that is absorbed and reemitted by dust grains in the circumstellar envelope. For optically thin shells, this technique will give an underestimate of $\dot{M}$ if and when scattering of the stellar light by the dust grains is more important than true absorption.

2.1 $\dot{M}_{dust}$ grains

The most comprehensive pre-IRAS analysis of far-infrared continuum emission from evolved giant stars is due to Sopka et al. (1985). They used the UKIRT to investigate the 400 μm emission from about one dozen objects for which they derived $\dot{M}_{dust}$ (or an upper limit). Advantages to using long wavelength emission (in the case of IRAS, 60 and/or 100 μm data) to derive $\dot{M}_{dust}$ include:

1) The Rayleigh-Jeans approximation applies so that the (uncertain) dust temperature appears only linearly in the expression for $\dot{M}_{dust}$;
2) The emission is optically thin;
3) In the region sampled by the IRAS beam, in the absence of secular variations in the mass loss rate, one can safely assume that the number density of dust grains is proportional to the inverse square of the distance of the grain from the star.

By way of contrast, these assumptions are probably not valid close to the star where warm dust grains emit near infrared radiation.

Sopka et al. showed that the typical gas-to-dust ratio in the outflowing envelopes was similar to that in the interstellar medium but with a substantial spread of values. It should be possible to obtain $\dot{M}_{gas}/\dot{M}_{dust}$ for hundreds of the brightest stars in the IRAS point source catalog. The total number of such stars would be limited, by millimeter wavelength state-of-the-art sensitivities, to those stars for which $\dot{M}_{gas}$ can be derived from CO observations as described below. For more distant objects, such as many OH/IR stars, in which CO cannot yet be detected, at least $\dot{M}_{dust}$ may be obtained from the IRAS data.

2.2 $\dot{M}_{gas}$

In the following we discuss results obtained to date on mass loss rates in gas as derived from CO observations. We do not discuss derivations of M_{gas} from OH observations of OH/IR stars in part because relatively few are available and, in part, because coming to Holland to lecture on OH/IR stars would be tantamount to traveling to the Vatican to lecture on Catholicism.

At this point in time (June 1985), approximately 130 evolved stars have been detected via either $J = 1 \rightarrow 0$ or $J = 2 \rightarrow 1$ rotational emission from CO molecules in their outflowing winds. Almost all these stars are listed in survey papers by Knapp and Morris (1985, hereafter KM), Zuckerman and Dyck (1985 and 1986, hereafter ZD 85 and ZD 86) and Zuckerman, Dyck and Claussen (1985, hereafter ZDC). The latter three surveys used the IRAS point source catalog as a guide for CO emission. Specifically, only catalog objects with fluxes greater than 100 Jy at 12 μm or greater than 50 Jy at 25 μm were searched. Attempts were made to discriminate against searching for CO in IRAS sources that were clearly young stars embedded in molecular clouds. In Table 3 we list stars that have been searched for CO about which very little is known. That is, these stars appear either only in the IRAS point source catalog or else, also, in the Revised AFGL catalog (Price and Murdock 1983) but, in either case, have not been characterized via previous ground-based observations. The CO line shapes and intensities permit a nearly unambiguous division into pre- and post-main-sequence stars. (The former have narrow, intense CO lines and the latter have broad, weak lines.) Also listed are some stars in which CO has been searched for but not yet detected. A few of these are very bright IRAS sources. All data are from ZD 85, ZD 86, and ZDC.

2.3 Mass Loss Mechanisms

The physical mechanism that drives mass away from red giant stars is still uncertain. Among suggestions appearing in the literature, one might list: 1) pulsations (shocks), 2) hydromagnetic waves, 3) the gravitational field of a close companion star and 4) radiation pressure on dust grains. If any of the first three mechanisms is responsible for mass loss, then no one really knows how to calculate, in a convincing fashion, an upper limit to MV_{∞} for a star of given luminosity (L_*). However, if radiation pressure dominates, then from momentum conservation:

$$\dot{M}V_\infty \sim \tau L_*/c$$

where, for very optically thick dust shells, $\tau \sim 2$ and, for shells optically thin in the near infrared, τ is the dust optical depth due to scattering and true absorption near the wavelength of maximum stellar emission.

To evaluate the idea that radiation pressure on dust grains drives the stellar wind, various authors (e.g., Jura 1983b, KM, ZD 85) have compared $\dot{M}V_\infty$, as estimated from CO emission profiles, with L_*/c. In particular, KM show that, for stars that are not too close to the Earth, $M_{gas} \propto T_B V_\infty^2 D^2$ where T_B is the peak CO line brightness temperature and D is the distance between the Earth and the star in question. In addition, $L_* \propto FD^2$ where F is the total infrared and optical flux received at the Earth. So a plot of $T_B V_\infty^3$ versus F yields a comparison of $\dot{M}V_\infty$ with L_*/c independent of the uncertain distances to the various stars. ZD 85 do this for approximately 80 stars and conclude that, for most stars, $\dot{M}V_\infty \lesssim L_*/c$. For $\lesssim 10$ stars, $\dot{M}V_\infty$ is apparently approximately three to ten times greater than L_*/c.

Because of errors in the measurements and model uncertainties and because τ can be larger than unity for very optically deep shells, it is not obvious that factors of approximately 3 should be regarded as inconsistent with radiation pressure-driven mass loss. Factors of 10 are more difficult to explain away, but two of these stars, NGC 7027 and AFGL 618, may have declined by a factor of 10 in luminosity subsequent to ejection of the observed CO molecular envelopes (e.g., Jura 1984; Spergel et al. 1983). Further analysis of line and continuum emission from the other stars with apparently large $\dot{M}V_\infty$ is required before a definitive conclusion can be reached.

3   OUTFLOW VELOCITIES (V_∞) AND C/O RATIOS

ZDC and ZD 85 find a fair number of carbon-rich stars with large V_∞ (up to 34 km/s) and apparently large $\dot{M}V_\infty$ located close to the galactic plane. This suggests that massive stars are involved but that close companion stars are not important in physically ejecting matter from red giants. The prototypical carbon star with large mass loss, IRC+10216, has a very modest outflow velocity (~ 15 km/s) which is less than the average of approximately 60 carbon stars plotted by ZDC. Therefore, it is likely that IRC+10216 is located substantially closer to the Earth than 290 pc since, at this distance, it would be a massive star at the very tip of the AGB.

ZDC find, in a total sample of more than 100 stars, that the dispersion in outflow velocities is similar in the oxygen- and carbon-rich subsets of this sample. The mean value of V_∞ for the carbon-rich subset may be slightly larger than the mean for the oxygen-rich set when known O-rich supergiants are excluded from the latter group.

IRC+10216 has a long period relative to optically bright Mira-type carbon stars. But OH/IR stars are concentrated toward the galactic plane and some have much longer periods (of order 1000 days) than does

IRC+10216. It will be interesting to measure the pulsation periods of the carbon stars with large V_∞ to see if they continue the trend of increasing V_∞ with increasing period plotted by ZDC. Olnon et al. (1984) argue that the reddest OH/IR stars, as determined from 12, 25 and 60 μm IRAS fluxes, pulsate only weakly or not at all. Again it will be of interest to determine if a similar trend holds for carbon-rich giants with large V∞.

KM have modeled mass loss from approximately 45 stars with detectable CO emission. They conclude that the total mass returned to the interstellar medium by AGB stars is in agreement, to within a factor of 3 or so, with the rate deduced in § 1 from the birth rate of white dwarfs. Both KM and ZDC find that this mass loss is roughly equally divided between carbon- and oxygen-rich stars. In particular, out of a total sample of 110 stars, ZDC find that 60% are either carbon-rich or of S-type (C/O ~ 1). Out of a total sample of 70 planetary nebulae (PN) Zuckerman and Aller (1985) find that 62% are carbon-rich or S-type. Therefore, to first order at least, the statistics are consistent with the idea that PN originate from very red, giant stars with large $\dot{M}$ detected in microwave CO emission.

This research was supported in part by N.S.F. Grant No. AST 83-18342 to UCLA.

REFERENCES

Bowers, P.F., Johnston, K.J., Spencer, J.H. 1983, Ap. J., 274, 773.

Jura, M. 1983a, Ap. J., 267, 647.

Jura, M. 1983b, Ap. J., 275, 683.

Jura, M. 1984, Ap. J., 286, 630.

Knapp, G.R., Morris, M. 1985, Ap. J., 292, 640 (KM).

Morris, M., Zuckerman, B., editors. 1985, Mass Loss From Red Giants, Reidel, Dordrecht.

Olnon, F.M., Baud, B., Habing, H.J., de Jong, T., Harris, S., Pottasch, S.R. 1984, Ap. J. Letters, 278, L41.

Price, J.D., Murdock, T.L. 1983, The Revised AFGL Infrared Sky Survey Catalog, AFGL-TR-83-0161.

Sopka, R.J., Hildebrand, R., Jaffe, D., Gatley, I., Roellig, T., Werner, M., Jura, M., Zuckerman, B., 1985, Ap. J., in press (July).

Spergel, D.N., Giuliani, J.L., and Knapp, G.R. 1983, Ap. J., 275, 330.

Zuckerman, B., Aller, L.H. 1985, Ap.J., in press.

Zuckerman, B., Dyck, H.M. 1985, Ap.J., submitted (ZD 85).

Zuckerman, B., Dyck, H.M. 1986, Ap.J., to be submitted (ZD 86).

Zuckerman, B., Dyck, H.M., Claussen, M.J. 1985, Ap. J., submitted (ZDC).

TABLE 1

Relevance of IRAS to What We Would Like to Know About
Mass Loss From Giant Stars

Property	IRAS	
	Direct Relevance	Indirect Relevance
1) $\dot{M}_{gas}$		
a) Individual stars		✓
b) Total over the Milky Way Galaxy for different stellar types (e.g., C-rich vs O-rich)		✓
2) $\dot{M}_{dust\ grains}$		
a) Individual stars	✓	
b) Total over the Milky Way as a function of stellar type	✓	
3) Underlying causes of $\dot{M}$		✓
4) Circumstellar chemistry		✓ (?)
5) Isotopic abundances		✓

TABLE 2

Infrared and Radio Techniques for Determining $\dot{M}$ in Cool Giant Stars

Technique	Quantity That is Measured	Remarks
1) Broadband IR Emission	$\dot{M}_{dust\ grains}$	IRAS data available for thousands of stars
2) Momentum Conservation $\dot{M}V_\infty = L_{IR}/c$	$\dot{M}_{gas}$	L_{IR} = "infrared excess" luminosity above black body flux extrapolated from short wavelengths
(Assumes that radiation pressure on dust grains drives mass loss)		IRAS data available for thousands of stars
3) Size of OH maser region (plus photodissociation model)	$\dot{M}_{gas}$ (oxygen-rich stars only)	Dozens of stars (e.g., Bowers et al. 1983)
4) Modeling CO emission ($J = 1 \rightarrow 0$, $2 \rightarrow 1$ and higher rotational transitions)	$\dot{M}_{gas}$	CO emission should be detectable from hundreds of stars with newest generation of radio telescopes and receivers

TABLE 3

IRAS Related Pre- and Post-Main-Sequence CO Emission

Detected in CO Emission			Not Yet Detected in CO Emission	
Broad CO Lines		Narrow Intense CO Lines		
"Pure" IRAS Sources	RAFGL/IRAS Sources		IRAS	RAFGL/IRAS
			0737−4021	5359
			0743−3750	1992
0215+2822	5102	IRAS 0423+5336	0801−3627	5369
0453+4427	5250	RAFGL 5206	1634−3814	5379
0713+1005	5254	RAFGL 5497	1646−4022	5384
0807−3615	6815S	RAFGL 5502	1700−4119	5146S
1610−4205	5416	IRAS 2155+5907	1744−4048	2088
2002+3910	2343	IRAS 2214+5206	1803−2201	2143
2131+5631			1809+2704	2298
2148+5301			1818−1623	2333
2155+6204			1824−0839	2350
2227+5435			1956+3423	2477
			2043+3825	2968
			2134+4508	
			2137+4540	

MODELS OF IRAS OBSERVATIONS OF CIRCUMSTELLAR SHELLS

M. Rowan-Robinson, A. Lock, D.W. Walker, S. Harris
Theoretical Astronomy Unit
Queen Mary College
London E1

We have modelled the IRAS observations of the circumstellar dustshells
around late-type stars previously studied by Rowan-Robinson and Harris
(1982, 1983a, b), using only high quality IRAS fluxes which do not
appear extended in the raw data. Figure 1a, b shows the IRAS 12-26-60
colour colour diagrams for late M and carbon stars, together with the
predictions of simple model sequences. Figure 2a, b show the spectra of
a sample of the stars.

The main results are:

1) The carbon star models of Rowan-Robinson and Harris (1983b) are a
good fit to the IRAS data, provided the outer radius of the dust shell
is taken as 1000 x the inner radius, consistent with the zone at which
the wind from the star is expected to interact with the interstellar
medium.

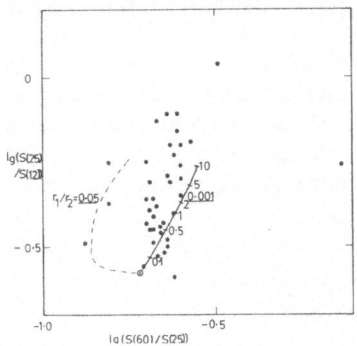 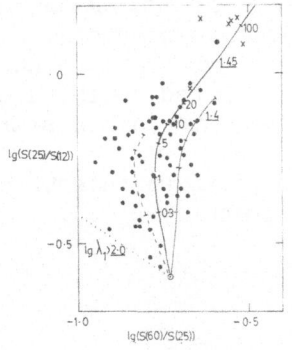

Fig.1a: 12-25-60 μ colour-colour diagram for carbon stars. Solid curve: model sequence for amorphous carbon grains, r_2/r_2 = 0.001 (labelled with values of τ_{uv}). Broken curve: r_1/r_2= 0.05.

Fig.1b: Some for late M stars. Crosses are selected OH-IR stars. Solid and dotted curves are model sequences for dirty silicate grains with $Q_\nu \propto \nu$ for $\lambda > \lambda_1$, with $\log \lambda_1(\mu)$ = 1.4, 1.45, 2, all with r_1/r_2 = 0.001. Broken curve: $\log \lambda_1$= 1.4, r_1/r_2 = 0.01

F. P. Israel (ed.), Light on Dark Matter, 101–102.
© *1986 by D. Reidel Publishing Company.*

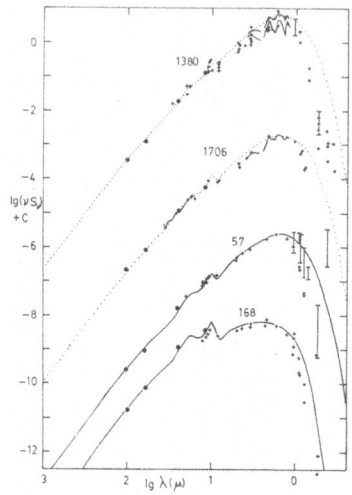 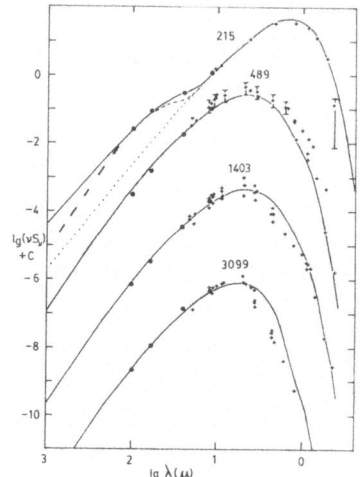

Fig. 2a: Models for selected late M stars, labelled with AFGL number.

Fig. 2b: Models for selected carbon stars. The far infrared excess in GL 215 has been modelled with black body grains, T = 125K (solid curve) and with $Q_\nu \propto \nu$ grains, T = 95K (broken curve).

2) The properties of the silicate grains around M stars have to be modified so that $Q_{\nu \text{ abs}} \propto \nu$ for $\lambda_1 \lesssim \lambda \lesssim 100$ μ, where $\lambda_1 = 28$ μ. Thus the grains formed in the atmospheres of both carbon- and oxygen- rich stars must be highly amorphous.

A few stars show excess radiation at 100 μ, which may be emission from the shell expected to build up where the stellar wind runs into the interstellar gas. The carbon star GL 215 shows excess radiation at 60 and 100 μ, consistent with a blackbody spectrum (Fig. 2). This could be the result of a previous injection which terminated recently or could be due to an Oort cometary cloud.

REFERENCES

Rowan-Robinson, M. Harris, S., 1982, Mon. Not. R. Astr. Soc. 200, 197.
Rowan-Robinson, M. Harris, S., 1983a, Mon. Not. R. Astr. Soc. 202, 767.
Rowan-Robinson, M. Harris, S., 1983b, Mon. Not. R. Astr. Soc. 202, 797.

LUMINOSITIES of OH/IR STARS

J.H. Burger
Sterrewacht Leiden, Postbus 9513, 2300 RA Leiden,
The Netherlands
J. Herman
ESTEC, SSD, Postbus 299, 2200 AG Noordwijk zh.,
The Netherlands

ABSTRACT

The stars that exhibit strong maser emission from the OH radical at 1612 MHz, are located in the Herzsprung-Russell diagram on the Asymptotic Giant Branch. The Mira variables show relatively weak masering. Their mass loss rate is moderate, a few 10^{-7} $M_\odot yr^{-1}$, and the central star is still optically visible. The maximum of the spectral energy distribution lies around 1 μm. The OH/IR stars, found in extensive surveys at radio wavelengths, are losing mass at extremely high rates, up to 10^{-4} $M_\odot yr^{-1}$, and the central object is surrounded by a thick circumstellar envelope of dust and gas, which can have as much as 100^m of visual extinction. The maximum of their energy distribution is found at 10 μm and for some of the extreme cases at even longer wavelengths. Over the past decade much attention has been given to these strong OH masers, that give valuable information on the very last stages in stellar evolution. However, many of the basic parameters, such as mass or luminosity, are still poorly known or estimated in a statistical way only.

In this paper we will determine the average luminosities for a number of OH/IR stars for which we have (a) geometrical distances, (b) knowledge of their time variability, and (c) observations shortward of 10 μm available. For those stars with only IRAS observations available a bolometric correction is derived.

From the correlations of the redness and the 9.7 μm absorption feature with the OH luminosity, a prediction can be made of the variation of the optical depth in the 9.7 μm feature. The expected variation is found for OH 26.5+0.6, not only qualitatively but also quantitatively. Whereas the optical depth of the feature varies by more than a magnitude over a period, the total energy absorbed remains constant. For that reason the silicate feature seems the best way to estimate the mass loss rates.

F. P. Israel (ed.), Light on Dark Matter, 103–104.
© 1986 by D. Reidel Publishing Company.

IRAS Field 67   α = 20^h00^m, δ = +30°, HCON-3, 24 μm. Field in
                   Vulpecula and Cygnus, showing the Galactic plane.
                   NGC 6820 is at bottom right. The bright object north
                   of the center is NGC 6857. At top left, NGC 6888 and
                   part of the IC 1318 complex can be seen.

OH/IR CATALOGUE AND CORRELATION WITH THE IRAS DATA BASE (ABSTRACT)

R. Breukers, W. van der Veen, P. te Lintel, M. Wiertz,
H. Habing
Sterrewacht
P.O. Box 9513
2300 RA Leiden
The Netherlands

ABSTRACT. After an extensive literature study we have produced a catalogue containing all known double peaked 1612 MHz OH-masers identified with stellar IR-sources out of the IRAS data base. For some of the strong IR-sources IRAS has taken a Low Resolution Spectrum (LRS).

The catalogue consists of a computer catalogue, containing all the numerical information of the over 500 sources (e.g. OH- and IR-positions, OH- and IR-fluxes and star velocities), and an index catalogue containing the same information as the computer catalogue extended with the OH- and LRS-spectra (if available). We plan to publish the index-catalogue at the beginning of 1986. This publication will contain:
OH- and LRS-spectra,
OH- and IR-fluxes,
OH- and IR-positions star velocities and the
OH light curves if available.

F. P. Israel (ed.), Light on Dark Matter, 105–106.

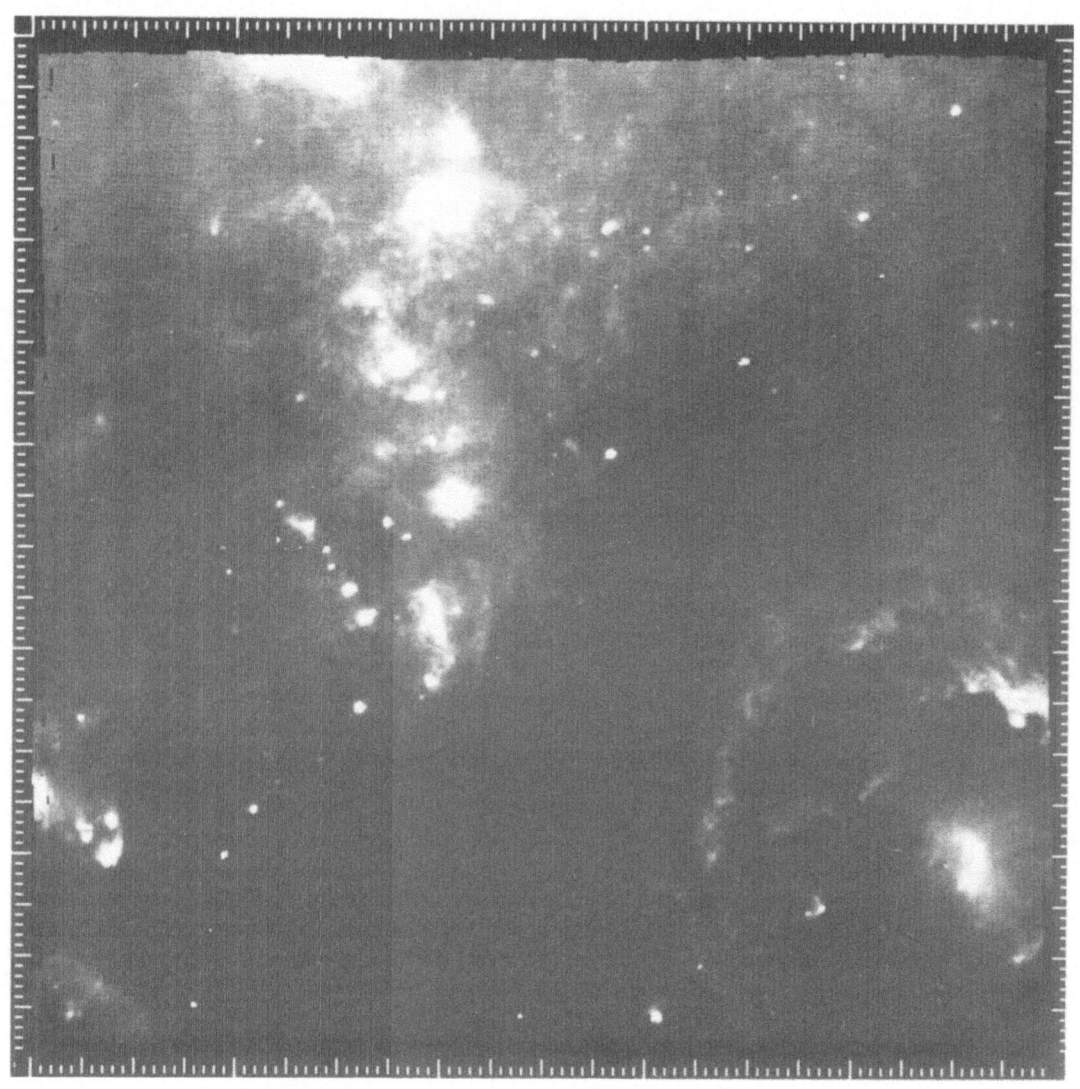

IRAS Field 77   α = 6^{h}00^m, δ = +15°, HCON-3, 60 μm. Field in Orion
and Monoceros, showing the λ Ori shell at lower
right. α Ori (Betelgeuse) is the bright object just
left of the λ Ori shell at bottom. As top right, IC
443 end going south NGC 2174/5, and the IC 2162
(S255) HII region complex; NGC 2264 (the Rosette
nebulae) is at lower left.

AGB STARS WITH HIGH MASS LOSS RATES IN THE BULGE OF OUR GALAXY (ABSTRACT)

W.E.C.J. van der Veen
Sterrewacht
P.O. Box 9513
2300 RA Leiden
The Netherlands

ABSTRACT. Using suitable criteria for the IRAS flux density ratios F_{25}/F_{12} and F_{60}/F_{25} a subset of AGB star candidates with high mass loss rates (OH/IR stars) were selected. These stars, which are at the end of their mass loss phase, clearly show a disk and a bulge component. The bulge component was studied by selecting those stars with $l < 10°$ and $2° < b < 10°$.

For all of these stars a total luminosity and a 'redness' were determined. Luminosities were determined by integrating under two Planck curves, one fitting the 12 to 25μm and one fitting the 25 to 60μm ratios and assuming a distance to the galactic centre of 8.7 Kpc. The observed luminosity histogram shows that the majority of the objects have luminosities between 2000 and 7500 L_o.

The 'redness', defined as the fraction of the total flux emitted at $\lambda > 25$ μm, is well correlated with the relative depth of the 9.7 μm silicate absorption as found from LRS spectra of over twenty nearby OH/IR stars, figure 1, which in turn is correlated with mass loss rates as model spectra show (Bedijn, priv. comm.). The redness of a star is thus a measure of its mass loss rate. Because the time scale for mass loss at the very end of the AGB is much smaller than the corresponding time scale for the growth of the core mass, both core masses and luminosities can be taken constant during this phase.

From the redness distribution within one bin in the luminosity histogram the mass loss distribution follows directly and thus the mass loss evolution for stars with the same initial ZAMS masses. The mass loss evolution at the end of the AGB appeared to be exponential in time ($M \sim \exp(t/\tau)$ in agreement with earlier results obtained from OH-data of OH/IR stars in the disk (Baud, 1983).

The duration of this phase was found to be a few times 10^4 years, during which the mass loss increases from 10^{-5} to 10^{-4} M_o yr^{-1}. Here it was assumed that 20% of the ZAMS mass was ejected in the preceeding low mass loss phase (Reimers wind).

For these calculations initial ZAMS masses of 1.5-2.5 M_o were assumed, corresponding with the observed luminosities and assuming solar abundances. The bulge stars, however, can be expected to be metal rich which will bring this mass range down.

107

F. P. Israel (ed.), Light on Dark Matter, 107–108.

   Whether we can get masses as small as 0.9–1.1 M_O and thus avoid
the conclusion of recent star formation in the bulge, is not yet
clear.

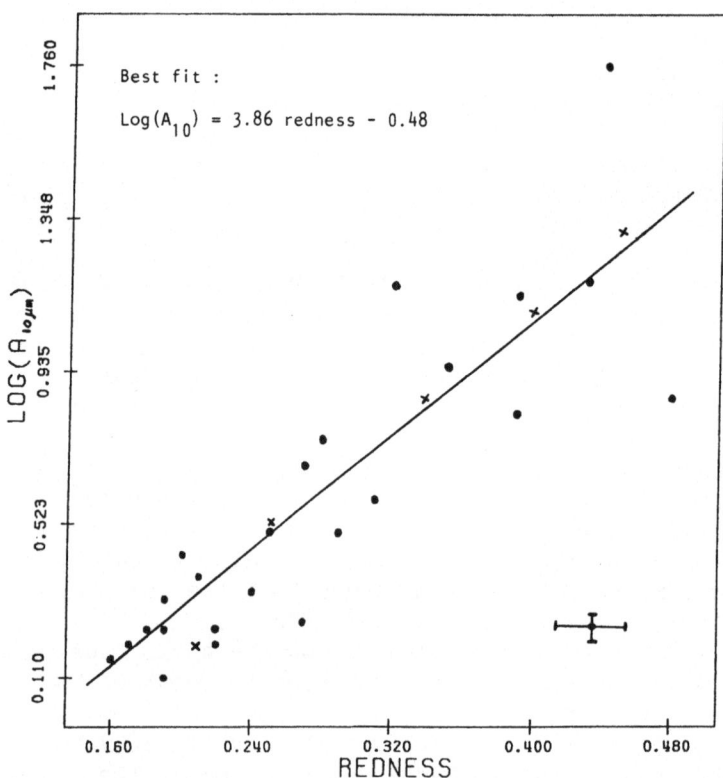

Figure 1: Relation between A_{10}(= log $(F_c/F_{10\mu m})$) and redness as
          obtained from over 20 spectra with 10μm absorption
          identified with known OH/IR stars (measurements indicated
          by dots). For comparison: absorption dips obtained from
          model spectra with accelerated mass loss (P. Bedijn, priv.
          comm.) are indicated by crosses.

Reference

Baud, B. Habing, H.J.: 1983, Astron. Astrophys. 127, 73.

THE CIRCUMSTELLAR ENVELOPE OF VX SAGITTARII

Jessica M. Chapman and R.J. Cohen
University of Manchester
Nuffield Radio Astronomy Laboratories
Jodrell Bank, Macclesfield, Cheshire SK11 9DL
United Kingdom

Radio maser emission from OH, H_2O and SiO masers provides a means of studying the circumstellar envelopes of long period variable stars. The OH 1612 MHz maser emission arises in the cool outer layers of the envelopes at distances $>10^{16}$ cm from the photosphere and is pumped by infrared photons (Elitzur 1982). The maser lines of the OH molecule at 1665/7 MHz and of H_2O and SiO molecules are valuable probes of the warm inner regions of the envelopes. Only a few sources have been mapped in these lines and physical conditions in the inner regions of circumstellar envelopes are not well understood. For the luminous M-type supergiant star VX Sgr, we have made MERLIN observations of the OH maser emission at 1612 MHz (LHC) and 1665/7 MHz (RHC) and the H_2O maser emission at 22 GHz (LHC). These have been combined with published data on the SiO masers (Lane 1984) to give a detailed description of the circumstellar envelope.

In Fig.1 the velocity separations of the masers from the stellar velocity are plotted against their projected angular offsets from the assumed stellar position. The dashed lines indicate the boundaries of the emission regions. The envelope radius at each boundary is given by the intersection of the dashed line with the line $V-V^* = 0$ and the expansion velocity at that radius is the velocity for which the projected separation is zero. The 1612 MHz masers lie in a 'thin-shell' region of angular radius 0.8 arcsec which is expanding at a terminal velocity of ~19 km s^{-1}. For an assumed distance of 1.7 kpc the linear radius of the 1612 MHz masers is 2.10^{16} cm. The H_2O and OH main-line masers at 1665/7 MHz occur in an accelerating 'thick-shell' region which has a mean radius of ~3.10^{15} cm and a mean expansion velocity of ~9 km s^{-1}. The SiO masers are closest to the stellar photosphere within a radius of about 10^{15} cm and have expansion velocities <6 km. s^{-1} (Lane 1984).

The H_2O emission region of VX Sgr is as extensive as the main-line OH region and the distribution of these masers is asymmetric. The H_2O masers lie approximately east-west and the main-line OH masers lie approximately north-south. The 1612 MHz OH emission also shows evidence of asymmetries. The maps at this frequency show shell-

F. P. Israel (ed.), Light on Dark Matter, 109–110.
© *1986 by D. Reidel Publishing Company.*

like structure which is strongest in the east and west and has gaps
in the north and south.

Our results conflict with theoretical models of maser emission
from circumstellar envelopes which assume spherical symmetry and
predict that the H_2O masers lie closer to the central star than the
main-line OH masers (e.g. Goldreich & Scoville 1976). We suggest that
the geometry of the mass outflow in the envelope of VX Sgr and
possibly many other late-type stars is influenced by the coupling of
charged circumstellar grains with the stellar magnetic field. In this
model the H_2O molecules are close to the equatorial plane where the
gas density is greatest. The 1612 MHz OH emission is also strongest
in the equatorial plane but occurs at a greater radial distance where
the gas density is sufficiently low for the photodissociation of H_2O
molecules to OH molecules (Huggins & Glassgold 1982) and where the
temperature is too low for the inversion of the main-line
transitions. The main-line OH emission originates from approximately
the same radius as the H_2O maser emission but along the direction of
the stellar magnetic axis where the gas density is least.

We have measured the Zeeman splitting of the OH spectral lines
from total power spectra at 1665/7 MHz. From the maps a Zeeman
pattern is confirmed by the coincidence in position and velocity of
OH masers at 1665 MHz and 1667 MHz. The stellar magnetic field is
approximately 5 mG at a radial distance of 3.10^{15} cm. For
a $1/r^2$ dependence this corresponds to a field at the stellar
photosphere of ~0.5 G comparable to the quiescent field strength of
the solar photosphere.

Elitzur, M., 1982. Rev.Mod.Phys., 54, 1225.
Goldreich, P. & Scoville, N.Z., 1976. Astrophys.J., 191, 93.
Huggins, G. & Glassgold, A.E., 1982. Astr.J., 87, 1828.
Lane, Adair P., 1984. IAU Symp.No.110, p329, eds. Fanti, R. et al.
    (Reidel)

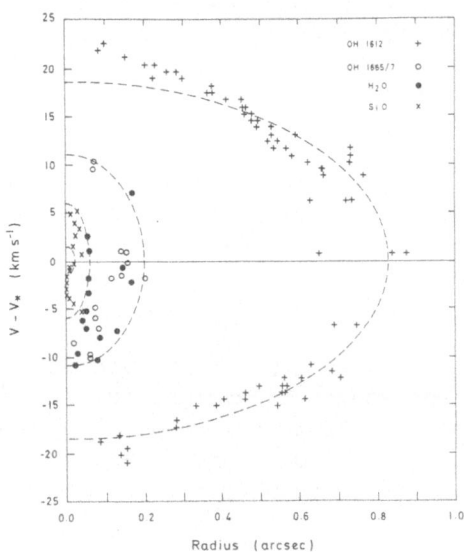

Figure 1. The velocities of
the SiO, H_2O masers of VX Sgr
are plotted against the
projected angular offsets of
the masers from the stellar
position. The dashed lines
show the boundaries of the
different maser regions.

THE INTERNAL RADIUS OF CS SHELLS AROUND COOL, OXYGEN-RICH STARS

R. PAPOULAR and B. PEGOURIE
SAP-CEN SACLAY
91191 GIF-sur-YVETTE, FRANCE

ABSTRACT. Using quantities derived from previous modeling of 23 objects of interest, we show that the internal radius of the shells varies approximately as T_*^3. This power-law is remarkably well reproduced by an equation derived from the thermodynamics of heterogenous condensation of magnesium silicates. It is also confirmed by diameter measurements at 10μm reported in the literature. These results also imply that dust wind is not sufficient to drive the mass-loss by itself. Finally, using the optical thickness, τ, derived from the models, and the above power-law, we deduce an approximate expression of the object distance in terms of T_*, τ and the near-IR flux of the star.

1. THE CONDENSATION OF GRAINS

Radiative models have been developed to describe the CS shells of cool giants and supergiants. Applying these models to the observed spectra of about 30 oxygen-rich stars, we derived the common optical properties of the CS grains in the range 8-30 μm, as well as the optical thickness and internal radius of the shells (A & A, 1983, 128, 355;1985, 142, 451). The ratio of the latter (r_1) to the star radius (r_*) is plotted in fig.1 as a function of T_* for 23 stars (filled circles). Here, we give a physical interpretation of this result. We assume the gas temperature, Tg, to decrease, with expansion, as $1/r^\delta$,where δ is to be determined empirically and should be between 0 and 0.5. The vapor of Mg_2SiO_4 is assumed to preexist in the expanding gas and to condense on nuclei of more refractory material, so that the required supersaturation is small (~ 1). The saturated vapor pressure is of the general form $10\uparrow(A-B/Tg)$, where A=38 and B=2.75x10^4K (from JANAF tables, for Mg_2SiO_4 and $MgSiO_3$). The radial distance, r_c, where condensation occurs, is then given by the implicit equation

$$T_* = \frac{B}{A+ \delta \log(r_c/r_*) - \log(T_* \, \eta_c)} \cdot \left(\frac{r_c}{r_*} \right)^\delta \qquad (1)$$

where η_c is the gas density at the same distance. The curve drawn through the points in fig. 1 corresponds to the particular set of

111

F. P. Israel (ed.), Light on Dark Matter, 111–112.

parameters δ =0.35 and T_* $n_c=10^{14}$K.cm^{-3}, which happens to be the best
fit. This relation is very nearly of the form $r_c/r_* \simeq (10\uparrow-8.9).T_*^{2.9}$.
From the same premises, we can also deduce : $n_c \simeq 3\times10^{10}$cm^{-3} and Tg($r_c$)$\simeq$
1100°K for all stars.

The crosses represent interferometric (α Ori, o Ceti), occultation
(RLeo, with error bar) and array-imaging (α Sco) measurements in the
10-μm band (from the literature).

It is clear that the condensation does not occur near enough to the
photosphere for the dust wind to be the sole mechanism responsible for
mass loss.

2. THE ESTIMATION OF STAR DISTANCES.

Assuming that all the available "silicate" vapor condenses instantly
at $r_c(=r_1)$, that the resulting grains are subsequently driven at
constant speed, v_d, by the star radiation pressure and that the mass
loss, $\dot{M}$(M⊙/y), is constant, then gas and grain densities vary as r^{-2}
beyond r_c. The optical thickness of the shell is then given by

$$\tau(10\mu) \simeq \left[3.10^{34}\ \frac{\dot{M}}{v_d r_*^2}\right].r_* T_*^{-2.9} \tag{2}$$

For our sample of objects, the quantity in bracket $\simeq 10^{-39}$. Now, r_*
can be expressed in terms of the star distance d_m and the flux F on
earth (outside the silicate band, e.g.λ= 3.6μm). Then,

$$d_m(\text{kpc}) \simeq 10^{-17}.\ \tau(10).T_*^{2.9}\left[BB(T_*,\lambda)/F(\lambda)\right]^{1/2} \tag{3}$$

This "model" distance is plotted in fig. 2 against the "measured" distance, d (from the literature), for 22 objects. Given the uncertainty on
d, the correlation is satisfactory, except for α Ori and α Sco, known
to have a small $\dot{M}$. On average, $\dot{M}/r_*^2 \simeq 2/10^{33}$ M⊙/y/cm².

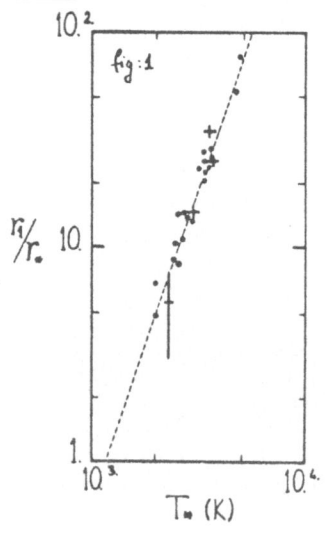

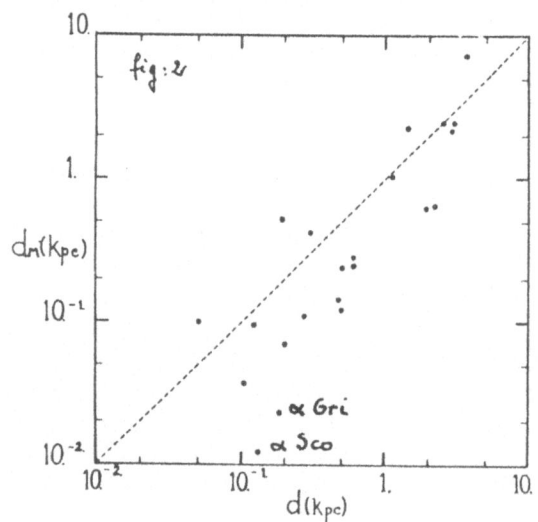

IRAS OBSERVATIONS OF CARBON STARS

F.J.Willems
Astronomical Institute 'Anton Pannekoek'
University of Amsterdam

1. INTRODUCTION

In view of the light shed by IRAS on dark matter one wonders what the
IRAS data add to our knowledge of carbon-rich dust. As the main factory
of this dust carbon stars deserve a thorough study. In this paper we
report on observations of cool carbon stars between 7 and 120 μm
obtained by IRAS. Cool carbon stars populate the asymptotic giant branch
(AGB) of the Hertzsprung-Russell diagram. In section 2 we describe the
sample investigated and report on the data available in the IRAS
databases. In section 3 we discuss the IRAS broad-band photometry of
the carbon stars in our sample. The LRS spectra are discussed in section
4. Section 5 contains some concluding remarks. A more detailed study
will be published elsewhere (Willems and de Jong 1985).

2. OBSERVATIONS

We studied a sample of cool carbon stars selected from the Catalog of
Cool Carbon Stars by Stephenson (1973). Out of this catalog we selected
all carbon stars outside 10 degrees from the galactic plane detected by
IRAS at 12, 25 and 60 μm. We ended up with 98 sources of which 68 have
an entry in the Low-Resolution Spectrograph (LRS) catalog.

3. THE BROAD BAND PHOTOMETRY

In Figure 1 we plot the log (S_{12}/S_{25}) versus log (S_{25}/S_{60}) color-color
diagram of the stars in the sample. The black body line is also given.
The error bars correspond to uncertainties quoted in the IRAS point
source catalog. Most stars fall below the black body line, where the 60
μm flux is in excess compared to the 12 and 25 μm flux. Five stars are
clearly located above the black body line. Three of those stars have an
entry in the LRS catalog and suprisingly all of them exhibit silicate
rather than carbon features as will be discussed in the next section.
This suggests that the color-color diagram can be used to discriminate
between carbon-rich and oxygen-rich circumstellar dust shells.
  A cross check with the General Catalog of Variable Stars (Kukarkin

F. P. Israel (ed.), Light on Dark Matter, 113–118.

et al. 1969) demonstrates that all the stars in our sample with a quoted
variability probability in the IRAS point source catalog larger then 85%
are Mira variables. Those stars form a small cloud around the 800 K
point on the black body line. The irregular variables (SRa,SRb and Lb)
are the ones responsible for the scatter in the diagram. Since figure 1
probes the circumstellar dust shells between about 7 and 120 μm we
conclude from this difference in appearance of the two groups that the
physical and chemical conditions in circumstellar dust shells are rather
similar for the carbon Mira variables and are subject to variations in
the other stars in the sample.

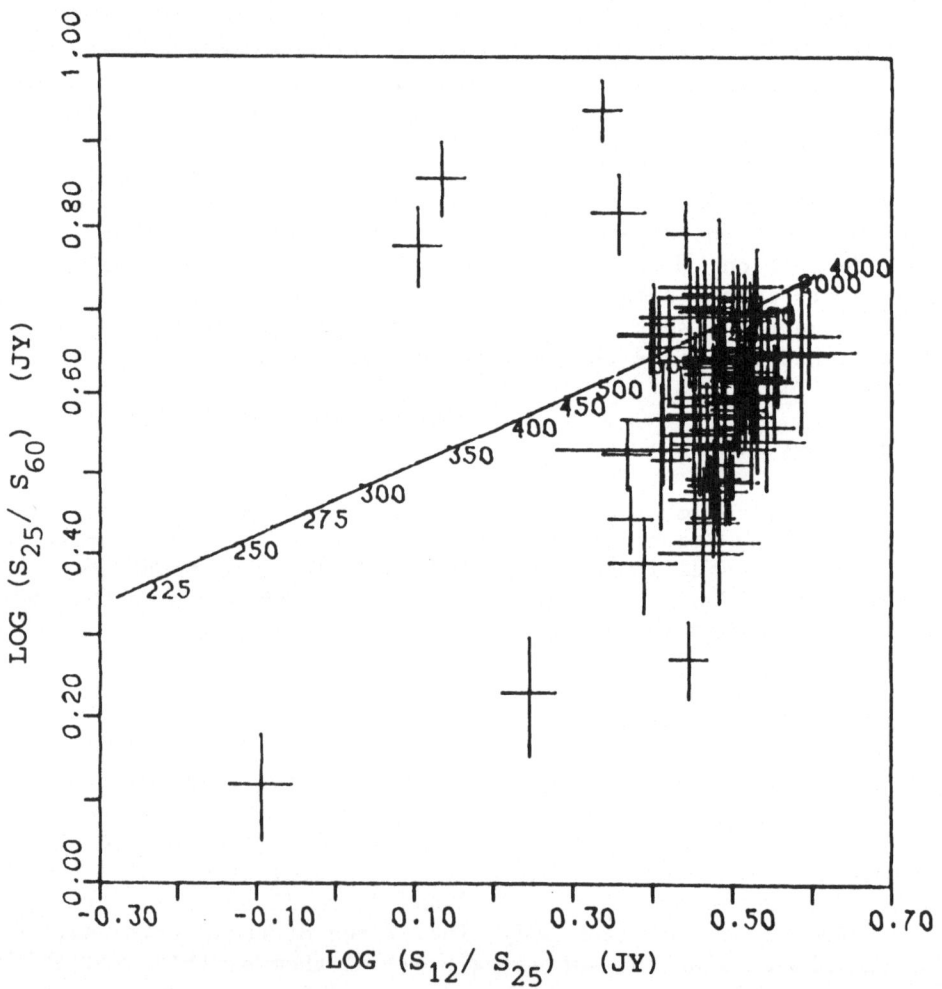

Figure 1.   The $\log(S_{12}/S_{25})$ versus $\log(S_{25}/S_{60})$ color-color plot for 98
carbon stars. The uncertainties quoted in the IRAS point source catalog
are indicated by the error bars. A black body line is also drawn.

Quantitatively this can be explained by a simple shell ejection model. Therefore we assume that the non-Mira variable carbon stars loose mass episodically. They eject distinct shells at large time intervals. Those shells are optically thin, isothermal and move with a constant speed and we assume that the dust obeys a λ^{-1} emissivity law. Thus some stars have shells close to the stellar surface and others have their shells more outward. Close to the star the shell will dominate all four IRAS bands. On its way out it will gradualy cool and dilute. After some time the shell becomes so cool that the stellar photosphere starts dominating at short wavelengths. It takes much longer before the photosphere will become visible at longer wavelenghts. This explains the elongated shape of the main cloud of carbon stars in the color-color diagram. The λ^{-1} emissivity law is choosen because the literature data (e.g Cambell et al. 1976) suggest to a λ^{-1} rather than a λ^{-2} emissivity of carbon-rich dust.

This simple model does not apply to carbon-rich Mira variables because they loose mass rather steadily and regularly, probably driven by the stellar pulsation.

4. LOW RESOLUTION SPECTRA

Of the sample under investigation 68 sources have an entry in the LRS catalog. The spectra will appear in print soon (IRAS Atlas of Low-Resolution Spectra 1985). Although all stars are classified in the optical as carbon stars, surprisingly three of them exhibit 9.7 and 18 μm silicate features. Figure 2 showes the spectra of C716, C1633 and C2919. These three stars belong to the group situated above the black body line in figure 1. This exciting result implies that we have now direct evidence for the transition of oxygen-rich to carbon-rich in the outer layers of AGB stars. Yamashita's (1972,1975) more careful classifiation of C716 and C2919 as carbon stars fully justifies their inclusion in Stephensons catalog.

The carbon-rich spectra show four features. Two narrow features, one centered at 8.8 μm and one at 11.3 μm and two broad features, one between 7 and 9 μm and one longwards of 13.7 μm. In Figure 3 examples of spectra showing the four features are given. The 11.3 μm feature between 10.2 and 13.7 μm is generaly described to SiC. The origin of the other features is still unknown.

Laboratory spectra of amorphous carbon show an excess at $\sim$ 8 μm (Koike et al. 1980). Cambell et al. (1976) showed that the far-infrared spectrum of IRC +10216 (CW Leo) could be fitted with a λ^{-1} emissivity law. This suggest that amorphous carbon may be a significant constituant of circumstellar carbon dust.

The feature longwards of 13.7 μm is clearly seen in three spectra and always in connection with the 8.8 μm feature. Although all spectra with the 13.7-20 μm feature also show an 11.3 μm feature, the fact that the majority of the spectra with an 11.3 feature have a featureless spectrum longward of 13.7 μm rules out a SiC origin. Perhaps this feature might be ascribed to amorphous carbon, but in laboratory spectra studied by Blea et al. (1970) and Koike et al. (1980) only one out of six samples show something like a 20 μm excess.

The broad 7 - 9 μm feature is always seen in spectra with a prominent 11.3 μm feature. The best examples of this feature are seen in the spectra of C1999 and C2476. C1999 (SS Vir) is a Mira and C2476 is a Mira according to the IRAS variability probability (see section 2). In a forthcomming paper (Willems and de Jong 1985) we show that these two Mira variables are surrounded by a thin isothermal shell. We suggest that during the growth of the mass loss rate the conditions in the shell change causing the feature to disappear. This explains the absence of the 7 - 9 μm feature in the majority of Mira variables.

On the basis of our data there is no way to rule out the simultaneous occurence of the 7 - 9 and the 8.8 μm feature. Our present understanding of the circumstellar dust shells and the state of laboratory experiments renders the identification of these features very difficult.

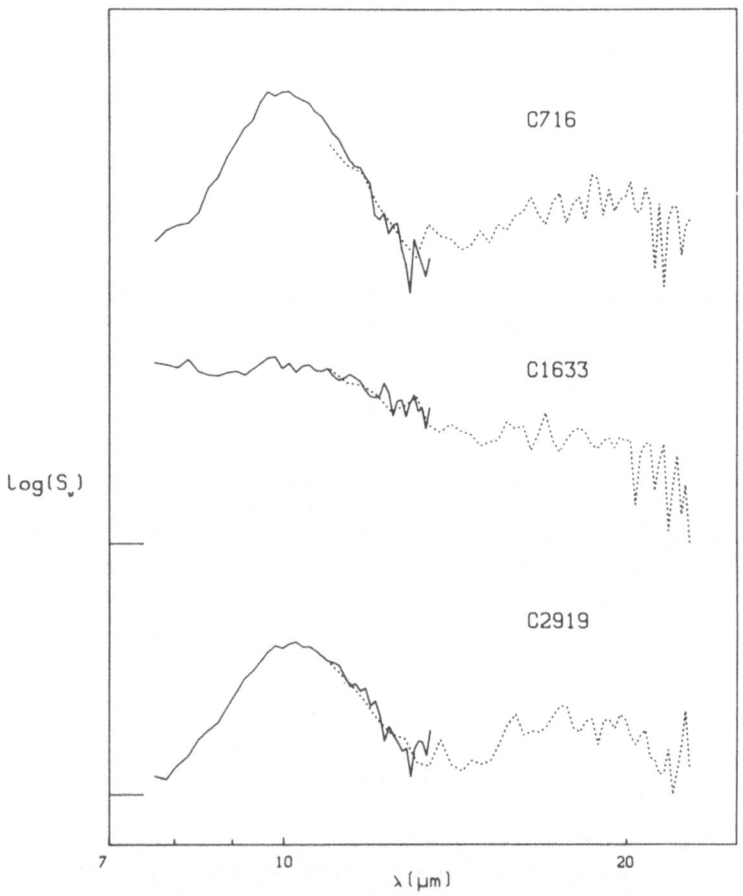

Figure 2. LRS spectra of the three carbon stars with silicate features. The tic marks on the vertical axis indicate a factor ten in flux density.

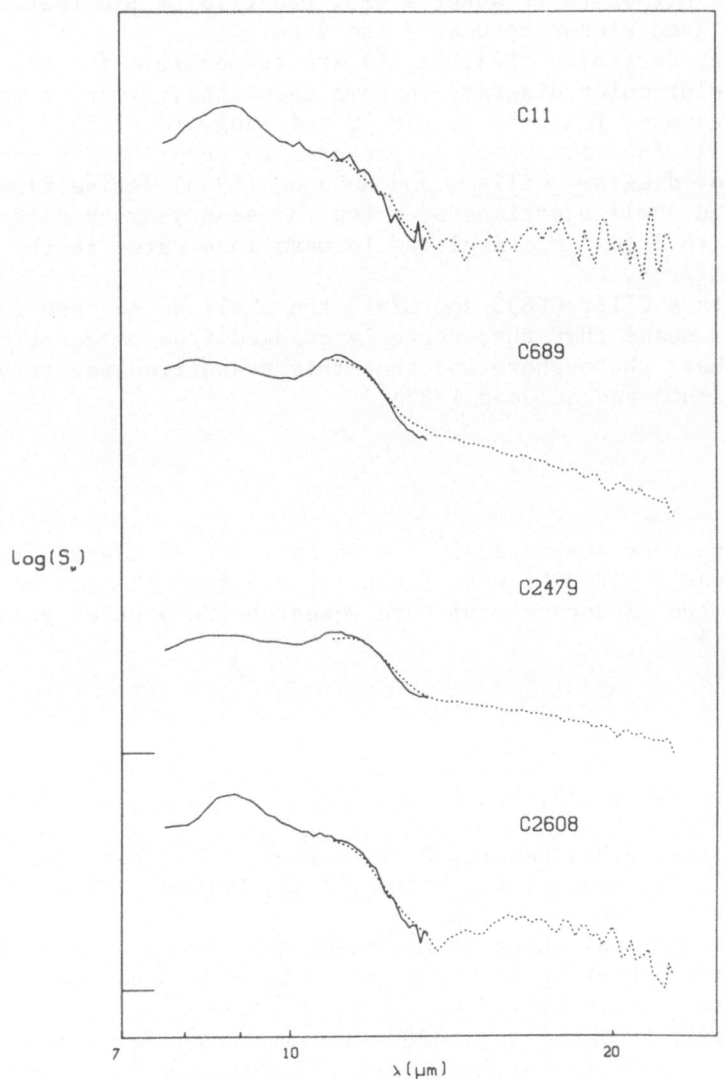

Figure 3. Four LRS spectra of carbon stars with carbon-rich
circumstellar dust shells. Four distinct dust features can be
recognized. The tic marks on the vertical axis indicate a factor
ten in flux density.

5. CONCLUSIONS

Infrared observations of carbon stars between 7 and 120 μm have revealed
some interesting facts. The $\log(S_{12}/S_{25})$ versus $\log(S_{25}/S_{60})$ color-color
diagram exhibit a large scattter. Four features in the LRS spectra can
be distinguished.

The regular variables (Miras) form a small cloud around the 800 K
black-body point. All of their spectra show the 11.3 µm SiC feature. A
few of them have some excess between 7 and 9 µm.

The irregular variables (SRa,SRb,Lb) are responsible for the large
scatter in the color-color diagram. In some cases their spectra show
beside the SiC feature, features at 8.8 µm and longward of 13.7 µm.

A simple shell ejection model is proposed to describe the scatter
in the color-color diagram. Willems and de Jong (1985) derive time
intervalls between shell ejections of a few thousand years and shell
masses of $5 \cdot 10^{-6}$ to $2 \cdot 10^{-4} M_\odot$ resulting in mass loss rates in the order
of 10^{-8} to $10^{-7} M_\odot yr^{-1}$.

For three stars C716, C1633 and C2919 the shell we see now is
oxygen-rich. This means that they recently changed from oxygen-rich to
carbon-rich in their photosphere and that this transition was very fast,
~ 1000 years (Willems and de Jong 1985)

ACKNOWLEDGEMENT

The investigations were supported by the Netherlands Foundation for
Astronomical Research (ASTRON) with financial aid from the Netherlands
Organization for the Advancement of Pure Research (ZWO) under grant
number 782-372-014.

LITERATURE

Blea, J.M., Parks, W.F., Ade, P.A.R. Bell, R.J. 1970 J.Opt.Soc.Am.
    60, 603
Cambell, M.F., Elias, J.H., Gezari, D.Y., Harvey, P.M., Hoffman, W.F.,
    Hudson, H.S., Neugebauer, G., Soifer, B.T., Werner, M.W.
    Westbrook, W.E. 1976 Ap.J. **208**, 396
IRAS Catalogs and Atlases. Atlas of Low-Resolution Spectra 1985, Joint
    Infrared Science Working Group, prepared by F.M. Olnon and E.
    Raimond Astron. Astrophys. Suppl. Ser. in press
Koike, C., Hasegawa, H. Manebe, A. 1980 Astrophys. and Space Sci.
    67, 495
Kukarkin, B.V., Kholopov, P.N., Efromov, Yu.N., Kukarkina, N.P.,
    Kurochkin, N.E., Medvedva, G.I., Perova, N.B., Fedorovich, V.P.
    Frolov, M.S. 1969 General Catalog of Variable Stars, Moscow
Stephenson, C.B. 1973 A General Catalog of Cool Carbon Stars, Publs.
    Warner & Swasey Obs. 1, no. 4
Yamashita, Y. 1972 Ann. Tokyo Astron. Obs. **13**, 169
                 1975 Ann. Tokyo Astron. Obs. **15**, 1
Willems, F.J., de Jong, T. 1985 Astron. Astrophys. submitted.

FROM MIRAS TO PLANETARY NEBULAE: A MODEL OF MASS LOSS

P. J. Bedijn
Institut für Theoretische Astrophysik
der Universität Heidelberg
Im Neuenheimer Feld 561
D-6900 Heidelberg, Federal Republic of Germany

ABSTRACT. Stellar pulsation in combination with radiation pressure on dust is proposed as the mechanism causing mass loss in Mira variables and an expression for the mass loss rate in terms of the basic stellar parameters mass and photospheric radius derived which leads to an acceleration of the mass loss rate in time as suggested by observations. Upper Asymptotic Giant Branch evolution with such a mass loss is presented. A model period distribution for classical Miras as well as a model 1612 MHz OH luminosity function for (OH/IR) Miras in the solar neighbourhood are fitted to observed ones. The time dependence of the past star formation rate and the average past star formation rate needed are in good agreement with those derived from totally different kinds of data.

1. INTRODUCTION

In the context of this paper the designation Mira variables stands for all large amplitude long period variables with more or less regular periods situated on the upper part of the Asymptotic Giant Branch (AGB). They encompass the classical, optically visible Miras with periods ranging from ~ 150 to ~ 600 days and most of the so-called OH/IR variables with even longer periods (up to 2000 days). They are stars with main sequence masses M_{ms} in the range $\sim 1\ M_\odot$ to $\sim 8\ M_\odot$, whose energy production is due to nuclear burning of H and He in very narrow shells around an electron-degenerate carbon-oxygen (C-O) core, and represent the final stages of stellar evolution in this main sequence mass range before becoming white dwarfs or supernovae (see Iben and Renzini, 1983, for a detailed review). The more or less regular luminosity variations are caused by radial pulsations of the hydrogen (H)-rich envelope.

Besides their variability these stars are also interesting because i) the He shell burning is unstable (thermally pulsing) which, in combination with (convective) mixing processes, may change the original chemical composition of the H-rich envelope (see Iben and Renzini, 1983); and ii) they all lose mass thereby creating circum-

119

F. P. Israel (ed.), Light on Dark Matter, 119–126.
© *1986 by D. Reidel Publishing Company.*

stellar (CS) shells of outflowing matter, observable through infrared
dust continuum emission and line emission of various molecules at
radio wavelengths. Modeling of this emission yields mass loss rates
($\dot{M}$'s) ranging from a few times 10^{-7} $M_\odot yr^{-1}$ for classical Miras to 10^{-4}
$M_\odot yr^{-1}$ or more for OH/IR Miras. The high $\dot{M}$ found for a number of objects suggests that mass loss may play an important, if not dominant
role in the evolution of upper AGB stars; instead of becoming supernovae most or all of the AGB stars may end up as white dwarfs.

The fact that $\dot{M}$ ranges over three orders of magnitude, whereas
stellar parameters (L_*, M_*) range over only one, is an indication for a
time evolution of $\dot{M}$ for a single star. Since Planetary Nebulae (PN)
masses, radii, and lifetimes suggest $\dot{M}$ of at least a few times 10^{-5}
$M_\odot yr^{-1}$ at the very end of AGB evolution leading towards PN and subsequent white dwarf formation this should be an evolution towards higher
$\dot{M}$.

At present two scenarios for the time evolution of $\dot{M}$ exist. In
the one due to Jones et al.(1983) a star of given M_{ms} evolves along
the AGB towards higher luminosity until at a certain L_* first overtone
pulsation starts. The star then becomes a classical Mira with modest
mass loss and with a low OH maser luminosity. Towards the end of the
first overtone phase $\dot{M}$ may have increased somewhat and the star may
have become a moderately strong type II OH maser source. Then a switch
to fundamental mode pulsation occurs, the pulsation period becomes a
factor ~ 2 larger, $\dot{M}$ suddenly increases by a large factor (100?), a
thick CS shell is formed, and the star becomes a high efficiency OH/IR
source.

The occurrence of OH/IR Miras with luminosities comparable to
those of classical Miras but with periods two times larger seems to
support this scenario. However, the luminosities used by the authors
are based on kinematic distances to the sources and these can be wrong
by an appreciable factor (see e.g. Herman et al., 1985). It is quite
possible that OH/IR stars lie on the average on the period-luminosity
relation for classical Miras extrapolated to larger periods and thus
may pulsate in the same mode (Feast, 1985).

In the scenario of Baud and Habing (1983) $\dot{M}$ increases smoothly in
time. From several unbiased galactic surveys it was found that the OH
luminosity function $\psi(L_{OH})$ (i.e. the number of sources with OH luminosity L_{OH} per unit interval of L_{OH}) for type II OH/IR sources is a power
law function of L_{OH} with slope $\alpha \sim -2$ over a range of at least a factor 100 in L_{OH}. The assumption that $\psi(L_{OH})$ reflects the time evolution
of L_{OH} for a single star combined with the "observed" correlation be-
tween L_{OH} and $\dot{M}$ ($L_{OH} \propto M^\beta$, with $\beta \sim 2$) leads to the following time
evolution of $\dot{M}$.

$$\dot{M}(t) = \dot{M}_{min} \left(1 - t/t_{OH} \right)^{1/2(1+\alpha)} \tag{1},$$

with $\dot{M}_{min} \sim 10^{-6}$ $M_\odot yr^{-1}$ and $t_{OH} \sim 10^5$ yr. The timescale on which $\dot{M}$ increases is proportional to $(1-t/t_{OH})$; this becomes increasingly smal-
ler towards the end of the evolution i.e. an acceleration of mass loss
occurs. The fact that $\dot{M}$ increases smoothly suggests the possibility of
mass loss evolution in a constant pulsation mode.

However, the physical mechanism which could cause such a time evolution remains unidentified. In the next section we shall propose such a mechanism and derive an expression for $\dot{M}$ in terms of basic stellar parameters which leads to just such a time evolution. In section 3 calculations of the evolution on the upper AGB using such a mass loss formula, calibrated to observed $\dot{M}$, are presented. In section 4 a theoretical period distribution for classical Miras and a theoretical OH luminosity function for the solar neighbourhood are compared with observations. Further possible tests of the model e.g. by using IRAS data are also mentioned.

2. MASS LOSS MECHANISM

We propose that the radial pulsations of the Miras in combination with radiation pressure on dust which condenses out in the upper part of the atmosphere is responsible for the mass loss experienced by these stars. This has already been suggested before (Wood,1979; Jones et al. ,1981). Though mass loss can be caused by either of these mechanisms alone (e.g. Willson and Hill, 1979; Wood, 1979; Kwok, 1975) their combined action may lead to substantially higher $\dot{M}$ (Wood, 1979). Due to the pulsation, shock waves periodically move outwards through the atmosphere. Numerical calculations of the structure of such periodically shocked atmospheres show that the density decreases markedly less steeply with radius than in a corresponding hydrostatic atmosphere. This is due to direct dynamical response to the shock waves, especially in the "isothermal" inner parts of the atmosphere (from the mean photospheric radius R_* to $R_t \sim 3\ R_*$), and indirectly through enhanced thermal gas pressure because the periodic passage of shock waves heats matter to a temperature higher than that for an atmosphere in radiative equilibrium. The latter process is especially important in the outer parts of the atmosphere, from R_t out to the dust condensation radius $R_c \sim 10\ R_*$.

Under the constraints $PM_*^{0.5}R_*^{-1.5} \sim$ constant for pulsating stars and $T_{eff} \sim$ constant for AGB stars, the hydrodynamical equations imply that the run of density, averaged over P, with radius is independent of M_* and R_* in the inner atmosphere, and only depends on the ratio M_*/R_* in the outer parts, to first approximation. Assuming that we can describe the run of density in the shock-heated outer atmosphere with a hydrostatic distribution with temperature distribution $T(R) = T_c\ (R/R_c)^{-\gamma}$, ($T_c \sim 1000$ K, γ is such that $T(R_t) \sim 5000$ K) and that the outflow velocity at R_c is equal to the thermal velocity v_T (~ 2 kms^{-1} for $T_c \sim 1000$ K) the mass loss rate can be written as

$$\dot{M} = \rho_* R_*^2\ B\ \exp(\ -A\ M_*/R_*) \qquad (2),$$

with

$$A = \frac{\mu m_h G}{(1-\gamma)kT_c}\ \frac{R_*}{R_c}\left[\left(\frac{R_c}{R_t}\right)^{1-\gamma} -1\right] \sim 2.5\ 10^3\ M_\odot^{-1}R_\odot$$

$$B = 4\pi v_T \left(\frac{R_c}{R_*}\right)^2 \frac{\rho\, t}{\rho_*} \left(\frac{R_c}{R_t}\right)^\gamma$$

ρ_* and ρ_t are the mean densities at R_* and R_t respectively, μ is the mean molecular weight of the gas, and m_h, G, and k have their usual meaning. The parameter combination A may depend on M_*/R_* and B is constant, to first approximation.

   Under the assumption that all quantities except M_* are constant during the AGB evolution of a star undergoing mass loss according to equation (2), this may be solved analytically to give the stellar mass as a function of time. Using this solution to eliminate M_* from equation (2) one gets an expression for $\dot{M}$ as a function of time which is the same as equation (1) for α =-1.5, provided $t_{OH} \equiv R_*/\dot{M}_{min}A$. With $R_* \sim 370\ R_\odot$ ($L_* \sim 10^4\ L_\odot$, $T_{eff} \sim 3000$ K) and $\dot{M}_{min} \sim 10^{-6}\ M_\odot yr^{-1}$ we derive $t_{OH} \sim 1.5\ 10^5$ yr. However approximate the above description of the proposed mass loss mechanism may be, we feel that it demonstrates the physical possibility that pulsating upper AGB stars may evolve under an accelerating mass loss in a constant pulsation mode.

3. EVOLUTION ON THE UPPER AGB WITH ACCELERATING MASS LOSS

   The upper (TP-)AGB evolution starts when hydrogen is reignited in a thin shell and the star begins thermal pulsation. The mass of the C-O core and the total luminosity (through the core mass-luminosity relation, see Iben and Renzini, 1983) at this moment are well-defined functions of M_{ms} and range from $M_{CO} = 0.54\ M_\odot$, $L_* = 2.6\ 10^3\ L_\odot$ for $M_{ms} = 1\ M_\odot$ to $M_{CO} = 1.06\ M_\odot$, $L_* = 4.1\ 10^4\ L_\odot$ for $M_{ms} = 8\ M_\odot$. The total stellar mass at this moment may be somewhat smaller than M_{ms} due to mass loss (via a Reimers wind) on the First Giant Branch and on the early AGB (of the order of $0.2\ M_\odot$ for small M_{ms}). The start of the TP-AGB is indicated by the solid line TP in figure 1.

   During subsequent evolution on the TP-AGB the core mass and the luminosity steadily increase (with $dlogL_*/dt \sim 3.5\ 10^{-7}\ yr^{-1}$ constant due to the fact that the main source of energy production is H burning and the existence of the core mass-luminosity relation) and the total mass decreases (due to a Reimers wind). The star may be stable against radial pulsations or pulsate in the first overtone. The evolutionary tracks (dashed lines) in figure 1 assume first overtone pulsation during this phase. Pulsation periods were calculated using the linear non-adiabatic pulsation models for standard disk stars of Fox and Wood (1983). This also presumes knowledge of T_{eff} as function of L_* and M_* (and metal abundance); we used a relation which gives a T_{eff} ranging from ~ 0.9 to $\sim 1.0\ T_{eff}$ of the Fox and Wood models.

   Upon reaching a certain L_* and M_* fundamental mode pulsation will begin. At which L_* this will happen for a star of a particular M_{ms} depends on the momentary stellar mass, effective temperature (and metal abundance). We used a transition locus derived from the pulsation models of Fox and Wood (1982) under the assumption that the growth rate for fundamental mode pulsation is equal to that for first

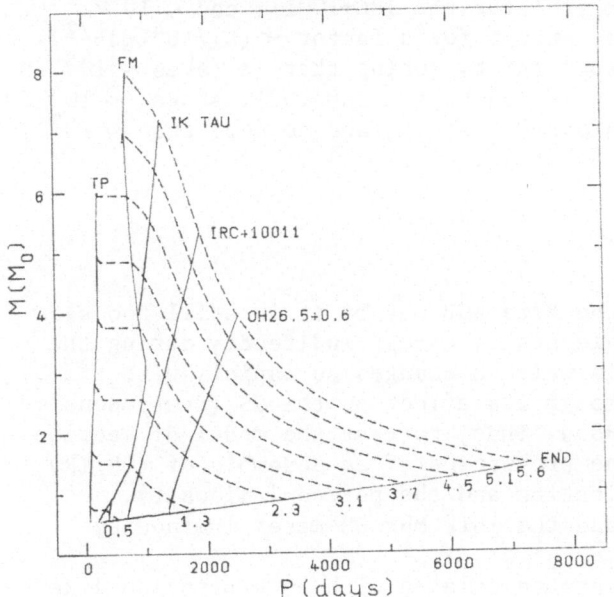

Fig. 1. Evolution on the TP-AGB for stars of different M_{ms} in a mass-period diagram. Dashed lines are evolutionary tracks with M_{ms} from 1 $M_\odot$ to 8 $M_\odot$ and they are labeled with L_* (in units 10^4 $L_\odot$) maximally reached. Solid lines indicate different evolutionary stages (see text for further explanation).

overtone pulsation. The solid line FM in figure 1 indicates this transition locus. In fact, fundamental mode pulsation already starts before the onset of thermal pulses in the most massive stars; the solid curve FM in figure 1 actually denotes the onset of the TP-AGB for $M_{ms} \gtrsim 6$ $M_\odot$. The duration from the onset of the TP-AGB to the start of FM pulsation is $7 \cdot 10^5$ yr for a star with $M_{ms} = 1$ $M_\odot$; in this time L_* increases to $4.5 \cdot 10^3$ $L_\odot$. This duration as well as the increase in luminosity continuously gets smaller towards larger M_{ms}.

During the subsequent evolution on the Mira-AGB mass loss according to equation (2) was adopted with $\rho_* = C \, M_*^2$, $A = A_0 + A_1 \, M_*/R_*$, and B constant; the values for the "free" parameters are $A_0 = 2.96 \cdot 10^3$ $M_\odot^{-1} R_\odot$, $A_1 = -1.45 \cdot 10^5$ $M_\odot^{-2} R_\odot^2$, and $B*C = 1.57 \cdot 10^{-7}$ $M_\odot^{-1} R_\odot^{-2} yr^{-1}$. These values were obtained by fitting equation (2) to some (OH/IR) Miras with reasonably well-known L_*, P, and $\dot{M}$, and by demanding that the most massive AGB stars can evolve towards a white dwarf as indicated by observations (Reimers and Koester, 1982). The evolution continues until the entire H-rich envelope is lost or only a small fraction remains; this stage is indicated with the solid line END in figure 1. Depending on M_{ms} the duration from stage FM to stage END varies from $1.7 \cdot 10^5$ yr (1 $M_\odot$), a mere $3.4 \cdot 10^4$ yr (2 $M_\odot$) to $6.5 \cdot 10^5$ yr (8 $M_\odot$). The pulsation period increases by a factor up to 7, reaching ~ 7000 days for stars with the largest M_{ms}. That objects with such large periods are not (yet) known and may never be found is not surprising since i) there are relatively few stars with large M_{ms} and ii) evolution of stars with such extreme P proceeds very fast.

The solid lines IK Tau, IRC+10011, and OH26.5+0.6 in figure 1 indicate different stages during Mira AGB evolution in which stars with different M_{ms} will look like the (OH/IR) Miras IK Tau,... in their infrared spectral appearance, since M is such that the CS dust shell optical depth is independent of M_{ms} in each particular stage ($\dot{M} \sim 10^{-6}$

$M_\odot yr^{-1}$ for the IK Tau, $\sim 10^{-5}$ $M_\odot yr^{-1}$ for the IRC+10011, and $\sim 10^{-4}$
$M_\odot yr^{-1}$ for the OH26.5+0.6 stage, except for a factor $\propto (L_*/10^4 L_\odot)^{1/2})$.
The acceleration of $\dot{M}$ becomes apparent by noting that it takes $\sim 10^5$
yr for a star to evolve from the IK Tau to the IRC+10011 stage, $\sim 10^4$
yr from the IRC+10011 to the OH26.5+0.6 stage, and no more than $\sim 10^3$
yr from the OH26.5+0.6 to the END stage.

4. OBSERVATIONAL VERIFICATION

However fast the evolution on the Mira AGB may be it is still too slow
to be directly visible for single stars; except indirectly during the
latest phases on the Mira AGB in which $\dot{M}$ changes so rapidly with time
that this may be observable through its effect on the CS (dust) density distribution (see Bedijn, 1985). Thus, to test the model of section
3 we must rely on explaining the properties of an ensemble of (OH/IR)
Miras such as the period distribution and the period-luminosity
relation for classical Miras, and the 1612 MHz OH maser luminosity
function of (OH/IR) Miras.

Theoretical distributions are calculated with an expression like

$$\frac{dn(P)}{dP} = \int_{M_l(P)}^{M_u(P)} \frac{BR_{AGB}(M_{ms})}{dP/dt(M_{ms},P)} dM_{ms} \qquad (3)$$

for the period distribution (number of Miras at period P per unit period interval per unit galactic disk surface area in the solar neighbourhood), and similar expressions for other distributions. The evolutionary rates dP/dt, dL_{OH}/dt (the latter through combining $d\dot{M}/dt$ and
$L_{OH}(\dot{M},L_*,..)$ such as given by Baud and Habing (1983)) are obtained
from the calculations presented in section 3. Furthermore, we need to
know the present-day local birth rate of AGB stars; this can be written as $BR_{AGB}(M_{ms}) = SFR(t_{AGB})*IMF(M_{ms})$. SFR(t) is the local star formation rate at time t ago, $t_{AGB}(M_{ms})$ the evolution time from the main
sequence to the upper AGB, and $IMF(M_{ms})$ the initial mass function. In
our calculations we used the Miller and Scalo (1979) IMF, normalized
such that $\int M_{ms} IMF(M_{ms}) dM_{ms} = 1$. In order that the theoretical period
distribution fit the observed one a time dependence of the SFR similar to that found by Twarog (1980) is needed; it peaked 8 10^9 yr ago,
was a factor two smaller 1.2 10^{10} yr ago, and is a factor three smaller at present. The average past SFR of 4.2 10^{-9} $M_\odot pc^{-2} yr^{-1}$ as implied
by the model fitting is in excellent agreement with the value 5 10^{-9}
$M_\odot pc^{-2} yr^{-1}$ Miller and Scalo derived from the luminosity function of
local main sequence stars.

In figure 2 the model (solid curve) and observed (dashed curve)
period distributions for classical Miras are shown. The observed distribution is the one of Wood and Cahn (1977) corrected for a period
dependent galactic z distribution (Ikaunieks, 1963). They agree very
well except at $P \gtrsim 450$ days. However, it should be realised that at
such periods a large fraction of the Miras have a CS dust shell with
such a large optical depth that they would not have been detected op-

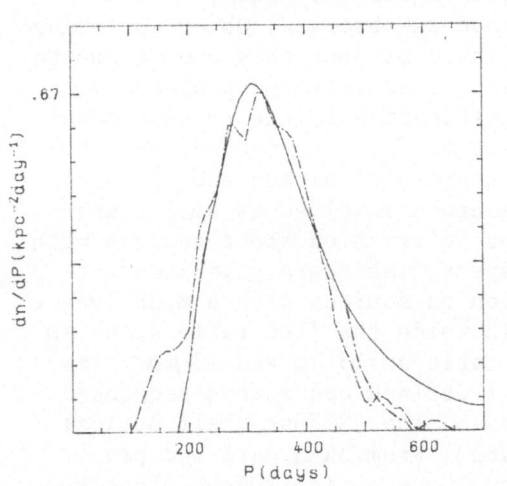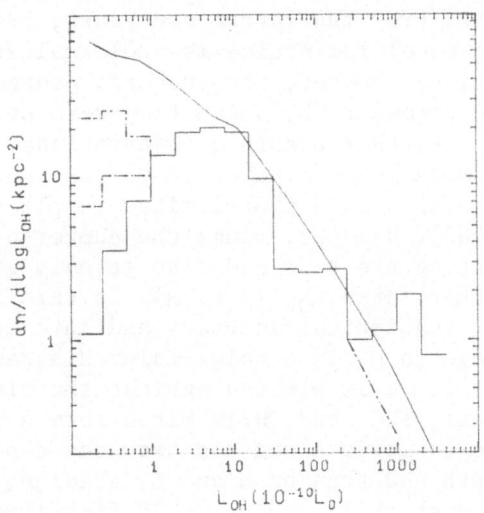

Fig. 2 (left). Model (solid curve) and observed (dashed curve) period distribution for solar neighbourhood, classical Miras. The latter stems from Wood and Cahn (1977). The dash-dotted curve is a model distribution excluding sources for which the CS $\tau_{10\mu m} \gtrsim 1.3$.

Fig. 3 (right). Model (solid curve) and observed (solid step function) 1612 MHz OH luminosity function for solar neighbourhood Miras. The latter stems from Herman and Habing (1985). Dash-dotted and dashed step functions are estimated upper and lower limits to the true luminosity function at small L_{OH} (see text). Dash-dotted curve at large L_{OH} indicates slope of luminosity function found in galactic surveys.

tically. Excluding those which have a CS dust shell optical depth $\gtrsim 1.3$ (at 10 μm) results in a model distribution (dash-dotted curve in figure 2) in good agreement with the observed one at all periods.

Figure 3 shows the model (solid curve) and observed (solid step function) 1612 MHz OH luminosity function for solar neighbourhood Miras. The latter stems from Herman and Habing (1985). The dash-dotted step function and the dashed one are upper and lower limits to the true luminosity function at $L_{OH} \leqslant 10^{-10}$ L_θ obtained by applying a correction to the distribution of Herman and Habing since the selected sample (all sources with flux $S_{OH} \geqslant 10^{-10}$ $L_\theta kpc^{-2}$ within a cylindric volume of 1 kpc radius and height 2 kpc in the z direction centered on the sun) on which the luminosity function is based misses sources with such low L_{OH}. Further, the dash-dotted curve in figure 3 indicates the slope of the luminosity function at large L_{OH} as found from galactic surveys. The model and observed distribution are in good agreement at $L_{OH} \geqslant 10^{-9}$ L_θ. The disagreement at lower L_{OH} is most probably due to the fact that at smaller $\dot{M}$ ($\leqslant 2$ 10^{-6} $M_\theta yr^{-1}$) the OH maser is unsaturated and L_{OH} is smaller than as implied by the relation between L_{OH} and M we used.

The model and observed period-luminosity relation for classical

Miras (for the latter see Feast, 1984) do agree very well; at each P
the model luminosity is $\sim$20% smaller than as observed. Observed luminosities, however, are probably overestimated by just this amount due to
the procedure by which they were obtained from infrared photometry.

Further possible observational verification of the present model
of Mira AGB evolution is to test whether it fits the period distribution of the OH flux-limited sample of sources of Herman and Habing
(1985). However, since the number of sources involved is small, statistics are poor and also it only gives information about sources with
rather large L_{OH} (i.e. $\dot{M}$). In this respect, IRAS data give much better statistical accuracy and information on sources with a much larger
range in $\dot{M}$. In a color-color diagram in which the flux ratio at 60 μm
and 25 μm is plotted against the flux ratio at 25 μm and 12 μm, classical, IRC, and OH/IR Miras form a well-defined continuous sequence.
The position along the sequence depends on the CS dust shell optical
depth and thus on $\dot{M}$ and L_* (Bedijn, 1985). From IRAS data and proper
CS dust shell models an IR flux-limited 10 μm optical depth distribution of Miras, i.e. the number of sources per unit 10 μm optical
depth interval as a function of τ_{10}, can be constructed; such a distribution can also be calculated as outlined in this paper.

ACKNOWLEDGEMENTS.   We thank Dr. D. Muchmore for reading the manuscript
and Drs. H.J. Habing and J. Herman for helpful discussions. We acknowledge financial support of the Deutsche Forschungsgemeinschaft (SFB 132).

REFERENCES

Baud, B, Habing, H.J.: 1983, Astron. Astrophys. 127, 73
Bedijn, P.J.: 1985, Astron. Astrophys., submitted
Feast, M.W.: 1984, Monthly Notices Roy. Astron. Soc. 211, 51P
Feast, M.W.: 1985, Monthly Notices Roy. Astron. Soc., in press
Fox, M.W., Wood, P.R.: 1982, Astrophys. J. 259, 198
Herman, J., Baud, B., Habing, H.J., Winnberg, A.: 1985, Astron. Astro-
       phys., in press
Herman, J., Habing, H.J.: 1985, Reports on Progress in Physics, in
       press
Iben, I. Jr., Renzini, A.: 1983, Ann. Rev. Astron. Astrophys. 21, 271
Ikaunieks, J.J.: 1963, Tran. Astroph. Lab. Acad. Scie. Latvian, SSR
       11, 58
Jones, T.W., Ney, E.P., Stein, W.A.: 1981, Astrophys. J. 250, 324
Jones, T.J., Hyland, A.R., Wood, P.R., Gatley, I.: 1983, Astrophys. J.
       273, 669
Kwok, S.: 1975, Astrophys. J. 198, 583
Miller, G.E., Scalo, J.M.: 1979, Astrophys. J. Suppl. 41, 513
Reimers, D., Koester, D.: 1982, Astron. Astrophys. 116, 341
Twarog, B.A.: 1980, Astrophys. J. 242, 242
Willson, L.A., Hill, S.J.: 1979, Astrophys. J. 228, 854
Wood, P.R., Cahn, J.H.: 1977, Astrophys. J. 211, 499
Wood, P.R.: 1979, Astrophys. J. 227, 220

GROUND-BASED AND IRAS OBSERVATIONS OF PROTO-PLANETARY NEBULAE

Sun Kwok, B.J. Hrivnak, E.F. Milone and R.T. Boreiko
Department of Physics
The University of Calgary
Calgary, Alberta, Canada
T2N 1N4

1. INTRODUCTION

Previous infrared sky surveys (IRC and AFGL) have revealed many asymptotic-giant-branch (AGB) stars which are totally obscured by circumstellar dust. These stars are characterized by the low colour temperatures (200-600K) and molecular emissions (in CO or OH) from the circumstellar envelopes. Subsequent infrared and millimeter observations suggest that they have mass loss rates of 10^{-5} - 10^{-4} $M_\odot$ yr^{-1} and have circumstellar envelopes which are optically thick as far as $\lambda \sim 10$ μm. Since the mass loss rates are suspected to increase as stars ascend the AGB, very late AGB stars with low colour temperatures can be discovered by the *IRAS* sky survey. These stars will soon evolve into planetary nebulae (PN), and therefore the IRAS results may provide valuable information on the evolutionary phase between AGB and PN.

2. OBSERVATIONS

Infrared photometric observations were made with the 3.6 m Canada-France-Hawaii Telescope in May and August of 1984 using the photometer with InSb and Ge bolometer detectors. Sixteen *IRAS* sources with colour temperatures ranging from 150 to 400 K were identified. Eight sources have no optical counterparts in the PSS near infrared (7550 - 8850 A) plates to a limit of $I \sim 19^m$. Five sources show silicate features in emission or absorption and are likely to be oxygen-rich OH/IR stars. The others show no obvious spectral features and could be carbon stars. An example of a carbon star spectrum is shown in Figure 1.

We also observed a number of compact ($\theta < 2''$) PN and found 8 to have definite far-infrared excesses. The colour temperatures of these nebulae are all < 200 K. In many cases the near-infrared (λ 1-4 μm) fluxes are much stronger than the expected emission from the cold-dust component. Extrapolation from the λ 6 cm radio fluxes (Kwok 1985) shows that the near-infrared excesses are consistent with free-free emission from the ionized gas. The infrared spectrum of the compact PN K3-62 is shown in Figure 2.

F. P. Israel (ed.), Light on Dark Matter, 127–128.
© *1986 by D. Reidel Publishing Company.*

3. DISCUSSION

At the end of the AGB
evolution, a star will
evolve to the left of the
HR diagram. During the
evolutionary phase where
the effective temperature
of the star is between
5000 and 30,000 K the
circumstellar nebula is
not ionized and this
period can be defined as
the "proto-planetary
nebula" phase. For a
star of ∿ 0.6 M☉, this
phase lasts ∿ 1,500 yr
(Schönberner 1983). Since
the remnant AGB envelope
is still there, a proto-
PN can only be identified
as a far-IR object (Kwok
1980). If the ∿ 200 K
dust component of compact
PN is the remnant AGB
envelope, then proto-PN
could be identified as
objects with colour
temperatures between
200-400 K, and some of the
16 *IRAS* objects observed
by us could be proto-PN.

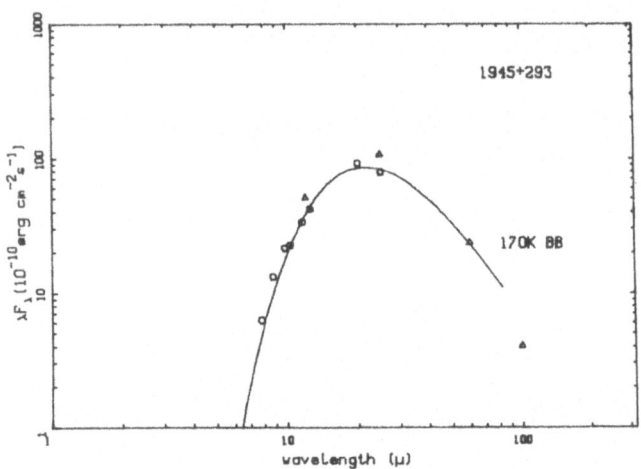

Figure 1.   Infrared spectrum of a possible
carbon star.

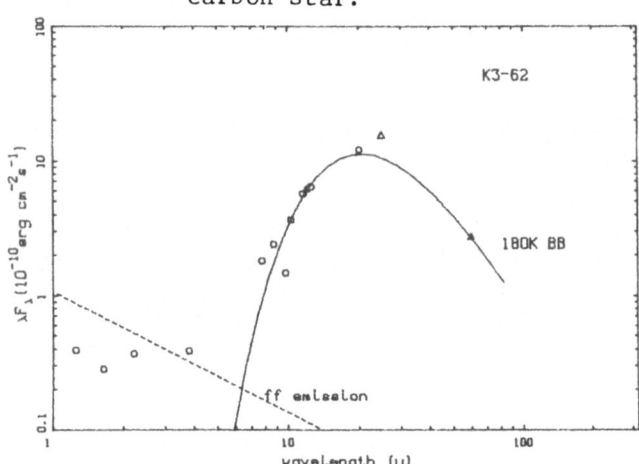

Figure 2.   Infrared spectrum of the
compact PN K3-62.

The CFHT is operated by NRC of Canada, CNRS of France and U. of Hawaii.

This work is supported by the Natural Sciences and Engineering Research
Council of Canada.

4.   REFERENCES

Kwok, S. 1980: *Astrophys. J.*, 236, 592.
Kwok, S. 1985: *Astron. J.*, 90, 49.
Schönberner, D. 1983: *Astrophys. J.*, 272, 708.

SPECTRA OF SOME IRAS SOURCES

J.W. Menzies, P.A. Whitelock and I.M. Coulson
South African Astronomical Observatory
P.O. Box 9
Observatory 7935 Cape
South Africa

ABSTRACT.   Three objects with peculiar spectra have been found in a small sample of IRAS sources identified from JHKL observations.

1. INTRODUCTION.

In a program aimed at investigating the detectability of IRAS sources at JHKL wavelengths, Whitelock and Feast (1984) observed a number of sources from the early lists (IRAS Circular Nos. 3, 4, 5, 8 and 9).   They found that the bulk of the sample were possibly OH/IR sources.   Three sources have near-infrared colours characteristic of stars with cool dust shells, while their IRAS colours place them on or near the blackbody line with dust temperatures of the order of 300K.

We have obtained spectra of seven of these objects covering the wavelength region $\lambda\lambda 5500\text{-}7400$ with a resolution of 3Å (FWHM) in what is a continuing program.

2. SPECTRA.

3 of the possible OH/IR sources (1705-022 P04, 1823+218 P08 and 1905-750 P08) show late-M spectra, 1705-022 having Hα emission and thus being a probable Mira.   The possible pre-main sequence object, 1952+279 P09 (Whitelock 1985), shows a very red continuum with no obvious molecular features and a very strong Hα emission line (Fig. 1a).   The emission line has two components, one broad, the other unresolved, while there appears to be a blue-shifted absorption component as well.

The 3 objects with near-IR dust shells were also examined.   1812+051 P08 has a late-K/early-M spectrum with no obvious peculiarities.   The optical identification may not be correct.   2005+185 P09 (Fig. 1b) has a moderately blue, relatively featureless continuum with very strong NaD lines in <u>emission</u>.   This suggests a very cool region associated with the radiation source.   1912+172 P09, which was observed in the blue as well as the red region (Fig. 2), appears to be a new planetary nebula of moderately high excitation class.   The emission lines (HI, [OIII], [NII], [NeII] and [SIII]) are superimposed on a stellar continuum.   The large Balmer decrement suggests fairly heavy reddening, probably mostly interstellar. The continuum appears to be that of a reddened B9 star which cannot be the nebula excitation source.   The central star may thus be a binary.

F. P. Israel (ed.), Light on Dark Matter, 129–130.

3. CONCLUSION.

Work on the peculiar objects is continuing.   Higher resolution and coverage of the blue spectral region should assist us in understanding the nature of these objects.

4. REFERENCES

Whitelock, P.A. & Feast, M.W., 1984.   M.N.R.A.S.,   **210,**   25P.
Whitelock, P.A., 1985.   M.N.R.A.S.,   **213,**   51P.

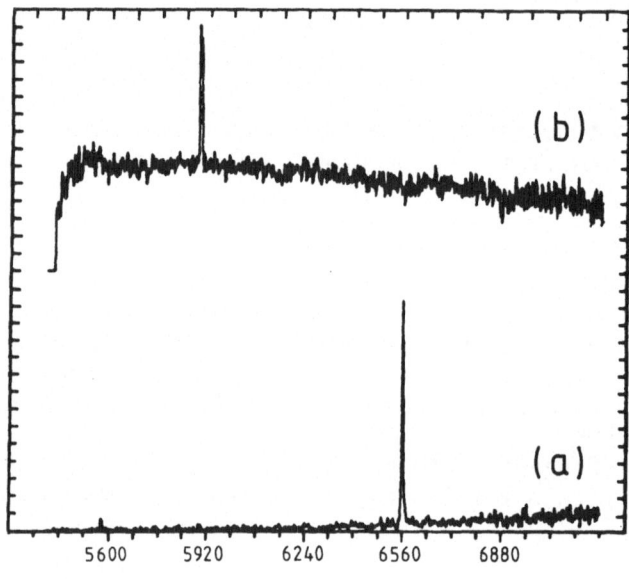

Figure. 1.
Spectra on a
relative flux scale:
(a)   1952+279 P09
(b)   2005+185 P09

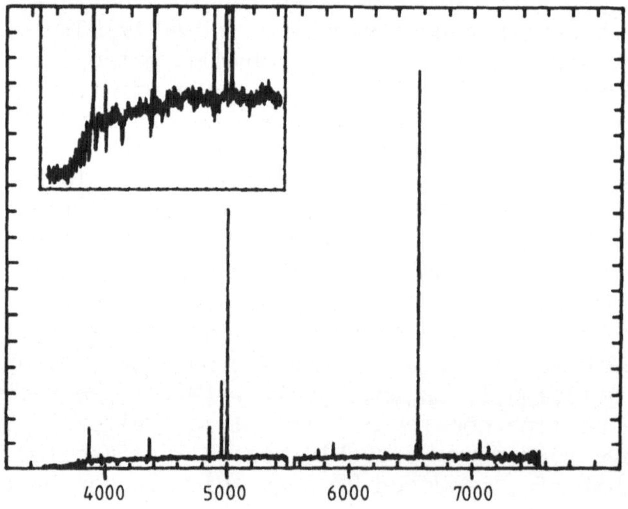

Figure 2.
Spectrum on a
relative flux scale
of 1912+172 P09.
The red and blue
spectra have been
joined.   The inset
shows the details of
the stellar
continuum in the
blue region.

IRAS MEASUREMENTS OF PLANETARY NEBULAE

S.R. Pottasch
Kapteyn Astronomical Institute, University of Groningen

SUMMARY:
    The low resolution IRAS spectra are described and a discussion of
their importance is given. The survey measurements are then discussed.
The relative importance of line and continuum emission in the survey
measurements is evaluated. The interpretation of the continuum emission
as dust radiation is considered. The energy balance is used as a probe
for the source of dust heating. An evolution of the dust 'temperature'
is shown to exist and simple interpretations are given. The gas to dust
mass ratio is presented and its evolution is discussed. Finally the
possibility of using the IRAS data to substantially extend the known
number of planetary nebulae is examined.

1. DESCRIPTION OF THE MEASUREMENTS

    Three sorts of measurements are available. First there is the point
source survey data which has been taken in four broad bands (Neugebauer
et al., 1984). These measurements are limited to nebulae with a diameter
less than about 2 arc minutes. If the size is larger the point source
reduction tends to give unreliable fluxes. Of the approximately 1000
objects which are identified as planetary nebulae with a reasonable
degree of certainty, 700 have been measured by IRAS. The positional
agreement is usually very good, better than 30". When the positional
disagreement is larger than this, it is usually because the optical
position is uncertain. Of these 700 nebulae, about 160 have been detected in all four bands, 230 in three bands, 220 in only two bands (almost
always 25 μm and 60 μm) and about 90 in only one band. The reason so few
are detected in only one band is because the nebulae usually have their
maximum flux in either the 25 μm or 60 μm bands and these fluxes are
often approximately equal, so that if it is not detected in both bands
it may not be detected at all. There has as yet been no systematic study
of the 300 nebulae which are undetected. It is possible that many of
these objects are not planetary nebulae at all. Some are nebulae which
are too large to be included in the point source catalogue. Furthermore
some of the objects listed as nebulae in the catalogue of Acker et al.
(1981) were not counted in the above. They were considered unreliably

F. P. Israel (ed.), Light on Dark Matter, 131–142.

identified on the basis of the IRAS fluxes. Two groups were suspect:
(1) those with peak flux at 100 μm and within 1° of the galactic plane.
They are probably HII regions, (2) those whose flux density at 12 μm was
a factor 1.3 or more greater that at 25 μm. These are probably stars or
extreme protoplanetary nebulae.

Secondly, maps can be made of those planetary nebulae which are
large enough (> 5' diameter) and sufficiently bright. These nebulae can
be studied to learn about the details of the radiation processes. Unfortunately there are less than 20 nebulae which fulfill the above requirements. Still these are enough to give a great deal of information
as will be seen from an example discussed below. Furthermore maps must
be made for all nebulae with a diameter of 2' or greater in order to
obtain more reliable fluxes than given in the point source catalogue.

Thirdly, Low Resolution Spectra (LRS) are available. These spectra
have a resolution of $\lambda/\Delta\lambda \simeq 30$ which is sufficient to separate line and
continuum emission and measure the most important lines. The LRS catalo-
gue contains the spectra of about 20 nebulae, but a special processing
of the data allows the extraction of many more spectra. A recent paper
(Pottasch et al., 1986a) shows the resultant spectra for about 75 nebulae. Because of the potential importance of line emission in the broad
band measurements, this will first be discussed.

2.   LINE RADIATION FROM PLANETARY NEBULAE IN THE IR

A.   LRS Spectra

Fig. 1 shows 13 examples of LRS spectra each covering the wavelength range between 7.5 μm and 22 μm. Both line and continuum emission
can be seen in each of the spectra; the emission comes from the entire
nebula. In general the very small nebulae (e.g. NGC 6790 and 6572, IC
418, BD +30 3639, Vy 2-2) show relatively fewer and less intense lines
than the larger nebulae, and stronger continuum.

The lines are identified on the figure. Four neon ions are seen,
NeII, NeIII, NeV and NeVI, sometimes all in the same nebula. In
addition, two sulfur lines, SIII and SIV ArIII lines as well. Some of
these lines can be measured from the ground (NeII, ArIII and SIV) and
one from a high flying aircraft (SIII). The rest of the lines have been
seen for the first time. The NeVI line was unexpected because of the
high ionization potential of the lower stage of ionization: 146 e.v. Its
wavelength, 7.65 μm, is just at the edge of the instrument sensitivity.
It has been seen in three nebulae.

These lines are interesting for several reasons. First, the neon
and sulfur abundances can be better determined than previously because
of the many stages of ionization present. Secondly, when combined with
lines of the same ions in the visible part of the spectrum, values of
the electron temperature T_e, and electron density, n_e, can be found. The
NeIII lines are especially valuable as an indicator of T_e, while the
SIII, ArIII and NeV lines are primarily sensitive to n_e. The value of n_e

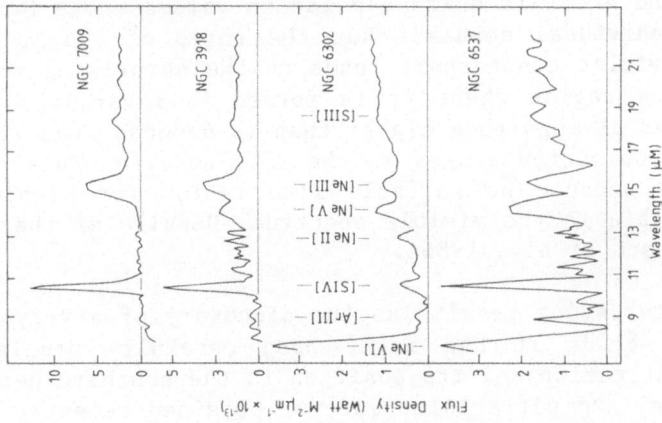

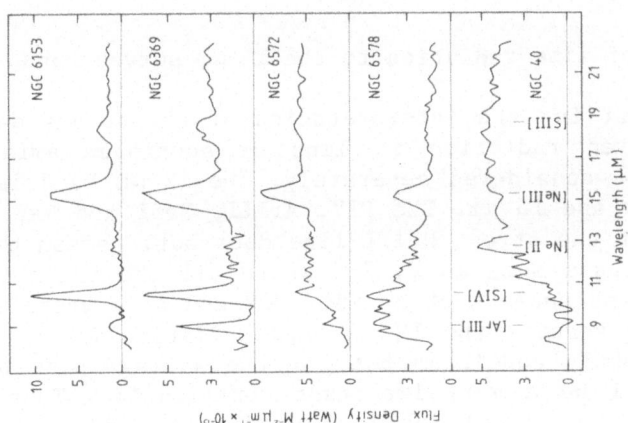

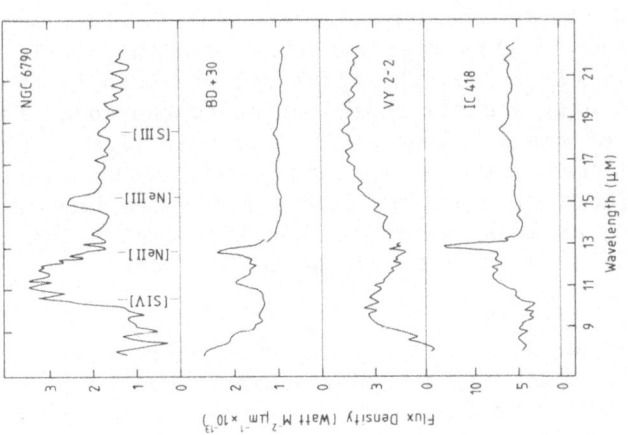

Figure 1. Low resolution spectra of 13 planetary nebulae in the wavelength range from 7.5 μm to 22 μm. The positions of the important line emission is marked.

found from SIII and ArIII is quite similar to values known from optical
line ratio for individual nebulae. But the ratio of the infrared NeV
line to its ultraviolet counterpart leads to the surprising result that
the density in the region where it is formed is invariably (with one
exception) an order of magnitude higher than is deduced from line ratios
of ions of lower ionization stages in the same nebulae. This result was
not known earlier because no suitable line ratios for highly ionized
species are available in the visible spectrum. Details of these results
are given in Pottasch et al., 1986a.

A further interesting result was the discovery of a very high neon
abundance in NGC 6153. It had never been carefully studied before
because of high extinction and its position in the southern hemisphere.
Analysis of optical and ultraviolet spectra obtained recently (Pottasch
et al., 1986b) show it to have extremely high abundances of almost every
major element and for this reason interesting for the evolution of
nebulae.

B.   Contribution of line radiation to the 12 μm survey band

It is important for the interpretation of the survey data to know
whether the observed radiation is line or continuum emission. Each
survey band will be considered separately. The 12 μm band is sensitive
between about 8 μm and 15 μm. The SIV, ArIII, NeII and NeV lines fall
within the band but the strong NeIII line does not. As can be seen from
Fig. 1 there are cases such as Vy 2-2 where all the radiation in this
band is continuum and cases such as NGC 7009 and 6153 where line radiation dominates. In general the 12 μm band radiation in the very small
nebulae will be mostly continuum but for the average size nebulae the
line radiation will be a very important contribution. Thus the 12 μm
survey band measurement should only be used with caution; the spectra
can be used in this wavelength range.

C.   Line radiation in the 25 μm survey band

The 25 μm band covers the wavelength range from 16 μm to 31 μm.
Four lines are present in this region: SIII (18.7 μm), ArIII (21.8 μm),
NeV (24.3 μm) and OIV (25.9 μm). The first two are seen in the LRS spectra and usually make only a small contribution to the total intensity.
Judging by strength of the NeV line at 14.3 μm, the line at 24.3 μm will
probably not be important either. This is confirmed by measurements of
this line made by Shure et al. (1983). The strongest line in this band
will be the OIV line, at least in medium and high excitation nebulae.
The importance of this line can be estimated by theoretical predictions
of its strength (Shure et al., 1983). From the measurement it is found
that about 17% of the IRAS 25 μm band flux is due to the OIV line in NGC
7662 and 7354 and about 8% in NGC 2392. The theoretical study (of 20
nebulae of various excitations) showed a similar result: in about 70% of
the nebulae the continuum is responsible for more than 80% of the measured survey flux. In a few cases, however, the OIV line can dominate the
25 μm band flux. A striking example of this is shown in Fig. 2. Here the
IRAS measurements of the large (15' diameter) planetary nebula NGC 7293

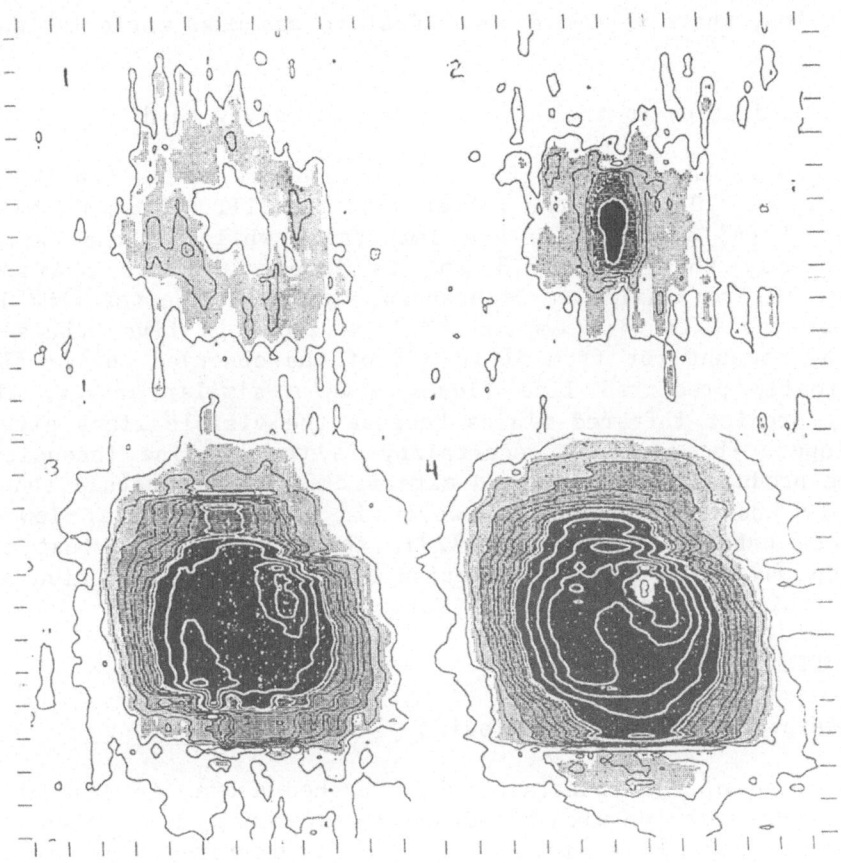

Figure 2. Maps of the survey emission from the planetary nebulae NGC 7293. The 12 μm map is at the top left, the 25 μm map is next to it while the 60 μm and 100 μm maps are on the bottom. Note the completely different distribrution of emission in the 25 μm map. Each map is 30 arc minutes in size.

are shown in the form of four maps, one for each of the IRAS bands (from Leene and Pottasch, 1986). Three of the maps are similar: the 12 μm, 60 μm and 100 μm maps. They are also similar to the pictures of this nebula in visible light, which originates mainly from the [OIII], [NII], Hα and hydrogen continuum. The IRAS map at 25 μm is dramatically different, indicating that it comes from a different radiation source than either the three other IRAS bands or the visible light. The most likely interpretation (Leene and Pottasch, 1986) is that the 25 μm map is dominated by and reflects the distribution of, the OIV line radiation. This is consistent with the expected intensity of the line and the fact that the more highly ionized material is near the center of the nebula.

This may also be true in other large nebulae excited by high temperature stars. The reason that this extreme effect is seen in large nebulae is that the dust temperature is low so that very little dust continuum is emitted in the 25 μm region. Furthermore, as will be discussed below, there may be a lower dust to gas mass ratio in the larger nebulae.

D.   Line radiation in the 60 μm and 100 μm survey bands

The 60 μm band extends from 45 μm to 80 μm, while the 100 μm band from 80 μm to 120 μm. In the former band the OIII (51.8 μm), NIII (57.3 μm) and OI (63.2 μm) lines are important, while in the latter band probably only the OIII (88.3 μm) is important. Here again limited measurements are available. Dinerstein, Lester and Werner (1985) report fluxes of the OIII 51.8 μm and 88.3 μm lines in five nebulae. These lines can account for from 5% to 30% of the observed survey flux. The theoretically predicted line fluxes give a similar answer; they can reliably predict infrared fluxes because the visible lines give sufficient input. The greatest uncertainty is the OI line intensity which could be produced from unionized material around the nebula about which little is known at present. However in the very large, low surface brightness nebulae such as NGC 7293, there is evidence that the line radiation may form a larger fraction of the survey band flux and even dominate this flux (see Leene and Pottasch, 1986).

3.   CONTINUUM EMISSION

A.   Interpretation as dust emission

In spite of the uncertainties, sketched above, in separating line and continuum emission with broadband measurements, it is clear that all nebulae show continuous emission in the far infrared. This emission is clearly in excess of hydrogen and helium free-free or free-bound nebular emission and its spectral distribution is such that there appears to be no alternative but to ascribe it to emission by dust heated to a temperature of approximately 100 K. The detailed properties of the dust, including its emission efficiency, is as yet very poorly known. In fact these properties are often used as free parameters in order to obtain agreement with the measurements.

In this summary we shall not try to derive the dust properties in detail because of the difficulty of disentangling the physical properties of the dust from temperature and density gradients in the nebulae. Instead we shall try to make some general deductions concerning the energy balance in the dust. One subject, the dust to gas mass ratio, which does depend somewhat on dust properties, will be discussed.

A parameter, the dust temperature T_D, will be used to describe the distribution of emission between 12 μm and 100 μm. It is that temperature for which a blackbody gives the best fit to the observed emission. The points at 25 μm and 60 μm measurements are given the greatest weight since usually most of the flux is found between these wavelengths. Furthermore comparison with weaker sources where only 25 μm and 60 μm measurements are available is simplified.

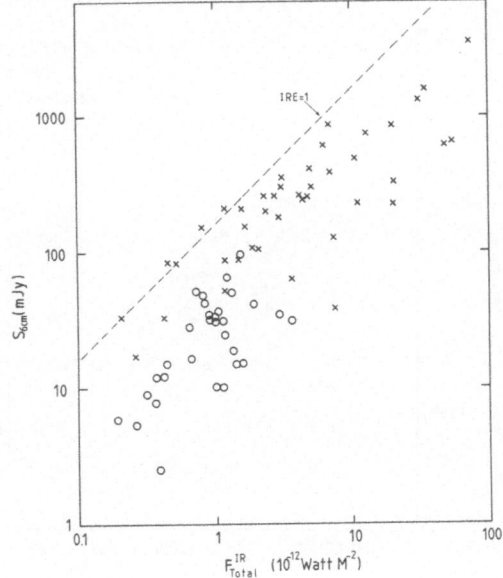

Figure 3. The total infrared flux of individual nebulae is plotted against its 6 cm radio flux density. The bright nearby nebulae are indicated as crosses and the galactic center nebulae as open circles. The line shows the infrared emission if all the Lyman α emission produced by the nebula heated the dust.

B.   Energy considerations

Fig. 3 is a plot of the total infrared flux against the 6 cm radio flux density for about 75 planetary nebulae. The objects marked as crosses are a selection of nearby, bright, well studied, nebulae for which it has been checked that the survey measurements are not strongly influenced by line emission. The circles are the selection of nebulae studied by Gathier et al. (1983). They are all within 1 kpc of the galactic center. With a single exception, all the nebulae detected with the VLA by these authors were also seen by the IRAS, usually in two bands. It can be seen that all nebulae fall in a wide band on the diagram.

The dashed line in the figure is the expected infrared radiation if all the Lyman α radiation produced in the nebula is absorbed by the dust and converted into infrared radiation. As can be seen, the observed infrared emission in about 10 to 15% of the nearby nebulae can be explained by this process. For the majority of the nearby nebulae, and all of the galactic center nebulae more energy, sometimes a factor 10 more, is required than is available in Lyman α. In this connection the term infrared excess (IRE) will be used to indicate the factor by which the observed infrared energy is greater than the Lyman α energy available.

The infrared excess depends on the properties of the nebula. Those with the greatest excess are usually small nebulae with bright, low temperature central stars. Their dust temperature T_D has a high value. As the nebulae increases in size and the central star decreases in brightness, T_D decreases and so does the IRE. The nebulae with IRE = 1 are large, low surface brightness nebulae excited by faint, hot stars. This suggests that the source of the infrared energy when the IRE is high is the central star of the nebula.

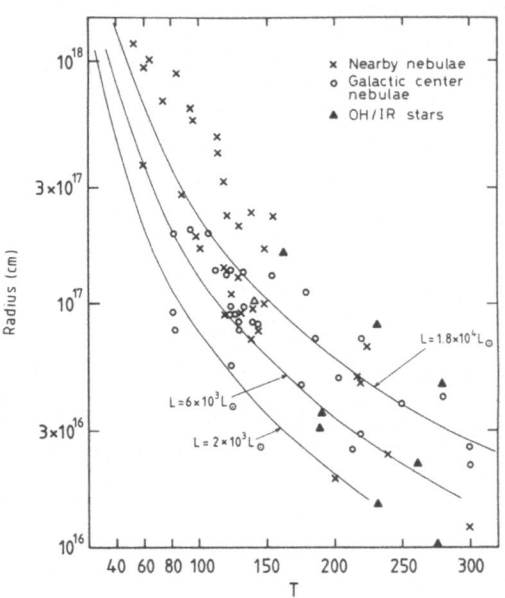

Figure 4. The dust 'temperature' T_D is plotted against the radius of the nebula. Again crosses are bright nearby nebulae and open circles galactic center nebulae. The lines are the theoretical predictions described in the text.

C.   Evolution of the dust 'temperature', T_D

Using the above definition, T_D is plotted as a function of the radius of the nebula. This is shown as Fig. 4. This illustrates very clearly the relation between dust temperature and size. Which can be explained as follows. Assume that the individual grain is in equilibrium and heated by radiation from the central star and by nebular Lyman α. Then

$$\pi a^2 \frac{R_s^2}{4r^2} Q_{UV} \sigma T_s^4 + \frac{E_{Ly\alpha} n_e n_p \Sigma\alpha_{2p}}{n_d} = 4\pi a^2 Q_{IR} \sigma T_D^4$$

where a is the radius of the assumed spherical grain, n_d is the density of grains, n_e the electron density, n_p the ionized hydrogen density, R_s the stellar radius, T_s the stellar temperature, Q_{UV} and Q_{IR} the absorption coefficient of the dust for ultraviolet radiation and infrared radiation, σ is the Stephan-Boltzman constant, $E_{Ly\alpha}$ is the energy of a single Lyman α photon, $\Sigma\alpha_{2p}$ is the sum of all radiative recombinations in hydrogen to the 2p level, and r is the distance from the star to the dust grain. For this simple calculation r is assumed to be 2/3 of the radius of the nebula, R_N. If the Lyman α radiation is neglected we find

$$T_D^4 = \frac{Q_{UV}}{Q_{IR}} \frac{R_s^2 T_s^4}{4r^2} = \frac{Q_{UV}}{Q_{IR}} \frac{4 \times 10^{33}}{16\pi\sigma r^2} \frac{L_s}{L_\odot} = \frac{Q_{UV}}{Q_{IR}} \frac{4 \times 10^{33}}{16\pi\sigma(\frac{2R_N}{3})^2} \frac{L_s}{L_\odot}$$

where L_s is the luminosity of the star in units of the solar luminosity, $L_\odot$.

With the assumption that $Q_{UV}/Q_{IR} = 100$, T_D may be computed as a function of the nebular radius R_N for a given value of $L_s/L_\odot$. The result is shown as a solid line in Fig. 4 for three reasonable values of $L_s/L_\odot$. The curves are seen to give an acceptable approximation to the observed points, especially for the higher values of T_D. For the nebulae with lower values of T_D, the larger nebulae, it is necessary to include the Lyman α heating. This can be done in a crude way by specifying n_e and the dust to gas mass ratio M_d/M_g. For a value of $n_e M_g/M_d = 2 \times 10^7$ cm^{-3}, the curve for $L_s = 2 \times 10^3 L_\odot$ changes as indicated by the dashed line in figure. This confirms that Lyman α heating plays an important role in the large nebulae, but that direct heating by the central star is often more important, especially for the smaller, younger nebulae. In a more detailed consideration the CIV resonance lines will also contribute to the heating.

Two further points are of interest regarding Fig. 4. Firstly, nine OH/IR 'stars' have been plotted on the diagram. They have been selected to have well determined distances from the phase lag method (Herman and Habing, 1984). They can be seen at the same position as the smaller planetary nebulae. This is expected if (1) the heating of the dust is due to central star radiation, and (2) the luminosity of the central star in the OH/IR sources is about the same as in young planetary nebulae. This strengthens the argument for an evolutionary association between OH/IR 'stars' and planetary nebulae.

Secondly, the relationship shown in the figure provides a means of determining the distance to planetary nebulae, since T_D can be determined independently of the distance, while the nebular radius is directly propertional to it. The nebulae which have been plotted have the best determined distances. The scatter in the diagram, much of which is intrinsic, limits the accuracy of this method to about a factor two.

D.   Comparison of stellar and far infrared luminosity

The questions now arise: (1) Does the dust absorb directly a substantial amount of stellar radiation? (2) If so does this affect the central star temperature derived by one of the standard methods?

The first question can be answered by comparing the luminosity measured in the infrared directly with the stellar luminosity. This can only be determined when the temperature is known from either the Zanstra method or the Energy Balance method and the radius is determined from the visual magnitude. In Table 1 a comparison is made for 16 nebulae where the temperature is usually taken from Preite-Martinez and Pottasch (1983) and Pottasch (1984), whose compilation of magnitudes has also been used. The nebular radius is also given in the table. As can be seen, the larger nebulae have a higher ratio of stellar to infrared luminosity (a factor of 10 for the nebulae with a radius of 10^{17} cm or large). For the smaller nebulae this ratio is only between 2 and 3.

When stellar radiation is directly absorbed by the dust the temperature determination can be effected. For example, the Zanstra temperature assumes that all Lyman continuum photons from the star ionize hydrogen or helium. If some are absorbed by dust, the derived Zanstra temperature will be lower than the actual temperature. But if a photon on the longwave side of 912 Å is absorbed, it will heat the dust without affecting the Zanstra temperature. It is very difficult to know

how strong the competition of the dust for hydrogen ionizing photons is
as long as the dust properties are not (better) known. But since especially the nebulae which are excited by low temperature stars have large
infrared excesses indicates that much of the stellar energy absorbed by
the dust is on the longwave side of Lyman α. The fact that the stellar
luminosity is considerably greater than the infrared luminosity means
that a possible correction to the stellar temperature will usually not
be greater than 10 to 15%. Even in the extreme case of He 2-131, where
the impossible situation exists that the infrared luminosity is greater
than the stellar luminosity, an increase in the stellar temperature by
25% (from 22,000 K to 28,000 K) will probably be sufficient.

TABLE 1
COMPARISON OF STELLAR AND INFRARED LUMINOSITY

NEBULA	NEBULAR RADIUS	$L_{IR}/L_\odot$	$L_{STAR}/L_\odot$	NEBULA	NEBULAR RADIUS	$L_{IR}/L_\odot$	$L_{STAR}/L_\odot$
IC 4997	0.12×10^{17}	120	300	N 3918	1.5×10^{17}	420	3000
BD+30	0.19 cm	350	1300	N 3242	2.3 cm	90	600
IC 418	0.47	200	550	N 2867	2.3	280	3000
He 2-131	0.5	525	450	N 2440	4.9	575	3000
N 6572	0.66	330	600	N 246	8.9	30	1100
N 6790	0.67	250	700	N 6072	9.5	150	600
N 6543	0.72	180	750	N 2438	10.2	50	400
N 7662	1.0	130	1300	N 6781	11.9	190	4300

E.   Determination of the dust to gas mass ratio

The dust to gas mass ratio may be written

$$\frac{M_D}{M_g} = \frac{d^2}{R_N^3} \; \frac{F_\nu}{\pi B_\nu(T_D)} \; \frac{a\rho}{Q_\nu} \; \frac{4.5 \times 10^{23}}{n_e}$$

if the dust is assumed to radiate as a blackbody B_ν. Here d is the distance to the nebulae, F_ν is the flux density at infrared frequency ν, Q_ν
is the dust emission efficiency and ρ its density (3 gm cm^{-3}). This
ratio depends on the properties of the dust and therefore is very uncertain. Surprisingly, the uncertainty in the way Q varies and its conse-
quent effect on the value of T_D, tend to compensate each other in the
determination of the ratio M_D/M_g. The average dust radius a is uncertain; the value 10^{-5} cm (Hildebrand, 1983) has been used.

The resultant values of M_D/M_g are shown in Fig. 5. They vary from a
value of about 4×10^{-2} (somewhat higher than the general interstellar
medium) to the low value of 10^{-4}. As in the earlier figures, the
'nearby' nebulae are shown as crosses and the galactic center nebulae as
open circles. No systematic difference between these two groups is seen.
A large scatter in the points is present. In spite of this Fig. 5 shows

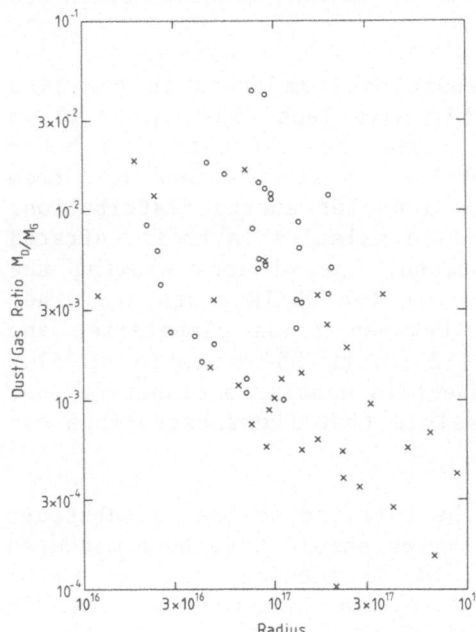

Figure 5. The dust to gas ratio is plotted against the radius of the nebula. Crosses and open circles as in the previous figures.

a definite evolution in the dust to gas ratio: it decreases as the nebulae increase in size. This is probably a real effect, because it extends over such a large range. Such a result is not unexpected: under the influence of the strong ultraviolet radiation field of the central star the dust probably will sputter and evaporate.

There are two further arguments for the correctness of these results. First, the abundances in the gas phase of Fe, Si, Mg, Al in the nebula BD +30 3639 have recently been determined (Pwa et al., 1985). These elements are found to be substantially underabundant compared to solar values. If they originally had the solar value and are depleted because they are in the form of dust, this already accounts for 20% of the dust mass computed from the above equation. Since these metals are probably in the form of oxides or carbides in the dust, it appears that this entirely independent way of computing the dust mass agrees (at least roughly) with the previous computation, arguing that it is probably not far wrong.

Secondly, as has been shown above, there is a good probability that no dust emission, and only line emission, is observed from the very large nebula NGC 7293. This indicates that very little dust is present in the nebula and the dust to gas ratio is very low. There are indications that this argument may be extended to other large nebulae.

F.   Can many new nebulae be discovered from the IRAS data?

Knowledge of the birthrate of nebula in different positions in the galaxy, a key factor in galactic evolution, is insufficiently known at present. This is primarily because only a very limited number of nebulae are known and their distance is a subject of controversy. The situation would be great improved if the number of known nebulae could be increased by a factor of two to three.

Planetary nebulae can be easily separated from stars in the IRAS survey data because they almost invariably have less flux in the 12 μm band than in the 25 μm band. There are a few exceptions to this behaviour (e.g. pre-main sequence stars). Nebulae can also be separated from galaxies because they almost always show a cooler energy distribution. The same is true of HII regions which mimic galaxies in their infrared energy distribution, with a few exceptions. The objects showing the greatest similarity to planetary nebulae are the OH/IR stars and other protoplanetary nebulae. The difference between these planetaries and OH/IR stars, etc can be found in the 12 μm to 25 μm ratio of flux densities. If it is less than 0.3 the object is usually a planetary, but there are exceptions to this. It is possible that these exceptions can be ignored for statistical purposes.

Nebulae are also bright enough in the infrared so that a substantial fraction of those present in the galaxies should have been measured by IRAS. It was already mentioned that of 35 nebulae selected to be close to the galactic center, all but one was detected by IRAS. In selected areas it will be possible to probe even considerably further by reanalysing the data and extracting sources 3 to 5 times weaker. In a few selected areas radio continuum measurements have been made with the VLA, confirming that planetary nebulae can be found with IRAS measurements. Further studies must be made to determine the selection effects which are inherent in this procedure.

REFERENCES

Acker, A., Marcout, J., Ochsenbein, F. 1981, Astron. Astrophys. Suppl.
    43, 265
Dinerstein, H.L., Lester, D.F., Werner, M.W. 1985, Astrophys. J. 291,
    561
Gathier, R., Pottasch, S.R., Goss, W.M., v. Gorkom, J.H. 1983, Astron.
    Astrophys. 128, 325
Herman, J., Habing, H.J. 1985, Astron. Astrophys. 145, 282
Hildebrand, R.H. 1983, Quart. J. Roy. Astron. Soc. 24, 267
Leene, A., Pottasch, S.R. 1986, in preparation
Neugebauer, G., Habing, H.J. et al., 1984, Astrophys. J. 278, LI
Pottasch, S.R. 1984, Planetary Nebulae (Reidel, Dordrecht)
Pottasch, S.R., Preite-Martinez, A., Olnon, F.M., Mo Jing-Er, Kingma, S.
    1986a, in preparation
Pottasch, S.R., Dennefeld, M., Mo Jing-Er, 1986b, submitted Astron.
    Astrophys.
Preite-Martinez, A., Pottasch, S.R. 1983, Astron. Astrophys. 126, 131
Preite-Martinez, A., Pottasch, S.R. 1986, in preparation
Pwa, R., Pottasch, S.R., Mo Jing-Er, 1985, submitted to Astron.
    Astrophys.
Shure, M.A., Herter, T., Houck, J.R., Briotta, D.A., Forrest, W.J.,
    Gull, G.E., McCarthy, J.F. 1983, Astrophys. J. 270, 645

IR Observations Of An Extended Planetary Nebula:
NGC 7293 - The Helix Nebula

A. Leene & S.R. Pottasch
Kapteyn Astronomical Institute
P.O. Box 800
9700 AV Groningen
The Netherlands

Generally infrared radiation is thought to be mainly caused by emission from dust. Only in a few sources line emission in the infrared has been observed, thanks to measurements form balloons and airplanes. This database has been greatly increased by observations of the Low Resolution Spectrometer (LRS) aboard the IRAS satellite. Most of this line emission has been observed in Planetary Nebulae and HII regions.

In this paper it is shown that the broadband fluxes as measured by IRAS in the nebula NGC 7293 are mainly due to atomic line emission; this is probably even true for the long wavelength bands. All the emission in these bands can be explained by line emission, but it is likely that some emission is also due to dust.

Comparing this nebula to the other ones observed by IRAS one sees that this nebula is more evolved. It has a small electron density and a large size (about 0.65 pc), which makes it one of the largest known planetaries. The 12-25 micron colour temperature is very high (245 K), which makes it one of the "hottest" known. The 60-100 micron colour temperature is on the other hand quite low (42 K), which makes it one of the "coolest". It must be emphasized that these temperatures do not say anything about the true dust temperature, because the broad band fluxes are mainly due to atomic line emission.

Assuming constant electron density in the nebula one can calculate the abundances of ions, emitting in the IRAS bands. This can be compared to known abundances. The main contributors to the 12 micron band are SIV (10.52 μm) and NeII (12.81 μm). For the 25 micron band these are the SIII (18.68 μm) and the OIV (25.87 μm) line. In the 60 micron band fall the OIII (51.71 μm) and the NIII (57.31 μm) line. The 100 micron band is mainly caused by the OIII line at 88.2 μm. Assuming that there is no continuum contributor (which is quite improbable) we can calculate the abundances needed to cause the IRAS broad band fluxes.

F. P. Israel (ed.), Light on Dark Matter, 143–144.
© *1986 by D. Reidel Publishing Company.*

The IRAS flux densities are 11.3, 18.3, 179 and 406 Jy in the 12, 25, 60 and 100 μm band respectively. The derived abundances are (assuming that one line causes the broadband flux): $X(SIV)=3.7 \ 10^{-6}$; $X(NeII)=4.8 \ 10^{-5}$; $X(OIV)=1.3 \ 10^{-5}$; $X(NIII)=1.2 \ 10^{-4}$; $X(OIII)=2.7 \ 10^{-4}$.

Comparing the abundances we need to explain the IRAS flux densities with what it is known about them; we then see that both the SIV and NeII can explain the 12 micron emission. For the 25 micron band the same is true for OIV and SIII. The 60 micron emission is mainly caused by NIII; OIII would require too large on abundance. The 100 micron emission can be explained by OIII. No exceptional abundances are needed to explain the infrared emission. This fact is already quite exceptional, because normally dust emission is a large contributor.

We also have information about the radial (on the sky) distribution of the emission. If we assume spherical symmetry we can deconvolve the observed emission in order to get the radial (in the nebula) abundance distribution. This will provide confirmation of the fact that the IRAS emission is caused by line emission. The 12 micron abundance profile shows a shell at a radius of 7 arcmin and a width of 4 arcmin. The 25 micron abundance profle is centered on the nebula and has a weak halo from 3 to 9 arcmins. The 60 micron abundance profile shows a shell centered on 5 arcmins radius and has a width of 4 arcmins. The 100 micron emission is very similar to the 60 micron abundance profile

If we combine this information with the possible ions then the 100 micron profile is due to OIII. The 60 micron profile is due to NIII, which has a similar ionisation potential as OIII. This can explain similarity in the profile. The 25 micron is due to OIV (in the center) and has the SIII halo. The 12 micron emission could be due to the SIV and NeII line. It peaks however at a larger radius than the 60 micron emission. SIV has a larger ionisation potential than OIII (and NIII), whereas NeII has a smaller ionisation potential. We probably see the NeII ion in the 12 micron band.

We conclude that line emission is an important contributant to the broadband IRAS fluxes. Any analysis of either planetary nebula or HII regions should include both dust and line emission.

NOVAE DETECTED IN THE IRAS POINT SOURCE CATALOG

Harriet L. Dinerstein and E.L. Robinson
Astronomy Department
University of Texas at Austin
Austin, Texas 78712

ABSTRACT. Five classical novae are found to have counterparts in the IRAS Point Source Catalog. Two are recent events measured during the first year; the others are old novae detected by IRAS more than 10 years after optical maximum. The IRAS measurements confirm the trend that slow novae have larger amounts of dust than fast novae and support the picture that dust grains condense in expanding nova shells.

1. INTRODUCTION

Infrared observations of novae over the last 15 years have shown the frequent appearance of thermal infrared emission characteristic of a blackbody at about 1000 K some 60 - 80 days after maximum visual light. This component appears simultaneously with a steepening or sudden decline in the visual light curve. Two models have been invoked to explain this behavior: 1) condensation of grains in the nova ejecta (the dust formation model); and 2) delayed heating of pre-existing grains (the "light echo" model). The dust formation model may more naturally explain the observed correlation between speed class and dust, such that slower novae appear to produce an optically thick dust shell while faster novae show little or no dust. One problem with this picture, however, has been presumed lack of dust in the slow nova HR Del (e.g. Bode and Evans 1983).

The IRAS survey provides the opportunity to examine a larger sample and to look at longer wavelengths for evidence of cooler dust. We have examined the Point Source Catalog (PSC) at the positions of 44 classical novae with well-determined optical positions, and found 5 correspondences. The objects are discussed individually below.

F. P. Israel (ed.), Light on Dark Matter, 145–148.

2. DISCUSSION

2.1. Novae in Early Phases: Nova Sgr 1982 and Nova Mus 1983

Nova Muscae 1983 was a fairly fast nova, fading by 3 mag within 40
days after visual maximum (1983 January 15). In the PSC, it is listed
as having a flux of 0.31 Jy at 25 μm, with upper limits of 0.31 and
0.45 Jy at 12 μm and 60 μm respectively. Ground-based observations by
Krautter et al (1984) at days 15 - 26 and 220 after maximum showed a
spectrum consistent with optically thin free-free emission at both
epochs, and no evidence for dust, although their later observations
extended only to 5 μm. The IRAS data are roughly contemporaneous with
the data of Krautter et al for day 220. They are consistent with a
decaying free-free continuum and little or no dust emission. Nova
Sagittarii 1982, although followed visually only one month past
maximum (1982 October 4), was clearly not a fast nova but rather
showed slow oscillations similar to those of HR Del (Iijima and
Rosino 1983). The variability flag value of 9 in the PSC indicates
varying fluxes; the catalog values are 22.6, 8.2, and 1.3 Jy at 12,
25, and 60 μm respectively, and an upper limit of 14 Jy at
100 μm. These values lie well above the free-free fluxes
corresponding to the last visual sighting (day 30). The long
wavelength data can be fit by a blackbody at roughly 1000 K, similar
to the temperatures for other novae observed near the time of dust
condensation, and the mass of dust implied is on the order
of 10^{-7} $M_\odot$, for d = 1 kpc, ρ = 2 gm cm^{-3}, and $(Q/a)25\mu m = .02\mu m^{-1}$.

2.2. Novae at Late Phases: FH Ser, HR Del and RR Pic

FH Ser (Nova Ser 1970), the first nova ever studied over a large
wavelength range, has been called "the Rosetta stone of nova
energetics" (Gallagher 1977). Between days 55 and 60, the optical
light declined abruptly and a hot (900 K) blackbody source appeared
(Geiser et al 1970). The 10 μm brightness increased until 110 days
after optical maximum, and then turned over and began to decay. The
total luminosity remained constant as the energy shifted to longer
wavelengths. This behavior was interpreted as an episode of dust
condensation in the nova ejecta. For an optically thin expanding dust
shell, the infrared flux should decline as time t^{-2} and the dust
temperature as $t^{-1/3}$ if the central luminosity source remains
constant (Gehrz et al 1980). The IRAS 12 μm flux of 0.33 Jy at time
log t (days) = 3.8 fits an extrapolated decay as t^{-2}. Since FH Ser is
the first nova for which infrared data over such a long time basis is
available, this consistency with the predicted time-dependence is of
interest.

HR Del (Nova Del 1967) was considered to pose a problem for the
dust condensation model because it was thought not to have formed
dust despite its extremely slow optical decline. The PSC quotes a
detection at 25 μm (0.36 Jy), and upper limits of 0.25 and 0.40 Jy at
12 and 60 μm. If we assume that HR Del, like other novae, formed dust

around 100 days after maximum at a temperature of about 900 K, we would predict that the temperature would be about 230 K by 1983. The IRAS data constrain the dust temperature to 130 K - 250 K for blackbody-like grains. The dust mass for these parameters is a few x 10^{-7} $M_\odot$; for a gas mass of 1-2 x 10^{-6} $M_\odot$ (Cohen and Rosenthal 1983), the dust/gas mass ratio is large, making HR Del a net source of dust to the ISM.

RR Pic (Nova Pic 1925) was a slow, luminous nova which did not show a sharp decline in its optical light curve. It is reported as a 12 and 25 μm source in the PSC, with fluxes of 1.2 and 0.3 Jy. However, any dust which formed in the 1925 outburst should have cooled to a temperature of about 150 K by 1983, while the dust detected by IRAS is much hotter, about 800 K. Therefore, either the dust must have been recently heated, or else the infrared emission arises from circumstellar dust.

REFERENCES

Bode, M.F. Evans, A. 1983, Q.Jl.R.Astr.Soc.,24,83.
Cohen, J.G. Rosenthal, A.J. 1983, Ap. J., 268, 689.
Gallagher, J.S. 1977, Astr. J., 82, 209.
Gehrz, R.D., Grasdalen, G.L., Hackwell, J.A., Ney, E.P. 1980, Ap. J., 237, 855.
Geisel, S.L., Kleimann, D.E., Low, F.J. 1970, Ap. J. Letters, 161, L101.
Iijima, T. Rosino, L. 1983, Pub. Astr. Soc. Pacific, 95, 506.
Krautter, J. et al 1984, Astr. Ap., 137, 307.

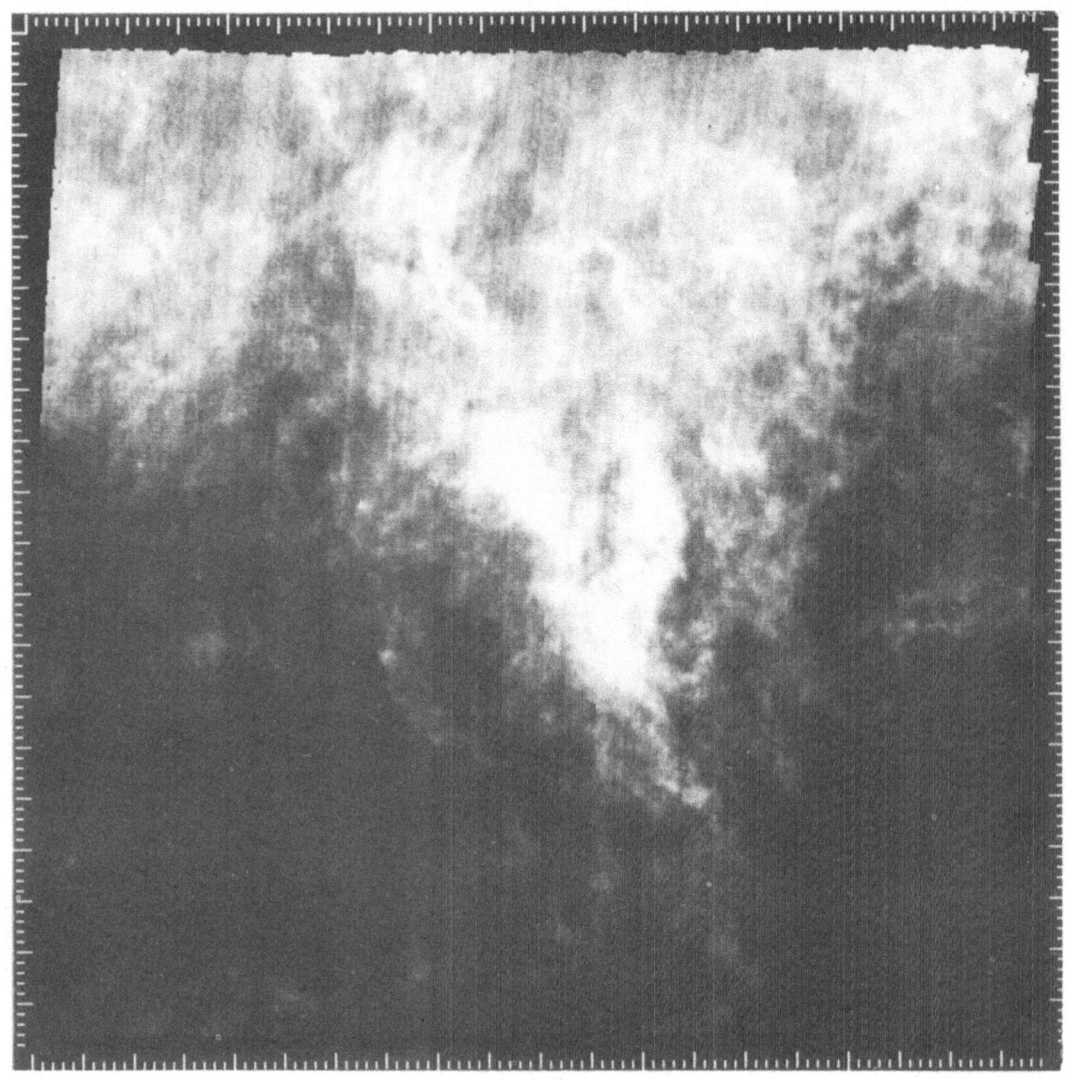

IRAS Field 99   α = 4^{h}00^m, δ = 0°, HCON-3, 100 μm. 'Empty' Field in
                 Taurus, showing large amounts of cool dust,
                 associated with the Taurus dark clouds.

IRAS OBSERVATIONS OF CLASSICAL NOVAE

C.M. Callus[1], J.S. Albinson[1], A. Evans[1] and M.F. Bode[2]

(1) Dept. of Physics, Univ. of Keele, ST5 5BG, U.K.
(2) Dept. of Astronomy, Univ. of Manchester, M13 9PL, U.K.

ABSTRACT. IRAS observations of classical novae are presented, with additional ground based IR photometry. Infrared emission from dust appears to arise from a wide range of conditions in the circumnova medium.

1. INTRODUCTION

Classical novae generally convert ejected material ($\sim 10^{-4}$ $M_\odot$) into dust with great efficiency[1]. IRAS observations of old novae can tell us about the long term evolution of the dust shells and of the remnants themselves[2]; while observations of recent novae can supplement ground based observations at shorter wavelengths. In this note we present a preliminary discussion of IRAS observations of recent novae.

2. OBSERVATIONS

Nova Sgr 1982 (V4077 Sgr) was observed by IRAS 374 days after outburst. The band I/II colour temperature was 800 K, consistent with ground based photometry carried out at SAAO. Assuming a distance of 1 kpc, the dust luminosity is ~ 24 $L_\odot$ and the corresponding mass of dust is $\sim 2.3 \times 10^{-10}$ $M_\odot$. Previous dusty nova shells have cooled to $\lesssim 700$ K within ~ 200 days[3], so the high dust temperature of V4077.Sgr is surprising. This can be understood if the grains in this nova were significantly smaller (~ 0.01 μ) than those in previous dusty novae (~ 0.5 μ), suggesting that significant depletion of the condensate may have occurred.

Nova Muscae 1983 (GQ Mus) was observed 225 days after outburst and was notable for the lack of evidence at short wavelength of dust formation[4,5]. We have definite detections in bands I and II and, while we cannot exclude the possibility that we may be seeing the integrated emission of fine structure lines, the implied colour temperature ~ 340 K leads us to suspect that the emission is due to dust. For a distance of 5 kpc[4] the dust luminosity is $L_d \sim 430$ $L_\odot$, and the corresponding mass of radiating dust $\sim 4.5 \times 10^{-7}$ $M_\odot$. The near infrared data [4] can be understood in terms of free-free emission from optically thin ejecta.

F. P. Israel (ed.), Light on Dark Matter, 149–150.

The possible existence of a cool excess suggests either (i) an infrared echo in an extensive circumstellar shell; (ii) collisional heating of pre-existing dust; (iii) condensation of dust having $T_{cond} \lesssim 800$ K. Of these (iii) seems unlikely but some support for the possibility of pre-existing dust (ii) comes from the detection of x-ray emission from GQ Mus[6], which may be interpreted in terms of shock-heating of gas in a pre-existing shell. The ratio $L_d/L_x \sim 16$, consistent with the ratio expected[7] for shock-heated dust.

The novae CK Vul (1672), Aquilae 1982 (V1370 Aql) and Serpentis 1983 (MU Ser) were not detected in the present survey.

ACKNOWLEDGEMENTS. The SAAO photometry was kindly provided by Prof. M.W. Feast. CMC, JSA & MFB are SERC supported.

References

1. Bode, M.F., Evans, A., QJRAS, 24, 83 (1983).
2. Evans, A., Observatory, 105, 6 (1985).
3. Bode, M.F. & Evans, A., MNRAS, 203, 285 (1983).
4. Whitelock, P. et al., MNRAS, 211, 421 (1984).
5. Krautter, J. et al., Astr. Ap., 137, 307 (1984).
6. Ögelman, H. et al., Ap. J. Lett., 287, L31 (1984).
7. Graham, J.R. et al., these proceedings, page 397.

COLLISIONAL HEATING OF DUST IN THE 1985 OUTBURST OF RS OPHIUCHI

J.S. Albinson, C.M. Callus & A. Evans,
Dept. of Physics,
University of Keele,
Keele, Staffordshire, ST5, 5BG, UK

Extended Summary. The recurrent nova RS Ophiuchi has undergone a
number of outbursts this century. Investigation of the interoutburst
state (Feast & Glass 1974) has shown that it is heavily reddened and
has a substantial circumstellar dust shell. That this shell is dusty
is suggested by a detection of RS Oph at 12 μm by the IRAS in 1983,
the flux density being well in excess of an extrapolation from near
infrared (JHKL) wavelengths. There is no doubt that the circumstellar
shell is an accumulation of material ejected in previous outbursts
and in a wind from the (evolved) secondary component. At outburst we
can expect a strong interaction between the radiation pulse from the
nova, the ejecta and the circumstellar shell. The most recent
outburst occurred in 1985 January (Morrison 1985). The outburst was
discovered before visual maximum, which occured on JD2446095.5 (Jan
29.5); visual maximum is taken as the origin of time hereafter. We
describe here a possible interpretation of the JHKL infrared
photometry obtained on the 0.75m and 1.9m telescopes at the South
African Astronomical Observatory (SAAO) since day 5, and made
available by Dr. P.A. Whitelock and colleagues.

Ultraviolet observations from IUE have led to a value of E(B-V)
= 0.73 ±0.07 (Cassatella et al. 1985). It is curious that, assuming a
value of 3.1 for the ratio $A_V/E(B-V)$, the resultant A_V is much less
than that during quiescence from the Balmer decrement (Dufay & Black
1964) or from JHKL photometry (Feast & Glass 1974), but consistent
with that derived from interoutburst IUE observations. For the
present we have dereddened the data using the standard interstellar
extinction curve (Savage & Mathis 1979) and E(B-V) = 0.73. The
assumed distance of 2 kpc is taken from the discussion of Cassatella
et al. (1985); the corresponding luminosity $L_* = 6.4 \times 10^{38}$ erg s^{-1}.

RS Oph had a strong infrared excess on day 5 (Laney 1985).
Infrared spectroscopy suggests that there were no strong emission
lines that might mimic a dust excess; we assume that the excess was
due to thermal emission by dust. The early appeerence of dust
precludes condensation in the new ejecta, as frequently occurs in
classical novae (e.g. Bode & Evans 1983a) and indicates that the
emission comes from dust in the pre-existing circumstellar dust

151

F. P. Israel (ed.), Light on Dark Matter, 151–152.
© *1986 by D. Reidel Publishing Company.*

shell. In order to identify the mechanism which supplies energy to the dust we need to determine (i) the dust temperature T_d and the dust luminosity L_d, and (ii), the respective time dependences.

We have estimated the dust temperatures by fitting blackbody functions, weighted by a frequency dependent dust emissivity, through the dereddend data at K and L.

The dust temperature is approximately constant between day 5 and day 60. This is very different to the behaviour seen in classical novae, where a rapid decline of dust temperature frequently occurs following grain condensation (.e.g. Bode & Evans 1983b). The dust luminosity is obtained by integrating the fitted blackbody. The resultant time dependence is in marked contrast to the behaviour of T_d.

That the infrared excess could be due to an 'infrared echo' can be rejected on the grounds that the dust temperature does not decline. In this model, $T_d(t)$ a t^{-173} for optically thin dust, and must decline more steeply if there is an appreciable optical depth.

Another possibility is that the pre-existing dust grains might be heated by the hot gas that results when the ejecta slam into the pre-existing shell. The presence of shocked material around RS Oph is indicated by the presence of coronal lines in the ultraviolet (Cassatella et al. 1985), and by the detection of RS Oph as an X-ray source by Exosat (Cordova et al. 1985). The deduced gas temperature is ~ 0.75 kev (T_{gas} ~ 8.5 x 10^6 k) if the gas radiates mostly by free-free emission; or KT_{gas} ~ 0.25 kev (T_{gas} ~ 3 x 10^6 K) if it radiates mostly in emission lines (Bode & Kahn 1985). Dust grains in the gas will be heated mainly by electron impacts: the equilibrium temperature is weakly proportional to both the gas temperature and number density. The survival time of the grains against sputtering is about a month, for plausible values of gas temperature and number density, ample to provide time for observable infrared emission.

The time dependence of the infrared luminosity is model dependent, but can be estimated to be ~ t^{-1} to $t^{-1.7}$, depending on the radial density gradient in the circumstellar shell.

We tentatively conclude that the infrared behaviour of RS Oph is consistent with that of shock heated dust in a pre-existing circumstellar shell. Much of the pre-existing dust will have been destroyed as a result of the present outburst and further obervations are required to find out how quickly the circumstellar shell of RS Oph is re-established.

REFERENCES
Feast, M.W. & Glass, I.S., 1974, MNRAS, 167, 81.
Morrison, W., 1985. IAUC 4030.
Cassatella, A., Harris, A., Snijders, M.A.J., Hassal, B.J., 1985.
    Proc. Bamberg meeting, 'Cataclysmic Variables'.
Dufay, J. & Black, M., 1964. Ann. Astrophys., 27, 462.
Savage, B.D., & Mathis, J.S., 1979. ARAA,17,73.
Laney, D., 1985. IAUC 4036.
Bode, M.F. & Evans, A., 1983a. QJRAS, 24, 83.
Bode, M.F. & Evans, A., 1983b. MNRAS, 203, 285.
Cordova, F.A., Mason, K.O., Bode, M.F., & Barr, P., 1985. IAUC 4049.

INFRARED OBSERVATIONS OF TYCHO USING IRAS

P.L. Marsden
Physics Department, University of Leeds, U.K.

Data obtained from survey scans using IRAS show that at 10, 25, 60 and 100 μm Tycho has a shell-like structure with an angular diameter of ∿ 8', comparable to the X-ray and radio maps of the remnant (Dickel et al, 1982). In Figure 1 the excess infrared flux density is plotted against frequency together with the extrapolated synchrotron radiation spectrum (Kellermann, 1969). The infrared excess is found to fit a Planckian spectrum with an emissivity varying as ν^2 and T = 70 K, or equally a Planckian spectrum with T = 90 K and emissivity varying as ν. The total infrared emissivity is ∿ 1000 $L_\odot$, where the distance to Tycho has been assumed to be 3 kpc.

REFERENCES

Dickel J R, Murray S S, Morris J, Wells D C, 1982, Ap.J. 257 145
Kellermann K I, 1969, Ap.J. 157 1

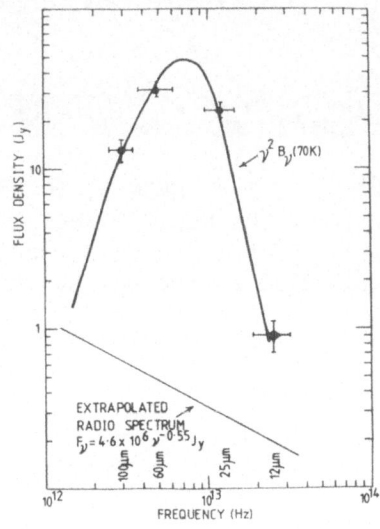

F. P. Israel (ed.), Light on Dark Matter, 153–154.

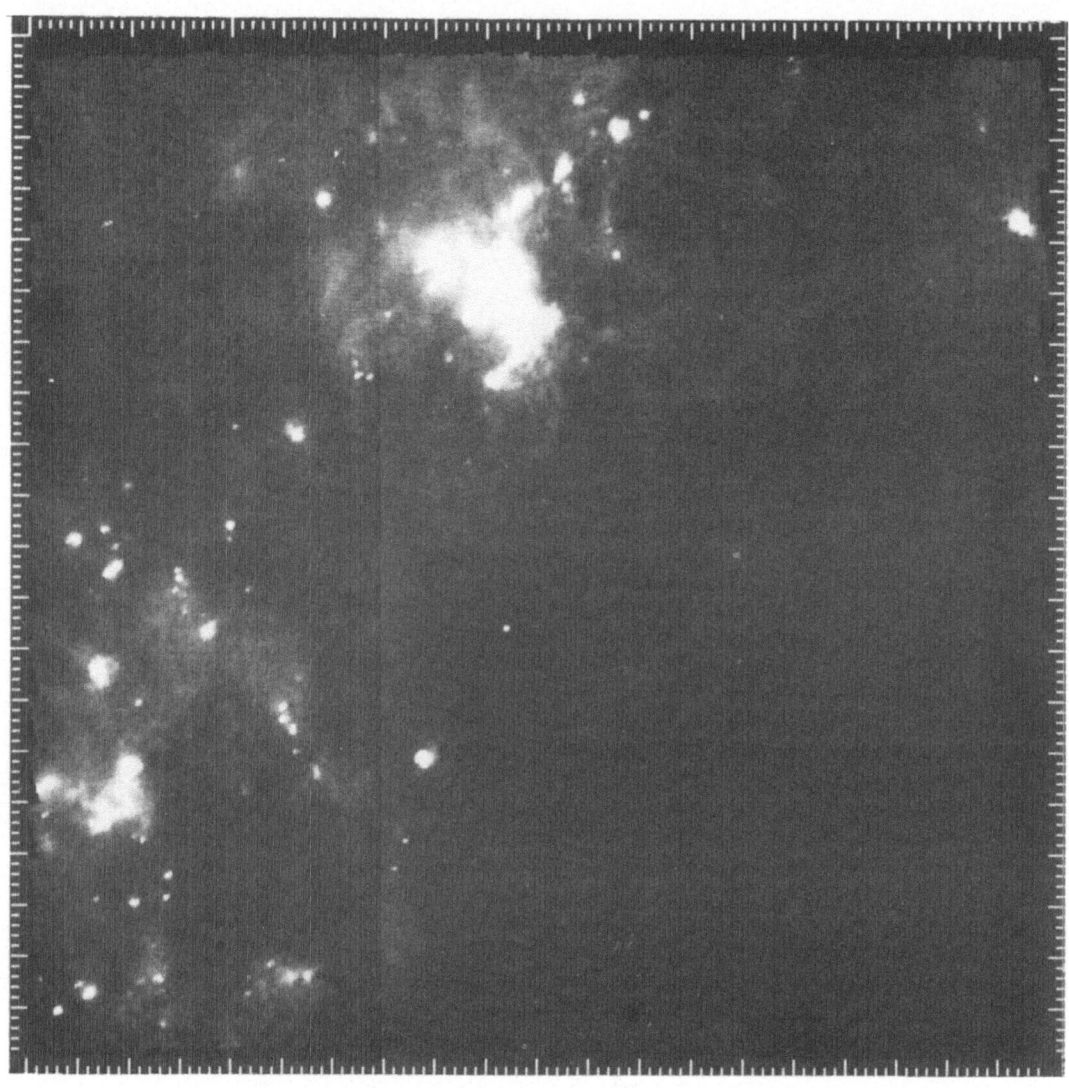

IRAS Field 126 $\alpha = 7^h00^m$, $\delta = -15°$, HCON-3, 60 µm. Field in Canis
Maior, showing the NGC 2327/NGC 2335/IC 2177 nebulae
associated with the CMa OB-1 association (top
center) NGC 2359 is the bright object SE of IC 2177.

SHOCK-HEATED DUST IN YOUNG SUPERNOVA REMNANTS

R. Braun, R.G. Strom, H. v.d. Laan, H. Greidanus
Sterrewacht
Postbus 9513
2300 RA Leiden
Netherlands

ABSTRACT. Infrared emission in young supernova remnants is interpreted as coming from shock-heated dust. Using models and data from other wavelength regimes, many physical parameters of the remnants can accurately be derived.

We have used IRAS observations of the young supernova remnants Tycho, Kepler and Cas A to further complete the description of these objects. Maps have been made in the 12, 25, 60 and 100 micron bands from the IRAS Skyflux data. Accurate (background corrected) total fluxes could be derived in all four bands, after removal of contributions from zodiacal light and cold dust, found by a spectral decomposition technique (Braun et al 1986). These fluxes were colour-corrected by comparison with a single temperature emission spectrum for graphite and silicate grains, based on the absorption efficiencies of Draine and Lee (1984) (Table I).

Table I. Infrared spectral parameters.

Source		Tycho	Kepler	Cas A
Nominal Flux Density (12 μm)		3.07	1.56	15.4
[Jy]	(25 μm)	23.7	11.7	191.
	(60 μm)	41.5	10.5	130.
	(100 μm)	11.3	2.52	31.3
Color Corrected Flux (12 μm)		2.73	1.69	17.6
[Jy]	(25 μm)	27.4	12.7	199.
	(60 μm)	36.3	8.50	102.
	(100 μm)	10.1	2.21	27.2
Color Corrected λI_λ (12 μm)		1.61×10^{-7}	5.29×10^{-7}	3.89×10^{-6}
[W m^{-2} sr^{-1}]	(25 μm)	7.74×10^{-7}	1.91×10^{-6}	2.11×10^{-5}
	(60 μm)	4.27×10^{-7}	5.32×10^{-7}	4.49×10^{-6}
	(100 μm)	7.13×10^{-8}	8.27×10^{-8}	7.19×10^{-7}
λ_{peak} [μm]		34 ± 1	$26.5\pm.5$	$25.0\pm.5$
λI_λ^{max} [W m^{-2}sr^{-1}]		1.01×10^{-6}	1.95×10^{-6}	2.11×10^{-5}

F. P. Israel (ed.), Light on Dark Matter, 155–158.

The derived spectral parameters were interpreted within the framework of the Draine (1981) model for IR emission from dust in shocked gas, after modification to include a filling factor and an incomplete cooling layer. Using furthermore: (1) other known source parameters such as the age, distance and shock velocity, and (2) an analytical representation of the results of Gull's (1973) numerical simulation of dynamical evolution of young SNRs, the physical parameters: density, filling factor of dense material, kinetic energy and total mass could be determined (Table II).

The three SNR were also mapped with the IRAS edge detector and as part of the CPC Additional Observations. A 2' FWHM gaussian beam was obtained by a linear deconvolution and spatial frequency taper in the fourier-transform domain. These maps were compared with maps of radio, optical and X-ray emission convolved to the same resolution (Fig. 1 - 3 for Cas A and Tycho).

The interpretation of these results (Braun 1985), using also X-ray derived masses, enables us to draw the following conclusions: (1) There exists both a diffuse and a clumpy (high density) component of the ISM, as well as of the SN ejecta. (2) Only the diffuse ejecta and the diffuse ISM are dynamically coupled. (3) A relatively small mass, $M < 0.4\ M_\odot$, of diffuse high velocity ejecta is responsible for driving the blast wave. (4) Most radio and X-ray emission is associated with the clumpy component of the ejecta. (5) The IR emission comes from shocked interstellar dust (as opposed to dust in the ejecta). (6) A large fraction of the progenitor mass must reside in slowly expanding ejecta.

Table II. Source parameters.

	Tycho	Kepler	Cas A
SN Date	AD 1572	AD 1604	AD 1680
Distance [kpc]	2.3	$4.1 \pm .9$	$2.9 \pm .1$
Initial Expansion Velocity [km s^{-1}]	$10,300 \pm 1000$	11,000[*]	$10,900 \pm 1000$
Diffuse Ejecta Mass [M$_\odot$]	$0.40 \pm .15$	0.27	$0.30 \pm .10$
E_k^o of Diffuse Ejecta [10^{50}erg]	1.8	1.4	1.5
Shocked ISM Mass [M$_\odot$]	$1.2 \pm .1$	1.5	$1.30 \pm .15$
ISM Density [cm^{-3}]	$0.20 \rightarrow 0.87$	$1.3 \rightarrow 3.6$	$0.42 \rightarrow 3.7$
Current Shock Velocity [km s^{-1}]	$3400 \rightarrow 1640$	$2510 \rightarrow 1490$	$3850 \rightarrow 2330$
E_k of Shocked ISM [10^{50}erg]	0.26	0.40	0.42

[*] assumed

References

Draine, B.T. 1981, Astrophys.J. 245, 880
Draine, B.T., Lee, H.M. 1984, Astrophys.J. 285, 89
Braun, R. 1985, Ph.D. Thesis
Braun, R., Strom, R.G., v.d. Laan, H., Greidanus, H. 1986, in
    these proceedings, page 47.
Gull, S.F. 1973, MNRAS 161, 47

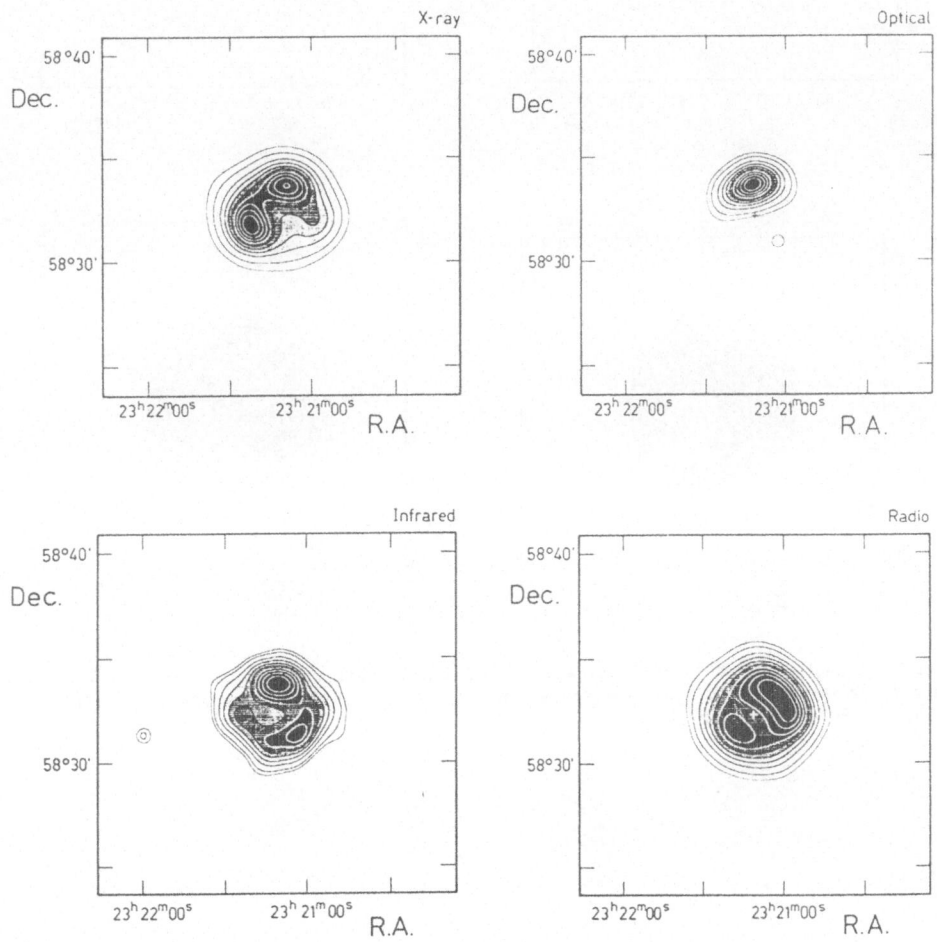

Figure 1. Cas A at X-ray, optical, infrared and radio wavelengths with the same (1.5 arcmin Gaussian FWHM) beam. Contours and grey-scale levels correspond to 5, 10, 20, 30, ... 90% of the peak brightness.

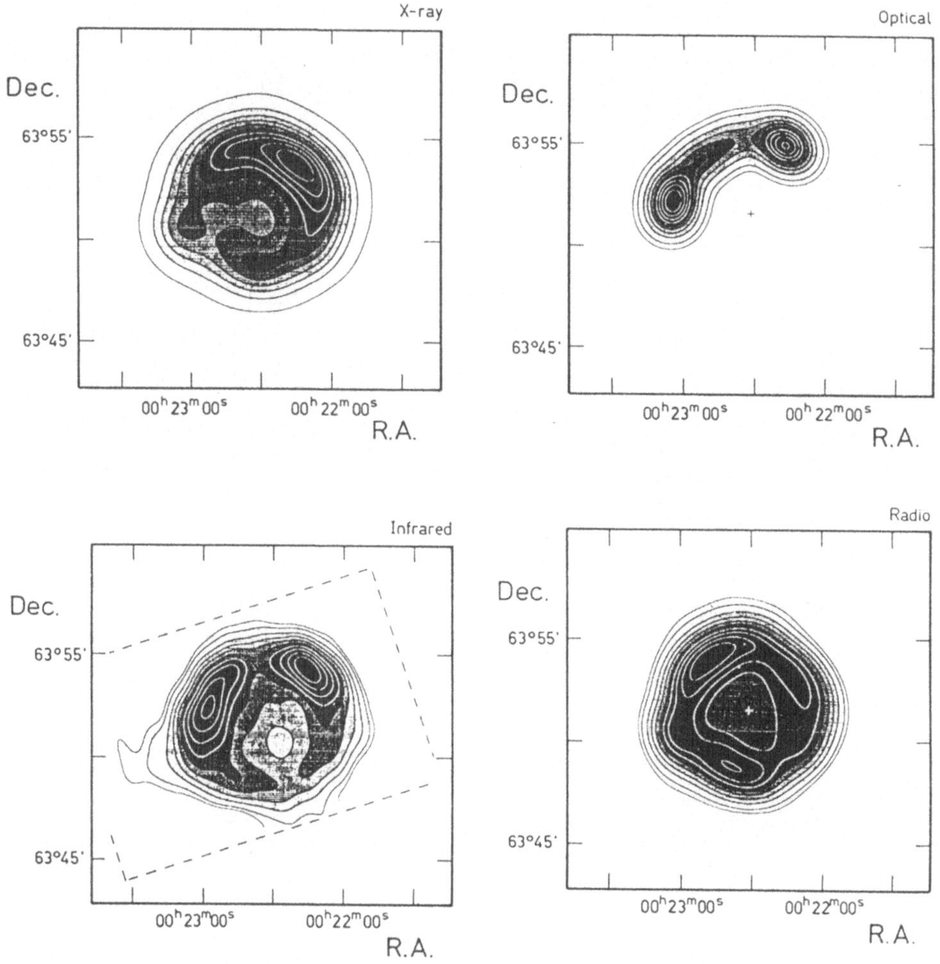

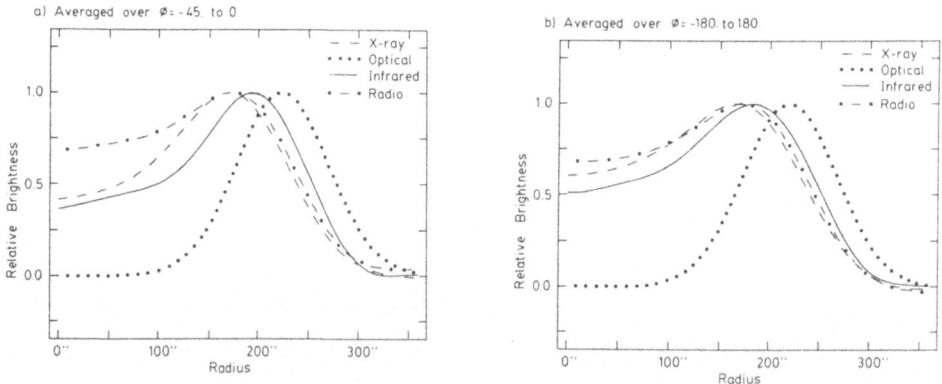

Figure 2. Tycho's SNR at X-ray, optical, infrared and radio wavelengths with the same (2 arcmin Gaussian FWHM) beam. Contours and grey-scale levels correspond to 5, 10, 20, 30, ... 90% of the peak brightness.

Figure 3. Angle averaged radial profiles of Tycho's SNR with respect to $\alpha_{50}=$ 0^{h}22^{m}31^s, $\delta_{50}=$ 63°51'36" through the images of figure 2. **a)** Averaged over position angles $-45°$ to 0°. **b)** Averaged over all angles.

SECTION 3.

DUST GRAINS AND THEIR PROPERTIES

GRAINS, WHAT DO WE KNOW?

H.C. van de Hulst
Sterrewacht, Huygens Laboratorium
Wassenaarseweg 78
Leiden, The Netherlands

1. THE HISTORICAL SETTING

This talk is an attempt to summarize the questions on which we think we know the answers and the questions on which we should like to have the answers, all in the area of interstellar grains. The review is personal, i.e., biased and incomplete. I cannot possibly compete with the review on grains over the past three years with 150 references, just written by John Mathis as chapter IV of the report of IAU Commission 34, although even that is admittedly incomplete.

I pay full tribute to the amazing technical developments which from time to time bring entirely new sources of information suddenly within reach. The IRAS satellite is a prime example. Yet I shall follow my own inclination and emphasize the issues of longer standing, corresponding to the Fourier components with a longer time period. Earlier (Van de Hulst, 1984) I have called this nanohertz astronomy.

Have you ever played this game? Take any issue of a good journal, say A + A or Ap.J., and find one of the many papers in which the author, a recent Ph.D., puts his introduction in the form: "Originally it was thought that ..., but now I tell you ...". The game is to guess from the 'original state' of scientific knowledge in what year the author started his graduate studies.

I was in that position 44 years ago, browsing without much inhibition through libraries of astronomy, chemistry and physics trying to understand what interstellar grains were, how we knew about them and what would happen to them.

The 'original knowledge', which I read in the literature at the time (top line of Table 1) was that the grains made themselves known by interstellar absorption and reddening, that they had temperatures of some 3 K, and that they were probably metallic. The idea of a metallic composition came from a hint at a continuous size distribution from interstellar grains to meteors (it was accepted as an established fact that many meteors had hyperbolic velocities) and it was helped in an odd way by the fact that Mie computations for metal spheres required less effort because a smaller size sphere leads to a reasonable fit of the observed slope of the extinction curve.

F. P. Israel (ed.), Light on Dark Matter, 161–170.
© *1986 by D. Reidel Publishing Company.*

TABLE 1   Ideas about interstellar grains

pre-1940	3 K	metal	constant	one environment
1945	10-20 K	dirty ice	growing	one environment
1985	10-20 K	core-mantle	stages in a cycle	several environments

The 'but now I tell you...' (next line of Table) was roughly this: the grains have temperatures of 10-20 K, which at the density of interstellar space is so cold that all atoms hitting their surface will freeze down permanently, except H (or H_2) and He, which evaporate. Consequently, a mantle will grow, in which H_2O is the most abundant molecule but otherwise a funny mixture. The term 'dirty ice' was invented by others later. This composition is independent of the initial nucleus of condensation, whose origin and composition we can only guess, and means a refractive index close to real, quite different from metallic. The mantle growth is somewhat fast. Oort and I speculated about what could be done to stop it well within the age of the galactic system.

I relate this story with some satisfaction because many of these conclusions still stand (third line of Table). Only I have used modern words like grain and mantle. At that time I advocated the use of 'smoke' instead of 'dust' because the chemists made the distinction between dust resulting from grinding bulk matter into small pieces and smoke resulting from the condensation of small solid (or liquid) particles in a vapor or gas. This crusade failed. Spitzer found the way out by avoiding both words and talking of grains.

The six times since then, that I was asked to summarize or review our knowledge of interstellar grains at a symposium were far enough apart to let the subject be brandnew again every time I looked at it. Today is another occasion of this kind. The average spacing of some 5 or 6 years is ideal to pick up the nanohertz component but obviously too long for an up-to-date review.

Remove old layers of paint, and all dirt and grease! What scrubbing and wiping do we have to do, and what sandpaper, soap or ammonia do we have to use, before our conception of what interstellar grains are or should be is ready to receive the new information offered from many sides and to integrate it into a new picture? We are up against some odds in this subject. In any subject in astrophysics, sharp "features" give the interpretation a better grip. Such features may exist in angle (a sharp image), in wave length (a spectral line) or in time (a pulse or sudden change). Interstellar clouds are diffuse in all three respects, which makes the subject evasive. Two astronomers talking about some star or galaxy with a catalogue name, may widely differ in their interpretation but they are sure that they talk about the same object. Two astronomers talking about interstellar clouds may not be so sure, in spite of the many adjectives invented to specify the kind of cloud more clearly.

Just for curiosity, I have listed a few words that once were coined or chosen in a specific context. But the context evolved or the model in which such a word fitted was superseded, so the use of such a word may inadvertently carry 'remnants' of old ideas and thus make present reasoning inconsistent.

Dust as a word is harmless because nobody is tempted to believe that the interstellar grains are the result of a grinding process.

Interstellar medium contains two danger words. They once had the strong connotation that the medium was homogeneous to as far away as we could see. One distinction, interstellar - circumstellar was made from the very start. Others: cloud - intercloud, arm - interarm, ionised - unionised, atomic - molecular, hot - warm - cool, quiescent - violent, were added at later dates. Recent additions to this list of distinctive names are: boundary regions of clouds, and the use of 'diffuse' as a distinctive term for everything that is not molecular or dense. Caution is clearly required in order not to take any (old or new) model for the reality.

H II region as a term has been a continual source of hilarity and confusion, but also what we mean by it has changed considerably since Strömgren introduced the concept and Spitzer the terminology.

Spiral arm often refers to a somewhat too eager identification of the observed data with a theoretical concept, which may in the long run form an undesirable restraint.

Standard cloud consists again of two danger words. When it comes to making reasonable estimates of extinction variations or of dynamical effects, randomly distributed homogeneous spherical clouds are a step better than a homogeneous 'medium'. But the model is still far from reality! Oort (1946) pointed out that in one case (Nova Persei 1901) the time at which successive features in the field light up could be used to complete the 2-dimensional image with the third dimension. This method merits more attention. Once in a century is not enough. Very curiously, X-ray diffraction haloes now actually observed around some X-ray sources (Rolf, 1983; Catura, 1983; Mauche et al. 1984) offer exactly the same opportunity, not only in principle but with closely the same scales: a field radius of some 10' and time scales of many months. Until such observations have actually been made, suggestive words like wisps, veils, and cirrus may serve as a reminder that we should not be sold on the model of nice spherical clouds.

2. ELEMENTS OF A SYSTEMATIC REVIEW

Reasonably complete reviews of most subjects in astronomy can be found in the tri-annual 'Reports on Astronomy' of the IAU. I know these reviews are no masterpieces of critical thinking but a lot of work of authors and editors has gone into them. I had the pleasure of composing two on interstellar matter myself, for Dublin 1955 and for Moscow 1958. In recent teaching I deliberately avoided the danger of working from old notes by using the Commission 34 report for Patras 1982. And just now the New Delhi 1985 report of this commission has arrived in manuscript.

The organization of these reports is by no means logical. Divisions may refer to modes of observation, or to physical processes in space, or to particular regions of space. This reflects the state of the subject rather than the clarity of mind of the reviewers. Grains occupy roughly 10 per cent by number of publications but even in that smaller area reviews differ greatly in emphasis and conclusions.

Naturally, I have tried to offer you the best systematic subdivision I could come up with. This is given in Table 2. Since many observed facts have not yet found their ultimate explanation, the division observation/theory, which we should aim to get rid of eventually, is still prominent. The table also gives a taste of history by stating when, in my subjective assessment, the question was first seriously addressed: c = classical, m = middle age, n = new. In two topics, where I remained in doubt how seriously I should reckon the attempts, I have left a question mark. Let me continue with a few comments rather than a lecture course.

Although the existence of grains through wide regions of the interstellar medium remained in doubt until 1930, we knew since Eddington 1923 that the energy of star light absorbed by any such grains should be re-emitted in the far infrared. IRAS now has succeeded in mapping this emission. Many theses are in the making but in the reference list this still seems a very modest topic.

The recent complaints that scattered light observed from reflection nebulae is hard to interpret because of ambiguous geometry echo those of 50 years ago. Diffuse galactic light does not have this problem to the same degree but the error range due to uncertainties in the calibration may be larger than sometimes has been suggested.

The extinction curve continues to hold its place as the one solid piece of evidence on which most of our ideas on grains are based. Its slight variations with position in space and with polarization form an important part of this evidence. Clearly, interstellar polarization is not a separate subject but should be treated as part of the extinction data.

Composition can, in principle, be inferred either from the obeserved infrared bands in absorption or emission or from the theoretical examination of the way grains form, grow, and are processed in the interstellar environment. Both ways are beset with ambiguities. A brief analysis is given in section 3.

3. DOES THE EXTINCTION CURVE REVEAL SIZE OR COMPOSITION?

Let us revisit the fundamental question which has been with us for over half a century: given the extinction curve, which shows the extinction E in magnitudes over a fixed path in interstellar space as a function of $1/\lambda$, what can we infer from it?

For the moment we simplify the problem by ignoring local variations. We do as if in all directions in which suitable background stars are available the same curve is found except for a scale factor proportional to the amount of dust (Figure 1). In many directions the curve splits in two, for two different linear polarizations. The maximum splitting is sketched in the figure. This interstellar polarization was discovered 40 years later than it might have been.

The necessity of using a background star whose distance and spectrum may not precisely be known, introduces greater uncertainties than we like. In the thirties the known part of the curve was practically limited to the range $1 - 3 \ \mu m^{-1}$ and it took long and clever research to arrive at the most reliable value of R, the ratio of total to selective extinction. This fixes the level where the line E = 0

TABLE 2   Methods to learn about interstellar grains and their effects

		first thorough studies		
I BASIC OBSERVATIONS		pre-1950	1950-1970	1970-1985
grand extinction	reddening, extinction curve	c		
	ratio extinction to reddening	c		
	extinction curve in UV		m	
	unidentified absorption bands	c		
	IR absorption bands		m	
	polarization	c		
grand scattering	brightness/colour of refl. nebulae	c		
	brightness/colour of diffuse galactic light	c		
	haloes around X-ray sources			n
grand emission	far IR emission continuum			n
	unidentified IR emission features			n
collection	in earth atmosphere, on moon or in space			?
II DIRECT THEORY				
temperature (in radiation environment)		c		
optical properties	absorption = emission	c		
	scattering	c		
+	extinction	c		
providing condensation nuclei		?	m	
mantle growth		c		
mantle processing				n
survival of refractory particle				n
electrical charge		c		
rotation			m	
III IMPORTANCE FOR RELATED TOPICS				
shielding of star forming regions			m	
mapping magnetic field			m	
modifying gas composition	depletion	c		
	molecule formation		m	

should be drawn and (since a 'neutral' extinction can be ruled out) this
zero line also fixes the point where the curve should start at $\lambda^{-1} = 0$.
Since the space age the curve can be measured from about 0 to 9 μm^{-1}. It
is known to have a conspicuous bump at 0.2175 μm and another rise in the
far UV. More prominent local variations are seen in the far UV curve
than in the visual range. They appear to be uncorrelated with those in
the visual part and for the moment we shall ignore both.

Extinction = scattering + absorption. In principle the measured
extinction contains no label telling which 'feature' is due to
absorption and which to extinction. Observations in arbitrary directions
are needed to separate these terms. Scattered radiation can be observed
in reflection nebulae, in the diffuse galactic light and in stellar
haloes, and emitted radiation in the far IR observed by IRAS. This means
that in discussing information based on the extinction curve alone, the
ratio

$$albedo = \frac{scattering}{extinction}$$

as a function of λ must be treated as unknown.

Although it is customary (and probably largely correct) to describe
enhanced extinction in certain narrow λ-intervals as 'absorption' lines
(or bands or features), we should at this stage realize that they are
details of the extinction curve, that should be studied as such with all
the methods that are appropriate (correlation studies, modeling, etc). A
well documented review would take pages but the scope of the information
hidden here can be understood from one table (Table 3).These 36 features
presumably form many signatures of the interstellar grains, or of the
interstellar gas. They can be read but are not understood. Less than 10
per cent are identified. The references may be consulted for details. I
have listed the (roughly averaged) total width at base, which for a
bell-shaped band is roughly twice the full width at half maximum. The
range of 30 over λ causes a range of 1000 in the conversion factor from
width in wavelength to width in wavenumber, so both are listed. However,
neither column seems to show a systematic trend. All features are very
broad when compared to atomic interstellar lines. Features whose
existence is observationally uncertain, like the 12 lines classed 'C' or
'G' by Herbig and one further very broad structure, have not been
included in table 3.

The art of interpreting the extinction curve obviously is to find a
model (or rather all models) matching it. A model is determined by
assumptions about number, size, size distribution, shape, composition,
distribution of composition, etc., of the grains. We may lump these
assumptions together in the words 'size' and 'composition'.

Clearly, any model extinction curve depends inseparably on both
size and composition, so it is not logical to request separate answers.
But is is a legitimate practical question to ask if certain properties
of the extinction curve are more revealing for size and others carry a
clear signature of composition. Many examples in other fields of
astronomy show that such a separation may be helpful: e.g., the inner
zodiacal light, the polarization maximum of Venus, an absorption line in
any particulate medium.

TABLE 3   Added-extinction bands

number of bands	wave length of represen- tative band	total width at base		identification	reference
	μm	in μm	in (μm)$^{-1}$		
1	0.218	0.10	2.5	'bump', tentative identification graphite	
1	0.55	0.10	0.4	'VBS' = very broad struc- ture, unidentified	a
5	0.443	0.004	0.02	unidentified, broad	
11	0.6284	0.0008	0.002	unidentified, medium	b
10	0.5797	0.00026	0.0008	unidentified, narrow	
8	6.0	0.05 to 1.2	0.002 to 0.05	water ice, silicate and unidentified	c

a: Van Breda and Whittet (1981). Most likely the VBS must be seen as decreased extinction = apparent emission. But the possiblity that it actually is added extinction in an adjacent band cannot be excluded, see Van Yzendoorn (1985).
b: Herbig (1975)
c: Wolstencroft and Greenberg (1984), De Graauw (1985)

My impression is that in interpreting the interstellar extinction curve a tendency has existed to jump too easily towards such a separation. Early authors have been content to assume a constant refractive index and explain everything by size. Others have discussed 'absorption' lines by composition effects only, which may lead to gross errors. Near the top of the extinction curve added absorption may even lower the extinction. Fortunately many authors are now aware of these complications.

My final assessment is that we are still in mid stream with very few firm effects to boast on. Table 4 shows what I consider the most important changes in our understanding over the last 40 years. We still need theoretical expectations to guide our choice of models as much as we did then. But the enormous progress, both in the observations and in the theoretical concepts , notably on mantle processing, gives us good hope that the next review wil be able to point at firmer results.

TABLE 4   Shifts in interpretation of extinction curve

	1943	1985
I Direct inference from observations		
Extinction curve	mostly size effect impossible to infer composition polarization unknown	also composition effect still impossible? polarization known but not of much use
Bands visual infrared	unidentified unknown	still unidentified much progress
Albedo	would be helpful, if we knew it well	critical assessment of present data needed
II Theory to complement inference from observations		
Nucleation	a gamble, but of no consequence	many ideas
Mantle growth	fair estimates	refined, not changed
Mantle processing	not even considered	good theories and firm experimental basis

References

Van Breda, I.G., Whittet, D.C.B., 1981, Mon. Notices Roy. Astr. Soc., 195, 79.
Catura, R.C., 1983 Astrophys. J. 275, 645.
De Graauw, T., editor, 1985 Infared Space Observatory, Alpbach Workshop, 1 non-serial ESA publication.
Herbig, G.H., 1975 Astrophys. J., 196, 129.
Van de Hulst, H.C., 1984 Nanohertz Astronomy in The early years of radio astronomy, W.T. Sullivan, ed., p.385 Cambridge Univ. Press
Oort, J.H., 1946 Mon. Notices Roy. Astr. Soc. 106, 159.
Rolf, D.P., 1983 Nature 302, 46.
Wolstencroft, R.D., Greenberg, J.M., editors, 1984 Hilo workshop, Occasional Reports Roy. Observatory, Edinburgh, 12.
Van Yzendoorn, L.J., 1985 Thesis Leiden University

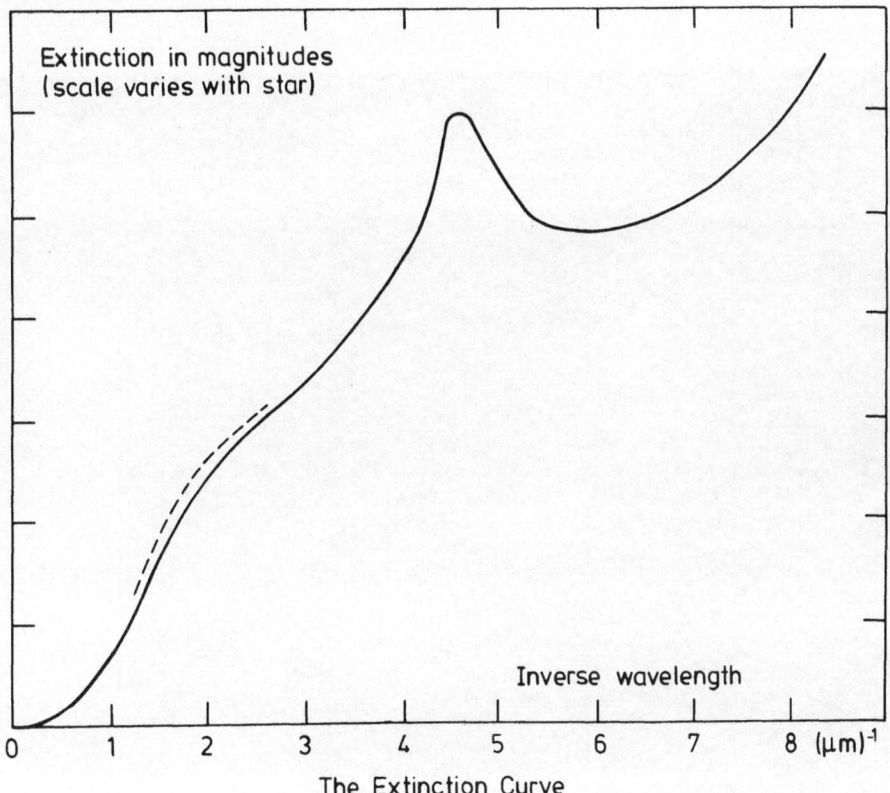

Fig.1 The standard extinction curve. The maximum polarization separation has been sketched in.

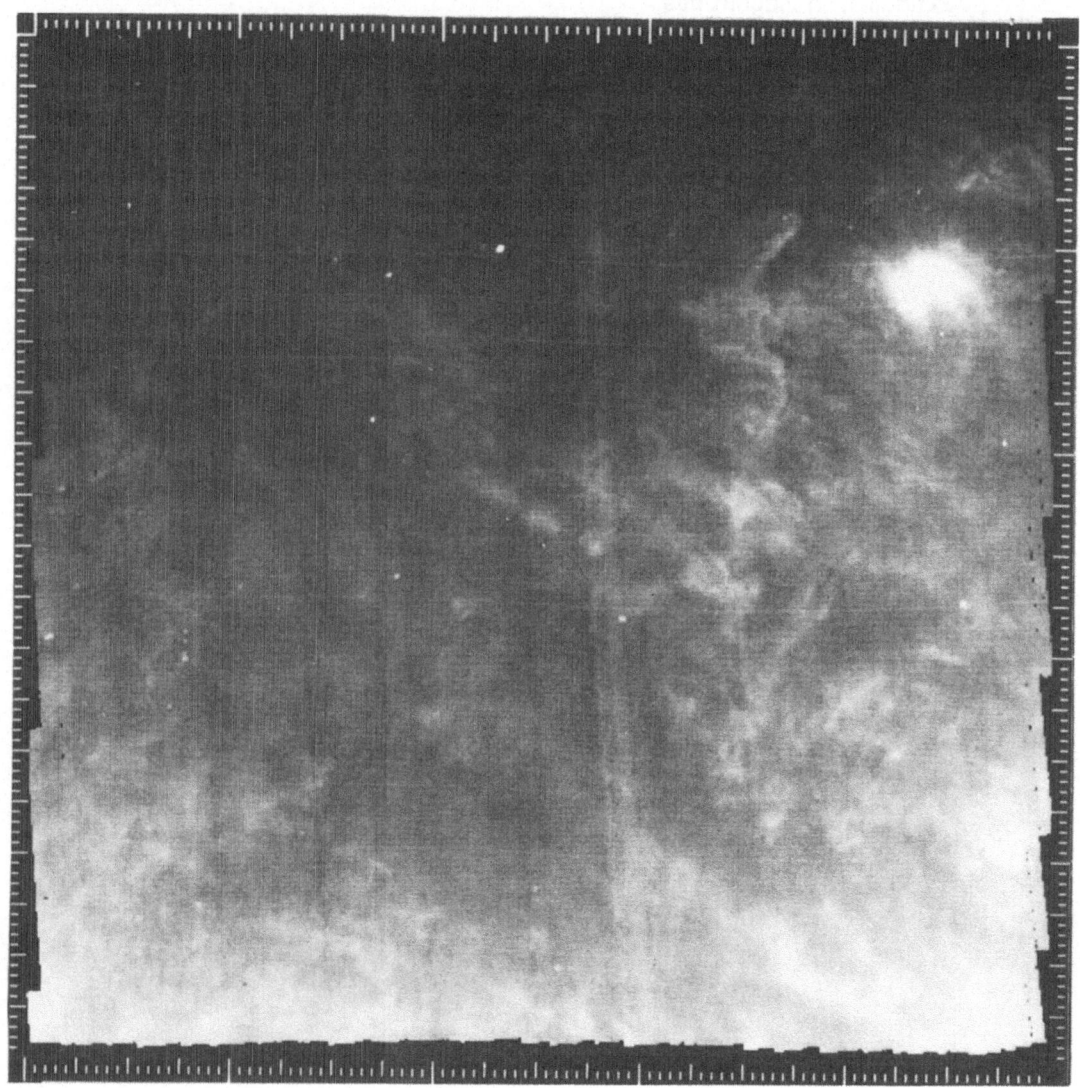

IRAS Field 136   α = 17^{h}00^m, δ = -15°, HCON-3, 60 μm. Field in
                  Ophiuchus and Serpens Cauda, showing extensive
                  cirrus and a very hot region near ξ Serpentis (top
                  right).

IRAS CIRRUS OBSERVATIONS AND THE NATURE OF DUST

John S. Mathis
Washburn Observatory
University of Wisconsin
475 N. Charter St.
Madison, WI 53706
U.S.A.

ABSTRACT. Two theories regarding the nature of interstellar dust are considered. One, "MRN", has bare silicate and carbon particles. The other, by Greenberg and his associates, has silicates with refractory mantles providing the extinction in the visible region of the spectrum. Some relevant observations are reviewed. The most important of these is the fact that linear polarization reaches a maximum at the same wavelength at which the circular polarization goes through zero; this fact implies that the polarizing grains, which apparently are silicates, are nonabsorbing. Since the albedo of real dust is well below unity, there must be another component besides the silicates which provides the absorption. The nature of the absorbing component, and of others, is discussed. The IRAS cirrus 60/100 μm color is that predicted by MRN. The 12/100 μm color is indicating the emission of the 11.4 μm band which is seen in many objects, and which has plausibly been attributed to polycyclic aromatic hydrocarbons. It seems to vary in strength from one region to another. The 20 μm band may be influenced by silicate emission in the 18 μm band.

Theories of Interstellar Dust: There have been countless suggestions regarding the composition of dust, but I know of only four "complete" theories which try to explain the entire wavelength dependence of interstellar extinction and polarization. Hoyle and Wickramasinghe (1982) prefer a biological origin for dust (see Jabir et al., 1982 for a description); they require an order of magnitude too much phosphorus (Whittet 1984). W.W. Duley, T.J. Millar, and their associates (e.g. Duley and Najdowsky 1983) use MgO to produce the λ2175 "bump" and other oxides for the extinction in other wavelengths. I feel that MgO cannot be a dominant absorpber because of the polarization of the 18 μm feature (Knacke and Capps 1979), which should be unpolarized if it is dominated by MgO. I will concentrate on the ideas of Mathis, Rumpl, and Nordsieck (1977, hereafter MRN), and of J.M. Greenberg and his associates (Greenberg (1983a,b, hereafter Gr). The theories are, of course, evolving. A very important recent development was the thorough investigation by Draine and Lee (1984, hereafter DL) of the optical constants of graphite and silicates. I strongly recommend using these constants in any applications of the MRN theory. I assume that in the next talk Prof. Greenberg will tell us of any modifications he might have had

F. P. Israel (ed.), Light on Dark Matter, 171–176.
© *1986 by D. Reidel Publishing Company.*

in his own thinking.

The MRN and Gr theories have the $\lambda 2175$ "bump" caused by graphite, although this identification is not essential to Gr. Both MRN and Gr have some or most of the visual extinction caused by silicates with a power-law distribution in sizes. The motivation for requiring silicates is the presence of the 9.7 μm and 18-20 μm features seen in both nebulae and in stars.

The differences between the Gr and MRN theories are mostly in the nature of the particles which absorb in the visual region. MRN has uncoated grains of silicates and a form of carbon, which is graphite in MRN but which may actually have much less crystal structure than graphite (see below). Gr has only one major type of particle in the visual-NIR region: silicates, with a fairly thick coating of a rather refractory material, "yellow-stuff", which can be produced in the laboratory. In the Gr model, most of the carbon is tied up in the mantles surrounding the silicate cores, but there is also some oxygen and nitrogen in the mantle as well.

It seems to me that there are observations which can decide whether there is only one major contributor to the visual extinction. They concern the polarization caused by interstellar grains. The wavelength dependence of the polarization is well known (Wilking et al., 1982); the wavelength of maximum polarization, λ_{max}, is 0.55 μm for the average, but has a range of about 0.4 μm to 1 μm. There is also circular polarization measured in several stars (Martin and Angel 1977); the interesting outcome of this measurement is that λ_c the wavelength at which the circular polarization is zero, is equal to λ_{max}, for stars with a fairly wide range of λ_{max}. Martin (1974) has showed that the condition $\lambda_{max} = \lambda_c$ implies that the polarizing material is dielectric, or nonabsorbing, if there is no strong wavelength dependence of the index of refraction, $\underline{m}(=\underline{n}-i\underline{k})$. It is possible to have the condition met if $\underline{n}$ and $\underline{k}$ depend on $\overline{\lambda}$ in a particular way, and in fact the substance magnetite (Fe_3O_4) happens to meet the requirements at $\lambda = 0.55$ μm. However, the condition is met for stars which have both small and fairly large values of λ_{max}. Thus, I see no way to avoid the conclusion that the polarizing grains are alsmost perfect dielectrics, with an albedo, ω, of unity.

What albedo is actually observed for the total population of grains? There is a qualitative difference between dust with $\omega = 1$ and, say, $\omega = 0.7$; if $\omega \approx 1$, then almost every photon eventually escapes, and the flux at the surface of a dusty object is just what has been generated within. Furthermore, the far-infrared (FIR) emission is correspondingly reduced by the lack of absorption in those wavelengths for which $\omega \approx 1$. The visual albedo of real dust is not easy to determine, but it seems clear to me that it is not very close to unity. The diffuse galactic light in the optical part of the spectrum (Witt 1968, Mathis 1973) and the reflection from dust clouds which are illuminated by the general interstellar radiation field (FitzGerald, Stephens, and Witt 1976; Mattila 1979) both suggest that $\omega \approx 0.6$. The Balmer decrement from HII regions would be unreddened if $\omega \approx 1$, but it is well known that most HII regions appear rather heavily reddened (e.g. Israel and Kennicutt 1980). The most extreme example of this phenomenon is the IRAS-discovered galaxies which are very bright at 100 μm but hardly visible in the optical. All of these considerations make me conclude that there is certainly a component besides the polarizing particles which is contributing to the optical extinction, and that the silicates (which we know do contribute the polarization, because of the strong polarization of the 9.7 μm silicate feature) are very good dielectrics. Even good absorbers, like graphite or amorphous carbon, have ω ~ 0.5. In order to bring the overall albedo down to ~0.6, we must have most of the extinction provided by the nonpolarizing material, which is not either bare or coated silicates.

One reason why I think that silicates are not the major component of the dust providing visual extinction is that about half of the dust being put into the ISM is from carbon stars (Jura 1985). Their spectra show some indication of a SiC band, but otherwise are featureless, as would be the spectrum of either graphite or amorphous carbon. There is not enough silicon for SiC to be the major component of the dust being ejected. The only carbon star for which the type of dust has been determined is R CrB (Hecht et al., 1984), for which the IUE obserbations showed that the bump is shifted to longer wavelengths. The dust is apparently amorphous carbon. Graphite, well-ordered crystalline carbon, is difficult to form in a stellar atmosphere or to produce by annealing amorphous carbon. These processes are even harder to imagine occuring in the ISM. However, there really is a continuum of degrees of crystallinity of carbon, with completely amorphous carbon on one end of the scale and graphite on the other. Probably some annealing does occur in the stellar atmosphere or while the grain is temporarily heated by passage through a shock in the ISM, and there may be some order, but not complete, in the carbon grains. The question really is, "over how large a region must the carbon atoms be well-ordered for the material to act like graphite as regards its absorption properties?" I suspect that the answer for the spectral region not including the λ2175 bump is "Not very large: a few dozen atoms in a graphite plane will do." It is remarkable that DL predicted the correct absorption in the 1-10 μm region using graphite constants, even though this region was not used in determining the MRN mixture. We shall see that MRN also predicts the correct IRAS 60-100 μm ratio, so that it seems successful over a wavelength range of a factor over 1000.

There is not very good agreement among various measurements of the optical constants of amorphous of poorly ordered ("glassy") carbon (Duley 1984, Koike et al., 1980, Williams and Arakawa 1972). One interesting feature is that small amorphous carbon particles, unlike graphite, show a steep rise with 1/λ between 0.15 μm and 12 μm.

I do not now think that graphite with a MRN size distribution is the carrier of the λ2175 bump. The main reason is that the position of the bump seems extremely constant from direction to direction, while theory (Gilra 1972; Savage 1975) requires that the position of the bump shift with particle size. It is possible, but not likely, that collisions between graphite particles establish the same size distribution in all directions. Recent laboratory studies (Sakata et al., 1983, 1984) have shown that the bump is found in the spectra of material produced in the laboratory by quenching the products formed from an electrical discharge in hydrocarbons. The nature of these "quenched carbonaceous condensates" QCDs) should be investigated vigorously. Another very interesting material very possibly found in QCDs, is a very near kin to graphite: "polycylic aromatic hydrocarbons", or "PAHs". Léger and Puget (1984), Puget et al. (1985), and Allamandola et al. (1985) have recently suggested that PAHs might be responsible for the emission bands which are commonly seen in the near infrared (NIR) in a wide variety of objects. I will be discussing these bands at some length later in this talk. The PAHs are, as their name implies, sheets of benzene rings, with H atoms bonded to some of the edge carbons. Graphite is a stack of a large number of such sheets of benzene rings. Thus, the statement that the λ2175 bump is caused by graphite might be actually rather accurate, if PAHs cause the bump. Of course, in PAHs there is no stack of benzene-ring sheets, and the sheets are not huge in extent. PAHs, which are really small portions of the graphite plane, might be more likely to force the bump position to be invariant. According to a study by E.L. Fitzpatrick and D. Massa (in preparation), the width of the bump does vary slightly but significantly from place to place; again, this is not what one would expect from graphite, but could be caused by a differing distribution of the PAH molecules.

It is clear from the presence of the 3.07 μm "ice" feature that mantles do form on grains deep within molecular clouds. I believe that mantles are not very important for the dust in the diffuse ISM because I suspect that they are destroyed rather quickly. In fact, a major problem in the theory of the destruction of grains is in under- standing how the silicates and graphite, which are more refractory than "yellow stuff", survive the rigors of supernova shocks.

In summary, I think that grains probably consist of bare silicates which provide the polarization, and which are very "clean" or nonabsorbing in the visual. Poorly ordered carbon probably provides the absorption and most of the extinction in the visual-NIR. Possibly very small carbon grains (but probably not silicates; see below) provide the extinction which is rapidly rising with $(1/\lambda)$ in the 0.12 - 0.15 μm range. This extinction does vary considerably from star to star. Perhaps even smaller PAHs, which are basically graphitic in nature, provide the λ1275 bump. Thus, I feel that there is a size distribution of carbon from fairly large (~ 0.25 μm) down to molecular sizes, with increasing order from large to small sizes. Probably all of the carbon grains have hydrogen atoms bonded to the edges. Possibly hydrogens inside the grains influence the properties of the larger grains significantly. There are doubtlessly many other substances present as well, but possibly in very small quantities; for instance, it is known that SiC is being ejected by carbon stars. magnetite, other oxides of iron, or metallic iron could also be present; in fact, I think that they have a major role in the alignment of interstellar grains (Mathis 1985).

Implications of the grain distributions for IRAS: The IRAS observations are a gold mine of information about the presence of grains. The IRAS cirrus 60/100 μm intensity ratio for a variety of MRN-type mixtures was predicted by Draine and Anderson (1985). The mixtures differed in the lower limit to the size distribution, which

was varied from 3 A to 100 A, and also in the relative numbers of
very small particles. The distribution of small grains is not well
determined by the extinction measures available even to IUE. Draine
and Anderson's prediction for the MRN mixture in the local
interstellar radiation field is that the 60/100 μm ratio is in the
range of 0.11 to 0.22 for the extreme range of mixtures he
considered. The observed ratio (de Vries 1985), on two well-observed
patches of cirrus, are 0.11 and 0.18. This is another correct
prediction of the MRN model. Most of the emission in the model is
from the carbon, except near the 9.7 μm band of silicates.

    I think that the 12 and 20 μm data are going to tell us about
one very important component of the grain population; the very tiny
grains (PAHs) which are heated by the absorption of a single photon.
There is an excess emission from several "reflection" nebulae, such
as NGC 2023 and NGC 7023 (Sellgren 1984; Witt et al., 1984). A very
recent paper (Sellgren et al., 1985) showing the 1-13 μm observations
of both nebulae has important implications for IRAS cirrus data. The
spectrum is dominated by the strong "unidentified" emission peaks ar
3.3, 3.4, 6.2, 7.7, 8.6, and 11.3 μm. This aspect is not too
surprising. The integral of the NIR emission is relative large, about
10% of the total energy emitted in the 60-150 μm range (Whitcomb et
al., 1981; Harvey et al., 1980). Thus, the carrier of the
unidentified bands is a major absorption component. The absorption
probably takes place in the ultraviolet, either the λ2175 bump or the
rise at λ < 0.16 μm, or both, because most of the energy is in that
spectral region (the central star of each nebula is an early B in
spectral type). The most plausible identification of the carrier is
PAH's (Léger and Puget, 1984). There is a relative minimum at 4.8 and
10 μm, so the carrier of these bands does not seem to be a silicate.
This is the reason why I suspect that the 0.15 μm-0.12 μm extinction
is not caused by silicates.

    The connection of the NIR spectra of reflection nebulae with
IRAS is obvious; the 11.3 μm feature is detected by the 12 μm IRAS
filter, and no doubt contributes much of the energy to that band in
the spectrum of cirrus. A very interesting feature of the cirrus (de
Vries and Le Poole, 1985) is that the 12/100 μm and 25/100 μm colours
are different for the two clouds which they studied, while the 60/100
μm colours are constant. This shows that the nature of the dust
varies from cloud to cloud. It would be interesting to contrast the
IRAS spectra from various reflection nebulae illuminated by stars of
differing spectral types, so that one can see how the IRAS colours
depend on the energy distribution of the star.

    It is possible that silicates contribute some emission to the 25
μm IRAS filter. The calculations of Draine and Anderson (1985), which
predict the correct 60/100 μm colour, suggest that the 20 μm emission
is appreciable if there is an enhanced number of very small silicates
(sizes > 30 A or so) over the standard MRN. I am not aware of any
unidentified emission bands within the 25 μm IRAS band pass.

    The IRAS observations can not easily distinguish between the MRN
and Gr theories. The 100 μm filter is close to the maximum of the
emissivity, so that it is not useful in determining how broad the
emission really is. However, IRAS is most useful for selecting bright

objects which can be studied from aircraft in the 50 - 400 μm region, so that the whole emission curve can be determined. All in all, IRAS observations tell us where dust is and provide powerful constraints on the interstellar radiation field within various places in the Milky Way and other galaxies. The 12/100 μm colors can give us useful insights into the relative amount of a very important grain component.

REFERENCES

Allamandola, L.J., Tielens, A.G.G.M., Barker, J.R. 1985, Ap. J., **290**, L25.
de Vries, C.P. 1985, Astr. Ap., in press.
de Vries, C.P., Le Poole, R.S. 1985, Astr. Ap., **145**, L7.
Draine, B.T., Anderson, N. 1985, Ap. J., **292**, May 15 issue.
Draine, B.T., Lee, H.M. 1984, Ap.J., **285**, 89 (DL).
Duley, W.W., Nadjowsky, I. 1983, Ap. Sp. Sci., **95**, 187.
FitzGerald, M.P., Stephens, T.C., Witt, A.N. 1976, Ap. J., **208**, 709.
Gilra, D.P. 1972, in Scientific Results from OAO-2, ed. A.D. Code (NASA
    SP-310), p. 295.
Greenberg, J.M. 1984a, Workshop of Interstellar Dust (Hilo, Hawaii), ed.
    J.M. Greenberg and R.D. Wolstencroft, several papers.
Greenberg, J.M. 1985, Les Houches Conference (preprint); see also these
    proceedings, page 177.
Harvey, P.M., Thronson, H.A. Jr., Gatley, I. 1980, Ap. J., **235**, 894.
Hecht, J.H., Holm, A.V., Donn, B., Wu, C.-C. 1984, Ap. J., **280**, 228.
Hoyle, F., Wickramasinghe, N.C. 1982, Ap. Sp. Sci., **86**, 341.
Israel, F.P., Kennicutt, R.L. 1980, Ap. J. Lett., **21**, 1.
Jabir, N.L., Hoyle, F., Wickramasinghe, N.C. 1982, Ap.Sp.Sci., **86**, 321.
Jura, M. 1985, NASA Workshop on the Inter-Relationships Among Circum-
    stellar, Interstellar, and Interplanetary Dust Grains, in press.
Knacke, R.F., Capps, R.W. 1979, A.J., **84**, 1705.
Léger, A., Puget, J.L. 1984, Astr. Ap, **137**, L5.
Martin, P.G. 1974, Ap.J., **188**, 517.
Martin, P.G., Angel, J.R.P. 1977, Ap.J., **207**, 126.
Mathis, J.S. 1973, Ap.J., **186**, 815.
Mathis, J.S. 1985, submitted to Ap.J.
Mathis, J.S., Rumpl, W., Nordsieck, K.H. 1977, Ap.J., **217**, 425 (MRN).
Mattila, K. 1979, Astr. Ap., **78**, 253.
Puget, J.L., Léger, A., Boulanger, A.: 1985, Astr. Ap., **142**, L19.
Sakata, A., Wada, S., Okutsa, Y., Shintani, H., Nakada, Y. 1983, Nature,
    301, 493.
Sakata, A., Wada, S., Tanabe, T., Onaka, T. 1984, Ap.J., **287**, L51.
Savage, B.D. 1975, Ap.J., **199**, 92.
Sellgren, K. 1984, Ap.J., **271**, 623.
Sellgren, K., Allamandola, L.J., Bregman, J.D., Werner, M.W., Wooden, D.H.
    1985, Ap.J., in press.
Whitcomb, S.E., Gatley, I., Hildebrand, R.H., Keene, J., Sellgren, K.,
    Werner, M.W. 1981, Ap.J., **246**, 416.
Whittet, D.C.B. 1984, M.N.R.A.S., **210**, 479.
Wilking, B.A., Lebofsky, M.J., Rieke, G.H. 1982, A.J., **87**, 695.
Witt, A.N.: 1968, Ap.J., **152**, 59.
Witt, A.N., Schild, R.E., Kraiman, J.B.: 1984, Ap.J., **281**, 708.

DUST IN DIFFUSE CLOUDS: ONE STAGE IN A CYCLE

J. Mayo Greenberg, Laboratory Astrophysics, Leiden University
P.O. Box 9504, 2300 RA Leiden, The Netherlands

ABSTRACT

An evolutionary point of view together with application of basic observational constraints suggest that the major solid particle constituents
of the diffuse cloud medium are: silicate core - organic refractory
mantle particles (~ 0.1 μm); small disordered carbon particles
($\lesssim$ 0.01 μm); small silicates ($\lesssim$ 0.01 μm). Implications of likely
distributions of such particles in the near infrared extinction and the
far infrared emission are considered. An extension of interstellar dust
to comet dust optical properties provides a basis for searching for a
hitherto unexpected source of infrared emission within the solar system.

1. INTRODUCTION

Rather than presenting any finished model of the solid particles in
diffuse clouds - an impossible task at this time - I shall try to place
the several alternative models in confrontation with the existing framework of established observational, theoretical and laboratory studies.
In some cases the conclusions are unequivocal, in others there remains
some room for question. It is only by combining all criteria together
that one can hope to arrive at a choice.

Although I will emphasize observed properties in diffuse clouds,
one of the keys, as I see it, is to consider this phase as a consequence
of the repeated cycling of grains between diffuse and dense clouds many
times. In an earlier paper I had looked at the dense cloud phase
(Greenberg, 1982b). Now I look at the diffuse cloud phase.

The astronomical observations providing the grain criteria in
diffuse clouds (D.C.) are:
(a) Wavelength dependence of extinction from the infrared to the
    ultraviolet
(b) Correlation of extinction with gas ($N_H/E(B-V) = 5.9 \times 10^{21}$ cm^{-2})
(c) Wavelength dependence of polarization; both linear and circular
(d) Maximum polarization relative to extinction: $P(V)/A(V) \leq 0.03$
(e) Scattering and absorption properties: albedo, phase function

F. P. Israel (ed.), Light on Dark Matter, 177–188.

(f) Spectroscopic properties: line and band absorption and emission
(g) Cosmic abundance of the condensable elements
    Astronomical observations relative to grain evolution are basically
all of the above with consideration of changes correlated with history
and present conditions such as:
(h) Star formation rate and cycling time for the interstellar medium
(i) Grain formation and growth - circumstellar, interstellar
(j) Grain destruction
(k) Coagulation vs. accretion. Effects on polarization and extinction.
    Wherever possible I will use simplified numerical results with
reference to detailed treatments elsewhere.
    Finally, I shall try to give some indications of how different
grain models may be related to the IRAS data. A brief extension of dust
modelling to comets, comet dust and dust in the solar system will be
made with the interesting result that the IRAS data on the zodiacal
light particles out of the ecliptic may be related to the interstellar
dust properties.
    The grain models to which most attention is given are the graphite
plus silicate model of Mathis, Rumpl and Nordsieck (1977, hereafter MRN)
as amended in later papers (Mathis, 1979; Mathis & Wallenhorst, 1981)
and the multimodal silicate core-organic refractory plus small bare
carbon and silicate particle model (hereafter referred to as C-M,b) of
Greenberg and colleagues (Greenberg 1978, Hong and Greenberg 1980;
Aannestad & Greenberg, 1983; Greenberg and Chlewicki, 1983; Chlewicki,
1985). Other proposed grain constituents will also be considered as
possible contributors to the interstellar dust population.

2. EXTINCTION

Both the MRN and the C-M,b models provide a reasonable match to the
average wavelength dependence of extinction between the near infrared
and the far ultraviolet. Some problems remain in the near infrared (NIR)
and the far ultraviolet (FUV). Recent observations of the far ultraviolet extinction variations even in diffuse clouds provides additional
criteria. These new criteria derive from the high degree of uniformity
of the shape of the far ultraviolet extinction simultaneous with large
variations in the amount of the FUV extinction relative to the strength
of the 220 nm (2175 Å) hump which, itself, correlates almost perfectly
with the visual extinction. It may be shown (Greenberg & Chlewicki,
1983; Chlewicki, 1985) that the MRN model predicts changes in the
position of the hump with varying FUV/hump ratios which is inconsistent
with the observed constancy of the hump position. Both the MRN and the
C-M,b models use graphite and silicate particles for the ultraviolet
extinction, and even for the average extinction neither of these is as
good as one would like. It is quite difficult to reproduce the FUV
extinction with silicates of known optical properties (Huffman & Stapp,
1973; Egan & Hilgeman, 1975).
    The difficulty with graphite for the hump is that the position
depends on both the shape and size of the particles (Gilra, 1972). The
latter is not a problem for the C-M,b model because it is already
required that all ultraviolet particles be very small. But, for the MRN

model the hump position depends on the particle size distribution which must be varied to account for the observed variations in the FUV extinction (Greenberg, 1985; also see Chlewicki, this volume). However, the observed degree of uniformity of this hump allows for little particle shape variation which in itself may not be an insurmountable problem but it is one which is bothersome. Also, graphite seems to be such a special form that less specialized forms of carbon have been proposed as alternatives: amorphous carbon (AC) (Borghesi, Bussoletti and Colangeli, 1985); quenched carbonaceous composite (QCC) (Sakata et al, 1983a). In addition, the O^{--} ion on magnesium oxide particles has been a suggested candidate to produce the 220 nm hump (McLean et al, 1982).

One critical difficulty with both amorphous carbon and QCC is that although they have peak absorptions at around 220 nm the positions differ substantially from the observed one. One might argue that the laboratory samples are not quite right in this respect but that in space a comparable type material with a better match to the hump is created. However there are two more basic constraints which are violated as seen in the sections on polarization and cosmic abundance.

Although we probably have to accept the possibility that neither graphite nor some more disordered form of carbon is suitable for the 220 nm hump, nevertheless the evidence exists that some sort of carbon is produced in stars and should appear in the I.S.M. (Czyzak & Santiago, 1973; Hecht et al, 1984). Furthermore the IRAS infrared data indicate the existence of dust particles which are substantially hotter than dielectric particles (de Vries and Le Poole, 1985). We suggest therefore that one should still consider AC or QCC as possible candidates for a portion of the I.S.M. particles.

Suppose that a fraction F of the extinction in the visual is produced, for example, by AC particles. Would this substantially modify our grain modelling at other wavelengths? To examine this let

$$\frac{c^{AC}(V)}{c^{CM}(V)} = F \qquad (1)$$

where $c^{AC} \equiv n^{AC} a_{AC}^2 Q_{AC}(V)$, $c^{CM} \equiv n^{CM} a_{CM}^2 Q_{CM}(V)$ are the extinction cross sections (per unit volume) of the AC and CM particles respectively. Using the data of Borghesi et al (1985) for $Q^{AC}(V)/a_{AC} = 6.43 \times 10^4 \, cm^{-1}$ and of Chlewicki (1985) for $Q^{CM}(V)/a_{CM} = 6.8 \times 10^4 \, cm^{-1}$ we get a volume ratio of AC to CM particles of $V^{AC}/V^{CM} \approx 1.06 \, F$. We now calculate the relative contributions of the AC and CM particles at, for example, 4 μm. If we assume a small value δ of the imaginary part of the index of refraction in the near IR for the organic mantles on the CM particles we get $Q^{CM}_{(4)}/a_{CM} \approx 2 \, \delta \, 10^4 \, cm^{-1}$ so that for $Q^{AC}_{(4)}/a_{AC} \approx \approx 8 \times 10^3 \, cm^{-1}$ we get

$$c^{AC}(4)/c^{CM}(4) = 0.4 \, F/\delta \qquad (2)$$

Since δ < 0.04 (Chlewicki 1985) we find that a large, or even, major part of the extinction in the near infrared (NIR) may be provided by the AC particles even if F is as small as 0.1. A similar result should apply also to QCC particles which may also contribute to the infrared emission features (Sakata et al, 1984). This would be a way of providing for the fact that the NIR extinction has always been difficult to match by dielectric grains either alone or with a small graphite component (Greenberg 1966, Hong & Greenberg, 1980) because the extinction cross sections of dielectrics tend to level off too rapidly at long wavelengths. The observed NIR wavelength dependence is possibly proportional to λ^{-1} (Davis et al, 1986) or perhaps $\lambda^{-1.85}$ (Landini et al, 1974) both of which are substantially steeper than the long wavelength dependence of $c^{CM}(\lambda)$ for weakly absorbing mantles. It should be pointed out that Jones and Hyland (1980) derive $A(\lambda) \sim \lambda^{-2.5}$ in the NIR which would be more representative of dielectric grains. Thus this important question of the actual functional form of $A(\lambda)$ in the near IR should be settled in order to provide an additional important criterion for grain models.

The contribution to the ultraviolet extinction by the AC particles is probably within acceptable limits (i.e. not too much distortion of the hump) for sufficiently small, but significant values of F. At $\lambda = 250$ nm one gets

$$\frac{c^{AC}(.25)}{c^{CM}(V)} = F \frac{Q^{AC}(.25)a_{AC}}{Q^{CM}(V)/a_{CM}} \simeq F \frac{1.5 \times 10^5}{6.8 \times 10^4} = 2.2 \, F \qquad (3)$$

Since the extinction at the hump relative to the visual is (Chlewicki 1985) $A(.22)/A(V) \simeq 3$ we see that the AC component with $F \simeq 0.1$ gives less than 20% contribution to the hump extinction and this is on the long wavelength side. Detailed calculations will determine the precise effects on the shape and position of the hump but it seems that a disordered carbon component in the ISM merits further consideration. I shall return to another possible reason for including disordered carbon particles in the discussion of grain temperature and emission properties in the FIR.

3. POLARIZATION

The average wavelength dependence of the polarization can be about equally matched by the MRN and the C-M,b models. The basic model differences show up in: (a) accomodation to variations in degree and shape of the polarization curve and (b) fraction of extinction particles providing the polarization; i.e. required degree of alignment of polarizing particles. Using the graphite and silicate power law size distribution, polarization is produced only by the silicate particles and to model the increase in the value of λ_{max} (position of maximum polarization) it is assumed that this results from particle coagulation which is represented by shifting the values of the maximum and minimum in the power law distribution towards larger values (Mathis, 1979; Mathis & Wallenhorst, 1984). The coagulation leads, at best, to an ill defined situation (spherical clumping, string clumping) in which increase in particle size is not necessarily accompanied by a change in the λ dependence of polarization (Greenberg 1985) and furthermore is a bit more complicated in its consequences than a mere change in upper and lower limits on the same power law (Aannestad and Greenberg 1984). In the C-M,b model the increase in λ_{max} is produced by accretion of extra

mantles of interstellar molecules which is automatically accompanied by
an appropriate change in the extinction shape. Furthermore it has been
shown by Aannestad and Greenberg (1984) that the variation of the width
of the polarization curve with λ_{max} more closely follows observation if
due to accretion than due to coagulation.

The circular polarization reversal at $\lambda = \lambda_{max}$ is a characteristic
of aligned dielectric grains. Both the pure silicates and the silicate-
organic refractory mantle particles have an absorption part of the index
of refraction which satisfies m" $\leq$ 0.05, a condition consistent with the
criterion of Martin (1974). We note here that the core mantle particles
with m"(550 nm) $\approx$ 0.05 have an albedo of α = 0.7 so that the total
albedo in the visual after dillution by the small graphite component
is α = 0.6 which is within the range 0.6-0.7 deduced from observation
(Savage & Mathis, 1979).

The degree of alignment of elongated dielectric particles - whether
pure silicate or silicate core - organic refractory mantle - required to
provide the observed polarization relative to extinction is given by

$$\frac{P}{A} = \frac{A_P R (P/A_P)_{max}}{A_{NP} + A_P} = 0.03 \qquad (4)$$

where A_P, A_{NP} are the extinction contributions of the polarizing and non
polarizing components at λ_{max}; $(P/A_P)_{max}$ is the maximum polarizing
capability of the polarizing particles; R is a reduction factor defining
the degree of disalignment. We assume that the only suitable alignment
mechanism is that of Purcell (1979) as shown by Cugnon (1985). A
reasonable estimate of $(P/A_P)_{max}$ is $\approx$ 0.12 (See Greenberg 1978; Mathis
and Wallenhorst 1981). Using $A_{NP} \approx 2A_P$ (Mathis and Wallenhorst 1981) one
finds R $\approx$ 0.75 suggesting the need for a high degree of alignment -
higher even than that which suprathermal (limited) spinup alignment
(SSA) provides (Aannestad and Greenberg, 1984). With no dilution by a
non polarizing component, SSA gives R $\approx$ 0.38 which gives (P/A) = 0.045.
This is a bit larger than the maximum observed but I find this a
comfortable margin because it allows for some randomness in the magnetic
field direction relative to a perfect 90° over a path length in the
cloud sufficient to reliably observe the amount of extinction producing
the polarization. It may be shown that the allowed randomness in B
direction is within about a 10° cone (Greenberg 1969). It is also worth
mentioning that cases of (P/A) > 0.03 have indeed been observed for
several (small) extinctions (Serkowski et al, 1975) but these apparent
anomalies were usually attributed to uncertainties in the determinations
of either A or P. Perhaps they were real after all, but in any case the
C-M,b model allows for a bit of imperfection in producing the required
polarization to extinction ratio.

4. COSMIC ABUNDANCE

It is only the abundant condensable atoms (O+C) which are adequate
to provide the visual extinction per hydrogen atom expressed by $N_H/E(B-V)$ = 5.9 x 10^{21} cm^{-2}. In the MRN model a major fraction ($\sim$ 80%) of the
carbon in the form of graphite provides the visual extinction while
silicates which involve the less abundant - by a factor of 10 - elements

(silicon, magnesium) provide the rest. In the C-M,b model the cores plus bare silicates consume all the silicates but the visual extinction is provided predominantly (in volume) by the organic refractory mantles which involve a bit less than half the available carbon (about half that of the MRN graphite). Including the graphite bare particles in the C-M,b distribution leads to a total carbon consumption not very different from MRN even though the two grain models are totally different in concept (Chlewicki, 1985). Thus, in both cases, carbon is the dominant element responsible for the visual extinction but in the MRN case it is in "metallic" form in the visual (m" ≃ 1) where in the C-M,b case it is in dielectric form (m" ≤ 0.05).

As already noted, materials other than graphite have been suggested for the hump. Both the AC and QCC materials have serious problems other than the lack of a proper position of the hump. The volume extinction efficiency factor for small (a ≤ 0.01 µm) graphite particles at the hump is $(C/vol)_G$ = 9 x 10^5 cm^{-1} which, for a mass density of 2.25 gm cm^{-3}, leads to a mass extinction efficiency of 4 x 10^5 cm^2 gm^{-1}. For AC one has $(C/vol)_{AC}$ = 1.2 x 10^5 cm^{-1} (Borghesi et al, 1985) and for Q.C.C. one has (C/vol) = 2.5 x 10^5 cm^{-1} (as deduced from Sakata et al (1983b) by assuming a mass density of 1.55 gm cm^{-3}). Just to provide the extra extinction (absorption) above the saturation level for the core-mantle particles at 220 nm requires 32% of the available carbon as graphite. Thus the minimum amounts of AC and QCC required are $\frac{4}{0.6}$ x 32% > 200% and $\frac{4}{1.6}$ x 32% ≃ 80% respectively, the first of which is clearly outside the cosmic abundance constraint and the latter is difficult to reconcile with the need for carbon to provide the CO in molecular clouds. This is perhaps not totally surprising for the reason that the extinction curves for both AC and QCC show, along with the hump, very significant contributions in the visual. Using the data by Borghesi et al (1985) gives $[C(hump)/ C(V)]_{AC}$ ≃ 2.5 and from Sakata et al (1983b) gives $[C(hump)/C(V)]_{QCC}$ ≃ 5 as compared with the observed diffuse cloud mean value of A(220)/A(V) = 3.2. For small graphite $[C(220)/ C(V)]_G$ ≥ 10. We see that for AC the entire visual extinction (even a bit more) is produced in order to give the hump and therefore allows for no dielectric component to produce the linear polarization. The Q.C.C. produces a very substantial contribution to the FUV extinction so that the decoupling of the hump and the FUV appears to be prevented. We are thus left at this time with graphite as the only viable candidate for producing the hump by small solid particles of carbon. However, recent work in Laboratory Astrophysics in Leiden indicates the possibility that certain types of carbon molecules are strong candidates for the hump not only because they provide a uniform hump position and shape (no size effects) but also because they require far less carbon than graphite for the same large strength (van der Zwet et al, 1985)

Referring back to the small amount of the AC or QC component which may be needed to provide the NIR extinction we find that the fractional cosmic abundance is ≃ 0.7 F x 2.00 ≃ 1.4 F. Thus an acceptable amount of carbon would be consumed by the AC particles, namely ≃ 15%. The molecular produced hump plus the AC particles together may consume no more (and possibly less) carbon than the "classical" graphite particles.

5. GRAIN EVOLUTION-CYCLING.

The spectrosopic evidence that grains in diffuse clouds are a product of sequences of stages involving residence and processing in both molecular and diffuse clouds has been building up over the past years. The analysis of these processes has been supported by laboratory studies on the photochemical processing (photoprocessing) of interstellar type ices by vacuum ultraviolet radiation. I shall limit myself here to a few key observations and experiments which provide the theoretical basis for the core-organic refractory mantle model of grains in diffuse clouds.

a) Growth of mantles in molecular clouds - the ice bands.

The growth of mantles in molecular clouds had long been limited to the example of solid ice in B.N. which is a special case not only because of the very deep extinction but also because it is a proto-stellar source. We now have many examples of "ordinary" molecular clouds in which grain mantles of ices are apparent (Whittet et al, 1983). The Taurus cloud exhibits H_2O ice bands at moderate extinctions which already indicate grain growth and evolution. The ice band for the star Elias 16 which stands behind the cloud is an example of a relatively unprocessed mantle (Van de Bult et al, 1985). It may be shown that it is representative of grains which have formed in moderately dense clouds leading to mantles whose major component is H_2O ($\geq 60\%$ is H_2O with at most ~ 10% NH_3), (d'Hendecourt et al, 1985). The star HLTau, on the other hand provides an example of grains which have undergone some form of heat processing, i.e. the ice mantle has been warmed above its formation temperature of ~ 10 K.

b) Photochemical processing of grain mantles.

The compact source W33A exhibits some very clear samples of photo-processing as well as heat processing. The laboratory studies of photo-processing of various mixtures which include oxygen, carbon and nitrogen always show a feature at 4.675 μm along side the CO stretch at 4.62 μm which can be attributed to the formation of a molecule containing the cyanogen ($C \equiv N$) group. The actual molecule is not yet identified except as $X-C \equiv N$. More recently, we have included sulfur in the form of H_2S in our mixtures and have shown that not only is S_2 formed (Grim and Greenberg, this volume) but also, in the warmed up sample we see a strong feature at 4.9 μm attributed to OCS (Geballe et al, 1985). This feature, along with a weak band at 3.9 μm (due to H_2S) has been detected in W33A and further confirms along with the $X-C \equiv N$ and CO absorption the concept of photoprocessed grain mantles at some stages of molecular cloud evolution (see d'Hendecourt, this volume).

c) End product of photoprocessing - organic residue mantles.

Already 15 years ago it was shown that large organic molecules were produced in ultraviolet irradiated ices (Greenberg et al, 1972, Greenberg, 1973) and this led to speculation that such complex molecules could provide a substantial portion of the grain mantle composition. In the Leiden laboratory we have produced organic residues which result from photoprocessing and warmup to room temperature of any mixture resembling interstellar type ices. As was demonstrated by the first crude experiments and now confirmed by many laboratory examples the rate of formation of organic residues is adequate to provide the observed thickness of grain mantles in the diffuse clouds. This has been demonstrated quantitatively for the galactic center source IRS7 as well as for other galactic center sources (Schutte and Greenberg, this volume). The volume of organic refractory is obtained by comparing the observed 3.4 μm optical depth to the galactic center of 0.22 with the absorptivity of our laboratory created "yellow stuff".

$$(\tau_{3.4}^{G.C.})_{obs.} = 0.22 = (\tau_{O.R.}^{G.C.})_{calc.} = V_{O.R.} \, s_{O.R.} \, [\frac{C(3.4)}{m_{O.R.}}]_{lab} \qquad (5)$$

where $V_{O.R.}$ = line of sight volume density of the organic residue mantles; $s_{O.R.}$ = mass density of the residue = 1.4 gm cm^{-3}; $[C(3.4)/m_{O.R.}]_{lab}$ = residue absorption per unit mass at 3.4 μm ≈ 440 cm^2 gm^{-1}. Substituting in Eq. 5 we get

$$V_{O.R.} = 3.57 \times 10^{-4} \text{ cm} \rightarrow \text{l.o.s. mass} = 5 \times 10^{-4} \text{ gm } cm^{-2} \qquad (6)$$

This is to be compared with the mass of silicates (in all forms) to the G.C., from the G.C. derived extinction of A(V) = 40 mag and a standard τ(9.7)/A(V). We get column density of silicon atoms to be $N_{Si} \approx 0.8 \times 0.32 \times 10^{-4} \times 10^{21} \times 40 = 24 \times 10^{17} cm^{-2}$ and a total mass of silicates (M = 116, s = 3.5 gm cm^{-3}), $M_{sil} = 4.5 \times 10^{-4}$ gm cm^{-2}, which is about that of the organic refractory. From a positive observational point of view it becomes difficult to conceive of a grain model which does not include the O.R. in the diffuse cloud dust. Even though the O.R. is made in molecular clouds, its observability is generally restricted to the diffuse cloud phase after all volatiles such as H_2O are evaporated or eroded away (Greenberg 1982a).

The rate of production of the O.R. in the M.C. phase is ≈ 1.5×10^{-41} gm $cm^{-3} s^{-1}$ (Schutte and Greenberg, this volume) which, when averaged over the M.C./D.C. phase, is about 0.8×10^{-41} g $cm^{-3} s^{-1}$ (assuming one half the total interstellar mass in M.C. and one half in D.C.). On the other hand the production rate for silicates based on the maximum amount of silicates (full cosmic abundance) accompanying a mass loss of 1.7×10^{-35} gm $cm^{-3} yr^{-1}$ from all M stars (Kwok, 1980) is $\frac{dm_{Sil}}{dt} < 10^{-45}$ g $cm^{-3} s^{-1}$ or about 10^4 times less than the O.R. mean production rate. The consequence of this difference on the maintenance of dust populations will be discussed in the next section.

6. STABILITY OF GRAIN POPULATIONS

It is generally accepted that such refractory particles as silicates and graphite are formed in stellar atmospheres as opposed to the formation of volatiles and organic refractory constituents which are created in the molecular clouds. All particle populations are eroded or destroyed in the diffuse cloud phase. The observed uniformity of the extinction in diffuse clouds implies a high degree of stability in the diffuse cloud grain population. Accepting the time scales for the destruction mechanisms proposed by Draine and Salpeter (1979a,b) and by others, one can show that <u>any</u> grain population is too strongly modified on time scales of the order of 10^8 yrs to maintain the required optical stability <u>unless</u> we are seeing a steady state situation; i.e. a continuous feed-in or replenishment of destroyed particles. Let us examine, in turn, the kinds of particles produced in stellar atmospheres and in the interstellar medium.

The maximum lifetime for silicate grains in the D.C. (DS 1979b) is $\tau_{Sil} \approx 4 \times 10^8$ yrs. which, converts to a mass loss rate (starting with a mean density of cosmic abundance $\rho_{Sil} \approx 6 \times 10^{-27} gm\ cm^{-3}$) of $dm_{Sil}/dt = -5 \times 10^{-43} gm\ cm^{-3} s^{-1}$. This is at least 100 times larger than the production rate. The question one may address to the MRN model is how is it possible to maintain the specific size distributions needed to produce the extinction or polarization during the lifetime of a diffuse cloud if the destruction rate is so high. The same question may be posed for the graphite population whether in the MRN or the C-M,b model. On the other hand if the hump is produced by very stable <u>molecules</u> the grain destruction mechanisms do not apply.

The mass creation rate for the O.R. is $\frac{d\rho_{O.R.}}{dt} \approx 6 \times 10^{-42} gm\ cm^{-3} s^{-1}$ (~ 1/3 of that obtained in section 5 where account was taken of the possibility of O.R. mantles using about 30% of the condensables). Since the mean mass density of the O.R. and the silicates is about the same, we see that it is possible to maintain a steady state of O.R. mantles allowing a destruction rate of ~ 10 times that for silicates. It had earlier been estimated that the erosion rate for organic residues is comparable with that of silicates (DS, 1979b; Greenberg, 1982b) so that in view of the remaining uncertainties in both the creation and destruction rates the balance required to achieve a steady state is not inconsistent with the data.

7. THE CARBON COMPONENT AND FAR INFRARED GRAIN PROPERTIES

The far infrared properties of the interstellar dust are determined by the relative volumes of the various grain components. One of the important differences between the MRN model and the trimodal model is in the relative proportion of dieletric to metallic (graphite) particles.

In the MRN model $n_C \approx [n_C]_{CA}$ and $n_{Si} \approx [n_{Si}]_{CA}$; i.e., carbon and silicon are fully accounted for in the dust (carbon being perhaps $\approx 80\%$ in dust). Since optical variations are to be accounted for by coagulation rather than accretion with rare exceptions (Mathis and Wallenhorst, 1981) we may deduce that in both diffuse and molecular clouds

$$\frac{V_C}{V_{Si}} = \frac{M_C}{M_{Si}} \frac{s_{Sil}}{s_C} \left[\frac{N_C}{N_{Si}}\right] \approx 1.3 = \left(\frac{metallic}{dielectric}\right)_{MRN} \qquad (7)$$

for reasonable values of the molecular weights and densities.

Quite a different result is obtained for the C-M,b model. We use the fact that the predominant dielectric component in the diffuse clouds is the mantle ($_{core-mantle} \approx$ mantle). From Greenberg (1982a) we deduce that $n_C \approx 0.27\ [n_C]_{CA}$ and $n_{O+C+N} \approx 0.22\ \{[n_O]_{CA} + [n_C]_{CA} + [n_N]_{CA}\}$. This gives

$$\frac{V_C}{V_M} = \frac{M_C}{M_{mantle}} \frac{s_{mantle}}{s_C} \frac{0.27\ [n_C]_{CA}}{0.22\ [n_{O+C+N}]_{CA}} \approx 0.1 = \left\{\frac{metallic}{dielectric}\right\}_{C-M,b} \quad (8)$$

and less in molecular clouds.

One of the problems which the IRAS data presents in the diffuse cloud interpretation is the existence of a high temperature (T ≥ 30K) component. It turns out that the graphite and silicate and C-M particles have equilibrium temperatures which are significantly less than this (Draine & Lee, 1984, Greenberg, 1971).

One way out of the dilemma, suggested by Draine & Lee (1984) is to ascribe the extra radiation at the shorter wavelengths as due to non equilibrium processes (Greenberg, 1968; Greenberg and Hong, 1976; Purcell, 1976).

As we have already shown the value of F = 0.1 chosen for the ratio of extinction in the visual by AC and C-M particles leads to a volume ratio of AC to C-M of ≈ 0.1. This is just about the same as the value we derived above for the graphite component in the C-M,b model so that small AC or QCC particles could indeed provide an important contribution to the FIR emission if they are as hot as the "graphite" metallic grains calculated by Greenberg (1971). It will be necessary to study the far infrared emissivity of various carbon constituents to resolve this problem but this is an example of where IRAS data will be helpful in establishing a particular carbon grain component.

8. OPTICAL PROPERTIES OF COAGULATED INTERSTELLAR DUST

Optical modelling of the zodiacal light was attempted for what were called birds' nest particles (Greenberg and Gustafson, 1981) and, more recently, Greenberg (1986) has derived some optical properties for a model of comet dust aggregates derived from meteor densities (Figure 1). The albedo of a porous structure made up of a loose tangle of submicron sized grains is much lower than that of the individual components. Using $\alpha = 0.6$ and $g \geq 0.8$ (> 0.8 for fully accreted grain mantles) as the individual interstellar grain albedo and asymmetry factor in the visual leads to a comet dust (Bond) albedo of $R_b \approx \alpha (1-g)/2 \approx 0.05$, which is smaller than normally assumed for zodiacal light particles. The implication of this is that comet dust as it first appears in the interplanetary medium hardly scatters visible light but it will emit relatively strongly in the infrared. Since comet dust is initially more spherically distributed in the solar system than the visible component of the zodiacal light one should expect to find a substantial infrared emission above and below the ecliptic and in the outer solar system which is not correlated with the scattered light. The IRAS data may possibly be used to look for this effect.

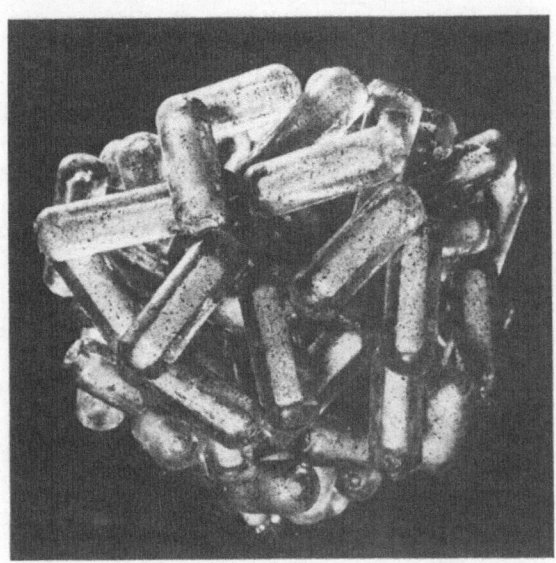

Fig.1: A model of randomly coagulated core-mantle dust grains. Shown is an ensemble of 91 particles incorporating within their outer icy mantles the very small particles of the interstellar dust population accreted in the final stages of condensation. The degree of volumetric packing is 0.4 (60 percent open space) and the overall dimension is scaled to about 4 microns.

REFERENCES

Aannestad, P.A. Greenberg, J.M. 1983, Astrophys. J. 272, 551.

Borghesi, A., Bussoletti, E., Colangeli, L. 1985, Astron. Astrophys. 142, 225-231.

Van de Bult, C.E.P.M., Greenberg, J.M., Whittet, D.C.B. 1985, MNRAS 214, 289.

Chlewicki, G.C. 1985, Ph. D. Thesis Leiden.

Cugnon, P. 1985, Astron. Astrophys. 152.

Czyzak, S.J., Santiago, J.J. 1973, Astrophys. Space Sci 23, 443.

Davis, D.S., Larson, H.P. and Hofmann, R. 1986, Astrophys. J. in press.

Draine, B.T., Salpeter, E.E. 1979a, Astrophys. J. 230, 106.

Draine, B.T., Salpeter E.E. 1979b, Astrophys. J. 231, 438

Draine, B.T., Lee, H.M. 1984, Astrophys. J. 285, 89.

Egan, W.G., Hilgeman, T. 1975, Astrophys. J. 80, 587.

Geballe, T.R., Baas, F., Greenberg, J.M., Schutte, W. 1985, Astron. Astrophys. 146, L6.

Gilra, D.P. 1972, in "The Scientific Results from the Orbiting Astronomical Observatory OAO-2", ed. A.D. Code, NASA SP-310, 295.

Greenberg, J.M. 1966, in "Spectral Classification and Multicolour Photometry", Willner Brothers, Birkenhead, 291.

Greenberg, J.M. 1968, in "Stars and Stellar Systems, Vol VII: Nebulae and Interstellar Matter", eds. B.M. Middlehurst and L.H. Aller, University of Chicago Press, 221.

Greenberg, J.M. 1969, Physica 41, 67.

Greenberg, J.M. 1971, Astron. Astrophys. 12, 240.

Greenberg, J.M., Yencha, A.J., Gorbett, J.W., Frisch, H.L. 1972, Soc. Roy. des Sciences de Liège 6e Serie tome III, 425.

Greenberg, J.M. 1973, in "Molecules in the Galactic Environment", eds. M.A. Gordon and L.E. Snyder (J. Wiley and Sons), 93.

Greenberg, J.M. Hong, S.S. 1976, in "Far Infrared Astronomy", ed. M. Rowan-Robinson, Pergamon Press, 299.

Greenberg, J.M. 1978, Interstellar Dust in Cosmic Dust, ed. J.A.M. McDonnell, Wiley, Chichester, 187.

Greenberg, J.M. Gustafson, B.A.S. 19881, Astron. Astrophys. 93, 35.

Greenberg, J.M. 1982a, in "Comets", ed. L. Wilkening (U. of Arizona Press) 131.

Landini, M., Natta, A., Oliva, E., Salinari, P., Moorwood, A.F.M. 1984, Astron. Astrophys. 134, 284.

Greenberg, J.M. 1982b, in "Submillimetre Wave Astronomy", eds. J.P. Phillips, J. Beckman (Cambridge U. Press) 261.

Greenberg, J.M., Chlewicki, G.C. 1983, Astrophys. J. 272, 563.

Greenberg, J.M. 1985, in "Birth and Infancy of Stars", Les Houches, Session SLI 1983, eds. R. Lucas, A. Omont and R. Stora, Elsevier Science Publishers BV, 139.

Grim, R., Greenberg, J.M. these proceedings, page 225.

Van der Zwet, G., de Groot, M., Baas, F., Greenberg, J.M. 1985, in preparation.

Hecht, J.H., Holm, A.V., Donn, B., Wu, C.C; 1984, Astrophys. J. 280, 2.

d'Hendecourt, L.B. 1984, Ph. D. Dissertation, University of Leiden.

d'Hendecourt, L.B., Allamandola, L.J., Greenberg, J.M. 1985, Astron. Astrophys. 152, 130.

d'Hendecourt, L.B. 1986, This volume.

Hong, S.S., Greenberg, J.M. 1980, Astron. Astrophys. 88, 189.

Huffman, D.R., Stapp, J.L. 1973, in "Interstellar Dust and Related Topics", eds. J.M. Greenberg and H.C. v.d. Hulst, Dordrecht, Reidel, 297.

Jones, T.J., Hyland A.R. 1980, MNRAS 192, 359.

Kwok, S. 1980, J.R. Astr. Soc. Canada 74 (4) 216.

Landini, M., Natta, A., Oliva, E., Salinari, P., Moorwood, A.F.M. 1984, Astron. Astrophys. 134, 284.

MacLean, S., Duley, W.W. and Millar, T.J. 1982, Astrophys. J. 256, L61.

Martin, P.G. 1974, Astrophys. J. 187, 461.

Mathis, J.S., Rumpl, W., Nordsieck, K.H. 1977, Astrophys. J. 217, 425.

Mathis, J.S. 1979, Astrophys. J. 232, 747.

Mathis, J.S., Wallenhorst, S.G. 1981, Astrophys. J. 244, 483.

Purcell, E.M. 1976, Astrophys. J. 206, 685.

Purcell, E.M. 1979, Astrophys. J. 231, 404.

Sakata, A., Wada, S., Okutsu, Y., Shintana, H., Nakada, Y. 1983a, Nature 301, 493.

Sakata, A., Wada, S., Tanabé, T., Onaka, T. 1983b, Preprint.

Savage, B.D., Mathis, J.S. 1979, Ann. Rev. Astron. Astrophys. 17, 73.

Schutte, W., Greenberg, J.M. 1986, these proceedings, page 229.

Serkowski, K., Mathewson, D.S. Ford, V.L. 1975, Astrophys. J. 196, 361.

de Vries, C.P., Le Poole, R.S. Astron. Astrophys. 1985, 145, L7.

Whittet, D.C.B., Bode, M.F., Longmore, A.J., Baines, D.W.T., Evans, A. 1983, Nature 303, 218.

INFRARED EXTINCTION IN MOLECULAR CLOUDS: THE FORM OF THE CURVE IN ORION

R. Hofmann[1], D. S. Davis[2], and H. P. Larson[1,3]

[1]Max-Planck-Institut für extraterrestrische Physik,
 8046 Garching
[2]Steward Observatory, University of Arizona
[3]Lunar and Planetary Laboratory, University of Arizona

Infrared extinction in dense molecular clouds is not well established by direct measurements. Moreover, there is little a priori justification for assuming that "standard" extinction curves should apply (i.e. those appropriate to the more diffuse ISM), or even that any two distinct dense regions should exhibit similar extinction properties. This situation may limit our understanding of the intrinsic spectral properties of some highly obscured IR objects. We have therefore determined an empirical extinction curve for the Orion molecular cloud in the near-IR spectral region using shock-excited H_2 emission lines as the "background" source. Comparison of the observed line intensities with theoretical predictions yielded the wavelength-dependent extinction measurements. This spectroscopic method has significant advantages over broadband photometric techniques: the intrinsic H_2 line intensities are easily modeled, and the H_2 emission is readily distinguished from independent sources of continuum emission. We used the Kuiper Airborne Observatory to minimize interference by telluric H_2O lines, and we used a Fourier spectrometer to provide simultaneous coverage of all H_2 lines. The observed spectrum contains 17 H_2 lines in the v=1-0 band covering a factor of 2 in wavelength (1.8-3.4 μm) at a spectral resolution of 1 cm^{-1}.

We represented the near-IR extinction curve in Orion as a power law of the form $A_\lambda \sim \lambda^{-\beta}$. The parameter β, which measures the curvature of the extinction law, was fit to the observed H_2 line intensity distribution in a weighted least-squares calculation. Other free parameters that were also fit in this calculation included the rotational temperature T=1600+100 K and the ortho/para H_2 ratio g_J=2.9+1.0. Our best-fit extinction curve is presented in Figure 1. The value of β, 1.04+0.08, indicates that the extinction law at near-IR wavelengths in Orion is not statistically different from a simple λ^{-1} curve. Two deviations from the curve cannot be explained by experimental error. First, the H_2 Q(6) and Q(7) lines near 4000 cm^{-1} are too intense at the best-fit temperature of 1600 K. This discrepancy may be removed by invoking a two temperature model in which a few percent of the excited H_2 is at a substantially higher temperature (~2100 K). The absence of the H_2 O(4) line at 3329 cm^{-1} presents a much more striking deviation. The excess extinction at 3 μm

F. P. Israel (ed.), Light on Dark Matter, 189–190.
© *1986 by D. Reidel Publishing Company.*

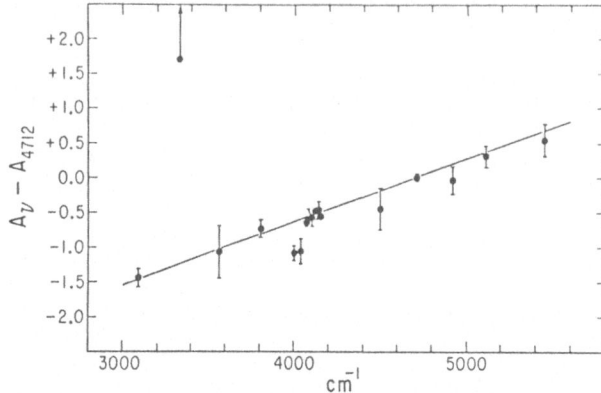

Figure 1. The IR extinc-
tion curve for Orion determined by airborne observations of H_2 line emission.

is most likely due to ice absorption in the foreground medium. Our lower
limit to the optical depth in the 3 μm ice absorprion is about 2.8,
which suggests that the H_2 source is at least as deep in the Orion molecular cloud as is BN.

The linear form of the near-IR extinction curve in Orion may be an
intrinsic property of the composition and size distribution of grains in
molecular clouds. This spectral signature is distinctly different from
that of the more diffuse ISM to which "standard" extinction curves apply.
This difference may be judged by comparing β values for representative
standard curves (β=1.5-2.0) with that for Orion (β=1.0). The astronom-
ically relevant grain core materials are silicate and graphite, with
mantles containing ices of condensed volatiles and products of secondary
chemical processes. It is significant that both silicates and graphite
show λ^{-1} extinction behavior for particle sizes of the order of 1 μm,
substantally larger than the interstellar mean (<0.1 μm; see Draine and
Lee, 1984, Ap. J. 285, 89 for discussion of the optical properties of
dust grains). We therefore suggest that the linear extinction curve in
Orion implies grain sizes much larger than in the diffuse ISM.

Extrapolations from our linear curve cannot be predicted with certainty.
However, the Orion cloud contains a large abundance of silicates and the
spectral properties of this component should influence the Orion curve.
The absorption cross-section for 1 μm silicate grains varies as λ^{-1} out
to 8 μm, where the "10 μm" silicate absorption begins, and as λ^{-2} beyond
about 11 μm. The possible influence of graphite includes linear behavior
to about 7 μm, but a simple power law is inadequate to describe its
spectral properties at still longer wavelengths. Thus, the linear curve
may apply throughout the near-IR, except in regions of strong molecular
absorption, and beyond about 10 μm the curve should acquire an approximate λ^{-2} character. From this expectation we estimate that the absolute
extinction in Orion at the H_2 S(1) line (2.12 μm) is 2-4 mag.

This research was supported by NASA Grant NGR 03-002-332 (HPL and DSD),
the Alexander von Humboldt-Stiftung (HPL), and by a NATO research grant
from the DAAD (RH).

ULTRAVIOLET EXTINCTION AS A KEY TO GRAIN OPTICAL PROPERTIES IN THE INFRARED AND ULTRAVIOLET

G. Chlewicki and J. Mayo Greenberg
Laboratory Astrophysics, University of Leiden,
The Netherlands

ABSTRACT. The paper summarizes the results of a large study of UV extinction based on IUE spectra. The analysis of variations in UV extinction between individual lines of sight is used to separate the contributions of individual grain populations and to constrain their properties. The temperatures of particles whose possible presence in the interstellar medium is indicated by grain models are compared with recent IRAS results.

1. Statistical analysis of extinction

In this paper we summarize the results of a large study of UV extinction carried out jointly by institutes in Italy and in the Netherlands, in which the analysis of the data concentrated on using variations in extinction rather than the average pattern to derive constraints on the properties of grain populations in the interstellar medium. In the final section, we shall also discuss the implications of grain models based on extinction for the interpretation of infrared data obtained by IRAS.

Following earlier investigations of extinction by the participants of this project (e. g., Aiello et al., 1982; Barsella et al., 1982; Greenberg and Chlewicki, 1983, hereafter Paper I), two samples of low-resolution IUE spectra of early-type stars were selected independently in Florence and in Leiden and were subsequently merged into a single, uniformly reduced data base, which consisted of ~120 extinction curves covering the wavelength range 1250 < λ < 3200A. A subset of ~40 objects located in dense clouds, including well-known nearby complexes (the ρ Oph cloud, the Orion Nebula and its surroundings), was identified in the sample on the basis of data other than extinction curves, such as radio observations of molecules or the presence of high surface brightness reflection nebulosity. Full details of data reduction and a discussion of the role played by accurate spectral classification will be given in forthcoming publications (Aiello et al., 1985; Chlewicki, 1985; Chlewicki and Greenberg, 1985).

Figs. 1a and 1b show the correlation between the visual extinc-

F. P. Israel (ed.), Light on Dark Matter, 191–196.
© *1986 by D. Reidel Publishing Company.*

tion and two main features in UV extinction curves: the 2200A hump
and the rise at λ < 1700A, both of which are characterized here by
colour excesses (extinction differences) at two selected wavelengths.
With errors in the UV indices estimated to be 1σ ≈ 0.15 mag for both
the hump and the FUV rise, the difference in the degree of correlation is immediately apparent from the scatter in each diagram. In the
diffuse medium, the correlation coefficient, r, exceeds unity for the
2200A hump, and is only 0.85 for the FUV rise, which confirms the
result obtained for a small sample of stars in Paper I. The scatter
is much greater for molecular cloud objects, presumably as a result
of differences in the processing of grains in individual clouds.

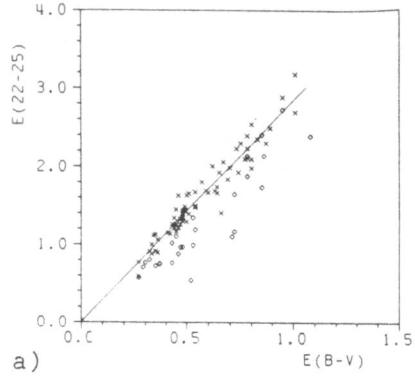
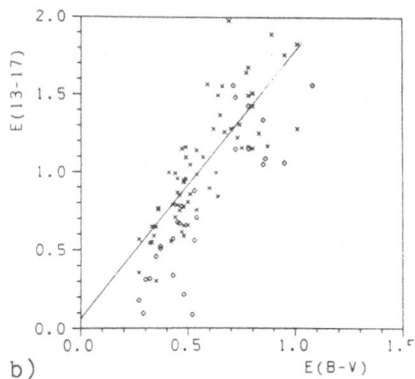

a)

b)

Fig. 1. Correlation of ultraviolet and visual extinction.
Crosses: diffuse medium lines of sight; diamonds: objects embedded in dense clouds. a) 2200A hump; the colour excess is defined
as the difference in extinction between the peak of the hump and
$\lambda = 2500A$: $E_{22-25} = A_{max} - A_{2500}$. b) FUV rise in extinction:
$E_{13-17} = A_{1300} - A_{1700}$.

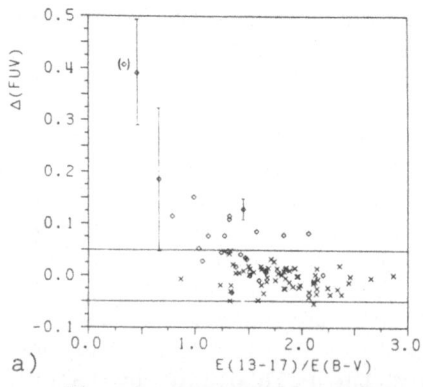
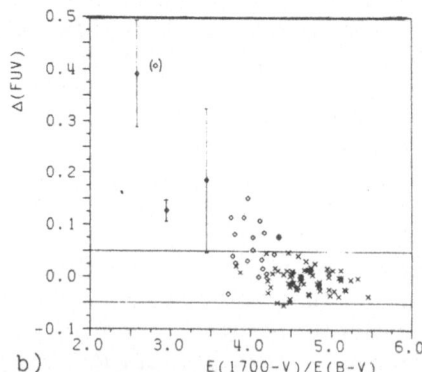

a)

b)

Fig. 2. Far UV curvature relative to the diffuse medium average
(see text). Symbols as in Fig.1. Horizontal lines indicate typical errors for diffuse medium objects. Individual error bars
and parentheses mark particularly uncertain positions of molecular cloud stars. a) Correlation with the FUV rise in extinction.
b) Correlation with extinction at the 1700A minimum.

More information about the nature of the particles responsible for the FUV rise in extinction and the 2200A hump can be derived from the shape of the curve at $\lambda < 1700A$, which becomes apparent in the FUV normalization, i. e., when $A(1300) = 1$ and $A(1700) = 0$ (see Figs. 3b and 5b in the following section). Figs. 2a and 2b show how the mean deviation of each curve from the diffuse medium average (both curves in the FUV normalization) at $1300 < \lambda < 1700A$, which measures the relative curvature at FUV wavelengths, correlates with other characteristics of FUV extinction. Within rather large errors, the curvature remains constant in the diffuse medium and only increases in those molecular cloud lines of sight which are characterized by very low values of extinction at the FUV minimum in the curve ($\lambda \approx 1700A$). As will be shown in the following section, the large scatter of points representing molecular cloud lines of sight in Fig. 2b is due to the fact that the FUV curvature reflects the properties of two grain populations responsible for the 2200A hump and the FUV rise. The interpretation of the apparently constant curvature in the diffuse medium is hampered by the uncertainties in both the observational measurements and the theoretical description of grain destruction in interstellar clouds. Within these uncertainties, the results shown in Figs. 2a and 2b are consistent not only with the stringent requirement that the particles responsible for the FUV rise must be almost pure absorbers (Paper I), but may also be explained if the FUV rise is partly due to scattering. However, indirect support for the concept of small, almost purely absorbing silicate particles (a < 0.01 μm) as a source of the FUV rise can be derived from the analysis of individual extinction curves (see below).

2. Wavelength dependence of extinction. Grain models.

Some of the most important extinction curves have been collected in Figs. 3a and 3b. The average curve for the diffuse medium subset of the sample represents the pattern seen in most lines of sight independent of the galactic location of the objects.

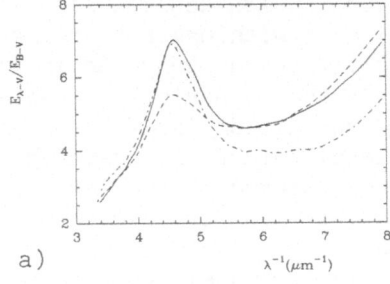 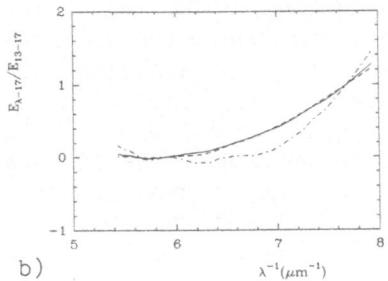

Fig. 3. Extinction curves for lines of sight with $A_V/E_{B-V} \approx 3.1$. Solid line: average for all diffuse medium objects; dashed line: HD 73882; dash-dot: average for 4 stars in Sgr OB1. a) Visual normalization: $A_V = 0$, $A_B = 1$. b) FUV normalization: $A_{1700} = 0$, $A_{1300} = 1$.

Figs. 3a and 3b also contain two apparently peculiar curves, which are particularly important in modelling grains. In spite of deviations from the average in the UV, in both cases the values of $R = A_V/E_{B-V}$, and consequently the properties of grains responsible for visual extinction, are normal (R ≈ 3.1). Therefore, the weak 2200A hump in HD 73882 and the reduced FUV rise in Sgr OB1 define the flexibility in setting the ratio of UV and visual extinction, which grain models must allow while preserving unchanged properties of grains responsible for visual extinction. The subtraction of the Sgr OB1 pattern from a curve characterized by strong FUV extinction directly determines the extinction curve for the population of particles responsible for FUV rise (Fig. 4). A nearly linear rise in the extinction due to the FUV population of grains is qualitatively consistent with the properties of small silicate particles (a < 0.01 μm) with the optical constants similar to those of olivine (Huffman and Stapp, 1973).

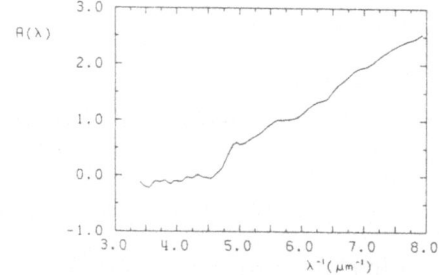

Fig. 4. Wavelength dependence of extinction due to particles responsible for the FUV rise obtained by subtracting extinction curves for two associations, Cep OB2 and Sgr OB1. Units of extinction are arbitrary.

Preliminary model calculations carried out as part of this study were intended to test the validity of the silicate + graphite concept of Mathis et al. (1977, MRN) and the three-population model (Greenberg and Hong, 1980), which in the modified version assumed here includes larger particles consisting of silicates with organic refractory mantles in addition to smaller silicate and graphite grains (Greenberg, 1982). The calculations indicate that spherical particles of graphite do not provide a convincing explanation for all the observed properties of the 2200A hump. A size distribution spanning a wide range of sizes of graphite spheres assumed in the MRN model leads to shifts in the position of the feature when its strength is reduced, which contradicts the observations. If only small particles of graphite (a ≈ 0.01 - 0.02 μm, three-population model) are present, their relatively strong FUV extinction makes it impossible to match the shape of curves with a reduced FUV rise, such as Sgr OB1.
    An alternative explanation for the hump may be offered by several other forms of carbon presumably present in the interstellar medium, which are reviewed below.

1. Non-spherical particles of graphite. Transitions which in small
spheres are responsible for the early onset of absorption in the FUV,
are only allowed for light polarized along the c-axis in strongly an-
isotropic crystals of graphite. Their contribution could be reduced,
if the particles had the shape of flakes or thin discs. Such a
change of shape, however, shifts the near-UV peak to ~2800A, and the
restoration of the observed position of the feature would require a
modification of the optical constants (possibly because of the existence of foreign atoms in the lattice), which has not been demonstrated experimentally.
2. Amorphous (Borghesi et al., 1985) or polymorphic (QCC, Sakata et
al., 1984) carbon. The absorption at the 2200A peak in a small graphite sphere is equivalent to a cross-section per atom of o_C =
$8.2*10^{-18} cm^2$; the corresponding numbers derived from published data
for amorphous carbon and the QCC (Sakata et al., 1985), are at least
3 times lower. In order to explain the strength of the 2200A hump,
amorphous or polymorphic materials would therefore require ~80% of
cosmic carbon, rather than ~25% needed for graphite, which exceeds
the observed upper limit of 50-60% (York et al., 1981), and may also
make it difficult to explain the abundances of molecules observed in
dense clouds (abundance calculations are based on numbers derived by
Greenberg, 1982 and Chlewicki, 1985).
3. Organic molecules in the gas phase. Although many species have
strong absorptions in the range 2000 < λ < 3000A, specific examples
of molecules that may explain the position and shape of the feature
have not been shown in the laboratory.

As a result of the stronly reduced FUV rise, the Sgr OB1 curve
(Fig. 3a) reveals the existence of a plateau centred at ~1700A, which
from the analysis of other curves appears to be associated with the
2200A hump and may therefore be important in identifying its carrier.
This new feature has no counterpart in any existing theoretical model
of the 2200A hump.

3. IRAS observations of interstellar grains

One of the most surprising early results derived from the
analysis of IRAS data is the consistently high 60/100 μm colour temperature in the diffuse medium with typical values $T_{60/100}$ > 25K (Low
et al., 1984; de Vries and le Poole, 1985; Walker, 1985). Here we
shall discuss only the hypothesis which attributes the 60 μm excess
to high equilibrium temperatures for one of the grain populations.

Table 1 illustrates the range of grain temperatures that can obtained
for particles most commonly included in grain models. For spherical
particles of both silicate and graphitic composition, the diffuse
medium temperature is below 20K. High temperatures required by IRAS
observations could, however, be achieved for thin (axial ratio of 10)
graphite discs. Table 1 also contains numbers calculated for a radiation field simulating the conditions in the cloud surrounding α Cam,
whose analysis by de Vries (1985) reveals the likely presence of two
grain populations with temperatures of 35K and 70K. It is evident
that non-spherical graphite grains do not provide an adequate expla-

nation for the 70K component. The 35K particles could be identified with large dielectric spheres (a > 0.1 μm), and it is important to note that these particles, which certainly contain a large fraction of the total volume of grains, will not be visible to IRAS in the diffuse medium as a result of their low temperatures (T < 15K).

Table 1. Grain temperatures for various particle populations.

| | Radiation field | |
Type of particle	Diffuse medium	α Cam
Silicate or core-mantle, a=0.15μm	12	35
Silicate, a=0.01μm	13.9	47.1
Graphite sphere, a=0.01μm	18.6	44.7
Graphite disc, a=0.01μm, a/b=10	25.7	49.1

Acknowledgements

This paper is largely based on research, to which an equal contribution was made by its Italian participants: S. Aiello, B. Barsella, M. Perinotto and P. Patriarchi.

References

Aiello, S., Barsella, B., Bonetti, A., 1982, Ap. Sp. Science, 87, 463.
Aiello, S., Barsella, B., Chlewicki, G., Greenberg, J. M.,
    Perinotto, M., Patriarchi, P., 1985, in preparation.
Barsella, B., Panagia, N., Perinotto, M., 1982, Astr. Ap., 111, 130.
Borghesi, A., Bussoletti, E., Colangeli, C., 1985, Astr. Ap., 142, 225.
Chlewicki, G., 1985, Ph. D. Thesis, University of Leiden.
Chlewicki, G., Greenberg, J. M., 1985, to be submitted to
    Astronomy & Astrophysics.
Greenberg, J. M., 1982, in "Submillimetre Wave Astronomy", ed.
    J. E. Beckman and J. P. Phillips, Cambridge Univ. Press, p. 261.
Greenberg, J. M., Chlewicki, G., 1983, Ap. J., 272, 563.
Hong, S. S., Greenberg, J. M., 1980, Astr. Ap., 88, 194.
Huffman, D. R., Stapp, J. L., 1973, IAU Symp. No. 52, ed. J. M. Greenberg
    and H. C. van der Hulst, Dordrecht: Reidel, p. 297.
Low, F. J., et al., 1984, Ap. J. (Letters), 278, L19.
Sakata, A., et al., 1984, Occasional Reports of the Royal Observatory,
    Edinburgh, 12, 128.
Sakata, A., Wada, S., Tanabe, T., Onaka, T., 1985, preprint.
de Vries, C. P., 1985, Astr. Ap., 150, L15.
de Vries, C. P., Le Poole, R. S., 1985, Astr. Ap., 145, L7.
Walker, H., 1985, private communication.

THE WAVELENGTH OF MAXIMUM POLARIZATION IN THE CHAMAELEON DARK CLOUD

D.C.B. Whittet
Lancashire Polytechnic

J.H. Hough
Hatfield Polytechnic

J.A. Bailey
Anglo-Australian Observatory

M.F. Rouse and T.M. Kirrane
Lancashire Polytechnic

1. INTRODUCTION

Observational evidence for grain growth in dark clouds is provided by variations in the interstellar extinction and polarization laws, and by infrared spectroscopy (e.g. Whittet and Blades 1980). Although anomalous extinction laws indicative of biasing of the grain size distribution towards larger sizes are now well-established for regions such as Orion, Rho Oph and R CrA, the extinction law in Chamaeleon is controversial. In a recent review, Hyland et al (1982) noted that estimates of the distance to the embedded T-association vary from 110 to 215 pc dependent on the value of $R = A(V)/E(B-V)$, the ratio of total to selective extinction. R values based on infrared photometry are often unreliable due to excess emission from circumstellar shells. This paper reports an attempt to resolve the problem by means of polarimetry, which provides an independent method of estimating R which is not hampered by such effects.

The observations were made on the Anglo-Australian Telescope in February 1984, as part of a more general investigation of the wavelength dependence of interstellar polarization. The Hatfield twin-channel polarimeter (Bailey and Hough 1982) was used in combination with the AAT infrared photometer-spectrometer, allowing simultaneous measurement through optical (B, V, R or I) and infrared (J, H or K) filters. Full details of the observations will be published elsewhere. A total of seven stars were observed in Chamaeleon. These are listed in table 1, with source designations from the catalogue of Kirrane (1985).

F. P. Israel (ed.), Light on Dark Matter, 197–200.

2. RESULTS AND DISCUSSION

Values of the wavelength of maximum linear polarization (λ_{max}) were determined by fitting the Serkowski law to the data; this takes the form

$$p/p_{max} = \exp[-K \ln^2(\lambda_{max}/\lambda)]$$

(Serkowski et al 1975), and K may be treated as a constant or a free parameter (Wilking et al 1980). In practice, this made little difference to our results; the values given in table 1 are based on the Wilking form with K = 1.7 λ_{max}.

TABLE 1

Star	λ_{max} (µm)	R(pol)	R(phot)
HD 97300	0.71 ± 0.02	4.0	5.1
Cha T21	0.67 ± 0.02	3.8	5.0
Cha F11	0.53 ± 0.02	3.0	3.6
Cha F16	0.62 ± 0.02	3.5	3.4
Cha F29	0.67 ± 0.02	3.8	(8.5)
Cha F30	0.61 ± 0.02	3.4	4.4
Cha F32	0.58 ± 0.05	3.2	5.4

Table 1 also lists R values deduced from polarimetry, using the relation deduced by Whittet and van Breda (1978):

$$R(pol) = 5.6\ \lambda_{max},$$

and from the colour excess ratio:

$$R(phot) = 1.1\ E(V-K)/E(B-V),$$

using photometric data from Rydgren (1980) and Vrba and Rydgren (1984). Non-systematic errors in R(phot) are typically ±0.3. It is clear that the results from these independent methods are in reasonable agreement for only two stars, Cha F11 and F16; in all other cases, R(phot) is higher than R(pol), as would be expected if infrared excess emission is present. The mean value of R(pol) for all seven stars is 3.5, only marginally greater than the normal interstellar value of 3.1, leading to a distance estimate (based on HD 97300, a B9 V star in reflection nebulosity) of 190 ± 15 pc.

3. CONCLUSION

We conclude from this study that either (i) five of the stars in our sample of seven have circumstellar infrared excess emission, or (ii) the normal correlation between the ratio of total to selective extinction and the wavelength of maximum polarization breaks down in the Chamaeleon cloud. The former explanation seems unlikely in view of the fact than none of these stars have emission lines in their optical spectra, precluding a hot gaseous shell. HD 97300 has a cool dust shell detected by IRAS (Wesselius et al 1984) but this produces negligible contamination at 2.2 μm. If the second explanation is correct, this could indicate an unusual grain chemistry in the Chamaeleon region (see McMillan 1978).

ACKNOWLEDGEMENT

We are grateful to the Science & Engineering Research Council for observing time and financial support.

REFERENCES

Bailey,J.A., Hough, J.H., 1982, Pubs.Astr.Soc.Pacific, 94, 618.
Hyland,A.R., Jones,T.J., Mitchell, R.M. 1982,
          Mon.Not.R.Astr.Soc., 201, 1095.
Kirrane,T.M.: 1985, Lancashire Polytechnic internal report.
McMillan,R.S.: 1978, Astrophys.J., 225, 880.
Rydgren,A.E.: 1980, Astr.J., 85, 444.
Serkowski, K., Mathewson,D.S., Ford, V.L. 1975,
          Astrophys.J., 196, 261.
Vrba,F.J., Rydgren, A.E. 1984, Astrophys.J., 283, 123
Wesselius,P.R., Beintema,D.A., Olnon, F.M. 1984,
          Astrophys.J., 278, L37.
Whittet,D.C.B., Blades, J.C. 1980, Mon.Not.R.Astr.Soc., 191, 309.
Whittet,D.C.B., van breda, I.G. 1978, Astr.Astrophys., 66,57.
Wilking,B.A., Lebofsky,M.J., Martin,P.G., Rieke,G.H., Kemp, J.C.
          1980, Astrophys.J., 235, 905.

IRAS Field 137   α = 18^{h}00^m, δ = -15°, HCON-3, 12 μm. Field in Scutum
                   and Sagittarius, showing the Galactic Plane. M17 is
                   the bright spot showing the effect of detector
                   hysteresis just left of the plane; to its north M16
                   can be seen.

THREE PRINCIPAL HEATING SOURCES OF DUST IN THE GALACTIC DISK

P. Cox, E. Krügel and P. G. Mezger
Max-Planck-Institut für Radioastronomie
Auf dem Hügel 69, 5300 Bonn 1, F.R.G.

The galactic infrared emission has been observed at various wavelengths from the near infrared 4 μm up to 900 μm. The spectrum of the central part of our galaxy is comparable to that observed towards external Sb/Sc galaxies: a broad emission peak centered around 100 μm and a distinct shoulder occuring shortwards of 20 μm. The presence of these features suggests that the infrared spectrum of normal spiral galaxies is a superposition of spectra originating in different classes of sources (Fig. 1).

We present here the results of a model of the galactic infrared emission where in all presently available observations can be accounted for. Our model assumes a cylindrical symmetry around the galactic center and adopts the MRN dust mixture of silicate and graphite grains, including the recent modifications as described in Draine and Lee (1984). The galactic infrared emission is a contribution from three main dust components with different temperatures and heating sources. We will in turn review the relevant characteristics of these three components and refer the readers to Cox et al. (1985) for a more detailed discussion.
(i) Cold dust (10-25 K) associated with atomic hydrogen and quiescent giant molecular clouds heated by the general interstellar radiation field (IRSF) to which old and young stellar populations contribute. It accounts for more than 90% of the emission at 900 μm where the HI and the GMCs contribute about equally. The luminosity content of this component is 5.7 10^9 L_0 and represents ∿40% of the total luminosity.
(ii) Warm dust (30-40 K) associated with ionized gas in extended low density HII regions and heated by O stars; and, with molecular gas heated by B stars. Hence warm dust traces the youngest stellar population. With the present star formation rate, as derived from Lyc photon counts O and B stars can account for the warm dust luminosity of 7.6 10^9 L_θ, which represents ∿50% of the total IR luminosity.
(iii) Hot dust (250-500 K) heated by post main sequence stars of a few solar masses undergoing mass loss (observationally known as OH/IR stars). The conspicious bump seen in the middle infrared (MIR) part of the galactic spectrum requires temperatures of a few hundred degrees Kelvin and contains 10% of the total IR luminosity 2 10^9 L_θ. One can argue that the main sequence stars cannot be the main heating sources for the hot dust

201

F. P. Israel (ed.), Light on Dark Matter, 201–202.

because the early stages of stellar evolution are very transient episodes and therefore have insufficient luminosity to account for the energy in the MIR. The OH/IR stars convert all intrinsic stellar radiation into MIR emission. These stars, near the end of their evolution, of a few solar masses have typical luminosities of 2 10^4 $L_\odot$ and a mass loss rate of 10^{-5} $M_\odot$ yr^{-1}. Their spectra peak around 10 µm and typical color temperatures are a few 100 K.

Since practically all stars in the mass range 2 to 8 $M_\odot$, which are the stars that build up the bulk of the galactic luminosity, evolve to M giants, only a few percent of the M giants are needed at a given time in the OH/IR phase to explain the MIR shoulder through their dust emission. This remark agrees in a quantitative way with estimates of the total averaged luminosity of these stars, based upon the OH masering stars survey, by Herman and Habing (1985). The derived total luminosity is of the order of 2 10^9 $L_\odot$ and accounts for the observed luminosity input in the MIR. More recent suggestions attribute the MIR shoulder to the presence of small dust particles (or big molecules). Puget et al. (1985) claim that all the MIR luminosity is contained in the broad emission features which characterize these polycyclic aromatic molecules. This far reaching and interesting argument has to be investigated furthermore e.g. by exploring the spectroscopic signature of these compounds in the galactic diffuse emission. However, OH/IR stars i.e. M giants with heavy mass loss, with their total number observed spectra and luminosities (based on observations) can accound for the MIR part of the spectrum and for the luminosity of 2 10^9 $L_\odot$; hence, they must contribute substantially to the galactic MIR emission.

REFERENCES

Cox, P., Krügel, E., Mezger, P. G. (1985) to appear in Astron. Astrophys.
Draine, B. T., Lee, H. M. (1984) Astrophys. J. 285, 89
Herman, J., Habing, H. J. (1985) preprint
Puget, J. L., Léger, A., Boulanger, F. (1985) Astron. Astrophys. 142, L19

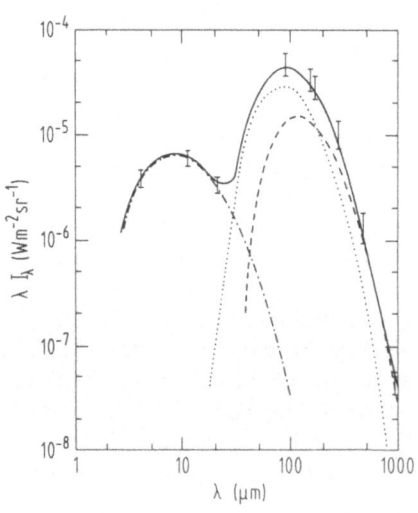

Fig. 1: Spectrum of the dust emission between 4 and 900 µm from the inner part (R ≲8 kpc) of our galaxy averaged over galactic longitudes 3-35° and latitudes |b|≤1°. Shown is a three-component fit to the observed spectrum: i) cold dust (10-25 K) - dashed curve; ii) warm dust (30-40 K)-dotted curve; iii) hot dust (250-450 K) - dash-dotted curve. The heavy solid line shows the superposition of the computed spectra.

MID-IR EMISSION OF THE INTERSTELLAR MEDIUM

F. Boulanger[1], M. Pérault[2], and J.L. Puget[2]
(1): Goddard Institute for Space Studies, 2880 Broadway,
New York NY10025, USA
(2): Ecole Normale Superieure, 24 rue Lhomond, 75005
Paris, France

Abstract. IRAS observations of the local interstellar medium and the
galactic plane are presented; for all components, the radiation
observed at 12 μm represents a large fraction of the far-IR emission.
The 12/25 μm color of the diffuse IR emission of the galactic plane
increases from the inner to the outer parts while the 60/100 μm ratio
decreases, a color-color anticorrelation observed also for giant H II
regions. The implications of these observations are discussed.

1. INTRODUCTION

Observations of several reflection nebulae (Sellgren et al.
1983, Sellgren 1984, and Sellgren et al. 1985) revealed unexpectedly
bright and extended near- and Mid-IR emission associated with the
unidentified IR bands at 3.3, 6.2, 7.7, 8.6, and 11.3 μm. Sellgren
(1984) proposed that this emission comes from very small grains
(~50 atoms) heated briefly to high temperatures each time they
absorb one UV photon. To explain the presence of the unidentified IR
bands, Léger and Puget (1984) and Allamandola et al. (1985) suggested
that the small grains are polycyclic aromatic hydrocarbon (PAH's)
molecules.

Puget et al. (1985) computed for different environments the IR
spectrum of a population of graphite grains with a size distribution
extending down to these molecules. Their prediction that the mid-IR
excess observed by Sellgren et al. (1983) should be observed in
various kinds of extended-IR sources was indeed confirmed by IRAS
observations (Beichman et al. 1984, Boulanger et al. 1985, Leene
1985).

This paper now presents observations which can be used to
constrain interstellar dust models built to account for the
ubiquitous mid-IR emission (Draine and Andersson 1985, Puget et al.
1985, and Désert **1986)**. Section 2 presents the spectral distribution
of the IR emission of the various components of the interstellar
medium in the Orion region (Boulanger et al. 1986). Large scale
properties of the mid-IR emission of our Galaxy are described in

F. P. Israel (ed.), Light on Dark Matter, 203–208.
© *1986 by D. Reidel Publishing Company.*

Section 3. The implications of these results on interstellar dust properties are discussed in Section 4.

2. THE LOCAL INTERSTELLAR MEDIUM

The diffuse galactic IR emission outside molecular clouds and H II regions is expected to come from dust associated with neutral atomic gas. The observed emission should be proportional to the column density of gas provided the interstellar radiation, the dust-to-gas ratio, and the dust properties are uniform along the line of sight.

The close correlation observed between H I integrated emission and the 100 μm brightness in the Orion region outside molecular clouds and H II regions (Fig. 1) confirms this idea. Figure 2 shows that the 12 μm emission also correlates with the gas column density; the ratio between the 12 and 100 μm emission $\lambda I_\lambda(12\mu m)/\lambda I_\lambda(100\mu m)$ is 0.45. This correlation, observed for one cloud by Boulanger et al. (1985), demonstrated from a larger set of data the existence of a significant number of dust grains with temperatures of several hundred degrees in the neutral atomatic interstellar medium.

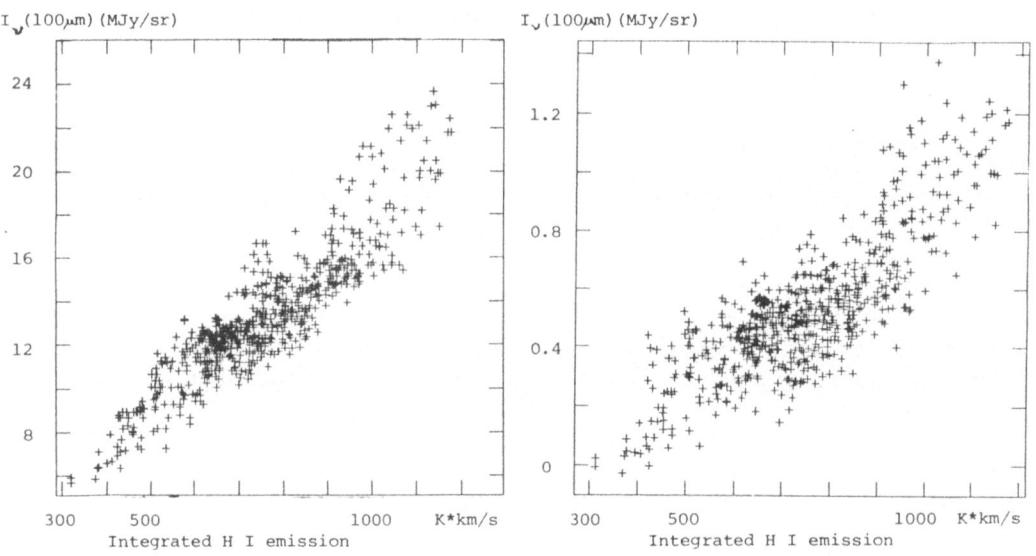

Figure 1: Correlation between 100μm brightness and H I emission integrated in velocity, observed in the Orion-Monoceros region outside molecular clouds and H II regions. Each cross represents a half degree pixel.

Figure 2: Correlation between 12μm brightness and H I emission integrated in velocity, observed in the Orion-Monoceros region outside molecular clouds and H II regions. Each cross represents a half degree pixel.

2.2. Molecular Clouds

The IR emission of molecular clouds can come from dust on the surface heated by the external radiation and dust around embedded stars. Both sources seem necessary to account for all features of the IR emission of the Orion-Monoceros clouds (Boulanger et al. 1986). Figure 3 shows that the 12 and 100 μm emission of these molecular

clouds are well correlated. The 12 to 100 μm emission ratio
$\lambda I_\lambda(12$ μm$)/\lambda I_\lambda(100$ μm$)$ is 0.4.

2.3. H II regions

λ Orionis excepted, the three significant H II regions in the
Orion constellation are Barnard's loop and Orion A and B (Reuch
1978). Barnard's Loop and the neutral diffuse interstellar medium are
pervaded by radiation fields of comparable energy densities and
present a similar mid- to far- IR emission ration (Fig. 4). In Orion
A and B, this ratio is 0.35 and 0.25 respectively; these slightly
lower values may be related to more intense radiation fields, as
discussed in the next section.

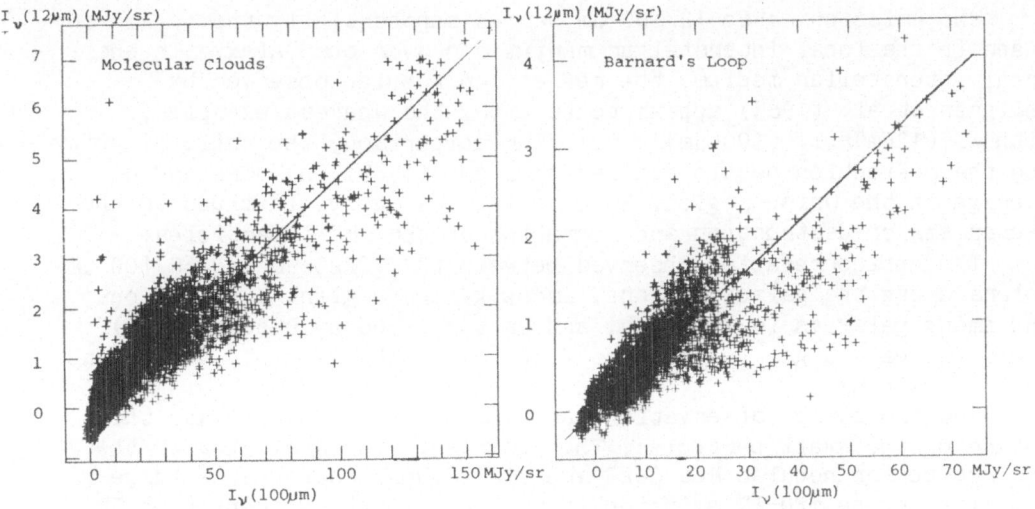

Figure 3: Correlation between the 12 and
100 m emission of the Orion-Monoceros molecular clouds. Each cross represents one
8 arcmin pixel. The straight line corres-
ponds to a $\lambda I_\lambda(12\mu m)/\lambda I_\lambda(100\mu m)$ ratio of
0.40.

Figure 4: Correlation between the 12 and
100 m emission of Barnard's Loop. Each cross
represents one 8 arcmin pixel. The straight
line corresponds to a $\lambda I_\lambda(12\mu m)/\lambda I_\lambda(100\mu m)$
ratio of 0.45.

3. INNER PARTS OF OUR GALAXY

The AFGL observations (Price 1981) revealed unexpectedly strong
mid-IR diffuse emission in the galactic plane. The IRAS results
(Sect. 2) show that the interstellar medium, its different components
all strongly emitting in themid-IR, is probably at the origin of this
emission.

Longitude profiles of the emission of the galactic plane at the
IRAS survey wavelengths 12, 25, 60 and 100 μm show that the emission
ratio 12/25 μm increases by a factor of 2 from the inner to the outer
parts of the Galaxy while the ratio 60/100 μm decreases. Bright IR
sources, typically giant H II regions, appear as holes and peaks
along the 12/25 μm and 60/100 μm color profiles respectively. First
noticed by Pajot et al. (1985) from a sample of galaxies, this color-
color anticorrelation is also observed between the different sources.

As the 60/100 µm ratio increases with the intensity of the
radiation that heats the grains, the anticorrelation implies that the
relative strength of the 12 µm emission decreases when the radiation
field becomes stronger.

4. DISCUSSION

The strong diffuse emission observed at 12 µm by IRAS in all
types of extended IR sources definitively demonstrates the existence
of very small grains throughout the interstellar medium. The fraction
of dust mass present in these small particles unfortunately cannot be
derived from the IRAS observations alone, because in many IR sources
big grains are expected to emit mostly at submillimetric wavelengths,
where more observations are needed.

The emission ratio 12/100 µm varies substantially from place to
place in the local interstellar medium and over our Galaxy. In the
local interstellar medium, the reflection nebulae observed by
Sellgren et al. (1983) appear to be among the weakest examples,
with $\lambda I_{\lambda}(12\mu m)/\lambda I_{\lambda}$ (100 µm) ~ 0.1. For comparison, the ratio is unity
for the reflection nebula studied by Leene (1985) in R Cra and 0.4 on
average in the Orion region. Such variations may yield clues to the
mechanisms of destruction and formation of the small particles.

The anticorrelation observed between the 12/25 µm and 60/100 µm
colors along the galactic plane, among galactic giant H II regions,
and among galaxies is striking, and as suggested by Desert **(1986),**
could indicate a gap in the size distribution between big grains and
small particles.

Spectroscopic observations are essential to investigate the
nature of the small particles. The near- and mid-IR spectra of the
two reflection nebulae NGC 2023 and 7023 demonstrate that a large
fraction of the mid-IR emission is radiated in the unidentified IR
bands. On the other hand the spectra do not show the 9.7 µm silicate
feature (Sellgren et al. 1985). These observations therefore indicate
that the small particles are mostly pregraphitic molecules and not
very small silicates (Désert et al. 1985).No evidence suggests that
this conclusion applies to all sources, notably those with a much
larger 12/100 µm emission ratio. Clearly, more spectroscopic
observations are needed.

ACKNOWLEDGEMENTS

We wish to thank E. Sarot for a careful reading of the manuscript. F.
Boulanger acknowledges support from a NAS/NRC Research Associateship.

REFERENCES

Allamandola, L.J., Tielens, G.G.M., Barker, J.R., 1985,Ap.J. 290,L25.
Boulanger, F., Baud, B., van Albada, G.D., 1985, Astr. Ap. 144, L9.
Boulanger, F., Maddalena, R.J., Thaddeus, P., 1986, these
     proceedings, page 293.
Désert, F.X., 1986, these proceedings, page 213
Draine, B.T., and Andersson, N., 1985, Ap. J. 292,494.

Desert, F.X., Boulanger, F., Leger, A., Puget, J.L., 1985, Astr. Ap., Submitted.

Leene, A., 1985, Submitted to Astr. Ap.

Léger, A., Puget,J.L., 1984, Astr. Ap. 137,L5.

Pajot, F., Boisse, P., Gispert, R., Lamarre, J.M., Puget, J.L., Serra, G., 1985, Astr. Ap., Submitted

Price, S., 1981, Astron. J. 86,193.

Pudget J.L.,Léger, A., Boulanger F., 1985, Astr. Ap. 142, L19.

Reich, W., 1978, Astr. Ap. 64,407.

Sellgren, K., Werner,M.W., Dinerstein, H.L., 1983, Ap. J. Let. 271, L13.

Sellgren, K., 1984, Ap. J. 277,623

Sellgren, K., Allamandola, L.J., Bregman,J.D., Werner, M.W., and Wooden D.H., 1985, Ap. J. in press.

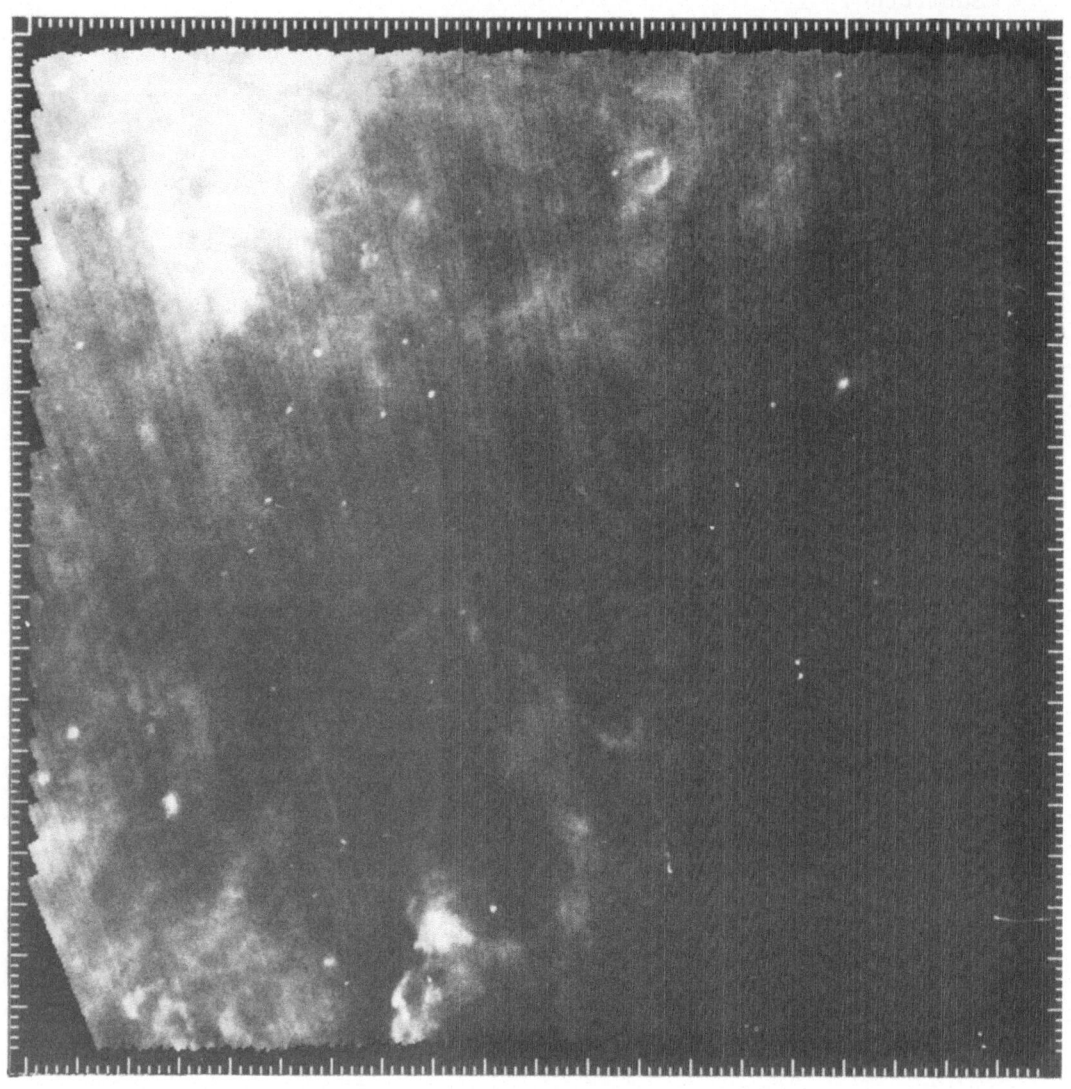

IRAS Field 150   α = 07^{h}00^m, δ = -30°, HCON-3, 60 μm. Field in Canis
Maior and Puppis. Note shell at top.

OPTICAL LUMINESCENCE FROM REFLECTION NEBULAE?

G. Olofsson
Stockholm Observatory
S-133 00 Saltsjöbaden
Sweden

It has been clearly demonstrated by Sellgren (1984) that the so-called
"unidentified" infrared emission bands plus a hot NIR continuum seen in
reflection nebulae cannot be caused by grains in thermal equilibrium
with the central stars. Instead she suggested thermal emission from very
small grains temporary heated to ~ 1000K by single UV quanta. Following
this idea, Leger and Puget (1984) found evidence that polycyclic aromatic hydrocarbons (PAH:s) are responsible for the IR emission bands.
These authors also noted that the optical properties for graphite cannot
be used for these tiny monolayer grains. This is in fact an important
statement - otherwise the model would not be able to explain the observed band to continuum ratio for e.g. the 3.3μm band: There is no doubt
that this band is caused by the C-H stretch vibration. The oscillator
strength depends on the detailed structure of the molecule, but the
absorption cross-section of a single C-H bond should not exceed
10^{-19} cm^2 (Duley and Williams, 1979). If we further consider carbon
grains with volume, V, and assume the number of CH bonds to be less than
the number of C atoms, we estimate the total cross-section of the grain
in the band to be $\leq 10^4$V cm^2. The absorption cross-section in the continuum is at least 1.5 x 10^4V cm^2 at 3.3μm for carbon grains (Koike et al.,
1980). Consequently we would expect to observe a band to continuum ratio
of at most 0.7. This is in clear conflict with the observations.

We are obviously discussing the limiting case particles/large molecules and it seems to be a matter of taste if we lable the NIR emission
as "thermal" or "fluorescence" (the latter is favoured in a recent paper
by Allamandola et al. 1985). I would personally prefer the lable
"luminescence" since it includes (in the case of hydrocarbons) both
transitions in the singlett and the triplett electronic states and it is
likely that both contribute.

It is well known that PAH:s exhibit very high quantuum efficiency
for optical luminescence and for this reason I have briefly looked into
the possibilities to observe PAH:s in the optical spectral region. In
Fig. 1 the basic mechanisms are summerized, and in Fig. 2 some examples
are shown. It is of interest to note that the main structure of the
fluorescence is similar for the three different molecules, which means
that their vibrational IR bands will approximately coincide. This means

F. P. Israel (ed.), Light on Dark Matter, 209–212.

that in a mixture these three molecules will tend to collaborate in the infrared (giving prominent bands) but they would be much easier to distinguish in the visual. In fact, Fig. 2 does not reflect the astronomical case (the laboratory spectra are taken in the liquid phase where the van der Waal forces smear out the energy levels) and Fig. 3 shows an example of a cooled sample: a large number of very distinct features appears and this fact gives us some hope for identifying specific PAH:s in their solid phase. Before rushing to the chemical litterature for PAH spectra we must however be aware of the fact that the frozen solvent (which hosts the PAH-molecules) causes a wavelength shift and that some details in the spectra are related to the specific solvent.

The astronomical litterature is scarce as regards spectra of reflection nebulae, but I would like to draw your attention to the remarkable emission spectrum found in the "Red Rectangle" by Schmidt et al. (1980). This spectrum has a general appearance which indeed resembles PAH luminescence. If so, it should obviously be possible to identify the molecule(s) - in particular the "ripple" near the band head at 5800Å should be useful. These features certainly deserves a detailed investigation, but it should be noted that the splitting of the electronic transitions between the ground vibrational states seen in laboratory spectra depends on the particular lattice (frozen solvent) which hosts the PAH molecule. This splitting, which gives rise to a ripple near the band head, may thus contain information on both the PAH:s and the dominating "grain" material which hosts the PAH:s.

As the "Red Rectangle" is known to exhibit the "unidentified" infrared emission bands (Russell et al., 1978), one would expect the main vibrational transitions as seen in the infrared to show up in the optical luminescence. Indeed, if we identify the sharp peak at 5800Å as the electronic trasition between vibrational ground states and simply "fold" the observed infrared spectrum into the optical (where $1/\lambda$ (optical) = $1/0.58 - 1/\lambda$ (IR)) we get the result in Fig. 4. I may be biased, but in my opinion this strongly suggests that the observed optical spectrum is luminescence emission from the same molecule(s) (presumely PAH:s) giving rise to the infrared emission. It is now very important to observe a number of nebulae to investigate the possible presence of emission bands in the optical. I have recently observed two reflection nebulae, NGC 2023 and the nebula surrounding HD 97300, but without finding any emission band. It may well turn out that the emission bands in the "Red Rectangle" are exceptionally strong, and forthcoming observations of reflection nebulae should aim at as good signal to noise ratio as possible in order to detect relatively faint emission bands.

References

Colmsjö, A. and Stenberg, U.: 1977, Chemica Scripta, 11, 220.
Lumb, M.D.: 1978, Luminescence Spectroscopy, Academic Press, London.
Russel, R.W., Soifer, B.T. and Willner, S.P.: 1978, Astrophys. J., 220, 568.
Schmidt, G.D., Cohen, M. and Margon, B.: 1980, Astrophys. J., 239, L133.
Leger, A. and Puget, J.L.: 1984, Astron. Astrophys., 137, L5.
Sellgren, K.: 1984, Astrophys. J., 277, 623.

Duley, W.W. and Williams, D.A.: 1979, Nature, 277, 40.
Koike, C., Hasegawa, H. and Manabe, A.: 1980, Astrophys. Space Sci., 67,
    495.
Allamandola, L.J., Tielens, A.G.G.M. and Barker, J.R.: 1985, Astrophys.
    J., 290, L25.

a)   b)

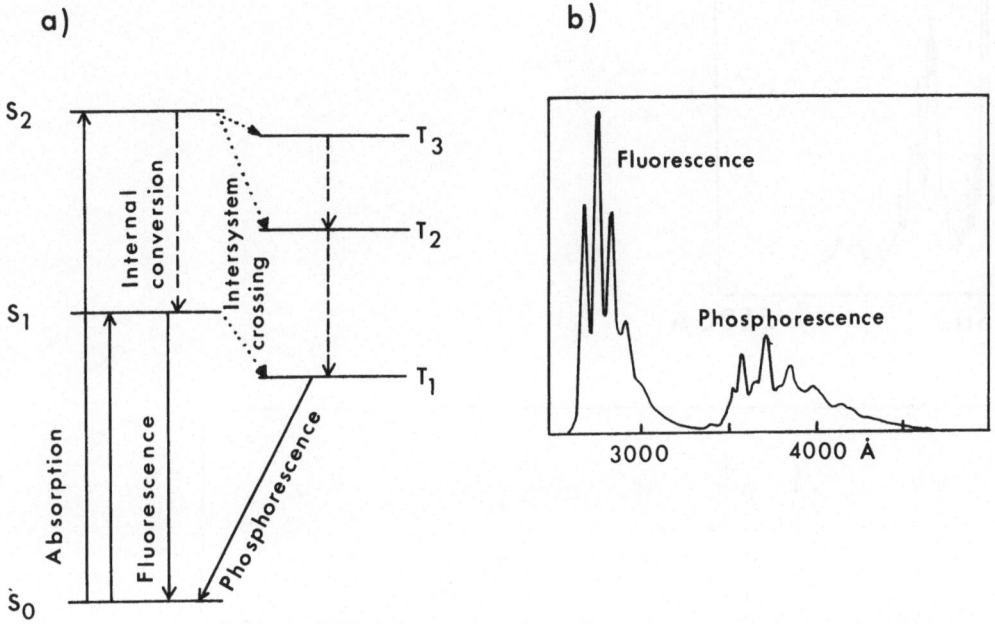

Figure 1. a) π-Electron luminescence processes. The full drawn lines
denote radiative transitions, the dashed lines internal conversions and
the dotted lines intersystem crossings.
        b) The fluorescence and phosphorescence spectrum of benzene
at 77K (Lumb, 1978).

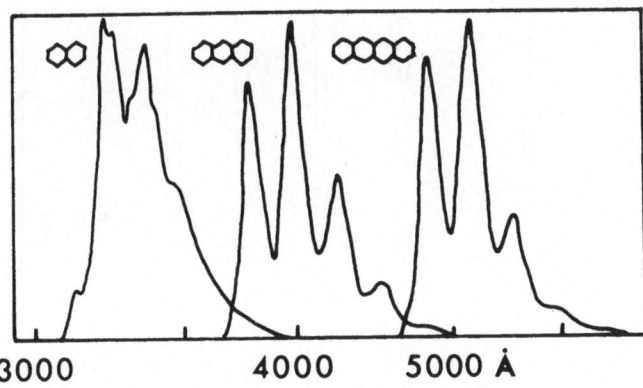

Figure 2. Fluorescence spectra of naphtalene, anthracene and tetracene
(Lumb, 1978).

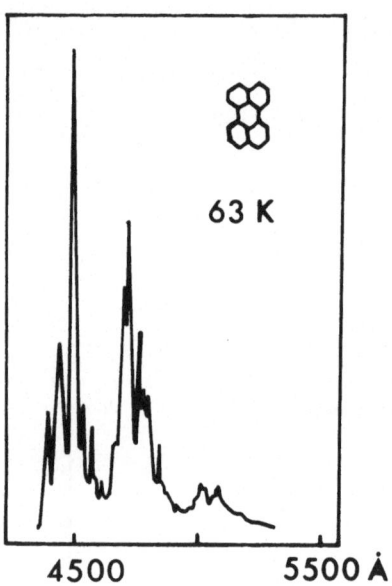

Figure 3. Fluorescence spectrum of perylene in n-heptane at 63K (Colmsjö and Stenberg, 1977).

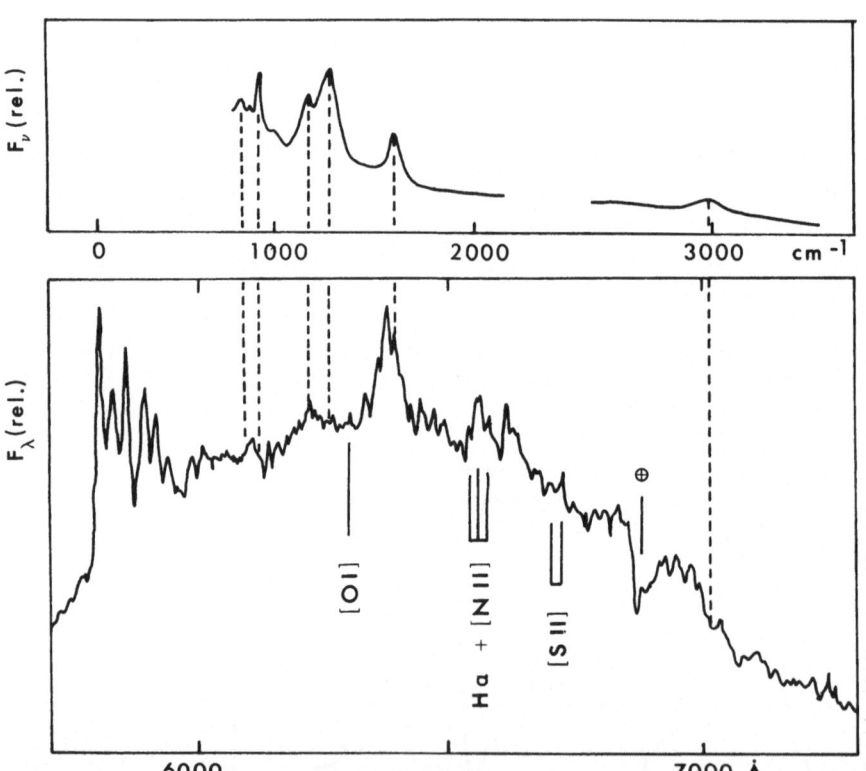

Figure 4. The optical emission from the "Red Rectangle" (Schmidt et al., 1980) is compared to the infrared emission (derived from Russell et al., 1978).

INFRARED SPECTRA AND DUST TEMPERATURE FLUCTUATIONS

F.X. Désert
Groupe de Radioastronomie de l'E.N.S.
24, rue Lhomond
75231 Paris Cedex 05
France

ABSTRACT. A model of dust emission is built which contains graphite, silicate grains and the recently discovered PAH molecules. The resulting infrared spectra are compared with IRAS observations. It seems likely that there is a gap in the size distribution between graphite grains and PAH molecules. Furthermore, this model can account for the anticorrelation seen in our Galaxy in the color-color diagram $F(12\mu)/F(25\mu)$ vs $F(60\mu)/F(100\mu)$.

1. INTRODUCTION

There is now large evidence that a new component of dust is necessary in the general interstellar medium, mainly to explain the emission observed at wavelengths lower than 20μm. For example, Boulanger et al. (1985a) observed this component in cirrus clouds (see also Pajot et al., 1985). As has been proved by Sellgren (1984), by studying reflection nebulae, the new component must be made of very small particles transiently heated to high temperatures by UV and visible photons so that they can emit a large fraction of their energy at shorter wavelengths (10-20μm) than the equilibrium temperature would indicate (20 K corresponds to 60 and 100μm). This new component has been identified with Polycyclic Aromatic Hydrocarbon (PAH) molecules (Léger & Puget, 1984) which emit their energy in the very characteristic midinfrared features at wavelengths of 3.3, 6.2, 7.7, 8.6, 11.3μm; Puget et al. (1985) have predicted that these molecules should be observed in the cirrus clouds.

2. THE DUST MODEL

To investigate the importance of those small particles, we have built a model of interstellar dust which consists in grains of graphite and silicate to which we add PAH molecules. As there is no evidence for small grains of silicate (Désert et al., 1985b) due to the lack of the 9.7μm emission features, we take for silicate grains a lower limit of 10 nm for their size distribution. Therefore, this model differs from Draine & Anderson's calculations (1985) by the composition of the very small grains. Table I resumes the parameters needed for the calculation. The infrared integrated emissivities of PAHs has been derived from laboratories measurements (Sadtler Standard

213

F. P. Israel (ed.), Light on Dark Matter, 213–216.
© 1986 by D. Reidel Publishing Company.

TABLE I

Grain type	Lower limit radius (nm)	Upper limit radius (nm)	Grain mass m(H)	Reference for Heat Capacity
Silicate	10	250	$6.0 \ 10^{-3}$	Léger et al. (1985)
Graphite	a	250	$3.4 \ 10^{-3}$	DeSorbo et al. (1953)
PAHs	0.4	1.5	$1.1 \ 10^{-4}$	Puget et al. (1985)

Spectra, 1959) on a typical PAH molecule, coronene, and an adjustment of emission line profiles to astrophysical observations has been allowed. We have not taken into account a possible variation of the lower limit of PAH radius with the intensity of the radiation field because it seems that evaporation of these molecules is less important (Omont, 1985) than estimated by Puget et al. (1985). For all grains, we adopt the size distribution found by Mathis et al. (1977) which corresponds to a power law with a -3.5 exponent. The emissivities of the big grains are Draine & Lee's ones (1984). To compute the emission of each grain and molecule, we use the method derived by Désert et al. (1985a) which takes into account the multiphoton processes in order to calculate its exact temperature distribution. Indeed, the fluctuations of the temperature T are important to consider for grains or molecules of low heat capacity $C(T)$: $T.C(T) < E$, where E is the energy of a typical incident photon.

3. SOME PRELIMINARY RESULTS

The infrared spectra emitted by interstellar dust are displayed in figure 1 corresponding to an interstellar radiation field (ISRF) ranging from X = 1 to 100 times the local one (Mathis et al., 1983), for 2 types of grain size distribution: a) a continuous one from PAHs to graphite grains (a = 1.5 nm) ; b) a bimodal one (a = 10 nm). We have calculated the corresponding fluxes in IRAS bands, using the spectral response of each band detector. We give the representative points of the different spectra we have obtained in the F(12)/F(25) versus F(60)/F(100) color-color diagram (figure 2). Two curves are shown, corresponding to the size distribution a) (continuous line) and   b) (dashed line). As the ISRF increases, the average temperature of all grains rises, implying an increase of the ratio of F(60)/F(100). In the same time, the PAH emission in IRAS band 1 increases linearly, because the cooling of these molecules occurs always at the same wavelengths, mostly in the 3.3 to 11.3μm emission features. On the contrary, the flux at band 2 increases non-linearly with the ISRF because the part of the emission coming from the smallest graphite grains shifts toward shorter wavelengths. On the same figure 2 are IRAS observations corresponding to emission peaks of 30'-averaged maps of the galactic plane (Boulanger et al., 1985b); filled circles correspond to the inner Galaxy, open circles to the molecular ring and crosses to the solar circle. The general anticorrelation in the color-color diagram seems to be qualitatively explained by the model, and suggests a possible gap in the size distribution between pregraphitic molecules as are PAHs, and big graphite grains. A further generalization of this model, including different ISRFs and interstellar environments for dust, could help in understanding the same anticorrelation observed in external galaxies.

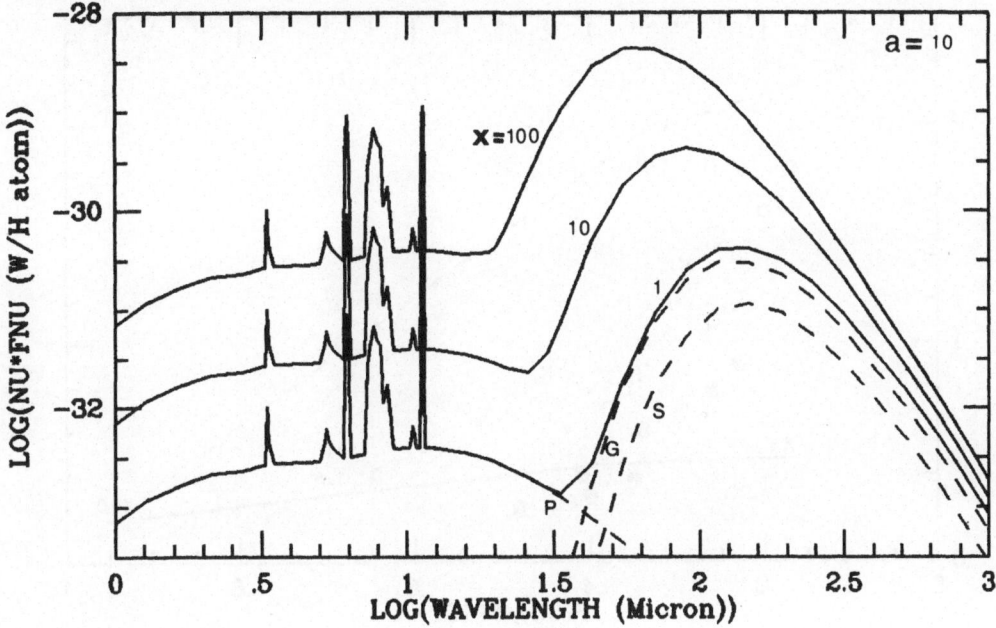

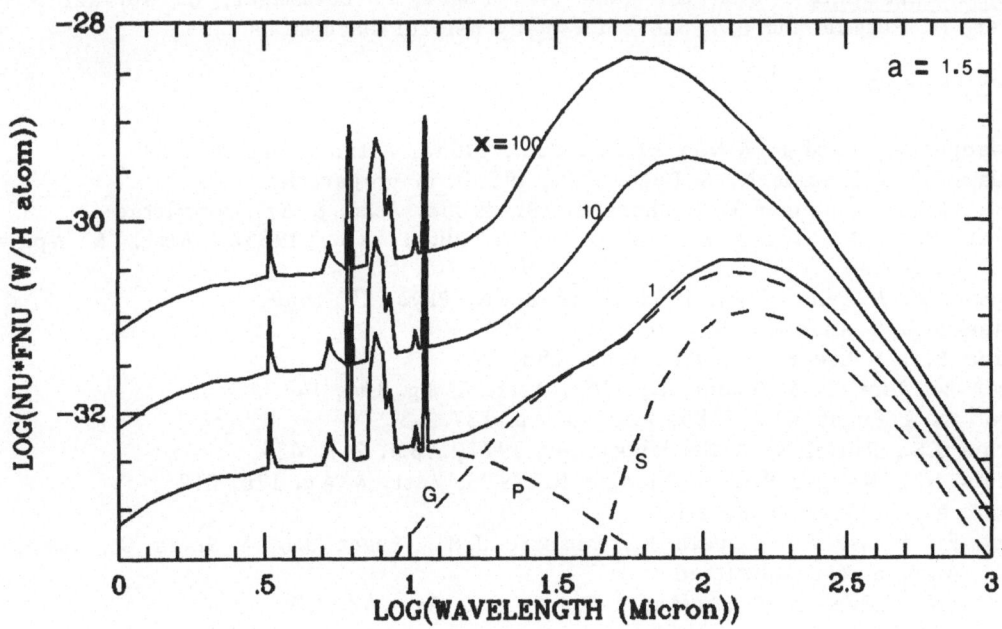

Figure 1 a) & b). Dust infrared emission. In dashed lines are the individual contributions of PAHs (P), graphite (G) and silicate (S) grains. X and $\underline{a}$ are defined in the text. The spectra are more bimodal in figure 1b, reflecting the gap in the size distribution between graphite and PAH molecules.

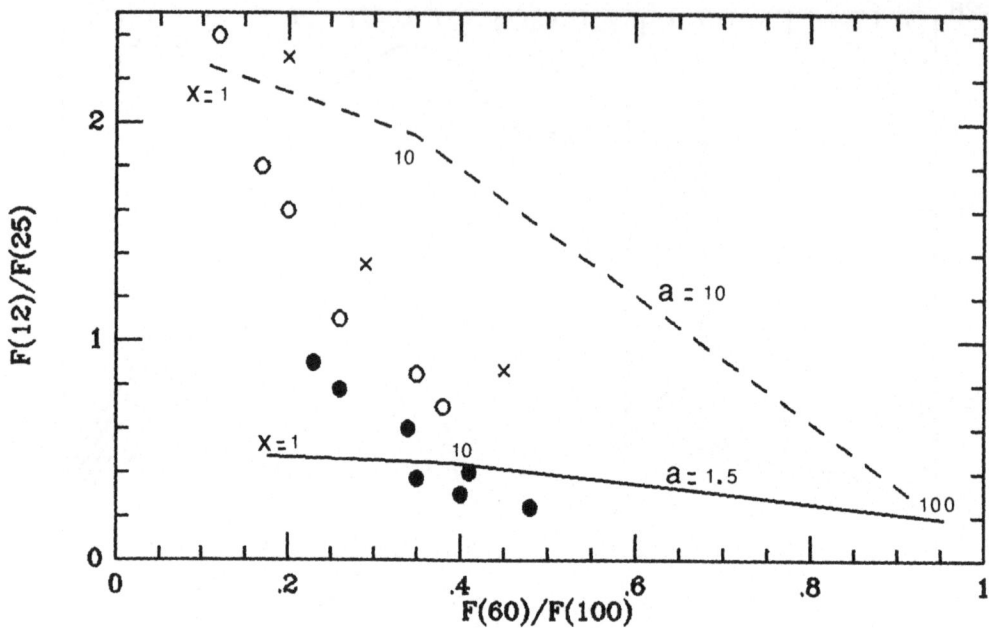

Figure 2. IRAS color-color diagram (see the text)

Acknowledgements I wish to thank J.L. Puget, F. Boulanger, C. Dupraz, A. Léger, M. Pérault and S.N. Shore for many helpful discussions.

REFERENCES

Boulanger F., Baud B. & van Albada G.D., 1985a, Astr. & Ap. **144**, L9
Boulanger F., Pérault M. & Puget J.L., 1985b, in preparation
Désert F.X., Boulanger F. & Shore S.N.S., 1985a, Astr. & Ap., submitted
Désert F.X., Boulanger F., Léger A. & Puget J.L., 1985b, Astr. & Ap., submitted
DeSorbo W. & Tyler W.W., 1953, J. of Chem. Phys. **21**, 1660
Draine B.T. & Anderson N., 1985, Ap.J. **292**, 494
Draine B.T. & Lee H.M., 1984, Ap.J. **285**, 89
Léger A., Jura M. & Omont A., 1985, Astr. & Ap. **144**, 147
Léger A. & Puget J.L., 1985, Astr. & Ap. **137**, L5
Mathis J.S., Rumpl W. & Nordsiek K.H., 1977, Ap.J. **217**, 425
Mathis J.S., Mezger P.G. & Panagia N., 1983, Astr. & Ap. **126**, 212
Omont A., 1985, in preparation
Pajot F., Boissé P., Gispert R., Lamarre J.M., Puget J.L. & Serra G., 1985, Astr. & Ap., submitted
Puget J.L., Léger A. & Boulanger F., Astr. & Ap. **142**, L19
Sadtler Standard Spectra, 1959, Midget edition
Sellgren K., 1984, Ap.J. **277**, 623

NON-EQUILIBRIUM EMISSION FROM SMALL PARTICLES

K. Sellgren
Institute for Astronomy
University of Hawaii
2680 Woodlawn Drive
Honolulu, HI 96822
U.S.A.

ABSTRACT. The idea of non-equilibrium thermal emission from very small particles, whether grains or molecules, has a growing interest for infrared astronomers as more and more examples of its potential applicability are found in the IRAS data set and other observations. This paper will review the observational development of these ideas, as well as discuss recent observations and suggest possible future directions for theory and observations of this phenomenon.

1. THE ORIGIN OF THE NON-EQUILIBRIUM THERMAL EMISSION MODEL

While the idea of temperature fluctuations in interstellar grains has been discussed theoretically for some time (Greenberg 1968; Duley 1973; Harwit 1975; Allen and Robinson 1975; Purcell 1976), the first observational evidence for the phenomenon was provided by the near infrared observations of visual reflection nebulae by Sellgren, Werner and Dinerstein (1983) and Sellgren (1984). The near infrared emission observed can be characterized by a 1000 K blackbody plus strong emission from the 3.3 micron unidentified emission feature, and is found to be extended over regions 0.4-0.9 pc in diameter. The spectrum of the emission is the same for three visual reflection nebulae observed and is independent of position within each nebula, while the surface brightness of the emission is found to closely follow the surface brightness distribution of visual reflected light. The need for a non-equilibrium thermal emission model was driven not by the presence of the 3.3 micron feature but by the inability of standard emission mechanisms to explain the continuum emission observed from 1 to 5 microns.

Several mechanisms for producing the emission of these sources can be immediately ruled out. The clusters of young stars associated with reflection nebulae fail by a factor of 10 to produce the observed surface brightness, as well as having very different near infrared colors. Radio observations of these nebulae, which are illuminated by middle B stars, show that free-free emission is unable to account for the emission (Sellgren, Becker and Pravdo 1983). The model of Allamandola,

217

F. P. Israel (ed.), Light on Dark Matter, 217–220.

Greenberg and Norman (1979) for UV-induced fluorescence in cold grain
mantles requires too high an efficiency of conversion of UV to infrared
photons. Reflected starlight can be ruled out by the inability of the
illuminating stars to produce the observed surface brightness, even for
the most favorable scattering parameters. This conclusion is supported
by recent polarimetric observations of NGC 7023 (Sellgren, Werner and
Dinerstein 1985), finding high polarization at short wavelengths (26%
at 1.25 microns) which steadily decreases to longer wavelengths (4% at
2.2 microns), as the fraction of reflected light compared to that of
the anomalous continuum emission decreases. Molecular hydrogen emission
excited by UV fluorescence has been recently discovered in NGC 2023
(Gatley et al. 1985), but these lines can account for at most 35% of the
2.2 micron broadband flux from this visual reflection nebula. Furthermore, Sellgren (1985) finds that 3 out of 6 visual reflection nebulae
having anomalous near infrared continuum emission show no evidence for
molecular hydrogen emission. Equilibrium thermal emission can be ruled
out, as middle B stars are unable to heat grains in radiative equilibrium, at distances of 0.1-0.2 pc, to the observed temperature of 1000K.

The model proposed by Sellgren (1984) to explain these observations was one in which small grains, whose heat capacity is small and
whose temperature therefore fluctuates, are briefly heated to high
temperatures by the absorption of single UV photons. The number N of
atoms in such a grain is found from $E = 3NkT$, where E is the energy
of the UV photon, T is the peak grain temperature, and 3Nk is the high
temperature limit of the grain heat capacity. For a 10 eV photon and
a grain temperature of 1000 K, the number of atoms is around 50,
implying a grain radius of about 10 A. This model has the advantage of
naturally explaining both the observed constancy of the color temperature with radius from the exciting star, as the peak temperature depends
only on the photon energy and not on the distance, and the similarity
of the spatial distributions of visual reflected light and near infrared
emission. These 10 A sized grains require only 0.3% of the total dust
mass, and account for only 1% of the UV dust absorption, in agreement
with the observation that about 1% of the stellar luminosity is re-
radiated by the 1-5 micron nebular emission.

2. FROM SMALL GRAINS TO BIG MOLECULES : THE PAH MODEL

The non-equilibrium thermal emission model of Sellgren (1984) was
intended to explain the continuum emission of visual reflection nebulae,
while leaving the question of the 3.3 micron emission feature largely
unaddressed. Léger and Puget (1984) and Allamandola, Tielens and Barker
(1985), however, soon applied the single UV photon excitation idea to
a specific material, polycyclic aromatic hydrocarbons (PAHs), in order
to explain the unidentified infrared emission features observed in a
wide range of astrophysical sources. Both groups found reasonable
spectroscopic agreement with the features. This was quite encouraging
progress in understanding the unidentified features, as previous work
had been unable to plausibly identify an emission mechanism and a
material simultaneously. The PAH model does not predict continuum
emission associated with the features, as is observed in visual

reflection nebulae, but Allamandola et al. and Puget et al. (1985)
argue continuum emission may be present, either from overlaps of weak
combination and overtone bands, or from an electronic continuum.

3. VISUAL REFLECTION NEBULAE BEYOND 5 MICRONS

The 1-5 micron observations of visual reflection nebulae lead to a
number of interesting observational questions, such as whether the
continuum emission extends past 5 microns, and whether the other unidentified features will be present. Sellgren et al. (1985) have recently
obtained spectra of two visual reflection nebulae from 4 to 13 microns,
to pursue these questions. These observations show that the continuum
seen at shorter wavelengths clearly extends to at least 13 microns,
while all 6 of the unidentified emission features are seen. Interesting-
ly enough, no silicate emission at 10 microns is observed, which in a non-
equilibrium thermal emission model has important consequences for the
composition of the small grains. The shape of the 1-13 micron continuum
agrees reasonably well with the predictions of a model of non-equilibrium
thermal emission from small grains, integrated over both the interstellar grain size distribution and the time dependence of the temperature of an individual grain as it cools. One important outcome of these
observations is that the IRAS 12 micron band, covering 8 to 15 microns,
includes several of the emission features as well as the underlying and
apparently associated continuum emission. If the material which emits
in reflection nebulae is ubiquitous, as seems likely from the appearance
of the unidentified features in objects as diverse as planetary nebulae,
reflection nebulae, H II regions, and galactic nuclei, then this
emission mechanism may have widespread effects on the IRAS 12 micron
band data.

4. FUTURE DIRECTIONS

The observations to date provide an exciting introduction to a new way
of looking at dust and the interstellar medium, but many unanswered
questions remain. The excitation of the non-equilibrium emission is not
well understood ; UV photons seem to be required for the features, but
the most effective wavelength is not known for either features or
continuum. This question is being studied by Sellgren et al. (1985) by
surveying the near infrared emission of a large sample of visual
reflection nebulae, and the same group plans to extend this survey to
IRAS observations. The spectrum of the emission needs further explora-
tion ; Houck and co-workers plan to obtain 20 micron spectra of visual
reflection nebulae, to search for further features, while near infrared
spectra of 12 micron excess sources such as the infrared "cirrus"
(Gautier and Beichman 1984 ; Boulanger et al. 1985) are needed to verify
that the emission mechanism is indeed non-equilibrium emission as these
authors propose and that the unidentified features are present. Laboratory studies of candidate small grain materials, such as PAHs, are
needed, to search for continuum emission associated with the features,

to obtain longer wavelength spectra, and to search for spectral differences between neutral, ionized, and de-hydrogenated species. The
question of survival of the small particles needs to be clarified both
observationally and theoretically. Evidence exists that the feature
emitting material is destroyed inside ionized regions (Sellgren 1981;
Aitken and Roche 1983), and Aitken and Roche (1985) have argued that
the lack of unidentified features in Seyfert galaxies compared with
the strength of these features in starburst galaxies implies the small
grains cannot survive the harsher environment of an active galactic
nucleus. This question could be explored by observations of the spatial
distribution of non-equilibrium emission, both of the features on small
spatial scales and of the 12 micron excess seen by IRAS on larger
scales. Finally, the questions of both survival and formation need to
be considered theoretically, particularly in light of the absence of
silicate emission from the small grains in reflection nebulae.

REFERENCES

Aitken, D.K., Roche, P.F., 1983: M.N.R.A.S. 202, 1233.
Aitken, D.K., Roche, P.F., 1985: M.N.R.A.S. 213, 777.
Allamandola, L.J., Greenberg, J.M. Norman, C.A., 1979: Astr. Ap.
     77, 66.
Allamandola, L.J., Tielens, A.G.G.M., Barker, J.R., 1985: Ap. J.
     (Letters) 290, L25.
Allen, M., Robinson, G.W., 1975: Ap. J. 195, 81.
Boulanger, F., Baud, B., van Albada, G.D., 1985: Astr. Ap. 144, L9.
Duley, W.W., 1973 : Nature Phys. Sci. 244, 57.
Gatley, I. et al., 1985 : in preparation.
Gautier, T.N., Beichman, C.A., 1984 : B.A.A.S. 16, 968.
Greenberg, J.M., 1968 : in Nebulae and Interstellar Matter,
     ed. B.M. Middlehurst and L.H. Aller (Chicago, Univ. of Chicago).
Harwit, M., 1975 : Ap. J. 199, 398.
Léger, A., Puget, J.L., 1984: Astr. Ap. 137, L5.
Puget, J.L., Léger, A., Boulanger, F., 1985: Astr. Ap. 142, L19.
Purcell, E.M., 1976 : Ap. J. 206, 685.
Sellgren, K., 1981 : Ap. J. 245, 138.
Sellgren, K., Werner, M.W., Dinerstein, H.L., 1983: Ap. J. (Letters)
     271, L13.
Sellgren, K., Becker, R.H., Pravdo, S.H., 1983, unpublished.
Sellgren, K., 1984 : Ap. J. 277, 623.
Sellgren, K., Allamandola, L.J., Bregman, J.D., Werner, M.W.
     Wooden, D.H., 1985 : Ap. J., in press.
Sellgren, K., 1985 : in preparation.
Sellgren, K., Werner, M.W., Dinerstein, H.L., 1985 : in preparation.
Sellgren, K., Werner, M.W., Allamandola, L.J., Dinerstein, H.L.,
     1985 : in preparation.

EVIDENCE FOR A 12 MICRON WATER-ICE ABSORPTION BAND IN THE IRAS LRS SPECTRA OF PROTOSTARS AND LATE TYPE STARS

M. de MUIZON (1), L.B. d'HENDECOURT (2) and C. PERRIER (3)
(1) Sterrewacht Leiden, Postbus 9515, 2300 RA, Leiden, THE
    NETHERLANDS and Observatoire de Paris, FRANCE
(2) Université Paris 7, 2 place Jussieu, 75251 Paris Cédex 05
    FRANCE
(3) Observatoire de Lyon, 69230 Saint-Genis Laval, FRANCE

ABSTRACT. We present evidence for a 12 micron absorption band due to water ice in some IRAS LRS spectra. This identification is based on the 3 micron spectrophotometry of the same objects showing the presence of the 3 micron water ice absorption band.

1. INTRODUCTION

Water ice presents in the mid infrared region an easily recognizable spectrum with three main absorption bands occurring at 3.1 μ (OH stretch), 6 μ (OH bend) and around 12 μ (hindered rotation or librational mode). Observations have revealed numerous absorption features in the infrared spectra of many heavily reddened sources (Allamandola, 1984). Among these features, two of them, at 3.1 and 6.0 μ, are attributed to water ice. The detection of the third important ice feature at 12 μ is more controversial, its presence in astronomical spectra being proved only in two objects (Gillett and Soifer, 1976; Roche and Aitken, 1984). Various physical reasons have been put forward to explain the presence or absence of this peculiar band in IR spectra of ice mixtures in the laboratory as well as in astronomical spectra (Hagen and Tielens, 1983; d'Hendecourt, 1984). The fact that the ice can be formed or evolve at different temperatures (10 to 150 K), leads to various forms of its structure, namely amorphous or crystalline phases whose IR spectra show notable differences in the 3 and 12 μ regions. The LRS instrument on board IRAS is particularly well suited for the study of this wide band. We have examined the LRS spectra where the 12 μ band can be tentatively assigned. Because of a deep absorption feature located at 9-10 μ, due to silicate absorption and because the precise shape of the silicate band is sensitive to model dependent parameters (Gillett et al, 1975), we have performed the 3 μ spectrophotometry of the objects where the presence of the 12 μ band is suspected. The goal of these observations is multiple: to definitely identify the 12 μ ice band in the LRS spectra and isolate the "true" profile of the silicate band; to study the correlation between the 3.1 and the 12 μ bands by measuring their respective

F. P. Israel (ed.), Light on Dark Matter, 221–224.
© *1986 by D. Reidel Publishing Company.*

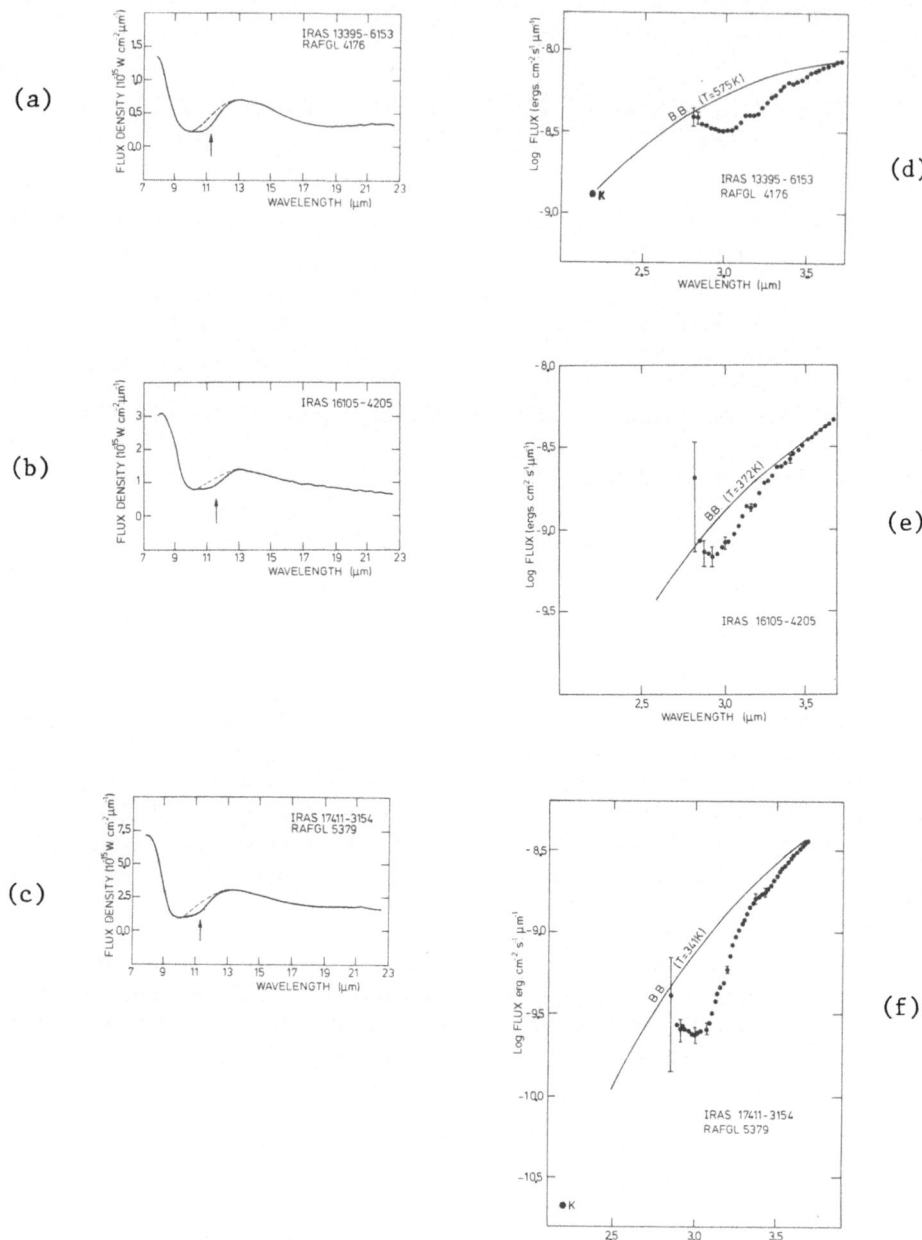

Figure 1:
a,b,c : IRAS Low Resolution Spectra of the three IRAS sources
        13395-6153, 16105-4205 and 17411-3154.
d,e,f : Three micron spectrophotometric data of the corresponding
        sources obtained at the ESO-3.6m telescope.

shapes and optical depths and to study the 3.1 μ band shape in order to look for the precise nature of the ice (degree of crystallinity).

2. RESULTS

Figure 1 a, b and c, shows three examples of LRS spectra displaying the 12 μ ice band on the side of the silicate band. The 3 μ spectrophotometry of these objects have been carried out on the 3.6 m telescope of ESO-La Silla in March 1985. The resolution of the CVF is 70. The results are shown on Figure 1 d,e and f . In order to derive the shape and depth of the band, a simple blackbody curve is used as the baseline. To derive the most correct baseline, the BB temperature is calculated in relation with the flux at 2.2 μ measured by photometry.

Table 1: Optical depths measured in the three bands of interest: 3.1 and 11.5 micron absorption of water ice, and 9.7 micron absorption of silicate.

Source	$\tau_{3.1\ \mu m}$	τ_{sil}	$\tau_{11.5\mu m}$
IRAS 17411-3154	1.29	1.93	0.36
IRAS 16105-4205	1.01	1.14	0.20
IRAS 13395-6153	0.53	1.51	0.32

3. DISCUSSION

As it is apparent from Figure 1, in each source where the presence of the 12 micron ice band is suspected, the 3 micron spectrophotometry reveals the presence of a broad and deep absorption centered on 3.1 micron and typical from the ice absorption band. All the spectra recorded in this region show a major absorption peak at 3.07 micron. The width and peak position show that this ice has an amorphous structure. A second absorption peak observed on the longwavelength side of the main peak, around 3.2 micron is quite typical of the presence of crystalline ice (Hagen et al, 1981). In the 12 micron region, the absorption peak is centered on 11.45 micron and again reveals the presence of annealed ice whose temperature has been raised to 120 K (Hagen and Tielens, 1983). Amorphous ice, whose vapor is slowly deposited at 10 K shows the librational mode at 13.3 micron (d'Hendecourt, 1984). These

spectra point to the presence of a mixture of both amorphous and crystalline ice in the line of sight towards these objects. This could partly explain why, in table 1, the measured optical depths in the two bands
(3.1 and 11.45 micron) are not well correlated. This lack of correlation can be due to different proportions of the two phases in these
objects. A two component model to explain these spectra including a
comparison with laboratory spectra is in preparation.

Finally, in at least two of the spectra of Figure 1 (d,f), we note
the presence of a small absorption dip around 3.4 micron. This dip
could be due to the absorption of a molecule containing a CH3 group
such as an alcohol or an hydrocarbon (d'Hendecourt, 1984).

REFERENCES

Allamandola, L.J.: 1984, in Galactic and Extragalactic Infrared Spec-
    troscopy, eds Kessler, M.F. and Phillips, J.P., Reidel, p. 5.
Gillett, F.C. et al.: 1975, Ap. J. 200, 609
Gillett, F.C., Soifer, B.T.: 1976, Ap. J. 207, 780
Hagen, W. et al.: 1981, Chem. Phys. 56, 367

Hagen, W. et al. : 1983, Astron. Astrophys. 117, 132
Hagen, W., Tielens, A.G.G.M.: 1983, Spectrochimica Acta 38A, 1089
d'Hendecourt, L.B.: 1984, PhD Leiden University, The Netherlands
Roche, P.F., Aitken, D.K.: 1984, M.N.R.A.S. 209, 33P

S_2 FORMATION IN INTERSTELLAR DUST; A DIAGNOSTIC OF THE MAXIMUM AGGREGATION TEMPERATURE FOR A COMET.

R.J.A. Grim, J.M. Greenberg and L.J. van IJzendoorn
Laboratory Astrophysics, Leiden University
Wassenaarseweg 78, 2300 RA Leiden, Netherlands

1. ABSTRACT

Ultraviolet photolysis of interstellar grain types of dirty ices containing H_2O, CO, CH_4 and H_2S produces S_2. At low temperature (10 K) this molecule can be trapped in the dirty ice for a time longer than the age of the solar system. The comet nucleus interior therefore retains the same composition as that of its initial material. The recent discovery of the S_2 molecule in comet IRAS-Araki-Alcock 1983d favors the theory that comets have accreted out of interstellar dust particles.

2. INTRODUCTION

The very close approach to the Earth (0.032 AU) of comet IRAS-Araki-Alcock 1983d made it possible to observe close to the nucleus where ultraviolet emission spectra obtained with the IUE show the presence of S_2 (A'Hearn et al. 1983). In their paper it was argued that the S_2 molecules were released directly from the nucleus and that the emissions resulted from resonance fluorescence by solar radiation. Another possible but excludable excitation mechanism, not discussed by A'Hearn et al., arises from the recombination of two ground state sulphur atoms. In the solid phase this has been demonstrated in chemiluminescence experiments by several groups (Lee and Pimentel 1979, Smardzewski 1978). However, in the gas phase the reaction has a low reaction rate because two bodies that recombine into a single one can not dissipate their energy of reaction. This means that the recombination is quickly followed by dissociation. A second argument that rules out the recombination as an excitation mechanism is that the energy released is not sufficient to explain the observed emission spectrum of S_2 in IRAS-Araki-Alcock. In a second paper A'Hearn and Feldman (1985) discussed possible formation mechanisms of S_2. They gave reasons to believe that the S_2 must have been formed in the icy mantles of interstellar dust particles either by ultraviolet photolysis or by cosmic ray irradiation. The recent observations of H_2S and OCS in the solid phase in the ice mantles of dust particles towards W33A (Geballe et al. 1985) support this idea. Both sulphur molecules are relatively abundant in the

225

F. P. Israel (ed.), Light on Dark Matter, 225–228.
© *1986 by D. Reidel Publishing Company.*

interstellar medium as is the case for SO_2 and H_2CS (Mann and Williams 1980), although the latter are still not detected in the solid phase.

It would be very useful if the observation of S_2 in comets not only provides evidence for the comet coagulation theory (Greenberg 1982) but also gives an upper limit to to the accretion temperature of a comet. This temperature limit restricts the birthplaces of comets. In this paper we will demonstrate that U.V. photolysis of H_2S inbedded in dirty ices produces S_2, and we will also discuss the implications for the maximum temperature of formation of the initial comet nucleus.

3. RESULTS

In the laboratory the formation of new species upon U.V. photolysis or C.R. bombardement of dirty ices is normally studied by means of infrared spectroscopy (Hagen et al. 1979, Moore et al. 1983). This technique, however, can not be applied to S_2 because the dimer shows no permanent dipole moment. Fortunately an even more sensitive technique, Laser Induced Fluorescence, exists for the identification of S_2. The S_2 molecules in the dirty ice are excited with a laser energy of 308 nm and the resulting emission spectra can be interpreted on the basis of the peak locations and line spacings.

Figure 1 shows the emission spectrum of a photolyzed $Ar/H_2O/H_2S$ = 1000/10/1 mixture. Due to the strong and narrow emission lines the emitting species can be positively identified. Both observed band systems are ascribed to S_2 and probably belong to two different excited states, $B\ ^3\Sigma_u^-$ and $B"\ ^3\Pi_u$ (Hamaguchi and Tasumi 1982).

In a dirty ice the emission bands are broadened and weakened due to the coupling of the S_2 molecules with their neighbours (figure 2). In this spectrum the two separate emission systems, $B\ ^3\Sigma_u^- \rightarrow X\ ^3\Sigma_g^-$ and $B"\ ^3\Pi_u \rightarrow X\ ^3\Sigma_g^-$, are not distinguishable. Efforts to determine the amount of S_2 molecules produced by means of ultraviolet absorption spectroscopy have not yet been successful.

The presence of S_2 in the cometary mantle not only requires the aggregation of previously photolyzed interstellar grains, but also, coagulation at low temperature. Grain-grain collisions at $V>40\ ms^{-1}$ will rise the temperature to approximately 30K, leading to diffusion of stored radicals and runaway reactions (d'Hendecourt et al. 1983). Under such conditions S_2 probably reacts with the diffusing radicals. Initial warm up experiments show that S_2 trapped in a photolyzed H_2O/H_2S = 10/1 mixture do not react at T<130K. However, we must emphasize here that the radical storage in such a mixture is too low to satisfy the conditions for runaway reactions, which are presumbaly characteristic of normal interstellar grain compositions. Above 130 K amorphous ice transforms to polycrystalline cubic ice (Hagen et al. 1981) and an increase of the S_2 emission is observed. Apparently sulphur atoms remained trapped until they were able to diffuse and recombine when the ice became more structured. At temperatures T>160K the S_2 molecules start to polymerize and/or evaporate so that emission rapidly drops to zero. Further experiments in which the reactivity of S_2 is studied in "explosive" dirty ices mixtures are currently underway.

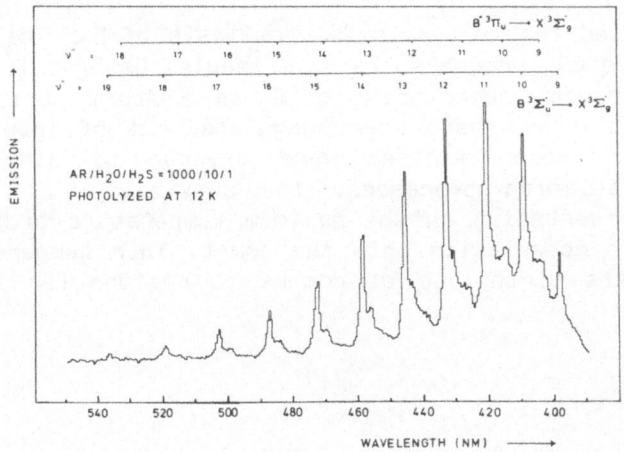

Figure 1: The B $^3\Sigma_u^-$ → X $^3\Sigma_g^-$ and B$''$ $^3\Pi_u$ → X $^3\Sigma_g^-$ emission system after excitation at 308 nm of a U.V. photolyzed Ar/H$_2$O/H$_2$S = 1000/10/1 matrix.

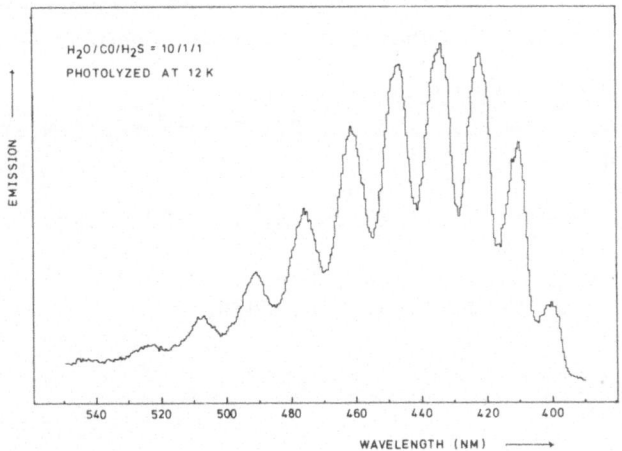

Figure 2: The B → X transition in a U.V. photolyzed H$_2$O/CO/H$_2$S = 10/1/1 matrix. The emission lines are broadened with respect to an inert matrix.

4. CONCLUSIONS

It has been demonstrated that ultraviolet photolysis of H_2S diluted in mixtures of H_2O, CO and CH_4 produces the S_2 molecule. Using this fact in connection with the recent observation of S_2 in a short period comet favors the theory that comets have been coagulated out of interstellar dust particles whose frozen mantles were created by simultaneous condensation and irradiation processes. The survival of S_2 in the cometary ice places a restraint on the maximum temperature of the dust particles during their coagulation into the comet. This temperature in turn   restricts the birthplace of comets to regions far from the sun.

REFERENCES

A'Hearn, M.F., Feldman, P.D., Schleicher, D.G., 1983, Ap. J. Letters, 274, L99.

A'Hearn, M.F., Feldman, P.D., 1985, in "Ices in the Solar System", ed. Klinger, J. Kluwer Acad. Publ. Group, Dordrecht.

Geballe, T.R., Baas, F., Greenberg, J.M., Schutte, W., 1985, Astron. Astrophys. 146, L6.

Greenberg, J.M., 1982, in "Comets", e.d. Wilkening, L.L., Univ. Ariz. Press, Tucson, p. 131.

Hagen, W., Allamandola, L.J., Greenberg, J.M., 1979, Ap. Sp. Sci., 65, 215.

Hagen, W., Tielens, A.G.G.M., Greenberg, J.M., 1981, Chem. Phys., 56, 367.

Hamaguchi, H., Tasumi, M., 1982, Chem. Phys. Lett., 91, 406.

d'Hendecourt, L., Allamandola, L.J., Baas, F., Greenberg, J.M., 1983 Astron. Astrophys. 109, L21.

Mann, A.P.C., Williams, D.A., 1980, Nature, 283, 721.

Moore, M.M., Donn, B., Khanna, R., A'Hearn, M.F., 1983, Icarus, 54, 388.

Lee, Y.P., Pimentel, G.C., 1979, J. Chem. Phys., 70, 692.

Smardzewski, R.R., 1978, J. Chem. Phys., 68, 2878.

FORMATION OF ORGANIC MOLECULES ON INTERSTELLAR DUST PARTICLES

W. Schutte and J. Mayo Greenberg
Laboratory Astrophysics, Leiden State University, Wassenaarseweg 78, 2300 RA, The Netherlands

ABSTRACT.
The 3.4 um absorption feature in the spectrum of several infrared
sources (Wickramasinghe and Allen, 1981; Willner and Pipher, 1982)
indicates the presence of organic molecules in the interstellar medium .
Several identifications were proposed such as Hydrogenated Amorphous
Carbon (Duley and Williams, 1983), and residues produced by proton
bombardment of ice mixtures (Strazzulla et al., 1984).

We show how an organic grain mantle is produced in a molecular
cloud. It is estimated that half of the volatile molecules containing O,
C, N are converted to organics in the lifetime of an average cloud. The
spectrum of an organic material made by laboratory simulation of this
process closely resembles the N.I.R. spectrum of the galactic center
source IRS7. We estimate the extinction towards the G.C.,finding a value
which agrees well with other determinations.

1. CHEMICAL PROCESSES ON DUST PARTICLES IN MOLECULAR CLOUDS, LABORATORY SIMULATION.

We simulated the chemical processes taking place on dust particles in
Molecular Clouds. In the cloud, an ice layer accretes on the grain which
is irradiated by the attenuated diffuse medium UV field given by:

$$F_{uv} = F_{uv}(0) \exp(-\tau_{uv}) \tag{1}$$

$F_{uv}(0)$ equals the Habing field of 10^8 fotons cm^{-2} s^{-1} (Habing,1968),
counting only photons of energy greater than 6eV, a typical threshold
energy for the photodissociation of chemical bonds. Furthermore, we take
$\tau_{uv} = 2A(V)$ (Hagen et al, 1979). In the laboratory we deposit
gas mixtures on an aluminum block cooled down to interstellar dust
temperatures (12K) with simultaneous irradiation by an UV lamp. For a
description of the setup see Greenberg (1982). The gas mixtures consist
mainly of H_2O and CO with some NH_3, CH_4, O_2 and N_2, representative of
the grain mantle composition as derived from a model of the gas and
grain surface chemistry in M.C.'s (d'Hendecourt, et al. 1985).

In the irradiated ices OH, HCO and COOH and other radicals were
detected by I.R. spectrometry (Hagen, 1982).

229

F. P. Israel (ed.), Light on Dark Matter, 229–232.

Table I. Average molecular cloud properties.

Cloud extinction	A(V)	4 mag	
Hydrogen density	n(H)	2×10^4 cm^{-3}	
Fraction n.c.c. in gas	f	0.6	(d'Hendecourt,1982)
Gas temperature	T_{gas}	20 K	
Cloud lifetime	τ_{mc}	3×10^7 year	(Blitz and Shu, 1980)
Dust to hydrogen ratio	n_d/n_H	5×10^{-13}	
Dust radius	a_d	0.19 µm	
Accretion time scale	τ_{ac}	0.39×10^6 year	
Irradiation density		1.8×10^{22} UV photons gr^{-1}.	
Depletion		0.4	

Upon warm up the sample explodes at about 27 K, due to a chain reaction of the radicals (d'Hendecourt, et al, 1982). The reactions form organic molecules, visible as a yellow residue remaining on the block after warm up to room temperature. This Organic Refractory material (henceforth O.R.), consists of a variety of molecules such as lactic acid, glycine and hexamethylenetetramine (Agarwal et al., 1985, Ferris, private communication). In space, grain mantle explosions with subsequent O.R. formation can be triggered by grain-grain collisions (d'Hendecourt, 1982).

2. FORMATION RATE OF THE ORGANIC REFRACTORY MATERIAL

Because the lifetime of a grain mantle against explosive disruption is much shorter than the lifetime of a M.C. (d'Hendecourt et al. 1982), we can assume that every molecule accreting on a grain will eventually take part in an explosion. The O.R. produced is then given by:

$$x(t) = 1 - \exp(-\gamma f t / \tau_{ac}) \tag{2}$$

$x = m(O.R)/(m(OR.) + m(n.c.c.))$
$m(O.R.)$ = mass of the O.R. per cm^3
$m(n.c.c.)$ = mass per cm^3 of none carbon chain molecules (n.c.c.) and atoms consisting of O, C and N completed with H (H_2O, CO, N_2, etc.)
f = fraction of the n.c.c. in the gas phase
τ_{ac} = accretion time scale for n.c.c.
γ = conversion rate n.c.c. to O.R. after explosion.

To calculate the amount of O.R. produced in a M.C. we estimate the average cloud properties (table I). The interstellar extinction curve gives an average radius of a_d = 0.15 µm for the diffuse medium grain (Chlewicki, 1985). With a visual extinction efficiency of 1.2 (Chlewicki, 1985), the N(H)/E(B-V) relation given by Martin (1978) yields the ratio of grains to hydrogen atoms. The depletion of O, C, N and the adopted value of f, taking ρ (ice) = 1 gr cm^{-3}, give the estimated radius of the M.C. grain, consisting of the diffuse medium grain with an ice mantle. The outcome agrees with observations of the

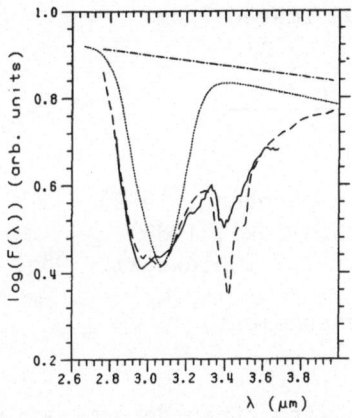

Figure 1. The spectrum of IRS7 (Solid line) compared with laboratory spectra of amorphous water ice (dotted line) and of an O.R. synthesised by irradiation of an ice of composition $H_2O:CO:CH_4:NH_3$ = 3:1:1:0.3 (dashed line).Dash dot line is the continuum.

B.N. object (Greenberg and Van der Bult, 1983). The accretion timescale was calculated for a molecule of 20 amu. Using a penetration depth in the ice mantle of 0.08 μm, similar to water ice (Okabe, 1978), we find from equation (1) the given irradiation density, that is the number of UV photons absorbed per unit accreted mass. Laboratory measurements then give $\gamma = 1.5.10^{-2}$. .Equation (2) now gives the conversion to O.R. in the cloud life: $x(\tau_{mc}) = 0.5$.

3. DETECTION OF ORGANIC GRAIN MANTLES FROM THE OBSERVATIONS OF IRS7

Star formation processes will eventually result in the return of the M.C. dust and gas to the diffuse medium. Here the grain mantle looses its smaller volatile molecules, due to processes like cloud-cloud collisions and crushing by shock waves of supernova remants (Greenberg, 1982). The resulting grain consists of a silicate core with an O.R. mantle.

Figure 1 compares the N.I.R. spectrum of the G.C. source IRS7 (Whittet, 1985) to the spectra of amorphous water ice and of an O.R. produced by irradiating an ice of composition $H_2O:CO:CH_4:NH_3$ = 3:1:1:0.3. The match between the O.R. and IRS7 was made by equating the spectra at 2.82 and 3.6 μm and multiplying the O.R. spectrum by a factor choosen to get a maximal overall resemblance . The continuum thus found agrees with the observations of Allen and Wickramasinghe (1981).

The water absorption indicates the presence of dense molecular clouds (Whittet, 1983). Obviously it is nearly absent, while the O.R. matches the observations rather well, implying a more tenuous medium where the grain mantles have lost their volatile molecules.

The features at 3.4 μm and 3.0 μm are respectively due to the C-H stretch and probably the O-H stretch vibration in organic molecules The O.R. matches the observed 3 μm feature, but its 3.4 μm feature is not so close. Furthermore, the depth ratio of the features differs from the observations. However, we did not yet account for the effect of photoprocessing of the O.R. in the diffuse medium.

From the depth of the 3.0 μm feature in the IRS7 spectrum the

Visual extinction towards the G.C. can be estimated from:

$$\frac{\tau_{G.C}(3.0)}{A_{G.C.}(V)} = \frac{Ab\,(3.0)\,f(O.R.)\,\frac{4}{3}\,\Pi(a_m^3 - a_c^3)}{1.086\,Q(V)\,\Pi\,a_m^2}. \qquad (3)$$

a_c = radius of the silicate core $\approx$ 0.05 μm (Greenberg, 1982)
a_m = radius of the O.R. mantle $\approx$ 0.15 μm (Chlewicki, 1985)
$Q(V)$ = extinction efficiency in the V band $\approx$ 1.2 (Chlewicki, 1985)
$p(O.R.)$ = 1.4 gr cm^{-3}
$Ab(3.0)$ = 6.7 x 10^2 cm^2 gr^{-1} (laboratory measurement)
$\tau_{g.c.}$ (3.0) = 0.47 (see fig. 1)

This yields $A_{g.c.}(V)$ = 34, in good agreement with other determinations.

REFERENCES

Agarwal, V.K., Schutte, W.A., Greenberg, J.M., Ferris, J.P., Briggs, R., Connor, S., van de Bult, C.E.P.M., Baas, F., 1985, Origins of Life, Vol. 16, 21.
Allen, D.A., Wickramasinghe, D.T., 1981, Nature, 294, 239.
Blitz, L., Shu, F.H., 1980, Ap.J. 238, 148.
Butchart, I., McFadzean, A.D. Whittet, D.C.B., Geballe, T.R., 1985, submitted to Ap.J.
Chlewicki, G., 1985, Ph.D. Thesis, University of Leiden, NL.
Duley, W.W., Williams, D.A., 1983, Mon. Nat. R. astr. Soc. 205, 67P.
Greenberg, J.M., 1982, Submillimetre Wave Astronomy, eds. Beckman, J.E. and Philips, J.P., Cambridge University Press.
Greenberg, J.M., Bult, C.E.P.M., van de, 1983, Proc. of the Hilo Workshop on Lab. and Obs. I.R. Spectra of Interst. Dust, 70.
Habing, H.J., 1968, Bul.. Astron. Inst. Ned., 19, 421.
Hagen, W., 1982, Ph.D. Thesis, University of Leiden, NL.
Hagen, W., Allamandola, L.J., Greenberg, J.M., 1979, Ap. Sp. Sci. 65, 215.
d'Hendecourt, L.B., Allamandola, L.J., Greenberg, J.M. 1985 Astron. Ap. 152, 130.
d'Hendecourt, L.B., Allamandola, L.J., Greenberg, J.M. 1982, Astron. Ap. 109, L12.
Martin, P.G., 1978, Cosmic Dust, Oxford University Press.
Okabe, H., 1978, Photochemistry of small molecules, Wiley Interscience.
Strazulla, G., Cataliotti, R.S., Calcagno, L., Foti, G., 1984, Astron. Ap. 133, 77.
Whittet, D.C.B., Bode, M.T., Longmore, A.J., Baines, D.W.T., Evans, A., 1983, Proc. of the Hilo Workshop on Lab. and Obs. I.R. Spectra of Interst. Dust, 61.
Willner, S.P., Pipher, J.L., 1982, Proc. of Workshop on the Galactic Center, Cal. Inst. Tech.

POLYCYCLIC AROMATIC HYDROCARBONS AND THE DIFFUSE INTERSTELLAR BANDS

G.P. van der Zwet[1] and L.J. Allamandola[1]
[1] Laboratory Astrophysics, Leiden State University, Wassenaar-
    seweg 78, 2300 RA Leiden, The Netherlands.
[2] National Research Council Senior Associate, NASA Ames
    Research Center, M.S. 245-6, Moffett Field, CA 94035, U.S.A.

ABSTRACT. The profiles, widths and positions of the Diffuse Interstellar
Bands (DIB's) are discussed. It is shown that the carriers of the DIB's
are molecular rather than impurities in grains. Polycyclic Aromatic
Hydrocarbons (PAH's) are attractive candidates for the DIB's, in view of
their likely abundance, their extreme stability, and the fact that these
molecules, and their corresponding radicals, show visible transitions.

1. INTRODUCTION-HISTORY OF THE DIFFUSE INTERSTELLAR BANDS PROBLEM

The DIB's form a series of approximately 40 visible absorptions
extending from about 4400 A into the near IR. The first bands were
discovered some 70 years ago, but their interstellar origin was first
recognized by Merrill (1934).

Over the years a number of identifications have been proposed for
the bands, which can roughly be divided into two groups: impurities
embedded in grains, and molecules. Up to date neither explanation seems
convincing (reviews: Herbig, 1975; Smith et al., 1977).

2. OBSERVATIONAL CONSTRAINTS

2.1. Profiles of the bands

The profiles of the DIB's are generally slightly asymmetric with a steep
edge on the blue (high-energy) side of the band. Some variation in the
degree of asymmetry occurs. With the resolution presently available no
fine structure is seen.

Impurity absorptions in large grains responsible for the visual
extinction (radius a ~ 0.12 µm), can be excluded as DIB-carriers,
because scattering theory predicts a much stronger asymmetry for the
bands than is observed (van de Hulst, 1949; Greenberg and Hong, 1976).
Small grains (a ~ 0.01 µm) can be excluded as well, because their low
scattering efficiency will result either in symmetrical bands, or in
asymmetric bands with a steep edge on the red side (van de Hulst, 1949;

233

Purcell and Shapiro, 1977). Therefore, the only remaining sizes to be considered are in the intermediate range (0.02 $\leq$ a $\leq$ 0.10 μm). It turns out that a range of sizes is needed in order to account for the observed variety of bandshapes, which seems highly unlikely (Chlewicki et al., 1985).

Electronic transitions in molecules show rotational fine structure. For large molecules however, this structure will be unresolved and specific rovibronic band contours arise (Ross, 1971). Like most of the DIB's the band contours are generally slightly asymmetric with a steep edge on the blue side (Danks and Lambert, 1976). The asymmetry arises from the difference between the rotational constants in the ground and excited electronic state of the molecule.

2.2. Widths of the bands

The widths of the DIB's range from 1 to 100 cm^{-1} (0.6 - 30 A).

Impurity absorption bands in grains are likely to be broad because of the "dirty" and amorphous nature of the grains. At least for the narrower DIB's a solid state origin seems to be an inadequate solution.

The widths of band contours of electronic transitions in molecules are determined by the rotational constants of the molecule, its temperature, and the type of transition involved, i.e. how the transition moment is oriented with respect to the molecular axes (Hollas, 1973). A typical width for a large molecule under interstellar conditions is a few cm^{-1}, in agreement with the narrower DIB's (van der Zwet and Allamandola, 1985).

Additional line broadening may arise for example from radiationless internal conversion. This is a process in which a particular vibronic level excited couples with nearby levels of a lower electronic state and subsequently relaxes nonradiatively. This causes a lifetime broadening of the absorption and linewidths up to 100 cm^{-1} are not unusual among large aromatic molecules (Byrne and Ross, 1971). The widths of some of the broader DIB's may be caused by this process (Douglas, 1977). It is worth noting that the broader DIB's tend to lie at shorter wavelengths (Herbig, 1975), which is consistent with the internal energy conversion idea because the density of states increases with energy.

2.3. Positions of the bands

No significant variation of the central wavelengths of the DIB's has been found. This requires a uniform chemical composition and constant temperature of the grains, if the DIB's are due to impurity absorptions. For molecules this constraint is naturally fulfilled.

3. POSSIBLE CARRIER: POLYCYCLIC AROMATIC HYDROCARBONS

The work that follows has been done independently by two groups of authors: Léger and d'Hendecourt (1985), and van der Zwet and Allamandola (1985).

A few examples of PAH's are shown in figure 1: they consist of

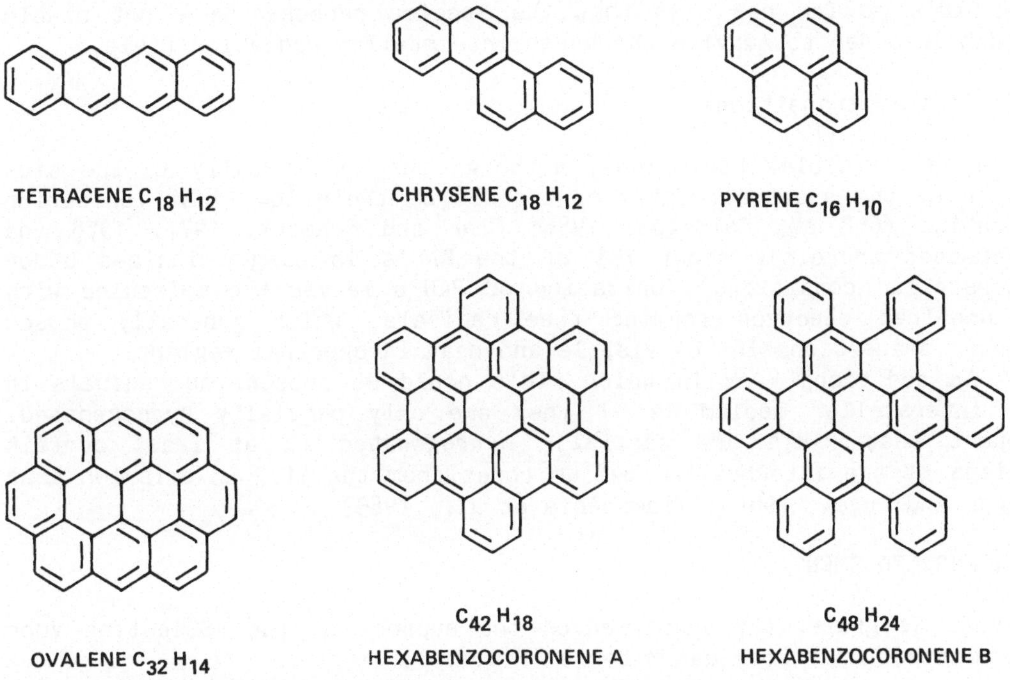

TETRACENE $C_{18}H_{12}$ CHRYSENE $C_{18}H_{12}$ PYRENE $C_{16}H_{10}$

$C_{42}H_{18}$ $C_{48}H_{24}$

OVALENE $C_{32}H_{14}$ HEXABENZOCORONENE A   HEXABENZOCORONENE B

Fig. 1. Some polycyclic aromatic hydrocarbons.

fused aromatic ring systems made up of carbon and hydrogen.

We believe that PAH's are likely candidates as carriers for the DIB's for the following reasons:

3.1. Abundance, evidence from the Unidentified IR Bands

Recently, it has been proposed that emission from (highly) vibrationally excited PAH-like species (or their corresponding positive ions) is responsible for the so-called Unidentified IR (UIR) emission features at 3.3, 6.2, 7.7, 8.6 and 11.3 μm (Léger and Puget, 1984; Allamandola et al., 1985). The abundance of these molecules needs to be about 2×10^{-7} n_H, in order to account for the observed flux in the UIR bands. The proposed presence of PAH's in a variety of objects points to their ubiquituous presence in the interstellar medium, an idea similar to Johnson's (1967) and Donn's (1968).

3.2. Stability

Large PAH's are among the most stable molecules known. Their stability is due to the delocalization of the electrons in the π molecular orbitals. Moreover, because these molecules are large, they have many internal degrees of freedom and are therefore able to get rid of energy quite easily by fast internal redistribution.

The extreme stability of these molecules is of key importance in the DIB-problem: a serious objection against any molecular carrier for the DIB's so far has been that the species proposed were not stable enough in order to survive the harsh interstellar radiation field.

3.3. Visible transitions

PAH's show visible transitions, although they are usually in the blue region of the spectrum. However, because of their low first ionization potential (6-8 eV, Gallegos, 1986; Clar and Schmidt, 1977, 1978 and references therein), about 75% of the PAH's is singly ionized under interstellar conditions. Ionization of PAH's leaves the molecules with an unpaired electron forming free radicals, which generally posses allowed transitions in the visible and near IR spectral regions.

An additional way in which PAH's could be present as radicals in the interstellar medium is if they are only partially hydrogenated. Support that PAH's are partially hydrogenated in at least certain regions of the interstellar medium comes from the 11.3 μm emission band (Léger and Puget, 1984; Allamandola et al., 1985).

4. ACKNOWLEDGEMENT

G. v.d. Z. gratefully acknowledges the support of the "Stichting voor Fundamenteel Onderzoek der Materie (F.O.M.)".

5. REFERENCES

Allamandola, L.J., Tielens, A.G.G.M., Barker, J.R.: 1985, Astrophys. J. Letters **290**, L25
Byrne, J.P., Ross, I.G.: 1971, Australian J. Chem. **24**, 1107
Chlewicki, G., van der Zwet, G.P., van IJzendoorn, L.J., Greenberg, J.M., Alvarez, P.P.: 1985, Astrophys. J. (accepted)
Clar, E., Schmidt, W.: 1977, Tetrahedron **33**, 2093
Clar, E., Schmidt, W.: 1978, Tetrahedron **34**, 3219
Danks, A.C., Lambert, D.L.: 1976, Montly Notices Roy. Astron. Soc. **174**, 571
Donn, B.: 1968, Astrophys. J. **152**, L129
Douglas, A.E.: 1977, Nature **269**, 130
Gallegos, E.J.: 1968, J. Phys. Chem. **72**, 3452
Greenberg, J.M., Hong, S.-S.: 1976, Astrophys. Space Sci. **39**, 31
Herbig, G.H.: 1975, Astrophys. J. **196**, 129
Hollas, J.M.: 1973, Chapter 2 in Molecular Spectroscopy: A Specialized Periodical Report Volume 1, The Chemical Society, London, p. 73
Johnson, F.M.: 1967, in Interstellar Grains, ed. J.M. Greenberg and T.P. Roark, NASA - SP-140, 229.
Léger, A., d'Hendecourt, L.: 1985, Astron. Astrophys. **146**, 81
Léger, A., Puget, J.L.: 1984, Astron. Astrophys. **137**, L5
Merrill, P.W.: 1934, Publ. Astron. Soc. Pac. **46**, 206
Purcell, E.M., Shapiro, P.R.: 1977, Astrophys. J. **214**, 92
Ross, I.G.: 1971, Adv. Chem. Phys. **20**, 341
Smith, W.H., Snow, T.P., York, D.G.: 1977, Astrophys. J. **218**, 124
van de Hulst, H.C.: 1949, Rech. Astr. Obs. Utrecht **11**, part 2
van der Zwet, G.P., Allamandola, L.J.: 1985, Astron. Astrophys. **146**, 76

IDENTIFICATION OF POLYCYCLIC AROMATIC HYDROCARBONS

A. Léger and L. d'Hendecourt
Groupe de Physique des Solides de l'Ecole Normale Supérieure
Université Paris 7, Tour 23 - 2 place Jussieu
75251 PARIS CEDEX 05 - FRANCE

ABSTRACT. The nature of the Very Small Grains evidenced by K. Sellgren (1985) is discussed. Their stability suggests that they are graphitic material and specifically Polycyclic Aromatic Hydrocarbons (PAHs). The expected IR emission of a typical PAH, coronene, gives an impressive spectroscopic agreement with the five observed "Unidentified IR Emission Features", leading to an unambiguous identification. Those PAHs are the most abundant organic molecules detected to this date ($f \sim 10^{-5}$). Several astrophysical implications are reviewed.

1. NATURE OF VERY SMALL GRAINS

1.1. Refractory grains...

Sellgren (1984, 1985) has shown strong evidence for quantum heating of Very Small Grains. To survive heating up to temperatures about 1000 K these particles must be refractory. Léger and Puget (1984) have shown

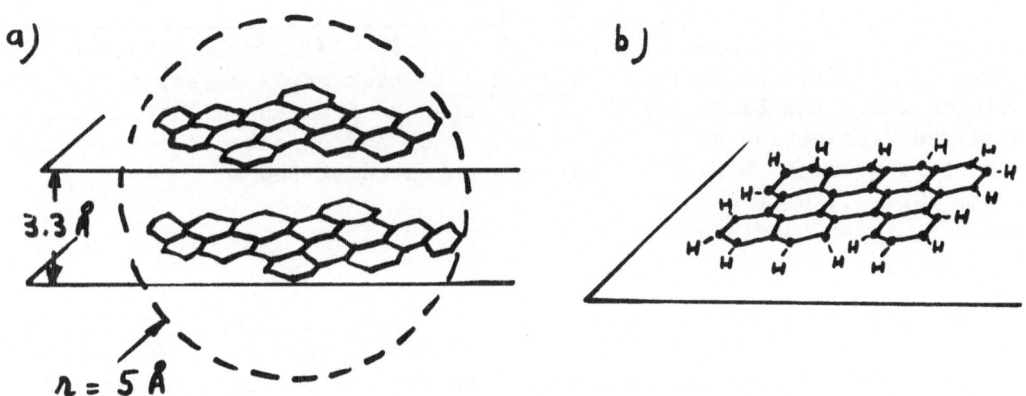

Figure 1 : a) What a 5 Å "sphere" of graphite would look like.
b) Proposed structure of an interstellar graphite cluster.

237

F. P. Israel (ed.), Light on Dark Matter, 237–240.
© *1986 by D. Reidel Publishing Company.*

that ices and silicates would be sublimated whereas graphitic grains could resist.

1.2. ... which are large molecules

Graphite is made of weakly bound planes. Therefore frequently heated graphitic clusters are likely planar. In addition, the presence of ambient H atoms and dangling bonds on peripherical C atoms suggest that they are large Polycyclic Aromatic Hydrocarbon (PAH) molecules (Fig. 1).

2. SPECTROSCOPIC IDENTIFICATION

Léger and Puget (1984) derived the expected IR emission spectrum of such free molecules using the laboratory measured absorption of a typical large PAH: coronene. They found an impressive agreement with the so-called "Unidentified IR Emission Features" at 3.28 - 6.2 - 7.7 - 8.6 - 11.3 μm (Fig. 2). This pro-position can also solve the puzzle of the high efficiency of the conversion - incoming UV/outcoming IR - because when the energy is absorbed by a free flyer molecule and degraded into vibrational energy it is entirely re-radiated in the IR, as this is the only de-excitation path way.

From the point of view of the Analytic Chemistry *the presence of some bands (3.28 - 6.2 and 11.3 μm) are highly characteristic of PAHs* (Bellamy, 1966) *and leaves very little doubt on their identification* in the objects where the bands are observed (reflection and planetary nebulae: NGC 2023, 7023, 7027..., H$^+$ regions: Orion..., star burst galaxies: M82, Arp 220...).

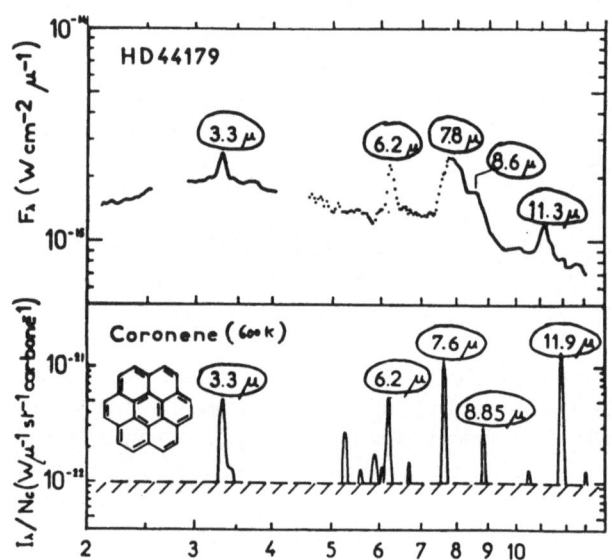

Figure 2 : Observed IR emission from HD 44179 (adapted from Russell et al., Ap. J. **220**, 568) and expected emission from coronene heated at 600 K.

3. THE MOST ABUNDANT ORGANIC MOLECULES DETECTED IN SPACE

The abundance of PAHs can be derived from the relative fluxes in FIR (large grains) and near-mid IR (transient heating of PAHs) as both species are submitted to the same irradiation flux and absorb according to their abundance and cross section. One finds that few percent of the cosmic C is needed in those PAHs molecules. This brings PAHs to the level of *very abundant detected cosmic molecules*, the most abundant

organic ones in the Universe (see
Table).

These molecules have been unrecogni-
zed for a long time whereas far less
abundant ones, as long cyanopolyynes,
were detected in radioastronomy. The
absence of a clear signature in the ra-
diofrequencies is at the origin of this
situation.

molecule	f = nb of atoms involved/n_H
H_2	~ 1
CO	10^{-4}
O_2, N_2	?
PAHs	10^{-5}
HCN, H_2CO...	$\} \lesssim 10^{-7}$
$HC_{11}N$	10^{-10}

4. IR EMISSION FROM PAHs IN THE GALAXY AND IN OUTER GALAXIES

Puget et al (1985) have computed the IR
emission of a whole population of carbon grains and PAH molecules when
submitted to an actual stellar irradiation field (see X. Désert and
F. Boulanger contributions in this Workshop). The optical properties
of grains and molecules were introduced taking into account their atomic
vibrations and electronic transitions contributions.

The IR emission (Fig. 3) mainly comes from: (i) grains (r > 15 Å)
at a steady and low temperature
(ii) small species (PAHs) emitting when transiently heated to
a high temperature by the absorption of a single UV photon.
The former emission peaks about
100 μm and was found in classical models of dust but the later
gives a completely new near and
mid IR flux. Process (ii) exhibits the vibrational bands of
the PAH molecules (3.3 - 6.2 -
7.7 - 8.6 and 11.3 μm) and a
continuum at shorter wavelengths
coming from the electronic transitions of the molecules. Contribution (ii) is cruxial and
can account for many observations, including :
- The formely called "Unidentified IR Emission Features"
- The 12 μm emission of the *IRAS*
"Cirrus" (prediction: Puget et al,

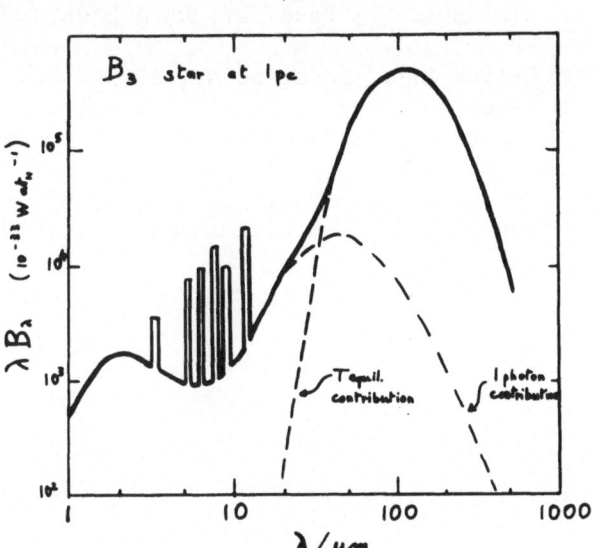

Figure 3 : IR emission of a population of grains and PAH molecules
irradiated by a B3 star.

1985, see also Draine and Anderson, 1985; observation: Boulanger et al,
1985, Beichman, 1986).
- The *extended* mid-IR (4-20 μm) emission extensively observed in AFGL
and IRAS surveys, far from heating sources.
- Possibly, the red emission excess in the Visual of the Red-Rectangle
and reflection nebulae.

5. CANDIDATES TO THE DIFFUSE INTERSTELLAR BAND (DIBs) MYSTERY

These recently recognized abundant molecules *might* also *be* at the origin of the half-century old DIBs mystery. Simultaneously Van der Zwet and Allamandola (1985) and Léger and d'Hendecourt (1985) (see also Van der Zwet in this Workshop) have shown that PAHs, probably in an ionized form, are good candidates for the DIB carriers, because :
    - they have absorptions in the Visual with possibly adequate line widths;
    - they can stand the UV of the Interstellar Medium;
    - they are abundant enough to account for the observed intensities. Other species fulfilling those requirements are very few indeed. But *this is only a suggestion* and a precise spectroscopic match is needed before this becomes a claim. *The situation is quite different from the IR*, from which, we think, that the presence of PAHs is already established.

REFERENCES
  Bellamy L.J., 1966 : IR spectra of Complex Molecules, Wiley & Sons
  Beichman C.A., 1986 : these proceedings, page 279.
  Boulanger F., Baud, B., Van Albada, G.D., 1985: Astr. & Astrophys. 144, L9
  Draine B.T., Anderson N., 1985 : Ap. J. 292, 494
  Léger A., Puget J.L., 1984 : Astr. & Astrophys. 137, L5
  Léger A., d'Hendecourt L., 1985, Astr. & Astrophys. 146, 81
  Puget J.L., Léger A., Boulanger F., 1985 : Astr. & Astrophys. 142, L19
  Sellgren K., 1984 : Ap. J. 277, 623; these proceedings, page 217.
  Van der Zwet G., Allamandola L.J., 1985 : Astr. & Astrophys. 146, 76.

SILICATE ABSORPTION STRENGTH; POLARIZED EMISSION AND ABSORPTION BY ALIGNED GRAINS

D. K. Aitken
Department of Physics (RAAF Academy)
University of Melbourne
Parkville
Victoria 3052
Australia

1. SPECTROSCOPY

It is well known (e.g. Willner, 1984; Aitken, 1981) that the observed astronomical silicate features are much better matched by amorphous materials (Day & Donn, 1978; Day, 1979; Stephens & Russell, 1979) than by any crystalline terrestrial silicate. It is also clear that the astronomical feature derived from molecular cloud studies, typified by the Trapezium region of Orion (Forrest, Gillett & Stein, 1975), differs significantly from that in the interstellar medium (Roche & Aitken, 1984; 1985), which in turn is well represented by the circumstellar shells of supergiants (Aitken, Roche & Spenser, 1980; Mitchell & Robinson, 1981) such as μDep (Russell, Soifer & Forrest, 1975). There is also evidence for differing silicate emissivities in various circumstellar environments (Papoular & Pegourie, 1983). Further spectroscopic studies of a wide range of late type stellar objects are needed to investigate these effects in more detail and with improved statistics.

It is not yet clear whether these differences can be attributed to variations in the chemical or physical form of the silicates, to the development of grain mantles, to temperature distributions and opacity effects, or to the presence of non-silicate components. It seems unlikely that the generally muted nature of the feature can be ascribed to a mixture of crystalline silicates, although this cannot be completely ruled out on the spectroscopic evidence alone.

In the interstellar medium (ISM) it seems that a correlation between A_v and τ_{sil} has been established, at least for the local region up to $\cong$ 3kpc, with A_v/τ_{sil} = 18.5 ± 1.5 (Roche & Aitken, 1984). However this relation does not hold over the longer path length to the centre of the Galaxy, where $A_v/\tau_{sil} \cong$ 8. The difference has been attributed to a lack of carbon based grains in the inner regions of the Galaxy (Roche & Aitken, 1985); this lends support to a model in which the visual and short wavelength extinction arises in a separate grain population from the silicates which give the bulk of the absorption at 10μ.

F. P. Israel (ed.), Light on Dark Matter, 241–244.
© *1986 by D. Reidel Publishing Company.*

2. SPECTROPOLARIMETRY

Spectroscopic studies do not reveal the absolute value of the absorption
coefficient of the grain material because the column density is unknown.
However if the grains are non-spherical the wavelength of a spectral
feature depends on the orientation of the E vector in a way which
involves the absorption coefficient. Thus spectropolarimetric studies
can provide independent constraints on the grain chemistry, and may also
give information on the grain distribution along the line of sight, and
sometimes unravel some of the complications of radiative transfer.
Although there is evidence that in some cases, for example in Orion
(Aitken et al, 1985a) and ηCar (Briggs & Aitken, 1985), mechanisms other
than magnetic alignment are operating, usually, whatever the alignment
mechanism, the Barnet effect (Dolginov & Mytrophanov, 1976) will be so
strong that the grain angular momenta will be distributed about the
local magnetic field. Consequently in most cases spectropolarimetric
studies also give information on the strength and direction of magnetic
fields in the grain environment.

Martin (1975) has shown that if the polarization is produced by
dichroic absorption in aligned grains, the ratio $p_a(\lambda)/\tau(\lambda)$ is related
to the band strength of the absorbing material. He has used this to
show that the observations of Dyck et al (1973) on the polarization of
the BN object in Orion imply that these grains have band strength much
less than crystalline silicates. In Fig 1a we present $p_a(\lambda)/\tau(\lambda)$ from
more recent observations of the BN object (Aitken et al, 1985a), and
compare with estimates for idealized silicate grains of band strength
3 and 9 x 10^3 cm^{-1} (Martin, 1975) in complete Davis-Greenstein alignment
(here Martin's estimates are reduced by a factor of 10). While $p_a(\lambda)$
is directly observed, the form of $\tau(\lambda)$ has to be inferred since the
absorption feature is influenced by the form of the underlying spectrum.
The observed spectrum is in fact well fitted by a combination of
Trapezium-like material in emission and absorption, and $\tau(\lambda)$ is taken
to be Trapezium-like with optical depth 3.3 at 10μm. Also shown is
$p_a(\lambda)/\tau(\lambda)$ for the interstellar polarization towards the Galactic
centre, as observed in IRS3 which has no intrinsic emission feature or
polarization (Aitken et al, 1985b). The absolute values of $p_a(\lambda)/\tau(\lambda)$
differ from Martin's estimates by scaling factors which reflect the
grain shape and alignment, but the form of the curves is similar and
suggests a comparable band strength for the ISM and molecular cloud
material, notwithstanding their different 10μm emissivities. Similar
polarization profiles have been observed in a number of heavily
obscured objects, and Lee & Draine (1985) consider that the form of
$p_a(\lambda)$ implies grain shapes which are predominantly oblate.

When we compare the 2 and 10μm polarizations in these two objects
we see that the ratio p_2/p_{10} for the line of sight to the Galactic
centre is approximately half that for Orion, and there is a position
angle change in the Galactic centre from 20° to 0° between 2 and 10μm.
The position angle change is confirmed by the observed polarization of
the Galactic centre [NeII] line emission, which at these wavelengths we
consider is uninfluenced by scattering and therefore representative of
the interstellar absorption. The polarimetry therefore reinforces the

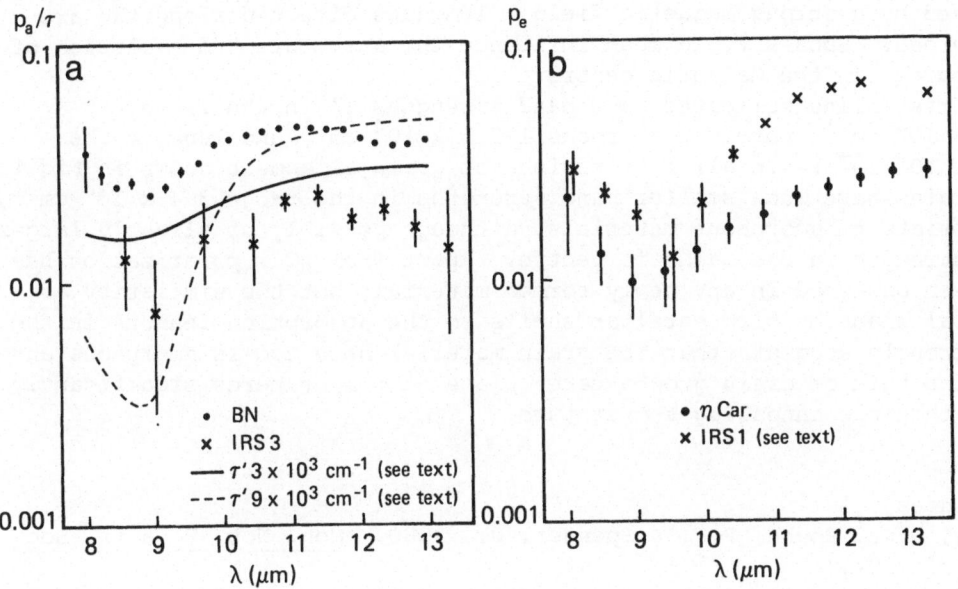

Fig 1a $p_a(\lambda)/\tau(\lambda)$ and Fig 1b $p_e(\lambda)$ (see text) for four regions of
grain alignment. The solid and broken lines denote computed values of
p_a/τ (taken from Martin (1975) and reduced by a factor 10) for the
band strengths indicated.

argument that, in the ISM, separate grains are responsible for
extinction at short wavelengths and 10µm (Mathis et al, 1977), and
further indicates that these components are fractionated along the line
of sight; both components must be capable of alignment.
       A number of infrared sources in ionized regions show evidence for
polarized emission from aligned grains, often overlaid by an absorptive
component. For emission the polarization, $p_e(\lambda)$ will have the same
spectral form as $p_a(\lambda)/\tau(\lambda)$ for the same grain composition; a
particular advantage here is that knowledge of $\tau(\lambda)$ is not required.
ηCar shows a silicate feature in emission which is polarized at
positions offset from the centre of the homunculus. In Fig 1b we show
$p_e(\lambda)$ for a representative offset position; $p_e(\lambda)$ compares well with
$p_a(\lambda)/\tau(\lambda)$ for the sources in Fig 1a. This dust formed in an outburst
only ∿ 100 years ago. In the Galactic centre there is also evidence
for polarized emission (Lebofsky et al, 1982), although here the effect
is complicated by the interstellar polarization. Assuming that the
interstellar polarization is well represented by that in the BN object,
the interstellar contribution may be unfolded from the data (Aitken et
al 1985b). The intrinsic component $p_e(\lambda)$ in IRS1 obtained in this way
is shown in Fig 1b. All the sources in the N-S arc (IRS1, 10, 5, 8 in
the nomenclature of Becklin et al, 1978) show similar large
polarizations with position angle directed closely normal to the line
of the arc, requiring the grain angular momenta to be directed along
the arc. Alignment of these warm grains in this environment can be

achieved by a strong magnetic field > 10mgauss directed along the arc
of sources.   Such a field must influence the structure and evolution of
the sources in the Galactic centre.

Crystalline silicates have band strengths τ' in the range
6-10 x 10^4 cm^{-1}, terrestrial rocks 1-2.5 x 10^4 cm^{-1} and lunar rocks
7-14 x 10^3 cm^{-1};   in all four regions of grain alignment shown in Fig 1
the grains have much smaller band strengths in the range 3-4 x 10^3 cm^{-1},
appropriate to amorphous materials, although possibly of slightly larger
band strength in the Galactic centre.   Apart from ηCar polarization has
not been observed in any newly formed material, but the similarity of
spectral shape of circumstellar shells to the absorption feature in the
ISM strongly suggests that the grain material here too is amorphous and
that the bulk of grain growth takes place at temperatures significantly
lower than the annealing temperature.

References

Aitken, D.K., Roche, P.F. & Spenser, P., 1980. Mon. Not. R. astr. Soc.,
        193, 207.
Aitken, D.K., 1981. IAU Symp. 96, Infrared Astronomy, ed.
        Wynn-Williams, C.G. & Cruikshank, D.P., Reidel, Dordrecht, 207.
Aitken, D.K., Bailey, J.A., Roche, P.F. & Hough, J.M., 1985a.
        Mon. Not. R. astr. Soc., in press.
Aitken, D.K., Roche, P.F., Bailey, J.A., Briggs, G.P., Hough, J.M. &
        Thomas, J.A., 1985b. Mon. Not. R. astr. Soc., in press.
Becklin, E.E., Mathews, K., Neugebauer, G. & Willner, S.P., 1978.
        Astrophys. J., 219, 121.
Briggs, G.P. & Aitken, D.K., 1985. Proc. astr. Soc. Aust., in press.
Day, K.L. & Donn, B., 1978. Astrophys. J., 222, L45.
Day, K.L., 1979. Astrophys. J., 234, 158.
Dolginov, A.Z. & Mytrophanov, I.G., 1976. Astrophys. & Sp. Sci., 43
        291.
Dyck, H.M., Capps, R.W., Forrest, W.J., Gillett, F.C., 1973. Astrophys.
        J., 183, L99.
Forrest, W.J., Gillet, F.C. & Stein, W.A., 1975. Astrophys. J., 195,
        423.
Lebofsky, M.J., Rieke, G.H., Deshpande, M.R., Kempe, J.C., 1982.
        Astrophys. J., 263, 672.
Lee, H.M. & Draine, B.T., 1985. Astrophys. J., 290, 211.
Martin, P.G., 1975. Astrophys. J., 202, 393.
Mathis, J.S., Rumpl. W. & Nordsieck, K.H., 1977. Astrophys. J., 217,
        425.
Mitchell, R.M. & Robinson, G., 1981. Mon. Not. R. astr. Soc., 191, 801.
Papoular, R. & Pegourie, B., 1983. Astr. & Astrophys., 128, 335.
Roche, P.F. & Aitken, D.K., 1984. Mon. Not. R. astr. Soc., 208, 481.
Roche, P.F. & Aitken, D.K., 1985. Mon. Not. R. astr. Soc., in press.
Russell, R.W., Soifer, B.T. & Forrest, W.J., 1975. Astrophys. J., 198,
        L41.
Stephens, J.R. & Russell, R.W., 1979. Astrophys. J., 228, 780.
Willner, S.P., 1984 in XVIth ESLAB Symposium, ed.Kessler, M.F. &
        Phillips, J.P., 37.

OPTICAL PROPERTIES OF SIMULATED ASTROPHYSICAL GRAINS AND THEIR
DYNAMICS IN THE NEAR-EARTH ENVIRONMENT   LAUR 85-1978

J. R. Stephens, T. D. Kunkle, and I. B. Strong
Los Alamos National Laboratory
P. O. Box 1663
Los Alamos, New Mexico 87545
USA

ABSTRACT. Plans are underway to prepare simulated interplanetary
grains and study their behavior when released into low-Earth orbit.
The program is twofold: preparing and characterizing, in the
laboratory, grains which simulate interstellar and interplanetary
grains, and monitoring the optical properties of grains released into
low-Earth orbit to study their dynamical behavior and interaction with
the fields and plasma near the earth. Such phenomena as grain
coagulation and alignment and also the effect of grains on the Earth's
plasma environment will be studied.

The properties of interstellar and circumstellar grains have been of
interest for many years. Observational astronomy is the major source
of information which is applied to elucidate the size, structure, and
composition of astronomical grains. In order to relate the
observations to properties of the grains, observations must be
compared with theoretical and experimental studies. We report here a
series of experimental studies which are aimed at providing data on
the spectral properties and dynamics in the near-Earth environment of
simulated interstellar grain materials.

Preparation of silicate, silicon carbide, and carbonaceous
materials is carried out by injecting volatile precursors (e.g.,
$Fe(CO)_5$, SiH_4, or CH_4) into an inductively-coupled plasma jet.
The high temperature plasma pyrolyzes the precursors to the elements,
which react and condense in the relatively low temperature plasma
tail, forming the grains of closely controlled composition. The
elemental composition, morphology, and crystalline phases present in
the grains are determined using electron microscope techniques.

Optical characterization of the grains will include measuring the
angle and wavelength-resolved intensities of visible light scattered
from the grains using a recently constructed scattering apparatus.
The scattering apparatus consists of a solar simulator light source, a
scattering chamber in which the grains are suspended, and a
monochrometer-optical multichannel detector. The angle and
wavelength-resolved scattering measurements will help characterize the

F. P. Israel (ed.), Light on Dark Matter, 245–246.

scattering efficiency, albedo, and absorption of irregular particles which are not amenable to theoretical scattering calculations.

Deployment of laboratory prepared grains in low-Earth orbit will be carried out using NASA developed Get-Away-Special (GAS) cannisters which allow ejection of a small (less than 1 meter^3) subsatellite from the Space Shuttle Orbiter. The subsatellite can be activated from the ground to release its contents during or after the Shuttle mission. The grains will be contained in a sublimating solid or other binder to reduce grain coagulation during storage. The grains will remain in orbit only a short time (less than one orbit) at typical shuttle altitudes (250 km).

Observations of the grain cloud will be carried out from the AMOS/MOTIF Facility located on Mount Haleakala on Maui in the Hawaiian islands. This facility houses telescopes which have the capability of tracking objects in orbit and performing spectroscopic analysis in various spectral regions. The total visible scattered light intensity and sunlight polarization will be monitored as a function of angle as the grain cloud passes overhead. Other observations including wavelength resolved infrared observations are possible. From the data obtained, the orbital decay due to atmospheric drag and other factors will be determined. Analysis of the polarization data will help elucidate mechanisms of grain alignment including streaming and interaction with the Earth's magnetic or electrical fields. From the scattering intensity of the grains in orbit relative to the scattering of the laboratory grains, the degree of coagulation can be studied.

The above program is an initial step in developing techniques to introduce into the near-Earth environment controlled quantities of materials for studying interaction of the materials with the space environment. Other possible experiments include injecting labeled grains into the atmosphere as tracers of stratospheric circulation and modeling the behavior of comets near perihelion by monitoring the behavior of simulated comets consisting of mixtures of grains and volatile material. Observations will intially be made from the ground, but observations of released materials from the Space Shuttle are desirable.

We are seeking input from the scientific community early in the program to ensure that the best scientific data are obtained from a limited number of launch opportunities. Collaborations in instrument design, operations, and observations are welcome. In addition, we are seeking careful scrutiny of the material releases with emphasis on possible adverse effects on other scientific programs, particularly those engaged in collecting and analyzing cosmic grains in exo and endoatmospheric environments. Our current plans are to tag the grains with an easily identified element or elements so that in the unlikely event that such grains are collected, they will be identified during routine analysis. Suggestions concerning uses of the orbital release capability and comments regarding possible interactions with other programs are encouraged.

ULTRAVIOLET PHOTOPROCESSING AND INFRARED SPECTROSCOPY OF LABORATORY SIMULATED GRAIN MANTLES

L.B. d' HENDECOURT
G.P.S. Université de Paris 7 4, Place Jussieu, 75251 Paris
Cedex 05 FRANCE and Laboratory Astrophysics, Leiden
University, 2300 RA Leiden THE NETHERLANDS

ABSTRACT. A series of experiments in which mixtures of ices are irradiated with ultraviolet light have been performed. The evolution of these simulated interstellar grain mantles is followed by infrared spectroscopy. The appearance of new bands has been observed and possible identifications of these bands are discussed. Implications for interstellar chemistry are discussed.

1. INTRODUCTION

The time dependent behavior of the composition, abundance and distribution of molecules in both the gas and solid phases in dense molecular clouds, using a method that takes atomic and molecular processes simultaneously into account have been calculated in a previous paper (d' Hendecourt et al., 1985a, paper 1). With such an approach, it is possible to compute not only the gas phase but also the accreted grain mantle compositions. A limited set of grain surface reactions was taken into account: only low temperature (10-20 K), diffusion controlled reactions have been considered and only reactions with a negligible activation energy have been included. Infrared spectroscopy is powerful tool for the study of molecular grain mantles in dense interstellar clouds (Allamandola, 1984) because the characteristic frequence of molecular vibrations lie in the mid infrared region of the spectrum (2-25 microns). Accretion mantles have been observed in many objects (Soifer et al., 1979; Aitken, 1981; Willner et al., 1982; Knacke et al., 1982) and various molecules such as water and ammonia firmly identified. More recently, solid CO has been discovered (Lacy et al., 1984; see also Whittet, this volume). Such accretion mantles can be easily simulated in the Laboratory by condensing gases, either pure or as mixtures of different cosmically abundant molecules, onto a cold (10 K) substrate and measuring their IR absorption spectrum. These simple ices can then be irradiated with ultraviolet light so that a more complex chemistry occurs. IR spectroscopy is again used to identify the new molecules formed. This semi-empirial approach allows a direct

247

F. P. Israel (ed.), Light on Dark Matter, 247–252.
© *1986 by D. Reidel Publishing Company.*

comparison with observed astronomical spectra in order to asses the importance of these irradiation effects on the chemistry of interstellar grains.

2. EXPERIMENTAL 'MANTLE' COMPOSITION

The experimental mantle composition has been chosen from the calculations described in paper 1. A classical gas phase ion-molecule reaction scheme describes the evolution of the gas phase molecular species which are able to stick onto the grains where they can react (Tielens and Hagen, 1982; paper 1) and evaporate back to the gas phase under certain circumstances (Grealy 1979) (d' Hendecourt et al., 1982). Naturally, mantle composition will differ according to various physical conditions in the cloud considered but basically, two different classes of mantles do arise as shown in Table 1: those in which water is dominant (60%) and those in which CO or CO2 and O2 are dominant. For the photolysis experiments, two mixtures were selected: mixture 1 = H2O/CO/CH4/NH3/6/2/1/1 and mixture 2 = H2O/CO/CH4/NH3/O2/N2/1/1/0.4/0.4/1/0.04 as representative of interstellar grain mantles at a time of their evolution (paper 1 and d'Hendecourt et al., 1985 b, paper 2).

TABLE I

	$A_v = 2$ $n_o = 10^3 cm^{-3}$	$A_v = 4$ $n_o = 2 \times 10^4 cm^{-3}$	$A_v = 8$ $n_o = 10^5 cm^{-3}$
H_2O	62	50	3
CO	18	3	30
CH_4	4	< 1	0.4
NH_3	9	< 1	0.4
O_2	–	3	43
N_2	–	7	8
CO_2	2[*]	32[*]	9

[*] The surface reaction $CO + O + CO_2$ has been taken into account for the computation of this table. If this is not the case the amount of CO_2 is reduced to about 1% for the standard case ($A_v = 4$, $n_o = 2 \times 10^4 cm^{-3}$).

Table I: Mantle composition obtained, as described in paper I, for various cloud parameters.

3. MANTLE MODIFICATION BY U.V. PHOTOPROCESSING

The IR spectra of the initial and irradiated mixtures are displayed in figures 1 and 2 for direct comparison.

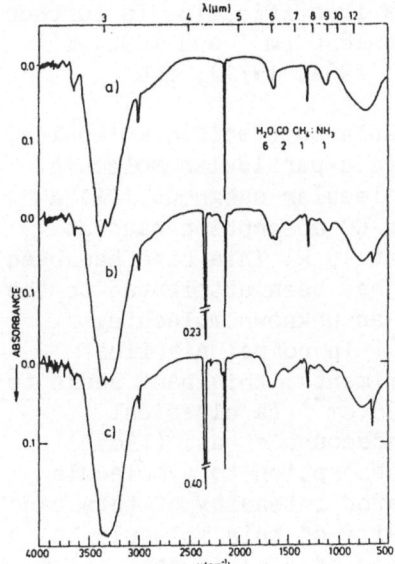

Figure 1: Infrared spectra of mixture 1:(a) no irradiation, b) and c) after 4 and 24 hours of irradiation respectively. See table III for the number of molecules corresponding to the absorptions.

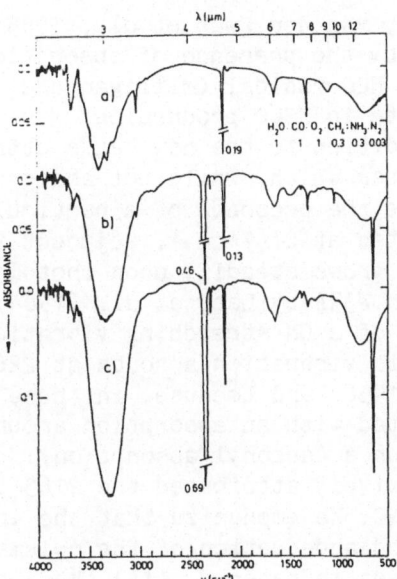

Figure 2: Infrared spectra of mixture 2: (a) no irradiation b) and c) after 4 and 24 hours of irradiation respectively. See table III for the number of molecules corresponding to the absorptions.

On these spectra we note that: the band charcterising methane, ammonia and, to a lesser extent, carbon monoxide show a steady decrease due to photolysis and subsequent reactions. The most intensive new bands appear at 2343 cm^{-1} (4.27 µm; FWHM = 12 cm^{-1}), 1470 cm^{-1} (6.8 µm; FWHM ≈ 100 cm^{-1}), 2167 cm^{-1} (4.61 µm; FWHM ≈ 25 cm^{-1}), 1850 cm^{-1} (5.41 µm; FWHM ≈ 15 cm^{-1}), 1370 cm^{-1} (7.3 µm; FWHM = 20 cm^{-1}), 1220 cm^{-1} (8.2 µm; FWHM ≈ 50 cm^{-1}) and finally 660 cm^{-1} (15.2 µm; FWHM = 22 cm^{-1}), in the shoulder of the so called "12 micron" ice band. The region 1800 - 1400 cm^{-1} shows many changes: new strong absorption bands, not mentioned above appear at 1720 cm^{-1} (5.81 µm), 1695 cm^{-1} (5.90 µm, a very strong and broad absorption, FWHM = 45 cm^{-1}), and 1585 cm^{-1} (6.31 µm a weak, broad and shallow absorption). Weaker bands are observed around 2960 cm^{-1} (3.38 m), 2900 cm^{-1} (3.53 µm), 1090 cm^{-1} (9.2 µm; FWHM = 15 cm^{-1}) and 1020 cm^{-1} (9.8 µm: FWHM = 20 cm^{-1}). The ratio and position of the strongest new lines at 2347 cm^{-1} (4.26 µm) and 660 cm^{-1} (15.2 µm), permit the unequivocal identification of the CO_2 molecule (Schimanouchi, 1972) Due to the high IR cross section of CO_2 at 2347 cm^{-1}, the isotope $^{13}CO_2$ can also be easily identified in the spectra. The next molecule which can be readily assigned with confidence is formaldehyde (H_2CO), with the lines situated at 1720 and 1500 cm^{-1}, corresponding to the CO strech and the CH2 scissoring modes respectively (Schimanouchi, 1972). This positive identification was made on the bases of the IR spectrum of H_2CO in a complex matrix

obtained by van der Zwet et al., 1985. This identification is further
supported by the presence of absorptions at 1850 cm^{-1} and 1090 cm^{-1}
due to the HCO radical (Milligan and Jacox, 1964, 1971), the
intermediate in H2CO production.

In addition to the new bands attributable to specific molecules,
others appear which, while not assignable to a particular molecule,
do indicate the presence of a particular molecular subgroup: (i) a
band, located at 2167 cm^{-1}, adjacent to the CO absorption band at
2143 cm^{-1}, grows steadily upon photolysis at 10 K. This band has been
observed in W33A by Lacy et al. (1984). It has been attributed to the
absorption of a CN stretching vibration in an unknown molecule.
Because this virbration absorbs at 2265 cm^{-1} in normal nitriles
(Bellamy, 1956) and because, in these experiments, this band seems to
be correlated with an absorption around 1695 cm^{-1} (a classical
position for a carbonyl absorption). d' Hendecourt et al. (1985b)
have tentatively attributed the 2165 cm^{-1} absorption to a molecule
like CH3CONC. We emphasize that the integrated intensity of this band
is large and a deduction of the column density of this molecule to
W33A is given in paper 2. (ii) the appearance of a broad absorption
band around 1500-1400 cm^{-1} ("6.8 μm) is also quite significant
because of the occurence of the same absorption in the spectrum of
W33A and in other protostellar objects (Willner et al., 1982). In
this region, the number of potential molecules is very large so that
a precise identification of the molecule responsible for this
absorption is meaningless. This occurs at a frequency identificable
with the deformation modes of saturated hydrocarbons, as pointed out
by Hagen et al., (1980). A more precise discussion of this
identification is given in paper 2 and comparison with observations
can be found in Tielens et al., (2984). The fact that, in these
experiments, this strong absorption at 6.8 micron is correlated with
a relatively week absorption in the 3.4 micron region (CH strech in
hydrocarbons), points to the formation of saturated aliphatic
hydrocarbons in which the methyl and methylene groups are adjacent to
unsaturated groups such as alkoxy and carbonyl groups (d' Hendecourt
and Allamandola, 1985).

In mixture 2, the evolution is rather similar: the UV photolysis
is more efficient and the original molecules are destroyed more
thoroughly. The 6.8 μm region shows a deeper absorption. However,
the 2167 cm^{-1} band as well as the band at 1695 cm^{-1} are totally
absent from this sample. Carbon dioxide is made in abundance as well
as formaldehyde and ozone (band around 1030 cm^{-1}). Other bands
attributed to nitrogen oxides are also observed, especially a band at
1878 cm^{-1} (NO) and at 1615 cm^{-1} (nitrogen peroxide).

4. ASTROPHYSICAL IMPLICATIONS

These implications are discussed extensively in papers 1 and 2
in terms of the role of the grains in interstellar molecule formation
and in terms of low temperature chemistry on the grain surface versus
"hot" atom chemistry initiated by UV radiation. Here, we focus on the

5-8 micron region observed in W33A as compared with various mixtures.
Figure 3 shows this comparison.

From this kind of comparison and the deductions made from the
experiments, it is possible to estimate the amount of material in the
line of sight to this objct: about 3 to 11% of the available cosmic
carbon is present in the line of sight to W33A in the form of solid
aliphatic hydrocarbons. This value is slightly lower than the one
derived by Tielens et al. (1984) but this is due to the fact that the
aliphatic hydrocarbons have larger IR cross-sections at 6.8 μm than
the sturated hydrocarbons.

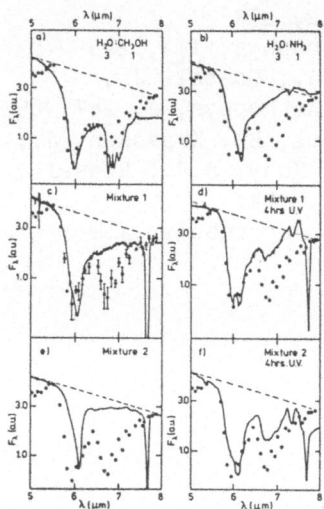

Figure 3: Comparison of the spectrum of W33A (dots) and various laboratory mixtures. Intensity units are arbitrary and the spectra are all
roughly normalized to the 1620 cm^{-1} (6.0 μm) water absorption. Observational error bars are shown only in frame (c).

REFERENCES

Aitken, D.K.: 1981, in Infrared Astronomy, IAU Symposium, n'96, eds.
    Wynn-Williams, G.G., Cruikshank, D.P., Reidel, 207
Allomandola, L.J.: 1984, in Galactic and Extragalactic Infrared
    Spectroscopy, eds. Kessler, ;F., Philips, J.P., Reidel, 5
Greenberg, J.M.: 1979, in Stars and Star Systems, ed. Westerland,
    B.E., D. Reidel, 173
Greenberg, J.M.: 1973, in Molecules in the Galactic Environment, eds.
    Gordon, M.A. and Snyder, L.E., John Wiley, p. 93
d'Hendecourt, L.B., Allamandola, L.J., Baas, F., Greenberg, J.M.:
    1982, Astron. Astrophys. 109, L12
d'Hendecourt, L.B., Allamandola, L.J., Greenberg, J.M.: 1985a,
    Astron. Astrophys. in press
d'Hendecourt, L.B., Allamandola, L.J., Grim, R.J.A., Greenberg, J.M.:
    1985b, Astron. Astrophys. in press

d'Hendecourt, L.B., Allamandola, L.J.: 1985, submitted for
    publication in Astron. Astrophys.
Knacke, R.F., McCorkle, S., Puetter, R.C., Erickson, E.F.,
    Kratschmer, W.: 1982, Astrophys. J. 260, 141
Lacy, J.H., Baas, F., Allamandola, L.J., Person, S.E., McGregor,
    P.J., Lonsdale, C.J., Geballe, T.R., van de Bult, C.P.E.M.: 1984,
    Astrophys. J. 276, 533
Milligan, D.E., Jacox, M.E.: 1964, J. Chem. Phys. 41, 3032
Milligan, D.E., Jacox, M.E.: 1971, J. Chem. Phys. 54, 927
Schimanouchi, T.: 1972 in Tables of Molecular Frequencies
    Consolidated, Volume 1, NSRDS-NBS no 39
Soifer, B.T., Puetter, R.C., Russell, R.W., Willner, S.P., Harvey,
    P.H., Gillett, F.C.: 1979, Astrophys. J. 232, L53
Tielens, A.G.G.M., Hagen, W.: 1982, Astron. Astrophys. 114, 245
Tielens, A.G.G.M., Allamandola, L.J., Bregman, J., Goebel, J.,
    d'Hendecourt, L.B., Witteborn, F.C.: 1984, Astrophys, J. 287, 697
Willner, S.P., Gillett, F.C., Herter, T.L., Jones, B., Krassner, J.,
    Merrill, K.M., Pipher, J.L., Puetter, R.C., Rudy, R.J., Russell,
    R.W., Soifer, B.T.: 1982, Astrophys. J. 253, 174
Van der Zwet, G.P., Allamandola, L.J., Baas, F., Greenberg, J.M.:
    1985, Astron. Astrophys. 145, 262

RADIATION EFFECTS ON GRAIN MATERIALS

Giovanni Strazulla
Observatorio Astrofisico ed Instituto di Astronomia
Citta Universitaria, I-95125 Catania, Italy

ABSTRACT. By comparing the carbonaceous and deduced organic components of the Murchison meteorite with MeV proton irradiated solid methane residues we have tried to suggest a method for making an evolutionary connection between protostellar and interplanetary material.

1. INTRODUCTION

In recent years it has become clear that whatever the site and history of their formation, grain materials evolve after birth under the action of external agents, mainly energetic photons and ions. The grain evolution is governed by physico-chemical processes whose nature has been and is presently being investigated in various laboratories.

The effects of photon irradiation on grain materials has been carefully investigated at Leiden (Holland) by Greenberg and coworkers (see e.g. Greenberg, 1984). The effects of fast ion irradation have been studied in our laboratory (e.g. Foti et al. 1984, Strazzulla 1985) at Catania (Italy). Similar studies have been made elsewhere.

Here I give a summary of the results obtained at Cantania by bombarding frozen carbon-containing targets with MeV ions. By comparing laboratory and observational data, I discuss the role they play in grain evolution. In particular I show that these results should be considered when a comparsion between grain properties is made for grains in different environment e.g., interplanetary vs. interstellar grains. In addition, I present a comparison between IR spectra obtained from laboratory synthesized organic residues and from the organic-carbonaceous component of Murchison meteorite.

2. EXPERIMENTAL ASPECTS

Experimental techniques and results have been discussed in detail elswhere (e.g. Foti et al. 1984, Calcagno et al. 1985).

The rationale of the experiments is the following: hydrocarbon

253

F. P. Israel (ed.), Light on Dark Matter, 253–260.

layers (mainly CH_4 and mixtures with H_2O) are condensed on a cold
(~10K) finger by admitting controlled vapours into a scattering
chamber. The layers are then bombarded by a keV-MeV ion beam which
can be also used to measure, in situ, the thickness of the target by
a backscattering technique. As a consequence of the bombardment,
solid complex organic materials have been obtained. The cross section
for the process converting volatile carbon containing molecules to a
refractory residue has been measured for some projective-target
combinations (Strazzulla et al. 1983a, 1984; Foti et al. 1984). The
synthesis of new materials occurs on a dose scale of ~ IE + 16 ions
cm-2.

The stable residues thus obtained, look like a yellow stuff, and
can be extracted from the scattering chamber and used for remote
analysis such as, e.g., IR studies. Moreover these organic materials
have been returned to the scattering chamber and their ion-beam
induced evolution (up to doses of ~ E + 17 ion cm⁻² has been
studied; in summary:

(i)   the residues as initially prepared with ion doses of ~ 2xE +16
      (1.5 MeV protons cm⁻²) are amorphous and fluffy (as tested by
      TEM images) and have low density (0.5 gr/cm³) and their IR
      spectra show strong signatures (2.9, 3.4, ~6., 6.9 µm) typical
      of polymere-like substances;

(ii) the additionally bombarded residues (up to ~1E+17 p cm-2) evolve
      towards an amorphous carbon-like material. This is noted by the
      observed preferential loss of H (mainly as H_2 molecules) with
      consequent density increase. As shown in Fig. 1 the IR signature
      decreases progressively in strength as the ion fluence
      increases. Reflectance spectra in the region 0.5-2.5 µm shows a
      similarity between our material and charcoal (Calcagno et al.,
      1985).

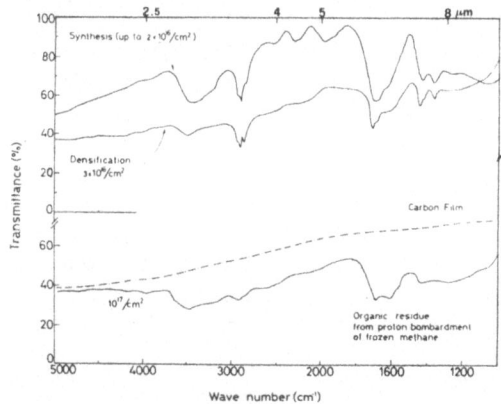

Fig. 1: IR transmission spectra of a residue obtained from CH_4 frozen
        after several 1.5 MeV proton fluences. For comparison the µgr cm⁻²)
        spectrum of an amorphous carbon film (20 µgr cm⁻²) is also
        shown.

TABLE 1: Short Summary of Typical Astrophysical Sources of Energetic
Protons

Proton Source	Energy (MeV)	Flux	Irradiation time (yrs)	Total dose (cm^{-2})
Solar Wind (at 1AU)	10^{-3}	$3 \times 10^8 cm^{-2} sec^{-1}$	4.6×10^9	4.4×10^{25}
Solar Flares (1 large event per yr at 1AU)	≥ 1	$10^{10} cm^{-2} yr^{-1}$	"	4.6×10^{19}
Modulated Galactic Cosmic Rays (E in MeV)	≥ 10	$6 \times 10^{-3} E^{-1.75} cm^{-2} yr^{-1}$	"	$3 \times 10^{22} E^{-1.75}$
Jovian Magnetosphere (R_j=Jovian radius)				
Io($6R_j$)	1	$2 \times 10^6 cm^{-2} sec^{-1}$	"	3×10^{23}
Europa($9.5R_j$)	1	3×10^7 " "	"	4.5×10^{24}
Ganimede($15R_j$)	1	3×10^6 " "	"	4.5×10^{23}
Callisto($26.6R_j$)	1	6×10^4 " "	"	9×10^{21}
Young Stars (T Tauri at 100/AU)	~1	$10^6 cm^{-2} sec^{-1}$	10^6	3.2×10^{19}
Low Energy Galactic Cosmic Rays	~1	$10 \ cm^{-2} sec^{-1}$	3×10^8	10^{17}

3.   GRAIN EVOLUTION

From our experiments we have learned that the targets, simulating to
some extent grains or planetary surfaces, undergo a deep modification
under the action of external radiation and evolve from simple
condensed gas to a comple polymer-like organic residue and, at higher
doses, to a carbonaceous amorphous layer.
       There is no question that in most astrophysical contexts ranging
from grains in diffuse or dense clouds to those in circumstellar
nebulae or discs to extended bodies in the Solar System, carbon-rich
cold surfaces are present. The problem is to see if a sufficiently
intense ion flux occurs to produce the laboratory simulated
processes. In Table 1 the proton fluxes at given energies, are
summarized as measured or inferred in representative astrophysical
situations. The last column in Table 1 shows that the total proton
dose is, with the possible exception of low energy cosmic rays, order
of magnitudes greater than required not only to build up organics
from simple gases but also to cause their evolution towards
carbonaceous material.
       The above results, I believe, have to be taken into
consideration wheñ comparing interplanetary versus interstellar
grains for example and in particular when evolutionary relations
between them are suggested.
       As an example of this let me suggest the following idealized
scenario. Suppose one starts with a cloud where grains have
collected, by radiation synthesis, organic mantles from the
3.4 µm feature. Suppose now that this cloud collapses to form a T Tau
star in the direction of which we no longer observe the
3.4 µm structure. We might conclude that the grains we are observing
in the circumstellar environment are different from those in the
parent cloud or on the basis of laboratory bombardment; we could also
say that the grains are the same and that because of the evolution of
the organic mantles into a carbonaceous material they have lost the
previously present IR signatures.

4.   COMPARISON WITH OBSERVATION

In the previous sections I have presented some reasons to suspect
that the experimentally simulated processes are effectively working
also in astrophysical scenarios.
       Many applications have in fact already been published. Among
others we have obtained the following results:

i)   Carbon containing molecules in icy mantles on interstellar
      grains can be totally polymerized by low energy cosmic rays in a
      time less than the estimated life time of dust clouds
      (Strazzulla et al. 1983a).
ii)   The polymerization process, activated by solar cosmic rays, is
      capable of building up substantial carbonaceous material on
      Pluto and Triton both of which are probably covered by methane

frost and/or surrounded by methane-rich atmospheres (Strazzulla et al. 1984).

iii) Ion irradiation is a very important mechanism to induce solid-state effects and cause chemical evolution of grains and planetary systems in the early (eruptive) stages of the Solar System (Strazzulla, 1985)

iv) The organic materials built up by ion fluence on interplanetary grains (cometary debris) could be the glue which cements submicron silicate particles to form a complex agglomeration whose density increases with increasing proton fluence (Strazulla et al. 1985).

v) Carbonaceous material of low reflectance could have been produced in large amounts by ion fluence on or in the icy surfaces of the Uranian satellites, Hyperion and the dark side of Iapetus (Calcagno et al. 1985).

In the following I describe a new application of the experimental results to the Murchison meteorite.

The spectra shown in Fig. 1 can be compared with those obtained from organic and carbonaceous components in some meteorites. In Fig 2 we compare the spectrum (3.2-3.6 μm) obtained for our carbonaceous material (shown in Fig. 1, lower section) with that of the carbonaceous component of the Murchison meteorite taken from Hoyle et al. (1984). The normalization has been chosen to give the same depth

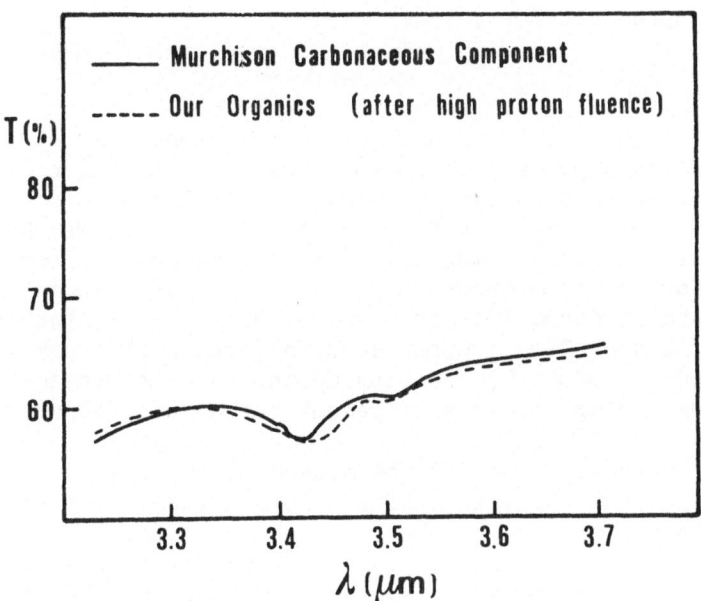

Fig. 2: Comparison between transmittance spectra from Murchison carbonaceous component (Hoyle et al. 1984) and from our organic after high proton fluence (see Fig. 1 lower section).

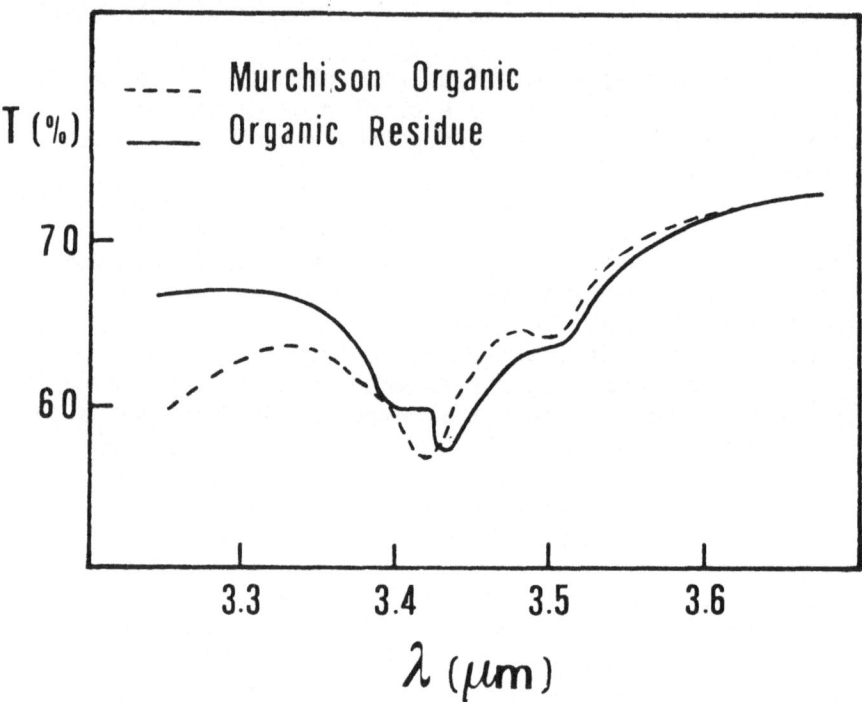

Fig. 3: Comparison between transmittance spectra from Murchison
        organic (after Hoyle et al. 1984) and from our organic (see
        Fig. 1 upper section).

at the minimum of the ~3.4 μm band, due to $-CH_2-$ and $-CH_3$ stretching
vibrations. The fit appears good so that the carbonaceous component
in Murchison looks, in this spectral region, similar to our material
obtained from and experiment at high (~E+17 p cm^{-2}) 1.5 MeV H fluence.
    Hoyle et al. (1984) estimated an originating organic component
for the Murchison spectrum shown in Fig. 3 (full line) where it is
compared with the spectrum (dashed line) of our organic synthesized
at low proton fluence (Fig. 1 upper section). Again the normalization
has been done to reproduce the maximum depth. The fit appears to be
not too bad considering the very different method used to obtain the
two spectra.
    It is thus suggested that if the meteoritic samples are really
among the most ancient components of the solar system it is then
possible that both purely organic and carbonaceous materials have
been synthesized by ion fluence on hydrocarbons which were frozen on
them in the first stages of solar system evolution when ion fluxes in
the T Tauri phase were much more intense than at the present time.

REFERENCES

Baratta, G.A., Strazzulla, G.: 1985, Astron. Astrophys, submitted.

Van de Bult, C.E.P.M., Greenberg, J.M., Whittet, D.C.B.: 1985, Monthly Notices Roy. Astron. Soc. (in press).

Calcagno, L., Foti, G., Torrisi, L., Strazzulla, G.: 1985, Icarus (in press).

Cohen, M.: 1983, Astrophys. J. Lett. 270, L69.

Foti, G., Calcagno, L., Sheng, K.L., Strazzulla, G.: 1984, Nature 310, 126.

Greenberg, J.M.: 1984, in Proc. of the Workshop on Laboratory and Observational Infrared Spectra of Interstellar Dust. R.D. Wolstencroft and J.M. Greenberg eds. p. 1.

Greenberg, J.M., van de Bult, C.E.P.M.: 1984, in Proc. of the Workshop on Laboratory and Observational Infrared Spectra of Interstellar Dust. R.D. Wolstencroft and J.M. Greenberg eds. p. 70.

Hoyle, F., Wickramasinghe, N.C., Al-Mufti, S.: 1984, Astrophys. Space Sci. 98, 343.

Strazzulla, G.: 1985, Icarus, 61, 48.

Strazzulla, G., Calcagno, L., Foti, G.: 1983a, Monthly Notices Roy. Astron. Soc. 204, 59p.

Strazzulla, G., Pirronello, L., Foti, G.: 1983b, Astrophys. J. 271, 255.

Strazzulla, G., Calcagno, L., Foti, G.: 1984, Astron. Astrophys. 140, 441.

Strazzulla, G., Calcagno, L., Foti, G., Sheng, K.L.: 1985, in Proc. NATO Workshop on Ices in the Solar System. (J. Klinger, ed) D. Reidel, The Netherlands.

Whittet, D.C.B., Bode, M.F., Longmore, A.J., Baines, D.W.T., Evans, A.: 1983, Nature, 303, 218.

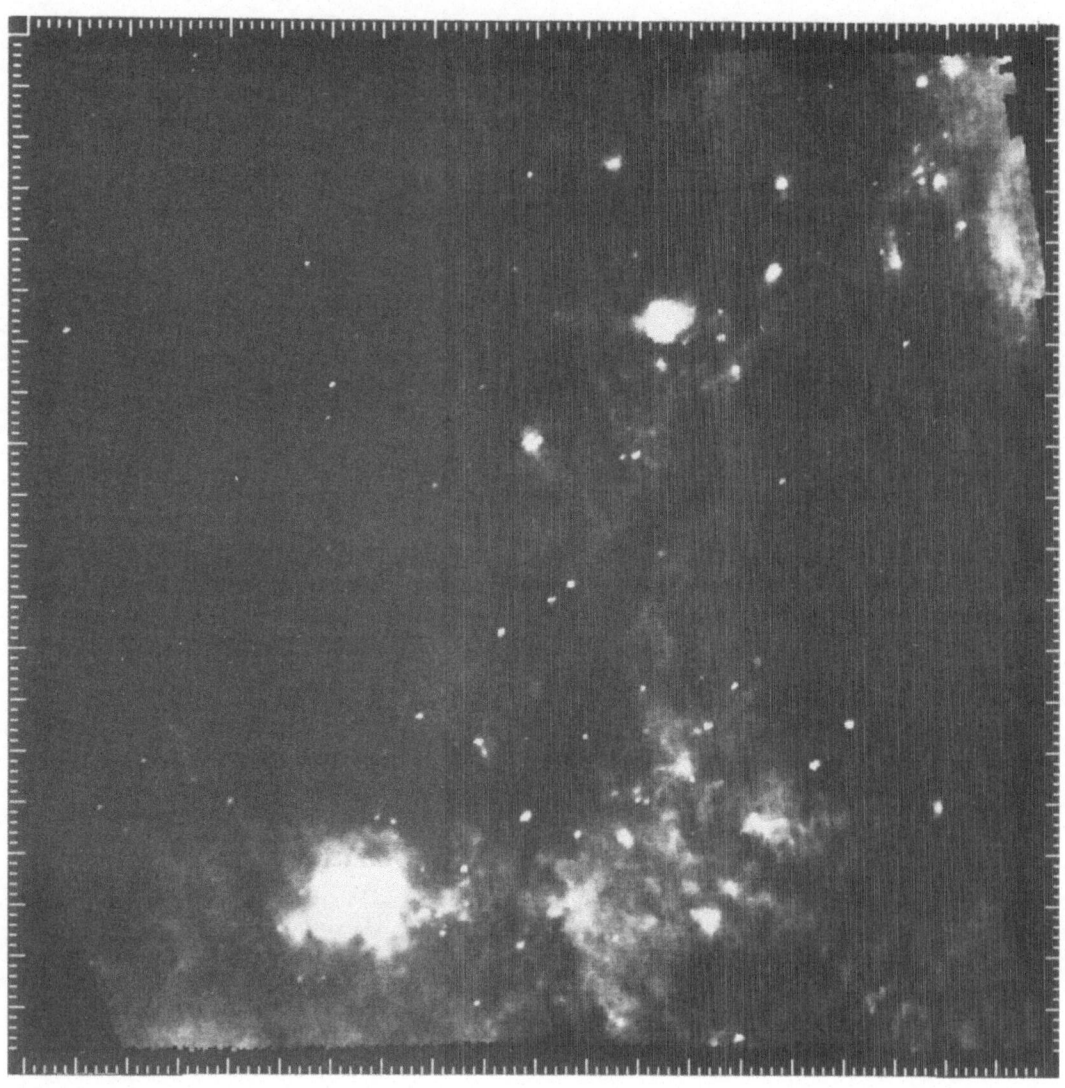

IRAS Field 151 $\alpha = 08^h00^m$, $\delta = -30°$, HCON-3, 60 µm. Field in
                   Puppis. Bright object north of center is NGC 2467;
                   NGC 2568 is at bottom.

REFLECTION NEBULAE, NON-EQUILIBRIUM THERMAL EMISSION, AND IRAS

K. Sellgren
Institute for Astronomy
University of Hawaii
2680 Woodlawn Drive
Honolulu, HI 96822
U.S.A.

L.J. Allamandola, J.D. Bregman, M.W. Werner, D.H. Wooden
NASA Ames Research Center, M/S 245-6
Moffett Field, CA 94035
U.S.A.

ABSTRACT. The 1-13 micron spectra of visual reflection nebulae are presented, and their implications for the IRAS 12 micron band discussed.

We present spectra (Fig. 1) of the visual reflection nebulae NGC 7023 and NGC 2023 from 1 to 13 microns (Sellgren et al. 1983, 1985), using the Kuiper Airborne Facility and the Infrared Telescope Facility. Six emission features, at 3.3, 3.4, 6.2, 7.7, 8.6 and 11.3 microns, are seen in addition to a smooth underlying continuum from 1 to 13 microns. There is no evidence for silicate emission at 10 microns. The emission features seen in reflection nebulae are similar in shape and relative strengths to the features seen in other sources, arguing for an universal emission mechanism and material.

Sellgren (1984) has suggested the 1-13 micron continuum in visual reflection nebulae, and possibly the features, are due to non-equilibrium thermal emission from small (10 A) grains, briefly heated to 1000 K by the absorption of single UV photons. The continuum shape agrees with model predictions for non-equilibrium thermal emission from small grains, integrated over both the interstellar grain size distribution and the time dependence of the temperature of an individual grain as it cools (Sellgren et al. 1985). Léger and Puget (1984) and Allamandola et al. (1985) have proposed the emission features are due to non-equilibrium emission from polycyclic aromatic hydrocarbons (PAHs), which give reasonable spectroscopic agreement with the features. The PAHs might also produce continuum emission such as is seen in reflection nebulae by an overlap of overtone and combination bands, or by an electronic continuum (Allamandola et al. 1985 ; Puget et al. 1985).

The IRAS 12 micron band covers 8 to 15 microns, and thus includes several of the unidentified emission features as well as the underlying and apparently associated continuum emission observed in reflection nebulae (Fig. 1). The observation of the unidentified features in

F. P. Israel (ed.), Light on Dark Matter, 261–262.
© *1986 by D. Reidel Publishing Company.*

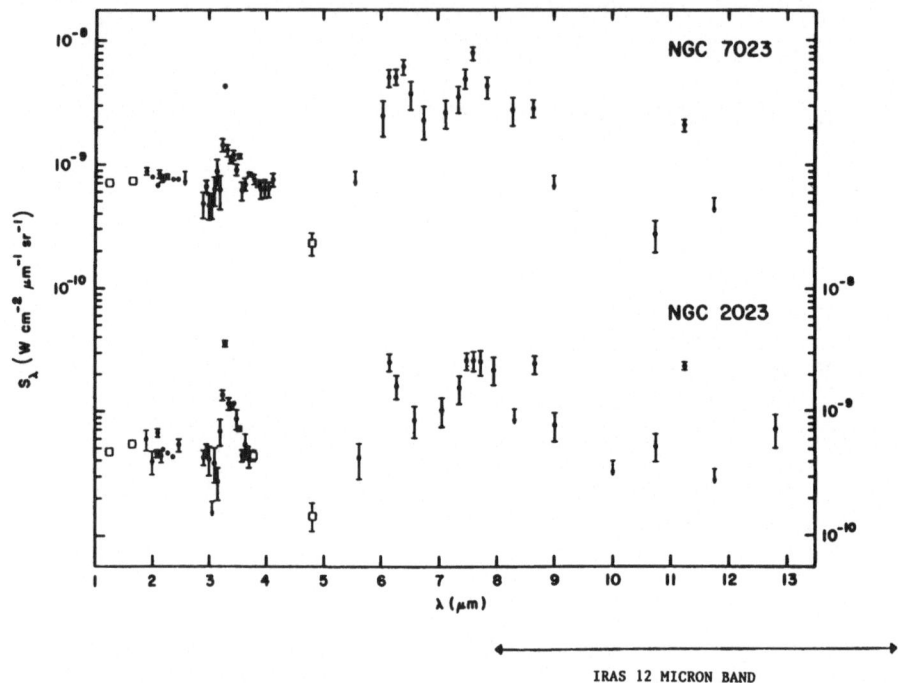

Fig. 1 : Spectra of two visual reflection nebulae, from Sellgren,
Werner and Dinerstein (1983) and Sellgren et al. (1985).

sources with as diverse dust histories as planetary nebulae and reflection nebulae implies the emitting material is ubiquitous, and is
either difficult to destroy or is easily replenished. The continuum and
feature emission observed in reflection nebulae might be expected to be
present wherever there is dust and UV radiation to excite the emission,
and be observable whenever other sources of 12 micron emission do not
mask its presence. IRAS may therefore discover 12 micron emission similar to that in reflection nebulae many other places in the Galaxy,
including the diffuse galactic medium.

REFERENCES

Allamandola, L.J., Tielens, A.G.G.M., Barker, J.R., 1985 : Ap. J.
    290, L25.
Léger, A., Puget, J.L., 1984 : Astron. Ap. 137, L5.
Puget, J.L., Léger, A., Boulanger, F., 1985 : Astron. Ap. 142, L19.
Sellgren, K., Werner, M.W. Dinerstein, H.L., 1983 : Ap. J. 271, L13.
Sellgren, K., 1984 : Ap. J. 277, 623.
Sellgren, K., Allamandola, L.J., Bregman, J.D., Werner, M.W.,
    Wooden, D.H., 1985, Ap. J., in press.

SECTION 4.

INTERSTELLAR MEDIUM AND STAR FORMATION

THEORIES OF STAR FORMATION CONFRONTED BY IRAS DATA

Bruce G. Elmegreen
IBM Thomas J. Watson Research Center
P.O. Box 218, Yorktown Heights, N.Y. 10598 USA

ABSTRACT -. The importance to star formation theories of IRAS
observations of the dust distribution around young stars, the
luminosities of protostars and embedded clusters, and the size
distribution of dust particles is discussed. The important role of
the smallest grains in determining the ionization fraction and
magnetic diffusion rate in dense molecular clouds is illustrated by
new solutions to the equations of charge equilibrum.

1. INTRODUCTION

IRAS can contribute directly to our understanding of the star
formation process by observing the distribution and density of
interstellar matter in the vicinity of star formation, the total
luminosities of newly formed stars and clusters, and the temperatures
of dust particles. The distribution of matter around a star-forming
core can reveal the mechanism of cloud formation, the triggering
process that might have initiated star formation, and the fractional
mass of the material that remains. The total infrared luminosities
can reveal the approximate masses of the embedded stars, and in
combination with the dust temperatures, the radii of extended dust
envelopes or disks. The stellar masses may be used to estimate the
efficiency of star formation in the cloud, and, for an embedded
cluster, the likelihood that the cluster will remain bound after the
gas disperses. The temperatures of dust particles in clouds with
known radiation fields might be used to determine the size
distribution and size range of these particles. The grain size
distribution is important for the ionization fraction, the ion-
neutral viscosity for magnetic diffusion and the total heating rate
of the gas by dust-gas collisions. Regions of high dust temperatures
could also correspond to protostellar disks, cocoons or dusty winds.
　　Most of the contributions by IRAS will ultimately be made in
conjunction with molecular-line and other observations. Of course,
the most obvious use of IRAS data is simply to find regions of star
formation that can be studied in detail with mm-wave and optical

F. P. Israel (ed.), Light on Dark Matter, 265–276.
© *1986 by D. Reidel Publishing Company.*

telescopes. Such a listing of sources has been one of the most
important first results from IRAS, leading to the conclusion that a
high fraction of dense globules contain stars, and therefore that the
collapse of globules to star formation is relatively rapid (Myers
1985). IRAS reveals much more than the mere locations of young stars,
however. Wesselius, Beintema and Olnon (1984) inferred from IRAS data
that the relatively high mass AO-type stars in the Cha I dark cloud
are forming with a greater residual of unused gas, and therefore
lower efficiency,than the lower mass, A- to K-type stars in the same
region. Beichman et al. (1984) found a young star in the core of the
dark cloud B5, and determined its stellar surface temperature (800 K)
and radius (2 A.U.). Emerson et al. (1984) suggested that an IRAS
protostellar source in the dark cloud L1551, designated as source NE,
could have been triggered by wind from the embedded star IRS5. Baud
et al. (1984) discovered 13 warm sources $[F(60\mu)>F(25\mu)]$ in the Cha I
cloud and suggested that they may originate in either warm gas that
surrounds low mass protostars, or ingravitationally collapsing
objects. IRAS observations of optically bright galaxies by De Jong et
al. (1984) revealed a correlation between the dust temperature and
the infrared-to-optical brightness ratio, which implies that higher
star formation rates correspond to both greater average dust
temperatures and greater dust luminosities from embedded stars. De
Jong et al. (1984) also estimted the total star formation rates in
galaxies. The IRAS discovery and observation of infrared-bright
galaxies by Soifer et al. (1984) implies that some galaxies can have
extremely large star formation rates, possibly triggered by
encounters with other galaxies.

     In these few initial publications, IRAS has already made an
enormous contribution to star formation theory. The purpose of this
review is to consider several general aspects of the additional
analysis that can be made, with a focus on the three topics listed
above: the determination of gas distributions, luminosities of
embedded stars, and grain properties. This focus is limited in range,
but illustrative of the type of information that could be obtained
from IRAS.

2. OBSERVATIONS OF THE DISTRIBUTION OF MATTER IN THE VICINITY OF STAR
FORMATION

IRAS has mapped the large-scale distribution of embedded stars, dust
and, therefore, gas (because the dust and gas should be well-mixed)
in a region of star formation. Such mapping has an advantage over
molecular-line maps in that IRAS could make a continuous detection of
gas from the molecular core to the atomic envelope, whereas molecular
line studies are sensitive to only the part of a cloud that excites
the observed transition. The angular resolution of IRAS limited its
direct perception of star formation to the largest scales, however.
These scales include the initial phases of cloud formation, and
fragmentation down to several tenths of a parsec for nearby clouds.
The initial phases are interesting because they may have been

triggered by the external perturbations. IRAS should therefore have determined how often star formation prefers the edges of cloud complexes, where such triggering may occur (Elmegreen and Lada 1977; Myers 1977; Cesarsky, et al. 1978; Gilmore 1980a,b), or if young starts frequently appear in the gas swept up by expanding HII regions, supernova explosions or stellar winds.

According to one triggering theory (Öpik 1953; Oort 1954; Elmegreen and Lada 1977), star formation propagates when the high pressure from one generation of stars forms the dense molecular core for the next generation. Core information occurs in two steps. First a shock front is driven into an adjacent low-density cloud by the first generation. The shock front compresses part of the cloud slightly, raising the density by a moderate factor of 5 to 10 in a typical case (e.g., for the ridge near W4 -- Lada et al. 1978). This density increase alone is not enough to trigger star formation, because the preshock molecular cloud typically has a density of only 10^2 cm^{-3} and the compressed ridge has an average density of only 10^3 cm^{-3}. The second step occurs when the ridge collapses by its own gravity into several cores with much higher density. The collapse supposedly occurs when the mass column density in the ridge, σ, exceeds $0.5(P/G)^{0.5}$ for shock-driving pressure P and gravitational constant G (Elmegreen and Elmegreen 1978). Then the core density will typically exceed 10^4 cm^{-3}, and the core mass will be $\pi^5 G^2 \sigma^3 t^4$ for shock age t (ibid.). For a 10^6 year old shock that moved 15 parsecs into a cloud with a density of 10^2 cm^{-3}, this characteristic fragment mass equals 4000 M_θ, which is typical for active molecular cores. The number of fragments equals approximately the total compressed mass divided by this characteristic fragment mass.

Star formation probably begins in the cores for the same reason it begins in nearly any molecular cloud with a density exceeding 10^4 cm^{-3}; the fact that the core is part of a shocked layer may be of no consequence. Star formation is 'triggered' because the cores apparently form stars much more rapidly than does the rest of the cloud. The increased rate may be simply because of the higher densities in the cores. The masses of the stars that form in the layer might also be larger than the stellar masses in the unshocked cloud, because of a higher temperature, lower specific magnetic field strength or lower specific angular momentum for gas in the layer.

In another theory of triggered star formation, the external pressure directly compresses pre-existing cloud fragments or globules past the threshold for stability (Bok 1955; Dibai 1958; see review in Klein 1985). Gravitational collapse to stars presumably follows such compression immediately. This theory has the attractive feature that the triggering of star formation is fast and direct. Observations should show star-forming globules surrounded by extremely high pressures from HII regions, supernova remnants or wind-swept bubbles.

The operation of either of these two triggering mechanisms should be evident from maps of the distribution of embedded stars relative to the overall distribution of gas (i.e., dust observed by IRAS) and pressure sources in an extended cloud complex. One

observable difference between the two theories is that the triggering
pressure in the first case should be located to one side of the most
recent generation of star formation, but the triggering pressure in
the second case should almost completely surround each star-forming
core. Cores that form from typical layers in the first theory are
also likely to be more massive than globules compressed by external
pressures in the second theory. Another possible difference is that
the velocity dispersions in the cores of the first theory should be
comparable to the virial theorem velocity for the observed core mass,
obtained by ignoring the external pressure term in the virial
theorem. This is because most of the binding of the cores in the
first case is from self-gravity. The velocity dispersions in the
compressed globules, however, should be much larger than the virial
theorem dispersions, again obtained by ignoring the external pressure
term. One should also be aware that both mechanisms can operate in
the same region if a core that formed by the collapse of a
pressurized layer also contains fragments that are compressed to the
point of instability by local pressures from embedded stars.

     IRAS observations of dust and therefore gas,located several
hundred parsecs from a star-forming core should show cloud envelopes
that are shaped by the ambient magnetic field (as determined by
starlight polarization) and by the gravitational force from the
galactic plane. Such low density envelopes could even be
interconnected by giant magnetic filaments resembling the Ophiuchus
streamer (Vrba, Strom, 1976), the filament in Orion and Monoceros
(Maddalena et al. 1985), and those in large scale HI maps (Cleary,
Heiles and Haslam 1979). Connections between neighbouring giant cloud
complexes may reveal a common origin for these clouds,as may be the
case for the Orion, Perseus and Sco-Cen-Oph clouds which are located
along the periphery of the Lindblad ring (Elmegreen 1982).

     Cloud maps might also reveal the mechanism of internal
fragmentation. A hierarchy of fragment structures could imply that
purely gravitational processes are involved (Hoyle 1953; Bodenheimer
1978). Alternatively, the presence of high pressures from embedded
stars could imply that the fragmentation is induced. Norman and Silk
(1980) proposed that star formation is initiated in cloud fragments
by pressures from embedded stellar winds. It could also be initiated
simply by the accretion of interfragment gas, with no direct
triggering (Zinnecker 1982). The collapse time can be very long if
the accretion is slow.

3. OBSERVATIONS OF THE TOTAL LUMINOSITIES OF NEWLY FORMED STARS AND CLUSTERS

The luminosities of pre-main sequence stars are poorly known as a
function of stellar mass and age because the pre-main sequence track
on an H-R diagram is virtually unchecked by observations. Stars may,
in principle, make excursions both up and down the Hayashi track.
They can go up if energy is transferred from stellar rotation or
magnetism into the photosphere, possibly during a FU Ori-type

outburst (Larson 1980), and they can go down as this excess energy is gradually released. For a large enough sample of IRAS protostars, the average relation between luminosity mass and age may be determinable, provided reasonable assumptions about the initial mass function and the stellar birthrates are made. IRAS luminosities of embedded stars should also be compared to wind luminosities obtained from moleculare line observations. One might expect the radiant and particle luminosites to vary in proportion (Bally and Lada 1983).

IRAS luminosities should also be used to identify embedded star clusters that will eventually become "open" clusters, like the Pleiades. A newly formed cluster will remain bound after the gas leaves if the ratio of the stellar mass to the total mass is sufficiently high, > 10%, depending on the magnetic field strength in the cloud and the rate of cloud dispersal (Hills 1980; Elmegreen 1983; Mathieu 1983; Wilking and Lada 1983; Lada, Margulis and Dearborn 1984; Elmegreen and Clemens 1985). The mass of an embedded cluster should be proportional to its total luminosity. Clouds with the largest rations of the stellar luminosity to total mass therefore have, on average, the highest star formation efficiencies. Such clouds are potential formation sites for open clusters.

4. OBSERVATIONS OF DUST GRAINS

Dust can be extremely important to the process of star formation. Dust regulates the ionization fraction, and therefore, the ion-molecule reaction rates and the ionic contribution to the viscosity of magnetic diffusion (Mestel and Spitzer 1956); it adds its own contribution to the viscosity of magnetic diffusion (Baker 1976; Elmegreen 1979; Nakano 1984); it contributes to the heating of gas (Goldreich and Kwan 1974) and therefore regulates the Jeans mass at high densities, and it determines the opacity of gas in a collapsing protostar (Gaustad 1963; Lynden-Bell 1973; Low and Lynden-Bell 1976; Silk 1977), and in the radiation-expelled envelope (Kahn 1974), thereby affecting the radii and masses of the pre-main sequence stars.

IRAS data for our galaxy and other galaxies should be examined for correlations between the dust-to-gas mass ratio and the IMF. Theory predicts that lower relative dust and metal abundances should correlate with higher gas temperatures (all else being equal, such as heating rates), and a greater proportion of massive stars (i.e., a shallower IMF), or a larger maximum stellar mass.

IRAS observations of the distribution of dust temperatures might also be used to determine a minimum grain size (corresponding to a maximum grain temperature), and the relative abundance of such small grains (from their relative luminosity). The relative density of very small grains is critically important for star formation. The distribution function for grain size given by Mathis, Rumpl and Nordsieck (1977) has an enormous number of tiny grains, with a grain number density $n_g(a) = n_o a^{-3.5}$ for grain radius a. Other evidence for a significant population of small grains (0.001 microns) comes from

Sellgren (1984). The importance of such tiny grains is assessed here.

Grains that are surrounded by free electrons and ions become negatively charged in the absence of photoionization. For large enough cloud temperatures, T, and large enough grains, the number of electrons on a grain, Z, is given by Z~2akT/e². Molecular clouds have such low T, however, that 2akT/e² is less than 1, even for large grains (a=1 micron). Thus the grains alternate in charge between -e and 0 (Elmegreen 1979). The fraction of grains of each radius that are charged at any one time Φ(a), depends on the overall ionization fraction. For ionization equilibrium between grains, electrons, molecular ions and atomic metal ions, with abundances relative to hydrogen nuclei equal to D(a) = n_g(a)/n,x,M, and A, respectively, and for mean thermal speeds of electrons,molecular ions and atomic ions given by c_e,c_m, and c_a, respectively, Φ(a) is given by the equation of steady state charge balance,

$$xc_e[1 - \psi(a)]\pi a^2 D(a) = (Mc_M + Ac_A)\psi(a)(1 + \frac{e^2}{akT})\pi a^2 D(a). \qquad (1)$$

The left-hand side of this equation equals the rate per hydrogen atom at which electrons hit neutral grains (having a relative abundance [1-Φ]D and collision cross section πa²), and the right-hand side equals the rate at which various ions hit negatively charged grains (having relative abundance [1-Φ]D and collision cross section πa²), and the right-hand side equals the rate at which various ions hit negatively charged grains (having relative abundance ψD and collision cross section πa²(1 + e²/akT)). For c_m~c_a and x~ M + A, ψ(a)~1 for all but the smallest grains (a>0.01 microns). The total charge locked up in grains per hydrogen atom is given by the expression

$$Q = \int_{a_{min}}^{a_{max}} \psi(a)D(a)da, \qquad (2)$$

where a_{min} and a_{max} are the minimum and maximum grain radii. For ψ(a)~1 and D_0=n_0/n, Q~0.4$D_0 a_{min}^{-2.5}$, which depends sensitively on a_{min}.

Numerical solutions to the charge-balance equations were obtained for the 4 species mentioned above, including charge exchange with rate coefficient β = 10⁻⁹ cm³ s⁻¹, dissociative recombination with rate coefficient α = 10⁻⁶ cm³ s⁻¹ and recombination on grain surfaces, with normalized rate πa²c_eD(a)(1-ψ(a)) for electrons, and πa²c_iD(a)ψ(a)(1 + e²/akT) for ions of type i= M and A. The cosmic ray ionization rate per atom is ξ(~10⁻¹⁷ to 10⁻¹⁸ s⁻¹), and the normalized cosmic ray ionization rate per atom is denoted by ξ = ξ/n. These solutions follow Oppenheimer and Dalgarno (1974) and Elmegreen (1979). The relative grain abundance is obtained from the ratio, ε, of the total grain mass to the total gass mass:

$$\int_{a_{min}}^{a_{max}} \frac{4\pi}{3}a^3 \rho_g D(a)da = \varepsilon m_H \qquad (3)$$

so that $D_0 = 2.6 \times 10^{-25}$ cm$^{-2.5}$, for $\rho_g = 1$ gm cm^{-3}, $\varepsilon = 0.01$, $a_{max} = 1$ micron, and $a_{min} \ll a_{max}$. The relative abundance of metal atoms is assumed to be 4×10^{-6}.

Figure 1 shows the resulting values x and Q as functions of ξ. The sensitivity of x and Q to a_{min} at low ξ is evident, as is the decrease in x below Q. In this limit, most of the negative charge is on the grains and not in free electrons. Very small values of ξ should occur in dense cloud cores and in collapsing protostars, where $n > 10^8$ cm^{-3}. The variation of x with n in this limit is extremely important for collapsing protostars (Black and Scott 1982).

Figure 2a shows the ratio of the viscous force density for magnetic diffusion from the charged grains,

$$F_g = \frac{4\pi}{3} c_m \rho \Delta v \int_{a_{min}}^{a_{crit}} n_g(a)\psi(a)a^2 da, \qquad (4)$$

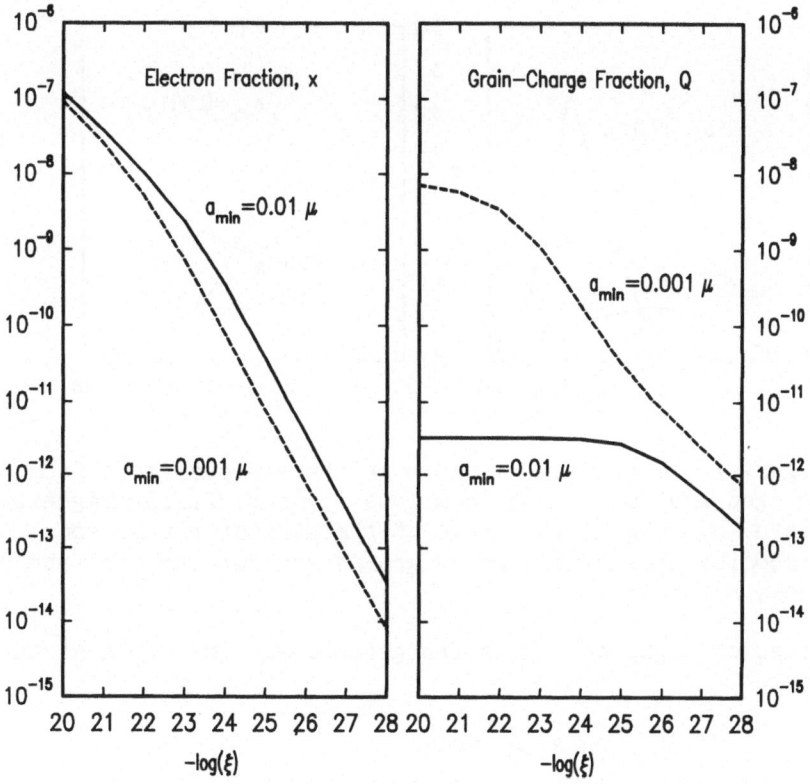

Figure 1 - The electron fraction, equal to the density of electrons divided by the density of all nuclei, n, and the grain-charge fraction, equal to the number density of charged grains divided by the density of nuclei, are shown as functions of the normalized cosmic ray ionization rate, $\xi = \xi/n$ for two minimum grain sizes.

to the viscous force density from the ions,

$$F_i = \rho \Delta v n (M + A) <\sigma_{in} c_i>, \tag{5}$$

where Δv is the relative drift velocity between charged grains or ions and neutrals, c_m is the mean thermal speed for all molecules, ρ is the mean molecular mass density, a_{crit} is the largest grain that can gryate around the field, derived in equation (6), below, and $<\sigma_{in} c_i> - 2.2 \times 10^{-9}$ cm^{3}s^{-1} (Spitzer 1978) is the ion-neutral collision rate. Charged grains are important for viscosity when F_g/F_i. The drift velocity for the grains equals that for the ions when the grains gyrate around the field, because then both grains and ions follow the field lines. Grain gyration occurs when the grain gyrofrequency exceeds the viscous dissipation rate:

$$\frac{eB}{m_g c} > \frac{4\pi c_m a^2 \rho}{3 m_g} \quad \text{or} \quad a^2 < a_{crit}^2 \equiv \frac{3eB}{4\pi c c_m \rho}; \tag{6}$$

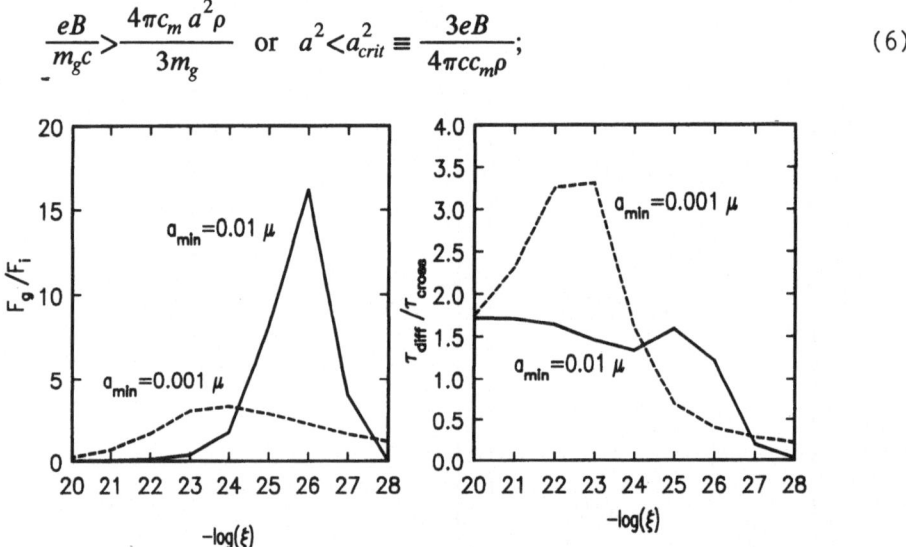

Figure 2 - a.(left) The ratio of the viscous force from charged grains to that from ions for a diffusing magnetic field. b. (right) The ratio of the diffusion time for viscosity including charged grains and ions to the internal crossing time.

c is the speed of light and m_g is the grain mass. The field strength is taken to be

$$B = 10^{-5} (\xi/10^{-20} \text{cm}^3 \text{ s}^{-1})^{0.5} \text{ Gauss}, \tag{7}$$

which corresponds to B = 10^{-5}(n/100 cm^{-3})$^{0.5}$ Gauss for c = 10^{-18}s^{-1} in a dense cloud core of protostar. Figure 2a shows that the force ratio depends sensitively on a_{min}. The increase in F_g/F_i at a large ξ is because M + A decrease with decreasing ξ, yet ψ is nearly constant until x-Q. At intermediate ξ, the force ratio is smaller for smaller a_{min} because x and therefore ψ are smaller (Fig. 1). The rapid decrease in F_g/F_i at very small ξ is because $a_{crit} \to a_{min}$.

Figure 2b shows the ratio of the magnetic diffusion time scale to the dynamic time scale, or crossing time, in a dense cloud. The diffusion time scale including both grains and ions is given by the following expression for a cloud of radius R_c in which the viscous force density from diffusion, $F_g + F_i$, balances the force density from self-gravity $4\pi G\rho^2 R_c/3$:

$$\tau_{diff} = \frac{R_c}{\Delta v} = \frac{3<\sigma_{in}c_m>}{4\pi Gm_H}(M + A)(1 + F_g/F_i). \qquad (8)$$

The internal cloud crossing time for rms velocity v_{rms} is given by the viral theorem,

$$\tau_{cross} = \frac{R_c}{v_{rms}} = 0.69(1 + \beta)^{0.5}(G\rho)^{-0.5}; \qquad (9)$$

β–1 is the ratio of the magnetic field pressure, $B^2/8\pi$, to the turbulent pressure, ρv^2_{rms} (see Elmegreen 1985). Figure 2b indicates that the diffusion time is comparable to the crossing time for large ξ, so the magnetic field is important for cloud dynamics in this limit. When ξ decreases below 10^{-24} cm^{3}s^{-1} or 10^{-26} cm^3 for a_{min} = 0.001 or 0.01 micron, respectively, τ_{diff}/τ_{cross} begins to decrease suddenly. The corresponding threshold density is 10^6 cm^{-3} or 10^8 cm^{-3} respectively, for $\xi = 10^{-8}$ s^{-1}. At higher densities, magnetic fields lose their importance for collapse retardation (Black and Scott 1982; Nakano 1984) and rotational braking (Gillis, Mestel and Paris 1979; Mouschovias and Paleologou 1980) of a protostar. Such a threshold density for collapse depends sensitively on a_{min} because it occurs approximately when x drops below Q.

The gas temperature at high density also depends critically on the minimum grain size. Collisions between warm grains and gas will heat the gas in molecular clouds with embedded stars. For high sticking probability, the heating rate is

$$\Gamma = \int_{a_{min}}^{a_{max}} nc_m\pi a^2 n_g(a)2k[T_g(a) - T]da \qquad (10)$$

for grain temperature distribution Tg(a) (Hollenbach and McKee 1979). Since $T_g(a)$ increases with decreasing a, the integral is over $a^{-\gamma}da$, with $\gamma > 1.5$, and the result scales with $a^{-\gamma+1}$, which depends primarily on a_{min}. Small grains contribute more to cloud heating than large grains because of the enormous abundance of small grains. Because the Jeans mass depends sensitively on the gas temperature, the characteristics mass for star formation should depend on a_{min}.

REFERENCES
Baker, P.L. 1976, Astron.Astrophys., 50, 327.
Bally, J. and Lada, C.J. 1983, Astrophys.J., 265, 824.
Baud, B., Young, E., Beichmann, C.A., Bientema, D.A., Emmerson, J.P., Habing, H.J., Harris, S., Jennings, R.E., Marsden, P.L., and Wesselius, P.R., Astrophys.J.(Letters), 278, L53.

Beichman, C.A., Jennings, R.E., Emerson, J.P., Baus, B.,·Harris, S.,
      Rowan-Robinson, M., Aumann, H.H. Gautier, T.N., Gillett, F.C.,
      Habing, H.J. Marsden, P.L., Neugebauer, G., and Young, E. 1984,
      Astrophys.J.(Letters), 278, L45.

Black, D.C. and Scott, E.H. 1982, Astrophys.J., 263, 696.

Bodenheimer, P. 1978, Astrophys.J., 224, 488.

Bok, B.J. 1955, Astron.J., 60, 146.

Cesarky, C.J., Cesarsky, D.A., Churchwell, E. and Lequeus, J. 1978,
      Astron.Astrophys., 68,33.

Cleary, M.N., Heiles, C. and Haslam, C.G.T. 1979, Astron.
      Astrophys.Suppl.36, 95.

de Jong, T., Clegg, P.E., Soifer, B.T., Rowan-Robinson, M., Habing,
      H.J., Houck, J.R., Aumann, H.H., and Raimond, E. 1984,
      Astroph.J. (Letters), 278, L67.

Dibai, E.A. 1958, SovietAstron.AJ, 2, 429.

Elmegreen, B.G. 1979, Astrophys.J., 232, 729.

Elmegreen, B.G. 1982, in Submillimeter Wave Astronomy, ed. J.E.
      Beckman and J.E. Philips, Cambridge University Press,
      Cambridge, p.1.

Elmegreen, B.G. 1983, Mon.Not.Royal.Astron.Soc., 203, 1011.

Elmegreen, B.G. 1985, Astrophys.J., in press (November 15 issue).

Elmegreen, B.G. and Lada, C.J. 1977, Astrophys.J., 214, 725.

Elmegreen, B.G. and Elmegreen, D.M. 1978, Astrophys.J., 220, 1051.

Elmegreen, B.G. and Clemens, C. 1985, Astrophys.J., in press (July 15
      issue).

Emerson, J.P., Harris, S., Jennings, R.E., Beichman, C.A., Baud, B.
      Bientema, D.A., Marsden, P.L. and Wesselius, P.R. 1984,
      Astrophys. J.(Letters), 278, L49.

Gaustad, J.E. 1963, Astrophys.J., 138, 1050.

Gillis, J., Mestel, L. and Paris, R.B. 1979, Mon.Not.Royal.Astron.
      Soc., 187, 311.

Gilmore, W. 1980a, Astron.J., 85, 894.

Gilmore, W. 1980b, Astron.J., 85, 912

Goldreich, P. and Kwan,J. 1974, Astrophys.J., 189, 441.

Hills, J.G. 1980, Astrophys.J., 235, 986.

Hollenbach, D., and McKee, C.F. 1979, Astrophys.J.Suppl. 41,55.

Hoyle, F. 1953, Astrophys.J., 118,513.

Kahn, F.D. 1974, Astron.Astrophys. 37,149.

Klein, R.I. 1985, in Protostars and Planets, II, ed. D.C. Black,
      University of Arizona Press, Tucson, in press.

Lada, C.J., Elmegreen, B.G., Cong, H.I., and Thaddeus, P. 1978,
      Astrophys.J.(Letters), 226 L 39.

Lada, C.J., Margulis, M., and Dearborn, D. 1984, Astrophys.J., 285,
      141.

Larson, R.B. 1980, Mon.Not.Royal.Astron.Soc., 190,321.

Low, C. and Lynden-Bell, D. 1976, Mon.Not.Royal.Astron.Soc. 139,221.

Lynden-Bell, D. 1973, in Dynamical Structure and Evolution of Stellar
      Systems, ed. G. Contopoulos, M. Heron, and D. Lynden-Bell,
      Geneva Observatory, Geneva, p. 131.

Maddalena, R.J., Morris, M., Moscowitz, J., and Thaddeus, P. 1985,
      Astrophys.J., in press.

Mathieu, R.D. 1983, Astrophys.J.(Letters), 267,L97.

Mathis, J.S., Rumpl, W. and Nordsieck, K.H. 1977, Astrophys.J., 217, 425.

Mestel, L. and Spitzer, L., Jr. 1956, Mon.Not.royal.Astron.Soc., 116,503.

Mouschovias, T.C. and Paleologou, E.V. 1980, Astrophys.J., 237,877.

Myers, P.C. 1977, Astrophys.J., 211,737.

Myers, p.C. 1985, in Nearby Molecular Clouds, Eighth European Regional Astronomy Meeting, ed. G. Serra, Springer-Verlag, Heidelberg, in press.

Nakano, T. 1984, Fundamentals of Cosmic Physics, 9, 139.

Norman, C. and Silk, J. 1980, Astrophys.J., 238, 158.

Oort, J.H. 1954, Bull.Astr.Soc.Neth., 12, 177.

Öpik, E.J. 1953, Irish Astron.J., 2, 219.

Oppenheimer, M. and Dalgarno, A. 1974, Astrophys.J., 192, 29.

Sellgren, K. 1984, Astrophys.J., 277, 623.

Silk, J. 1977, Astrophys.J., 214, 152.

Soifer, B.T., Rowan-Robinson, M., Houck, J.R., de Jong, T., Neugebauer, G., Aumann, H.H. Beichman, C.A., Boggess, N., Clegg, P.E., Emerson, J.P., Gillett, F.C. Habing, H.J. Hauser, M.G., Low, F.J., Miley, G., and Young, E. 1984, Astrophys.J. (Letters), 278, L71.

Spitzer, L. Jr., 1978, in Physical Processes in the Interstellar Medium, Interscience, New York, p. 25.

Vrba, F.J., Strom, S.E., and Strom K.M. 1976, Astron.J., 81, 958.

Wesselius, P.R. Bientema, D.A., and Olnon, F.M. 1984, Astrophys.J.(Letters), 278, L37.

Wilking, B.A., and Lada, C.J. 1983, Astrophys.J., 274, 698.

Zinnecker, H. 1982, in Symposium on the Orion Nebula to Honor Henry Draper, ed. A.E. Glassgold, P.J. Huggins and E.L. Shucking, New York Academy of Sciences, New York, p. 226.

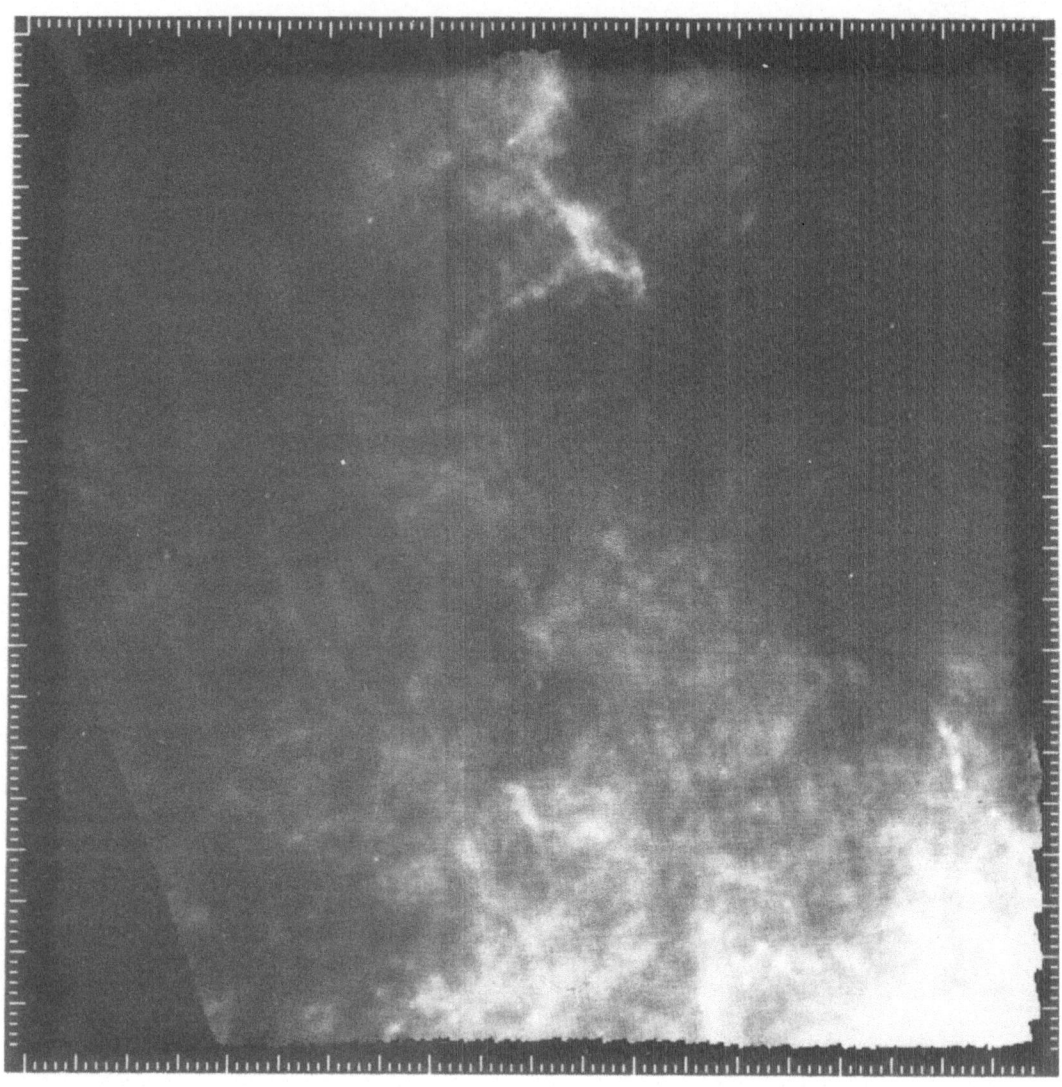

IRAS Field 152   α = 09ʰ00ᵐ, δ = -30°, HCON-3, 100 μm. Field centered
on Pyxis, showing appreciable cirrus in Vela
(bottom)

IMF IN STARBURST REGIONS

Hans Zinnecker
Royal Observatory,
Edinburgh EH9 3HJ
SCOTLAND

1. OBSERVATIONS

There is growing evidence that the Initial Mass function (IMF) appears to be strongly weighted towards high-mass OB stars in certain regions where star formation occurs in a burst. Here, I define a "starburst region" to be a star formation region with very high infrared surface brightness ($\gtrsim 10^{10}L_\odot/kpc^2$), a definition independent of distance. Starburst regions reveal themselves by unusually small mass-to-light ratios (M/L<1). Ratios as small as 0.04 and 0.03 have been inferred for the central 0.5 kpc diameter of M82 and NGC253 (Rieke et al. 1980), and 0.05 for the central 3 kpc diameter of NGC 1068 (Telesco et al. 1984). M/L-ratios of order 10^{-1} to 10^{-2} are typical for interacting and merging galaxies (Joseph et al. 1984, Joseph and Wright 1985) respectively. These galaxies turn out to be among the most luminous IRAS sources detected (e.g. Arp 220, Soifer et al. 1984). In another case (Mk 171 = NGC 3690 + IC 694 in collision) all the properties of the system are accounted for by the formation of massive stars in the range $20-35M_\odot$ only (Augarde and Lequeux 1985, Telesco et al. 1985).

2. MODEL CALCULATIONS

Telesco (1985) has constructed models of starburst regions aimed at explaining the small M/L ratios. While he concludes that cases where M/L ~ 0.1 could still be compatible with a normal IMF, there is no way to explain M/L ~0.01 other than with a truncated IMF (i.e. no stars with M $\lesssim 3M_\odot$). The question arises : what is the physical reason for such a truncated IMF and whereupon does the cut-off depend?

I have made simple analytic calculations which may offer a hint as to answer this question. The calculations try to follow the evolution of a collection of gaseous clumps of unit mass meant to represent a dense interstellar cloud. The key condition to truncate the lower IMF at a relatively high mass is to have a very large number of clumps (N~10^6 or more). This is seen in the following way:

277

F. P. Israel (ed.), Light on Dark Matter, 277–278.
© *1986 by D. Reidel Publishing Company.*

The mean mass $\overline{M}$ to which a clump will grow is given by the initial clump mass M_O times the average number of collisions that occur before the cloud will have undergone its free-fall collapse. Now the average number of collisions is given by the ratio between the crossing time of the cloud and the clump-clump collision time. This is equal to the cloud radius divided by the mean free path of the clumps. For a volume filling factor f, and assuming a geometrical cross-section for clump-clump coalescence, this then leads to

$$\overline{M} = 3N^{1/3}f^{2/3} M_O$$

(cf. Section II in Pumphrey and Scalo 1983).
$\overline{M}$ is identified with the lower cut-off mass of the IMF. For M>$\overline{M}$ the final mass distribution tends to be a power-law independent of initial conditions (cf. Section III in Pumphrey and Scalo 1983). Starting with $M_O \sim 1 M_\odot$ one requires $N \sim 10^6$ to obtain $\overline{M} \sim 30 M_\odot$ (taking f ~0.03). Other values for $\overline{M}$ are readily obtained depending on the choice of the parameters (N,f,M_O): $\overline{M} \sim 3 M_\odot$ for $N \sim 10^5$ and f ~0.003 (keeping $M_O = 1 M_\odot$). One realises that for a given M_O a larger cut-off mass $\overline{M}$ requires, by implication, a larger total cloud mass (=NM_O). A problem implicit in the above picture of collisional growth is that the clumps must somehow be supported against collapse for a time longer than their own free-fall time (by a factor $f^{-1/2}$). Turbulence fed by the collisions may provide the support.

3.   CONCLUSION

From the foregoing I conclude that the formation of very massive (clumpy) clouds may be a necessary condition for a special IMF without low-mass stars. Once some high-mass stars form, their radiation induces the implosion of other massive clumps while evaporating the lower mass clumps (LaRosa 1983). For a single cloud the prediction, therefore, is a highly coeval burst ($\Delta t \sim 10^5$ yr).

I acknowledge discussions with B.G. Elmegreen and C.M. Telesco.

REFERENCES

Augarde, Lequeux (1985), Astron. Astrophys. (in press).
Joseph et al. (1984). M.N.R.A.S. 209, 111.
Joseph , Wright (1985). M.N.R.A.S. 214, 87.
LaRosa (1983). Astrophys. J. 274, 815.
Pumphrey , Scalo (1983). Astrophys. J. 269, 531.
Rieke et al. (1980). Astrophys. J. 238, 24.
Soifer et al. (1984). Astrophys. J. 283, L1.
Telesco et al. (1984). Astrophys. J. 282, 427.
Telesco et al. (1985). Astrophys. J. 15 Dec. 1985 issue.
Telesco (1985).   In "Extragalactic Infrared Astronomy".
    RAL Workshop held in May 1985 (ed. P. Gondlekhar).

Point Sources in the Orion Complex

C. A. Beichman
Infrared Processing and Analysis Center
California Institute of Technology 100-22
Pasadena, CA 91125   USA

ABSTRACT. A 667 sq. deg. region in Orion contains thousands
of IRAS sources, including stars, embedded sources,
galaxies, as well as clumps in more extended emission
(cirrus). Two particular types of sources, visible stars
with infrared excesses characteristic of T Tauri stars and
invisible, cool, embedded objects (L~25 L-sun) are probably
related to the formation of low mass stars in Orion.

1. INTRODUCTION

The proximity of the Orion Molecular Cloud (~500 pc) as well
as the richness and diversity of the astrophysical phenomena
displayed there have made Orion the premier region for the
observational study of star formation and giant molecular
clouds. The purpose of my talk is to describe the point
sources that IRAS has measured toward Orion, giving an
overview of how one might use the IRAS catalog in a complex
region of this kind. Some 54 new objects are identified
that are probably low mass stars in early stages of
formation, either embedded in the surrounding molecular
cloud or in a subsequent T Tauri-like phase.

2.   THE IRAS SAMPLE

Fig. 1 shows the large scale 100 micron emission from the
Orion region. A rectangular $32^{\circ} \times 47^{\circ}$ area between
$6^{h}15^{m} > \alpha > 4^{h}45^{m}$ and $15^{\circ}00' > \delta > -15^{\circ}00'$, comprises some 667 sq.
deg, and encloses most of the extended emission. In galactic
coordinates this region extends from $220^{\circ} > l > 185^{\circ}$ at galactic
latitudes that range from $-3^{\circ} > b > -26^{\circ}$. The dominant clouds
in this region include Orion Molecular Clouds (OMC) 1 and 2,
Mon R2 and the clouds associated with λ Ori. A description
of the molecular gas emission in this same region is given
by Maddalena et al. (1985). The content of the IRAS Point

279

F. P. Israel (ed.), Light on Dark Matter, 279–292.

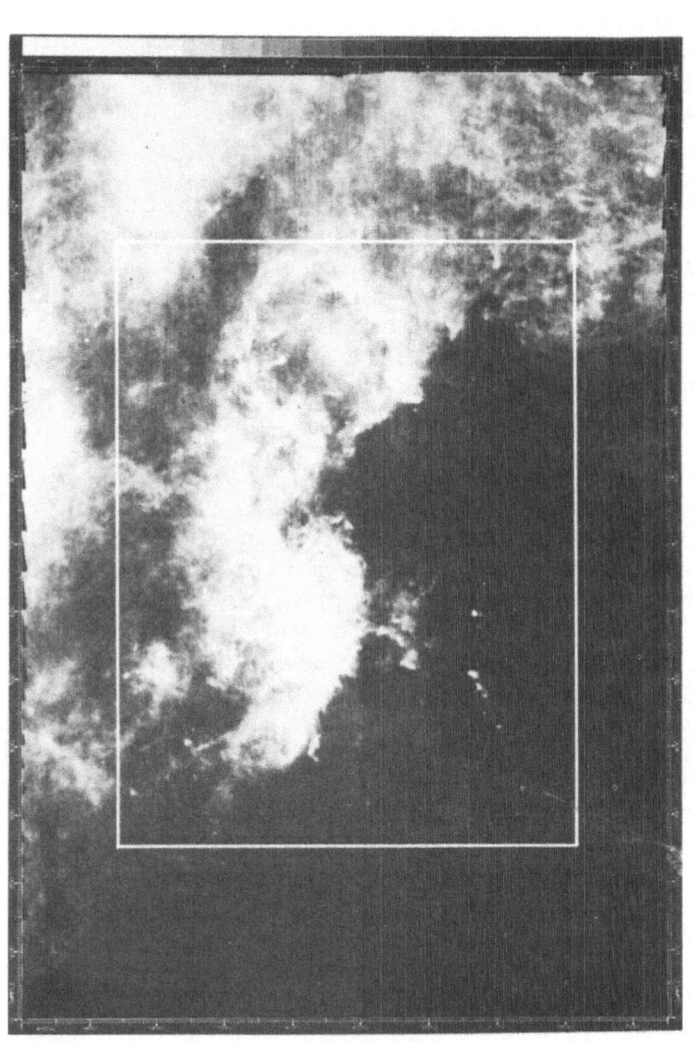

Figure 1. The total 100 micron emission from the Orion region. The image is centered at $5^h30^m, 0°02'$. The pixel size corresponding to one tick mark is 2'. The white rectangle outlines the region discussed in this paper.

Source Catalog within this region forms the sample discussed in this paper. The details of the IRAS data processing have been described in the Supplement to the IRAS Catalogs and Atlases (1985).

For the purpose of comparing source counts it is convenient to define a rectangular region in Gemini comprising 191 sq. deg. and extending between $205^{\circ} > 1 > 195^{\circ}$ and from $6^{\circ} > b > 26^{\circ}$ (corresponding to $8^{h}08^{m} > a > 6^{h}40^{m}$ and $25^{\circ} > \delta > 10^{\circ}$). This comparison region is at the same average l and |b| as Orion, so that disk populations symmetric in reflection about the galactic plane should appear in similar numbers in the two regions. Differences in the density of sources between the two regions should thus be related to populations specific to the vicinity of a giant molecular cloud.

Table I lists the numbers and surface densities of some of the various types of sources to be found in the IRAS catalog for both Orion and the comparison region.   Several types of source are defined below:

1) All sources:
Sources were only required to appear in the Point Source Catalog (Fig. 2).

2) Sources with various kinds of associations:
Many astronomical catalogs were searched for possible counterparts to IRAS sources (see p. V-65 in the Supplement). Categories of objects include stars, galaxies and other objects (HII regions, planetary nebulae, IRAS Small Extended Sources).

3) Classes of object predominantly due to cirrus emission.
Throughout Orion, most of the sources detected only at either or both of the two long wavelength channels (60 micron or 100 micron) are probably due to small scale structures, called cirrus (Low et al. 1984), within the overall emission from the molecular clouds.

4) Sources detected at either 12 or 25 micron with no associations:
Much of the interest in Orion centers on the nature of sources with no previously known counterparts. Objects were selected for having no associations in the major astronomical catalogs described in the Supplement. To decrease problems with structured extended emission at the long wavelengths, sources in this category had to be detected at either 12 or 25 micron. Sources associated with confirmed small extended sources were excluded.

Table I. Point Sources in Orion and Gemini

Source Type	------Orion------		-----Gemini-----	
	Number	Density (sq. deg.)$^{-1}$	Number	Density (sq. deg.)$^{-1}$
All sources	5035	7.54	589	3.08
Assoc. Type:				
Stars	810	1.21	260	1.36
Galaxies	115	0.17	36	0.19
Other/multiple	839	1.26	63	0.33
None	3271	4.90	230	0.83
Cirrus				
100 μm only	879	1.31	99	0.52
60 μm only	464	0.69	28	0.15
60 and 100 μm	1263	0.80	61	0.32
Cirrus Free:				
12 and/or 25 μm (no assoc.)	1022	1.53	105	0.55

3. NATURE OF THE IRAS SOURCES

Sources detected at 60 and/or 100 micron abound in Orion having a surface density three to six times greater than that seen in the Gemini region. A map of the distribution of the Orion point sources (Fig. 2) shows that the source density follows the total 100 micron emission. On the other hand, in the comparison region in Gemini the surface density of sources detected at 60 micron or at 60 and 100 micron, 0.47 (sq. deg.)$^{-1}$, is close to the value found for infrared bright, spiral galaxies at high galactic latitudes. (Chester, this conference).

Two arguments suggest that the majority of the long wavelength sources in Orion are cirrus and not distinct, self-luminous objects such as galaxies. First, the correlation coefficient quoted in the IRAS catalog is a measure of whether a source is truly point-like. For most of the long wavelength sources in Orion the correlation coefficient is poor, suggesting that these objects are extended. Second, the Orion sources are cooler than those found in Gemini. Fig. 3 gives histograms of the 60-100 micron flux density ratio for the two regions. In Orion the average color temperature is 31 K versus 45 K for the comparison field. The latter temperature is characteristic

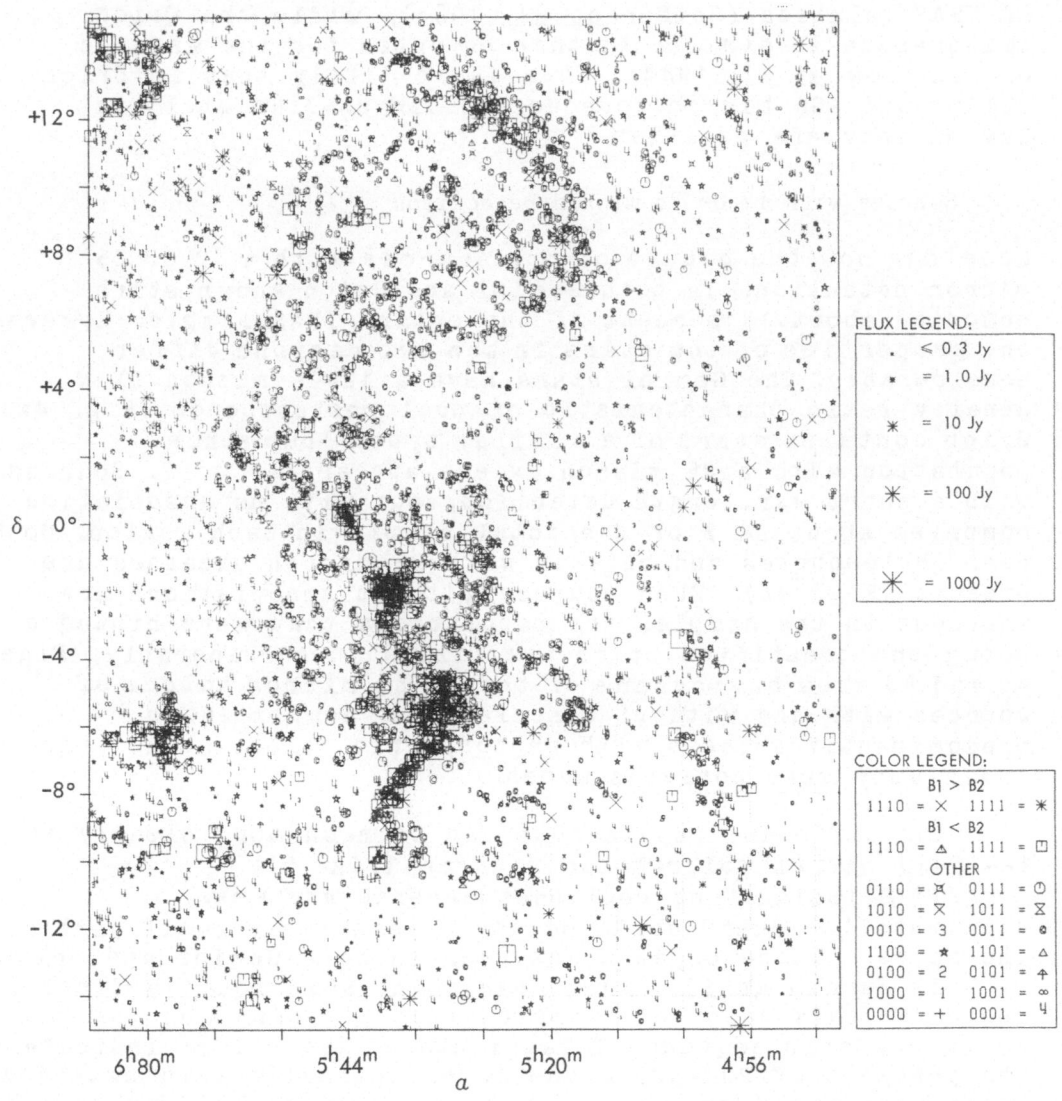

Figure 2. All of the IRAS point sources in Orion. The type
of symbol plotted denotes the wavelengths at which IRAS
detections were made as described by a four digit number in
the legend. In the four digit number a '1' indicates that a
band was detected. Bands are encoded 12, 25, 60 and 100
micron from left to right.

of IRAS galaxies (Soifer et al. 1984), while the Orion
temperature is similar to that reported for the warmest
cirrus (Low et al. 1984). Thus it is likely that in Orion
the 60 and 100 micron sources are cirrus-like while in
Gemini they are galaxies.

3.2. Sources with 12 or 25 micron Detections

Consider now the nature of the sources with a 12 or 25
micron detection. In both Gemini and Orion known stars
comprise about 15 percent of the entire IRAS sample. However
the properties of the stars in the two regions differ
considerably. The Gemini stars have a 12-25 micron flux
density ratio characteristic of cool photospheres (Fig. 4a).
Orion contains stars of a similar type, but a new
population with a 25 micron excess appears as well. Sources
with a short wavelength detection, but with no association
comprise about 20 % of the total sample in each region. Both
cool photospheres and objects with 25 micron excesses are
present (Fig. 4b). When sources without associations are
included in the sample, the populations that were hinted at
among the identified stars are filled in considerably. Figs.
4c and 4d show histograms of the 25-60 micron colors of
sources with and without associations. Gemini shows
predominantly objects with photospheric colors while Orion
shows many more cooler sources.

      The different populations can be seen more clearly in a
12-25-60 micron color-color diagram (Fig. 5):
      1) Emission from cool photospheres and weak
circumstellar dust shells occupy a relatively small part of
the space. These objects are seen in both Gemini and Orion.
      2) Harris et al. (in preparation) have found a distinct
region in this color-color space that contains most of the
known pre-main sequence T Tauri stars. The colors indicate
the presence of 200-300 K shells of optically thin material.
Orion contains many more of this type of source than Gemini.
      3) A few stars have a photospheric 12-25 micron color,
but a pronounced 60 micron excess. These stars resemble
α Psa and Vega and may be stars associated with the Orion
region and having circumstellar disks (cf. Aumann et al.
1984). The number of these sources is small.
      4) Some objects are very red with colors similar to the
objects studied by Beichman et al. (1985) and Myers (this
conference). These may be newly formed or still forming
stars that are embedded within their parent molecular
material. Many more of these objects are found in Orion than
in Gemini. Bright galaxies occupy a similar part of the
color-color space, accounting for some of the Gemini objects
with these colors (Chester, this conference).
      5) A group of objects located in the extreme lefthand

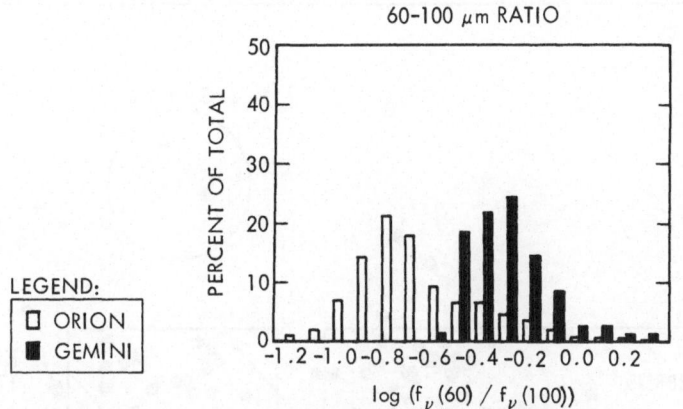

Figure 3. A histogram showing the 60 micron to 100 micron flux density ratios for Orion (white) and Gemini (black).

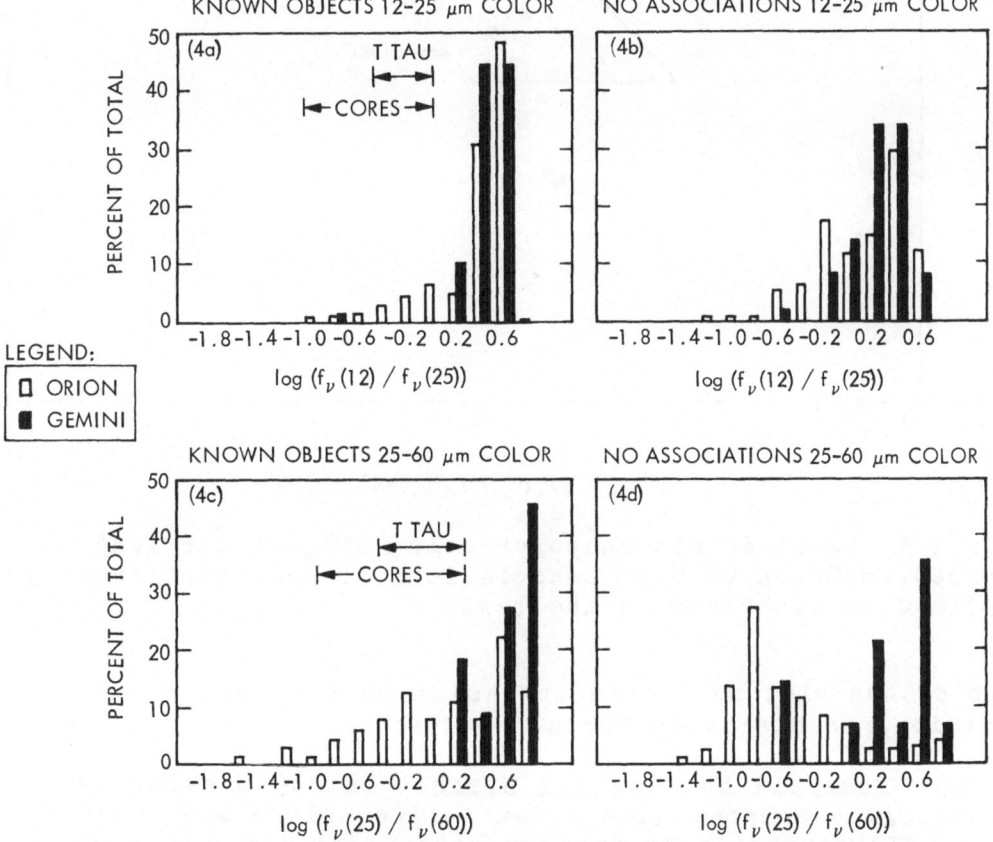

Figure 4(a-d). Histograms of the number of sources with various ratios of 12 to 25 and 25 to 60 micron flux densities for both Orion (white) and Gemini (black). Histograms are given for sources with and without associations with known objects.

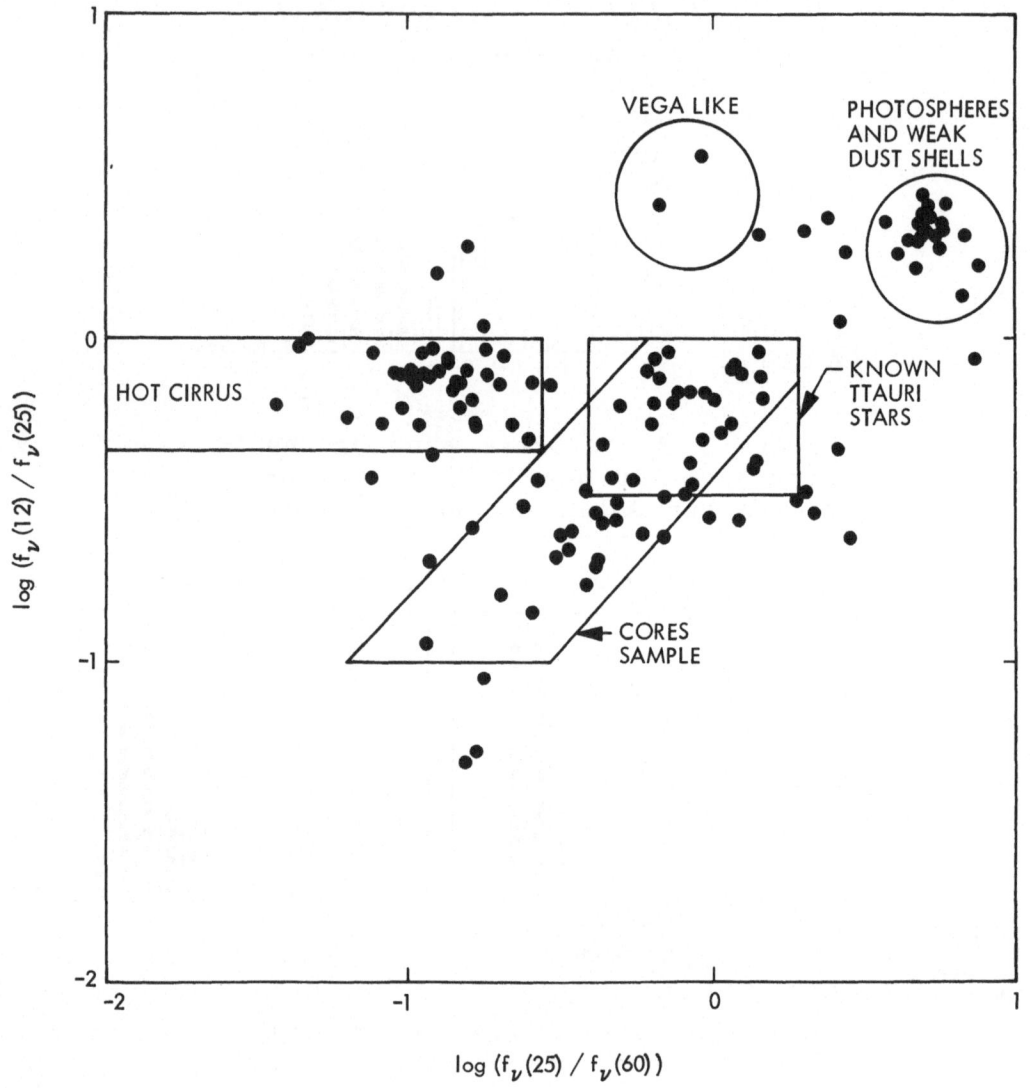

Figure 5. 12-25-60 micron color-color diagram for IRAS sources in Orion with no associations. A number of areas are outlined as discussed in the text.

side of the plot may represent emission from hot cirrus emission, as discussed further below.

3.2.1.   <u>Possible New T Tauri Stars</u>.   Thirty previously uncatalogued sources with T Tauri-like colors are found (Table II) clustered in regions containing already known T Tauri stars (Fig. 6) near the Trapezium and NGC 2024. Comparison with the Palomar Observatory Sky Survey (POSS) prints shows that two thirds of the IRAS sources are associated with visible stars, some with reflection

nebulosity. A few positions are either blank or impossible
to classify without better positions or optical images.
Confirmation of the premain sequence nature of these objects
awaits spectroscopic observations.

Table II. Sources with T Tauri-like Colors

| Position | | Color Corrected Flux Densities (Jy) | | | | POSS Type |
α(1950)	δ(1950)	12 μm	25 μm	60 μm	100 μm	
04h45m06.6s	-05°39'31''	0.4	0.5	0.6	<3	Two stars
05 14 13.5	-01 54 37	1.2	1.3	1.0	1.2	Red star
05 20 22.2	-01 43 19	0.4	0.7	0.5	<7	Faint star
05 20 37.3	+00 52 01	0.8	1.1	1.2	1.6	Red star
05 20 57.8	-01 07 08	0.7	1.7	1.9	<2	Red star
05 22 13.4	-08 44 39	0.8	2.1	3.0	3.4	Star
05 23 34.3	+11 29 01	6.1	8.5	14	23	Red star
05 23 53.1	-06 26 31	0.9	1.7	1.1	<2	Star
05 24 30.9	+00 22 40	1.5	3.2	2.3	1.6	Star
05 25 41.5	+01 07 44	0.5	0.6	0.4	<1	Red star
05 28 10.0	+02 57 26	0.7	0.7	0.6	<2	Red star
05 28 49.5	+09 49 46	0.4	0.7	0.7	<2	Blank ?
05 29 54.9	+12 19 02	0.4	0.5	0.7	<13	Crowded
05 30 26.7	+02 26 05	0.8	0.8	0.5	<2	Star
05 35 05.7	-02 10 00	1.0	1.1	0.8	<30	Star
05 35 43.8	-06 50 56	3.0	3.2	5.0	32	Red star
05 36 20.1	-07 14 20	2.8	3.0	2.4	<70	Star+Refl
05 37 56.0	-08 15 37	0.6	1.1	2.4	<9	Blank ?
05 38 30.1	-09 07 33	1.2	2.0	1.8	<7	Blank
05 39 22.3	-08 22 57	0.5	0.6	0.9	<50	Blank
05 40 00.7	-08 00 13	0.3	0.8	1.6	9.3	Blank?
05 40 21.5	-10 26 29	0.6	1.5	1.6	<5	SAO Star ?
05 40 27.8	-08 39 44	0.8	1.1	2.2	<20	Blank ?
05 40 35.6	-09 52 10	0.6	1.2	1.3	<20	Blank
05 46 27.1	+01 06 40	0.9	1.5	2.4	11	Star+Dust
05 51 21.0	-10 24 34	20	25	28	52	Star+Refl
05 59 26.7	-09 25 32	1.6	2.0	2.3	<3	Faint star
06 06 52.3	-10 10 57	0.3	0.5	0.6	<2	SAO Star ?
06 11 10.0	-06 24 07	0.5	0.5	0.6	<10	Red star
06 11 24.6	-09 19 32	0.6	1.5	2.6	<10	Blank

3.2.2. <u>Possible</u> <u>Low</u> <u>Mass</u> <u>Proto-stars</u>. Twenty four sources
(Table III) have colors like the objects found by Beichman
<u>et al.</u> (1985) in molecular cloud cores in Taurus and
Ophiuchus. More than half of these are located within
obvious dense clouds and have no optical counterparts on the
POSS prints. These may be embedded young stars or protostars
of about 1-100 L_0 (0.5-2 M_0). Some of these objects are
found in well known molecular clouds, others in new regions
(Fig. 7).

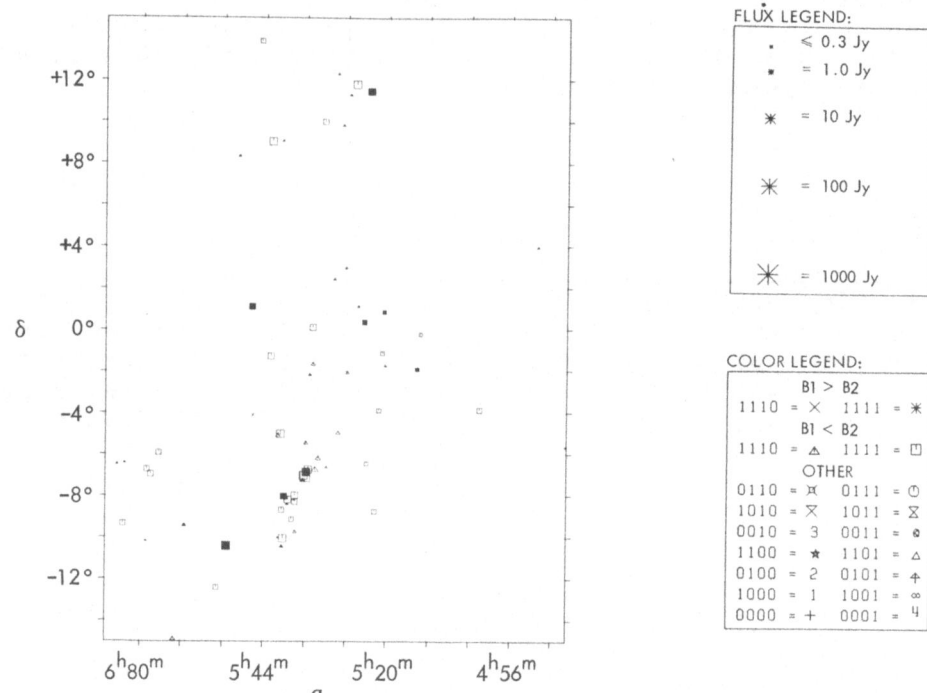

Figure 6. A map of Orion showing the distribution of
IRAS sources with 12-25-60 micron colors similar to
T Tauri stars. Sources associated with known stars
are shown as filled symbols, those without
associations as open symbols. The legend is as
described in Fig. 2.

Larson (1985) has argued that giant molecular clouds like
Orion which are hotter and denser than, say, the Taurus
clouds, should form more massive stars than those found in
cool clouds. The average luminosity of the sources in Table
III without optical counterparts is about 25 L-sun compared
with about 8 L-sun for comparable sources described by
Beichman et al. (1985) in Taurus and Ophiuchus. While a more
careful study of the selection effects must be made before
arguing for differences in the initial mass functions in the
two types of molecular clouds, it is intriguing to note that
although there is low mass star formation in Orion, the
average mass of the embedded young stars may be higher than
those in cooler, less dense regions like Taurus.

3.2.3. Hot Cirrus. The population alluded to earlier in the
extreme left of the 12-25-60 micron plot has a relatively
hot 12-25 micron color temperature of about 300 K and a cold
25-60 micron color temperature of about 70 K. Examination
of the Point Source Catalog shows that these sources have
very poor correlation coefficients and are thus likely to be

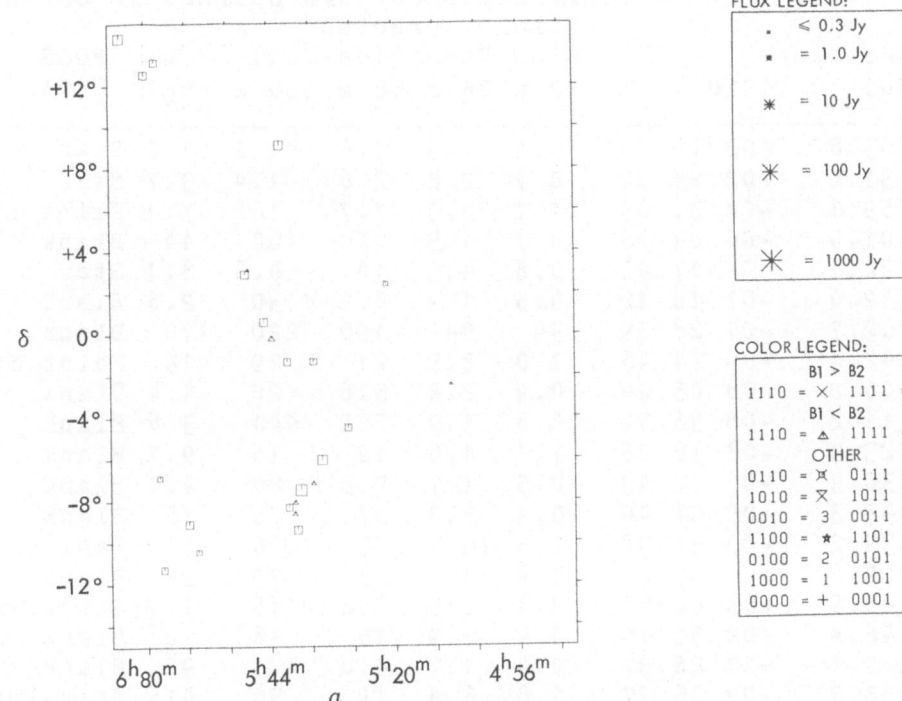

Figure 7. A map of Orion showing the distribution of
IRAS sources with colors characteristic of deeply
embedded stars or protostars. The legend is as
described in Fig 2.

compact structures in the cirrus. The location of these
sources in the hottest, most intense regions of OMC-1 and
NGC 2024 support this interpretation.

4.   CONCLUSIONS

The infrared emission in Orion extends over hundreds of
square degrees and contains many thousands of IRAS sources.
While many of these, particularly at 60 and 100 micron can
only be associated with emission from wisps and
condensations within the molecular clouds, there are also
many discrete point sources that merit further observation.
Some types of source are seen toward Orion, but not toward a
comparison region located at a similar galactic coordinates.
Populations of object exist that are unique to the environs
of the molecular clouds.

        Objects that may be previously unknown T Tauri stars
or younger, embedded protostars can be seen throughout the

C. A. BEICHMAN

Table III. Possible Embedded Core Sources in Orion

Position (1950)	(1950)	Color Corrected Flux Densities (Jy)				L (L$_0$)[1]	POSS Type
		12 m	25 m	60 m	100 m		
05^{h}09^{m}03.8^s	-02°26'25''	0.4	1.3	1.6	<1.2	1.7	Star
05 21 31.6	+02 25 05	0.9	2.8	2.6	1.4	3.7	Star
05 28 59.1	-04 30 09	1.1	3.3	7.7	16	7.1	Faint Star
05 34 01.5	-06 03 06	0.7	1.5	12	100	14	Blank
05 35 22.1	-01 17 03	0.8	4.3	10	8.6	8.1	Star
05 35 42.0	-07 10 12	0.5	1.2	4.2	<130	2.6	Blank
05 38 02.7	-07 28 59	34	94	190	220	170	Blank
05 38 42.4	-09 24 46	1.0	3.5	21	29	13	Faint Star
05 39 06.2	-08 05 04	0.9	2.4	5.8	<20	4.4	Blank
05 39 11.2	-08 36 51	0.6	1.9	5.2	<40	3.6	Blank
05 40 23.3	-08 18 26	1.1	4.6	12	16	9.7	Blank
05 40 30.3	-01 17 43	0.5	1.7	3.5	20	4.4	Blank
05 41 45.3	+09 07 40	0.4	3.3	27	75	18	Blank
05 43 34.2	-00 11 08	2.5	10	27	<30	19	Refl HH-24
05 45 07.8	+00 37 41	4.4	14	28	39	26	Blank
05 48 15.9	+03 06 57	0.4	0.5	1.6	<15	1.1	Star+Refl
05 48 46.4	+02 55 10	0.4	1.9	16	38	10	Blank
05 58 07.1	-10 26 37	0.4	1.7	4.0	4.3	9[2]	Blank
05 59 53.3	-09 06 27	1.8	6.4	19	30	41[2]	Star+Refl
06 04 47.8	-11 17 26	0.5	1.5	6.3	11	13[2]	Blank
06 05 32.3	-06 53 16	0.6	2.2	3.7	4.5	10[2]	Blank
06 05 58.4	+13 09 13	0.7	3.0	6.9	16	6.4	Faint Star
06 08 00.8	+12 33 24	0.8	4.6	22	25	14	Crowded
06 12 44.8	+14 18 04	4.9	32	110	140	78	Crowded

[1] 12-100 μm luminosity assuming a distance of 500 pc except as noted. [2] Assumes a distance to Mon R2 of 830 pc.

region. A study of the luminosities and masses of the low mass stars forming in the vicinity of a giant molecular cloud like Orion may have important consequences for theories of how the initial mass function depends on the size and temperature of the parent molecular cloud.

References:
Aumann, H.H., et al. 1984, Ap.J.,278, L23.
Beichman, C. A., Myers, P.C., Emerson, J.E.,
        Harris, S., Mathieu, R., Benson, P.J.
        and Jennings, R. 1985, preprint.
Explanatory Supplement to the IRAS Catalogs and Atlases
        eds. Beichman, C.A., Neugebauer, G. Habing, H.J.
        Clegg, P.E. and Chester, T.J. 1984,
        Government Printing Office.
Larson, R. J. 1985, Mon. Not. R. Astr. Soc., 214, 379.
Low et al. 1985, Ap.J., 278, L19.
Maddalena, R. J., Morris, M., Moscowitz, J. and
        Thaddeus, P. 1985, preprint.
Soifer et al. 1984, Ap.J.,278, L71.

The Infrared Astronomical Satellite was developed and
operated by the Netherlands Agency for Aerospace Programs
(NIVR), the US National Aeronautics and Space Administration
(NASA) and the UK Science and Engineering Research Council
(SERC). Thanks are due to R. Dumas for assistance in the
preparation of this manuscript. The research described in
this paper was performed by the Jet Propulsion Laboratory,
under contract with NASA.

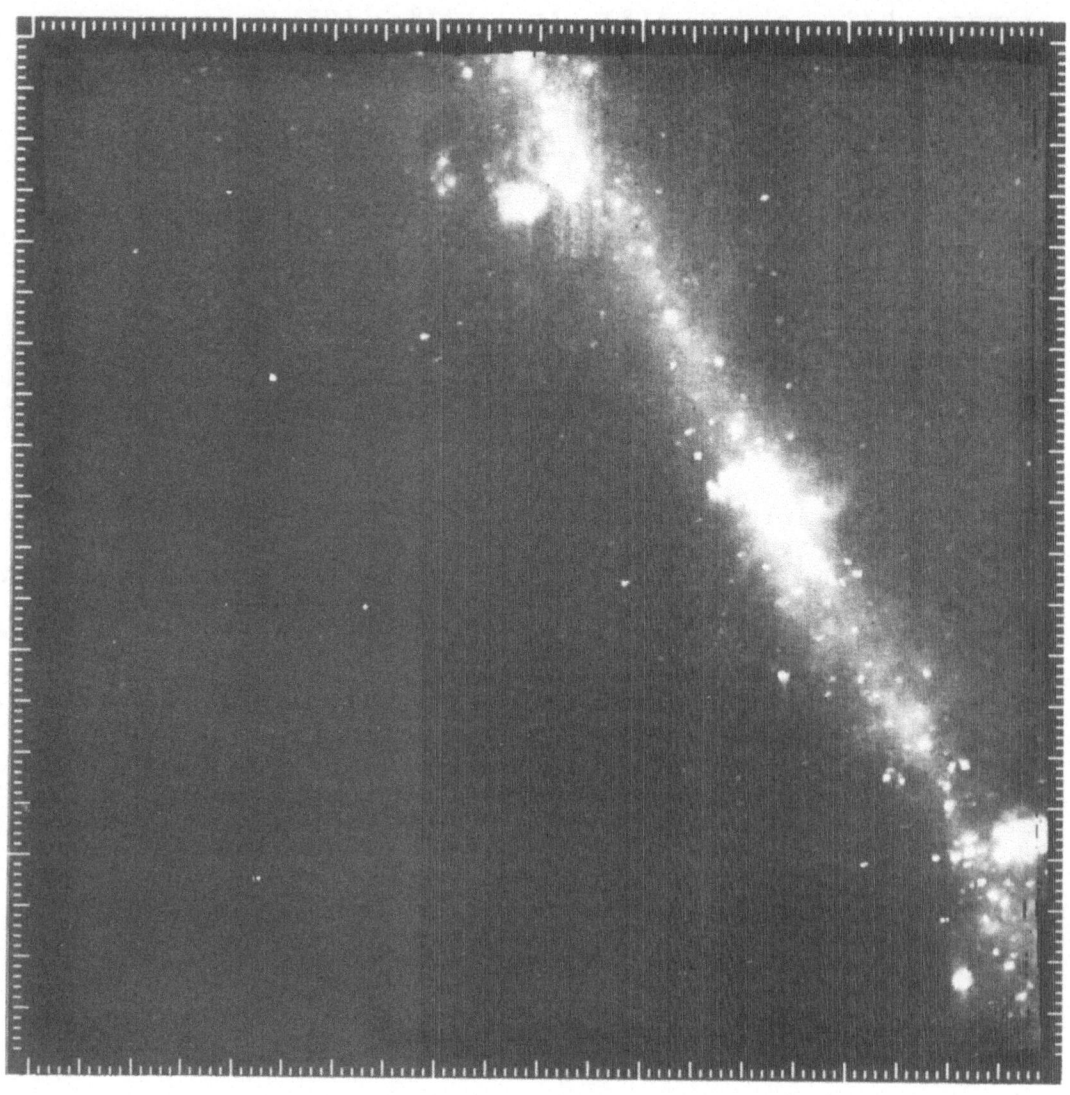

IRAS Field 161   α = 18^h00^m, δ = -30°, HCON-3, 24 μm. Field in
                    Saggitarius, showing the Galactic Center (center
                    right), M8 and M20 (top) and NGC 6357 (bottom
                    right).

THE IR EMISSION OF THE ORION-MONOCEROS MOLECULAR CLOUDS

F. Boulanger, R.J. Maddalena, and P. Thaddeus
Goddard Institute for Space Studies
2280 Broadway, New York, NY10025

1. OBSERVATIONS AND DATA REDUCTION

The present investigation is based on both IRAS and Columbia CO observations of a region ($200° \leq 1 \leq 225°$ and $-25° \leq b \leq -7.5°$) which includes all molecular clouds observed by Maddalena et al. (1985), except the one around λ Ori. The IR emission of the sky comes from dust in molecular clouds, H II regions, the diffuse atomic interstellar gas, and the solar system. Consequently, to get the IR emission radiated by the Orion-Monoceros molecular clouds and HII regions, the zodiacal and atomic gas contributions must be subtracted from the IRAS data. Since outside the molecular clouds the IR emission correlates well with the integrated H I emission (Boulanger et al. 1986), the atomic gas contribution is modelled from the HI data. The zodiacal light is modelled by a wedge in ecliptic coordinates fitted over all map pixels outside the molecular clouds.

2. EMISSION OF MOLECULAR CLOUDS

The total IR luminosity of all molecular clouds and of associated H II regions after bolometric correction is $6.5 \ 10^5 \ L_\Theta$. Most of this luminosity comes from the brightest parts of the Orion A and B clouds; except for a few known sites of star formation, the three other clouds, the northern and southern filaments, and Mon R2 are much fainter. Of the total luminosity, 60% is radiated by the Orion A and B nebulae alone, while 90% of the projected surface of all molecular clouds emits only 20%.

The Orion A and B clouds have a ratio of IR luminosity-to-mass of 1.5 in solar units, a value comparable to that of the surrounding atomic gas. The ratio is about 5 times lower for the other three clouds. Both sets of clouds differ also in color temperature, the faint clouds being much colder than the bright ones.

From the stellar statistics of Reeves (1978), the total luminosity of the Orion OBI association stars of type earlier than B2 can be estimated to B $10^5 \ L_\Theta$, these ionizing starts are probably the dominant source of the IR emission of these clouds. On the other hand, the low IR brightness of the other clouds may be accounted for

F. P. Israel (ed.), Light on Dark Matter, 293–294.
© *1986 by D. Reidel Publishing Company.*

by the energy absorbed by their external parts from the average local
Inter-Stellar Radiation Field (ISRF). Then, for all clouds, embedded
low mass stars do not represent a large fraction of the total
luminosity.

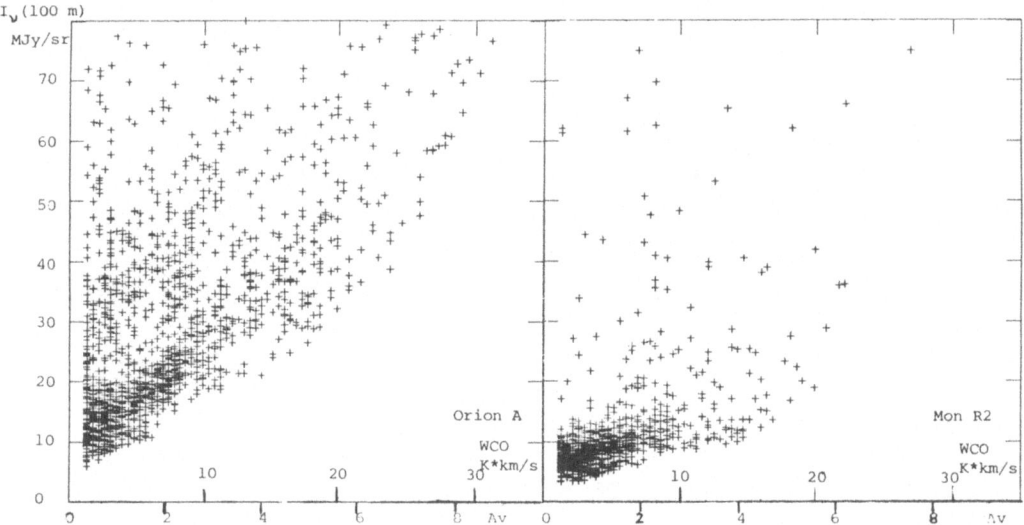

Figure 1-a,b: Pixel by pixel correspondence between the 100 m brightness and the CO integrated
emission (WCO). The CO integrated emission has been converted to extinction assuming: Av = 0.28*WCO.

3. IR-CO COMPARISON

Figure 1-a,b shows the correspondence pixel by pixel between the CO-
integrated emission and the 100 μm brightness for Orion A and Mon R2,
chosen here as examples of bright and faint molecular clouds.
        For Mon R2, the IR brightness, basically uniform over most of
the clouds does not follow the variations of the integrated CO
emission. For Orion A, the points are scattered because many bright
point sources are observed over the cloud, but a clear lower envelope
in the points distribution excludes Mon R2.
        The Mon R2·case is easily understood if the IR emission at
wavelengths shorter than 100μm comes almost exclusively from the
external parts of the cloud heated by the ISRF. On the other hand,
the continuous increase of emission observed in Orion A and B up to
high oprecities suggest that part of this emission comes from dust
inside the clouds heated either by a large number of embedded stars
or by the near IR emission of the OB association.

REFERENCES:

Boulanger, F., Pérault, M., Puget, J.,L., 1986, (ed.), These
proceedings, page 203.
Maddalena, R.J., Morris, M., Moscowitz, J., Thaddeus P., 1985, Ap. J.
    in press.
Reeves, H. 1978, in T. Gehrels (ed.), Protostars and planets,
    University of Arizona Press, p. 399.

YOUNG STARS AND HIGH DENSITY CONDENSATIONS IN THE HORSEHEAD REGION

G. Sandell
Observatory, Univ. Helsinki, SF-00130 Helsinki, Finland
B. Reipurth
Copenhagen Univ. Obs., DK-1350 Copenhagen K, Denmark
C. Menten, M. Walmsley
MPIfR, Auf dem Hügel 71, D-5300 Bonn 1, FRG
H. Ungerechts
Institut for Space Studies, 2880 Broadway, NY 10025, USA

ABSTRACT. We have mapped the Horsehead in ammonia and made partial maps in the CO and 13CO lines, both with a resolution of 30". We find strong molecular emission close to the ionization front, and several NH_3 condensations both in the head and down at the base of the Horsehead. Within our ammonia map there are four IRAS sources. Two are deeply embedded, one of them close to a NH_3 condensation. The other two are identifiable with visible stars, one with the nebulous star B33-1 in front of the Horsehead, the other with V615 Ori, which we have found to be a luminous G-type emission line star associated with several Herbig-Haro objects.

1. INTRODUCTION

The recent optical and near-infrared study of the Horsehead (Reipurth and Bouchet, 1984) shows that it is an active site of low mass star formation. In the Horsehead region there is also two Herbig-Haro objects and four IRAS sources, two of which are deeply embedded low luminosity PMS objects without optical or near-infrared counterparts. We have therefore undertaken a more complete study of the region to see how these PMS objects relate to and interact with the surrounding dark cloud, and to study the shock effects close to the ionization front.

2. MOLECULAR OBSERVATIONS

We have made an almost complete map with 30" spacing of the Horsehead in the NH_3 (1,1) and (2,2) lines with the Effelsberg 100 m telescope. We have also made strip maps in the J=1-0 CO and ^{13}CO lines with the same resolution using the Onsala Space Observatory 20 m telescope, as well as some limited maps around some of the PMS stars found by Reipurth and Bouchet.

F. P. Israel (ed.), Light on Dark Matter, 295–296.

Our NH_3 map reveals several condensations in the Horsehead and hardly any emission in the lower portion of the neck. Both the NH_3 and the ^{13}CO emission have a distinct peak just behind the rim. This maximum is also seen in CO, although not as pronounced as in ^{13}CO. We therefore have quite a sharp increase in the density when we reach into the Horsehead and also an enhancement in the kinetic temperature behind the ionization front.

Otherwise the CO and the NH_3 emission appears quite similar to that of normal dark clouds. The lines are narrow, ^{13}CO 1 km s^{-1} and NH_3 0.5 km s^{-1}, although the kinetic temperature is somewhat higher than in a typical dark cloud, 15K. The radial velocity is 10.7 km s^{-1}, with a weak gradient along the axis of the Horsehead. There is a distinct velocity shift and the lines get considerably broader as we reach into the L1630 molecular cloud. We see no tendencies for wings or line enhancement near any of the PMS sources, except for perhaps close to V615 Ori, the IRAS source thought to be responsible for the exitation of the Herbig-Haro objects (see below).

3. IRAS SOURCES AND HERBIG-HARO OBJECTS

The IRAS satellite detected four point sources within the boundaries of our ammonia map of the Horsehead region. IRAS 05383-0228 is identified with the little nebulous star B33-1, which is partly embedded in the very front of the Horsehead (Reipurth and Bouchet, 1984). IRAS 05384 -0229 is situated at the edge of an ammonia condensation in the middle of the Horsehead itself, and is possibly identical to the very faint star B33-7. IRAS 05386-0229 appears to have no optical counterpart, but may be related to a one arc-minute long luminous ray, which seems to emanate from behind the Horsehead. All of these sources are of low luminosity.

The fourth source, IRAS 05388-0224, is identified with the variable nebulous star V615 Ori. Our spectrum of the star shows only H in emission, a steep red continuum and a few absorption lines, which suggests a G-type spectrum. Using interference filters we have found 5 Herbig-Haro knots on a line from V615 Ori. Some of the knots were independently found by Malin (1985). Searches at optical, near-infrared and IRAS wavelengths have not uncovered other sources near the HH objects than V615 Ori. Since this star is visually very bright (after dereddening), and is fairly luminous for a T Tauri star and since it has a G-type spectrum, we speculate that it may be in an advanced fading phase after a FU Ori type eruption.

REFERENCES

Malin, D.: 1985, AAT Newsletters
Reipurth, B., Bouchet, P.: 1984, Astron. Astrophys. 137, L1

MAPPING OF THE CORONAE AUSTRINAE STAR FORMING REGION

A. Evans[1], J.S. Albinson[1], M.F. Bode[2], D.C.B. Whittet[3]

(1) Dept. of Physics, Univ. of Keele, Keele, Staffs.
(2) Dept. of Astronomy, Univ. of Manchester, Manchester.
(3) Dept. of Astronomy, Lancashire Polytechnic, Preston.

ABSTRACT. We describe observations of the Coronae Austrinae (CrA) star forming region. With the exception of R CrA, TY CrA and HH100 no infra-red sources were detected in the region surveyed.

1. INTRODUCTION

The CrA dark cloud, a nearby (126 pc) region of star formation, has been extensively studied at near infrared[1,2] and millimetre[3] wavelengths. The region around the Herbig Ae/Be stars R CrA and TY CrA has also been mapped in the far infrared[4,5].

We describe IRAS observations of a 12' x 70' region centred on R CrA and TY CrA. The observations consisted of multiple scans using edge detectors in the four bands of the survey array. Two separate observations of the region, taken 2 weeks apart, were combined to produce the maps discussed here.

2. RESULTS

Infrared sources associated with the regions around R CrA, TY CrA were detected in all four bands; in addition a source apparently associated with the Herbig-Haro object HH100 was certainly detected in bands I, II, while in bands III, IV the separation of R CrA and HH100 is less than a beamwidth. In each case the objects detected are essentially point sources.

No other sources were detected, in any of the four bands, in the entire field surveyed. An examination of the background in each band leads to upper limits on the strengths of any sources of 0.3 Jy, 2.0 Jy, 47 Jy and 27 Jy in bands I-IV respectively. In particular we note that, with the exceptions of the detections noted above, there are no IRAS sources, to the limits cited, as the positions of (i) the sources in the JHKL Glass-Penston survey[1]; (ii) the sources in the 2.2 µm survey of Vrba et al.[2]; (iii) the regions of high ($\sim$ 20 K) CO antenna temperature in the survey of Loren[3]. Further there is no detectable emission from the region of the Herbig-Haro object HH101, which lies $\sim$ 6' SW

F. P. Israel (ed.), Light on Dark Matter, 297–298.

of HH100. There is no definite evidence of any diffuse emission (particularly in band IV) which may be associated with cold ($\sim$ 30 K) dust distributed throughout this region, although there is an indication of some extended, low brightness emission in bands III, IV around R CrA and TY CrA.

3. DISCUSSION

The upper limits given above lead us to conclude that there are no embedded sources having L $\gtrsim$ 2 L$_\odot$ in the region surveyed as such sources would have been detected in at least one band. In particular, there can be no embedded obscured sources that can account for the peak in the CO distribution S of R CrA[3]; presumably this peak must, as discussed by Loren[3], arise as a result of heating by B8 stars and by TY CrA.

The colour temperatures at the position of TY CrA range from $\sim$ 335 K (I–II) to $\sim$ 80 K (III–IV), and from $\sim$ 315 K (I–II) to $\sim$ 70 K (III–IV) for R CrA. The total infrared luminosity $L_{IR}/L_\odot \sim$ 20.5 ± 5.0 for TY CrA and $\sim$ 30 ± 7 for R CrA, corresponding to spectral classes A4 ± 1 and A3 ± 1 respectively. This classification is consistent with previous work in the case of R CrA but not in the case of TY CrA, for which a B2 class is indicated[3]. This discrepancy may be resolved if only a fraction of the radiation of TY CrA is reprocessed by dust, lending support to the suggestion that this object lies in front of the molecular cloud.

The far infrared flux distributions of R CrA and TY CrA, derived from IRAS observations, are similar to those described in ref. 5, although the present band I,II fluxes for TY CrA are $\sim$ 10 times higher than those given in ref. 5; we cannot rule out variability at these wavelengths.

Indeed if TY CrA does lie in front of the could, the dust temperature and stellar luminosity lead us to expect a phase difference $\sim$ days between variability in the optical and at 20 μm; such variations could easily be monitored in future.

References

1. Glass, I.S. & Penston, M.V., MNRAS, 172, 227 (1975).
2. Vrba, F.J., Strom, S.E., Strom, K.M., AJ, 81, 317 (1976).
3. Loren, R.B., ApJ, 227, 832 (1979).
4. de Muizon, M et al., A & A, 83, 140 (1980).
5. Cruz-Gonzalez, I., McBreen, B., Fazio, G.G., ApJ, 279. 679 (1984).

A. ANALYSIS OF POINT SOURCES IN THE OPHIUCHUS AND PERSEUS CLOUDS

B. CPC OBSERVATIONS OF NGC 1333

R.E. Jennings, W. Cudlip, C.J. Hirst and D.H.M. Cameron

Department of Physics and Astronomy
University College London

ABSTRACT: Sources have been chosen from the IRAS point source
catalogue which are associated with particular molecular clouds, in
this case the Ophiuchus and Perseus clouds. The special relationship
that the infrared cirrus detected by IRAS appears to have with these
molecular clouds is discussed while the distribution of embedded
sources shows that the number having a luminosity in the middle of
the range appears to be relatively low.

There is considerable interest in studying the distribution of
the luminosities of dust embedded sources in molecular clouds, as
many of the sources will be stars at an early stage in their
development. Sources in the Ophiuchus and Perseus molecular clouds
were selected from the IRAS point source catalogue - unfortunately,
the situation is not ideal as for both clouds there are problems of
confusion as well as problems connected with the cirrus. Optical
depth contours generated from the IRAS flux maps were used to define
the boundary of each of the clouds.

It was important that the sources should have a band three flux
(B3) showing that they were dust embedded. All combinations of two or
more (adjacent) IRAS bands will include B3 except for B1+B2. Sources
only registering in B1+B2 were typically foreground stars - spread
uniformly over the whole field - and were ignored. Sources
registering in only one band were also ignored. The infrared
luminosity of each of the sources was obtained by integrating over
a $1/\lambda$ Black Body curve fitted to the observations, the aim being to
include the flux outside as well as inside the IRAS bands.

Of particular interest was the group in which a signal was
recorded at 60 and 100 μm only (B3+B4), as a large proportion of
these sources are warm infrared cirrus. These sources cover the whole
field in a fairly uniform manner, but if a very modest threshold
luminosity is imposed it is found that almost all of the sources
above the limit lie inside the boundary of the respective cloud. This
result is significant as it shows that the luminosity of the point
sources associated with the cirrus is dependent on the local density

F. P. Israel (ed.), Light on Dark Matter, 299–300.
© *1986 by D. Reidel Publishing Company.*

and that there is a smooth transition from low density cirrus to
molecular clouds.
The distribution of luminosities of the sources within the boundary
of each of the clouds was obtained and, in the case of the Rho
Ophiuchi cloud, was compared with the distribution obtained by Lada
and Wilking (Ap.J.,287,610,1984) for the central core of the cloud.

It appears that molecular clouds and cirrus are related, one being a
denser form of the other.

Sources of large luminosity (100Lo+) form in optically denser regions
while sources of ~ iLo form throughout the clouds.

Sources of luminosity 5 - 1000 Lo are deficient in number,
possibility due to a lack of middle range density in the case of the
Rho Ophiuchi cloud.

NGC 1333: The region lying south of the reflection nebula NGC
1333 in the Perseus molecular cloud is known to be associated with
active star formation and an area of about 25 arc min. square has
been scanned at 50 and 100 μm with the Chopped Photometric Channel
(CPC) on IRAS, this instrument giving higher angular resolution than
could be obtained directly with the survey detectors. The area
scanned has a number of Herbid Haro (HH) objects and embedded
infrared sources. As well as observing sources in the region of HH12
(for map see, for example, Moran,J.M.,1983, Rev. Mexicana Astron.
Astrof.,7,95) and the energising source for HH7-11, other sources
were seen, such as the one at 3h 25 32, 31° 03.3. This source could
be prestellar, particularly as it is known from CO observations that
it has a significant outflow of gas (Glenn White et al., to be
published).
Another interesting source is coincident with the Herbig-Haro objects
HH6. While this source shows up clearly on both
the 50 μm and 100 μm maps, the temperature map shows that it is not
significantly warmer than its surroundings. Either this region is
simply a condensation which radiates more flux because of its greater
density, or it is a prestellar object at the beginning of its
evolutionary sequence whose surroundings dust shell has not become
appreciably hotter than its surroundings. The latter appears to be
more likely, not only because of the presence of HH objects but also
because a water maser has recently been found there (Sandell, G.and
Liseau, R....IAU Conference, Toulouse, 1984)

These findings will need confirmation when the fully processed CPC
maps have been analysed.

ACKNOWLEDGEMENT: The data used here were obtained from the IRAS
project, the success of which was the result of much hard work on
both sides of the Atlantic.

STAR FORMING LOOPS IN THE IRAS SKY IMAGES

P.R.Schwartz
E.O.Hulburt Center for Space Research
Naval Research Laboratory
Washington DC   20375-5000

ABSTRACT. Loops of heated dust and embedded young stars are seen in IRAS images of the galaxy. Some of these loops are undoubtedly OB association "bubbles": a conclusion which revives the possibility of sequential star formation.

1.   SHELLS AND LOOPS IN THE INTERSTELLAR MEDIUM

The IRAS sky images, particularly at 60 and 100 um are sensitive to both interstellar material, in the form of extended emission from heated dust, and to embedded stars which produce localized discrete sources. This data set affords us a uniquely unobscured view of the relationship of recently formed stars to the interstellar medium. Inspection of IRAS galactic fields reveals a large number of loops in the extended emission. Some of these loops are stellar bubbles, ionized bright rims, or are associated with HI shells and supershells. In these cases, the loops are limb brigtened shells of interstellar material. Several loops are not only defined by extended emission but contain discrete sources with the characteristic structure, spectrum and luminosity of embedded young stars or even significant HII regions.

2.   TWO EXAMPLES

Figure 1 is an image of the approximately 80 pc diameter loop centered on the old open cluster Tr3 studied in CO by Blitz (1978). The extended loop of emission is associated with the dark clouds L1605 and L1624. The discrete sources to the northwest are optical HII regions (IC446, 2169, NGC 2245, 2247 and 2259). The eastern source(s) are associated with NGC2264 and are Allen's (1972) infrared source (IRS1) and its companion (IRS2, Schwartz et al. 1985). These objects, including Allen's source, are early B stars (S Mon which is probably also associated with the structure is an O7). The overall structure of the loop supports Blitz's suggestion that it is a stellar wind bubble which has triggered star formation activity. Figure 2 is an image of a structure found in TMC1. The five discrete sources are associated with a rotating quasi-stable molecular ring (Schloerb and Snell 1984) and are likely to be embedded low luminosity stars in this ring. The scale of the ring is approximately 1 pc.

F. P. Israel (ed.), Light on Dark Matter, 301–302.

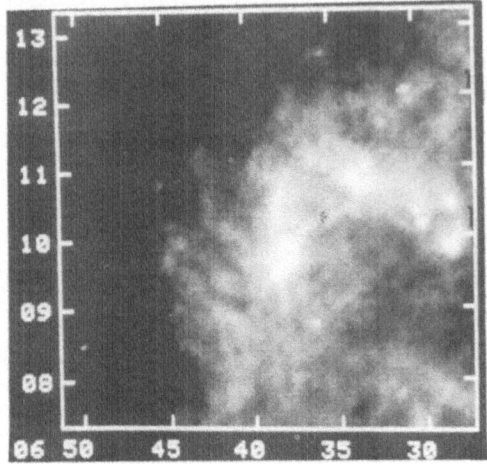

Figure 1 (left): 100 μm image of the L1605, L1624 region sometimes called Mon OB1. The bright source near the center is NGC2264 IRS 1 and 2.

Figure 2 (right): 100 μm image of the TMC1 ring. The maxima are also peaks in CO.

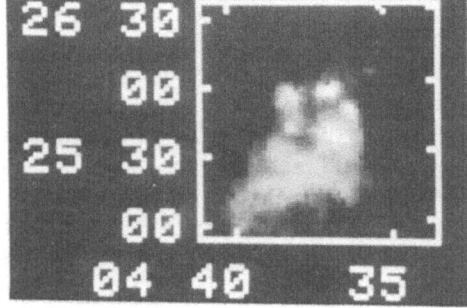

3. DISCUSSION

The examples above are but two of many cases of star forming loops that can be seen in the IRAS images. Other than the fact that the NGC2264(IRS1 and 2) region also represents a pc scale quasi-stable ring similar to the TMC1 ring (Schwartz et al. 1985) they bear only morphological similarity. The Mon OB1 phenomenon is worthy of particularly intensive future study. It and cases like it represent the possible signature of a in which stellar wind bubbles can compress the interstellar medium and trigger sequential star formation.

REFERENCES

Allen, D.A. 1972, Ap.J.(Letters) 172, L55.
Blitz,L., 1978, NASA TM79708.
Schloerb,F.P., Snell, R.L., 1984, Ap.J., 283, 129.
Schwartz, P.R., Thronson, H.A., Odenwald, S.F., Glaccum,W.,
    Loewenstein, R.F., Wolf,G. 1985, Ap.J. 292,231.

MODELS FOR IRAS OBSERVATIONS OF GALACTIC HII REGIONS

J. Crawford and M. Rowan-Robinson
Theoretical Astronomy Unit
Queen Mary College
London E.1.

We have investigated the transfer of radiation through a spherically symmetric distribution of dust surrounding a centrally located hot star. Comparison with observation is made via IRAS colour-colour diagrams for compact HII regions.

1. MODEL PARAMETERS

The star melts a cavity into the surrounding dust cloud, out to a radius r_1, at which the grain temperature falls below the sublimation temperature T_1. The parameters of the model are:

T_s The temperature of the illuminating star, chosen to be 40000 K in all models, although the emergent infrared spectrum is insensitive to variations in T_s for $T_s \gtrsim$ 10000 K (Rowan-Robinson 1980)

T_1 The grain sublimation temperature. Extensive work on modelling hot centred clouds indicates T_1 = 1000 K, the value adopted here (e.g. Rowan Robinson 1980, 1982a, 1982b)

β The grain number density distribution index, $n(r) \propto r^{-\beta}$. No successful models were found with β = 1 or 2. The most satisfactory fit was for β = 0 and this is the value used in all the models here.

r_1/r_2 The ratio of the inner cavity radius to the radius of the cloud r_1/r_2 was varied between 10^{-3} and 10^{-4}. .

τ_{uv} The optical depth due to dust of the star at ultraviolet wavelengths. τ_{uv} was varied from 100 to 200.

The grain properties are as used by Rowan-Robinson (1982b).

2. SELECTION OF DATA

To ensure the data selected are those which are most relevant to the simple model adopted, several selection criteria were employed.

i)   FQUAL = 3 in all four bands: Only those objects with the highest quality fluxes in all bands were chosen.

303

F. P. Israel (ed.), Light on Dark Matter, 303–304.
© *1986 by D. Reidel Publishing Company.*

ii) <u>SES2 = 0 in all four bands</u>: To ensure correct balance of colour
     in the flux ratios, only fully resolved objects were accepted.
iii) <u>High far infrared surface brightness</u>: It has been shown (Rowan-
     Robinson 1982a) that HII regions with far infrared surface
     brightness $\gtrsim 10^{-31\cdot5}$ $L_\odot/cm^2$ are the most likely to satisfy the
     assumption of spherical symmetry (this can be understood in term
     of an evolutionary sequence, where deviations from spherical
     symmetry develop as the HII region grows and bursts out of the
     ambient molecular cloud). For our IRAS HII region sample, this
     requires $S(100\mu) \gtrsim 2.8 \times 10^3$ Jy.

RESULTS AND CONCLUSIONS

     The IRAS comparison set of HII regions and the relevant models
can be seen plotted on colour-colour diagrams in Figure 1. The
agreement of the models with the IRAS data is excellent supporting
the hypothesis that the high far infrared surface brightness clouds
are approximately spherically symmetric.

REFERENCES

Rowan-Robinson M. 1980, Ap.J.Supp. <u>44</u>. 403-426.
                   1982a, Submillimetre Astronomy, ed. Philips and
                   Beckman.
                   1982b, Mon.Not.R.Astr.Soc. <u>200</u>, 197-215.

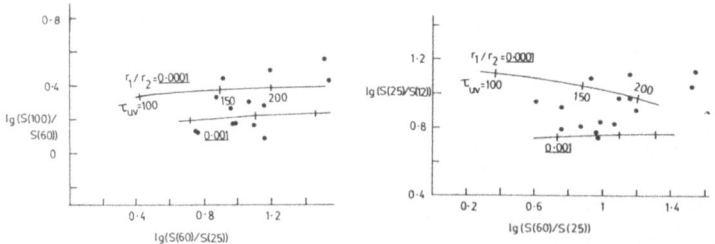

Fig. 1. The colour-colour diagrams for the selected IRAS compact HII
         regions (black circles) are shown. Superimposed are lines of
         constant r_1/r_2 where on each line τ_{uv} increases from left to
         right from 100 to 200, indicated by the tick marks.

FAR-INFRARED (100-200 µm) PHOTOMETRY OF HII REGIONS WITH A 1M BALLOON BORNE TELESCOPE

R.R. Daniel, S.K. Ghosh, K.V.K. Iyengar, T.N. Rengarajan,
S.N. Tandon, and R.P. Verma
Tata Institute of Fundamental Research
Homi Bhabha Road, Bombay 400005
India

ABSTRACT. A 1m balloon-borne telescope has been developed at TIFR for far-infrared (FIR) astronomical observations. The telescope was successfully flown on March 2, 1985 from Hyderabad, India. During the flight RCW 57, RCW 108, RCW 122, W 31, G 351.6-1.3 and Carina nebula were observed. This paper describes the telescope, its performance and very preliminary results from this flight.

A f/8 Cassegrain 1m telescope was fabricated at TIFR for FIR observations. The telescope mirrors are made of aluminum alloy. The plate scale is 26 arc sec per mm and the field of view 2.6 arc min. The FIR radiation is chopped at a frequency of 20 Hz by vibrating the secondary mirror. The chopper throw is 3.5 arc min.

The telescope can be oriented towards any direction in the sky using a three axis orientation system. The coarse orientation, in azimuth and elevation is done using local gravity and the geomagnetic field as reference. Fine orientation is done using a star tracker and rate gyros. The star tracker is sensitive to stars brighter than fifth magnitude. The rms pointing stability of the telescope is about 15 arc sec. The main telescope can be offset with respect to the star tracker by an angle upto $4°.5$. An area around a source is raster scanned to map the source. The FIR signal is detected by a liquid helium cooled Ge(Ga) bolometer. The effective wavelength of the photometer is ~ 160 µm for a black body source with a temperature of 80K. The sensitivity of the telescope is ~ 150 Jy Hz$^{\frac{1}{2}}$.

The telescope was flown on March 2, 1985 from Hyderabad, India at an altitude of ~ 30 km. During the flight we observed HII regions RCW 57, RCW 108, RCW 122, W 31, G 351.6-1.3 and Carina nebula. Saturn was observed for calibration. Fig. 1 shows a part of the signals recorded during raster scan across Saturn for illustrating the pointing capability of the telescope. Using peak signals recorded during scans across various sources and the peak signal for Saturn, we have given in Table 1 very preliminary estimates of the peak flux densities within a 2'.6 beam for the observed sources.

F. P. Israel (ed.), Light on Dark Matter, 305–306.

TIME →

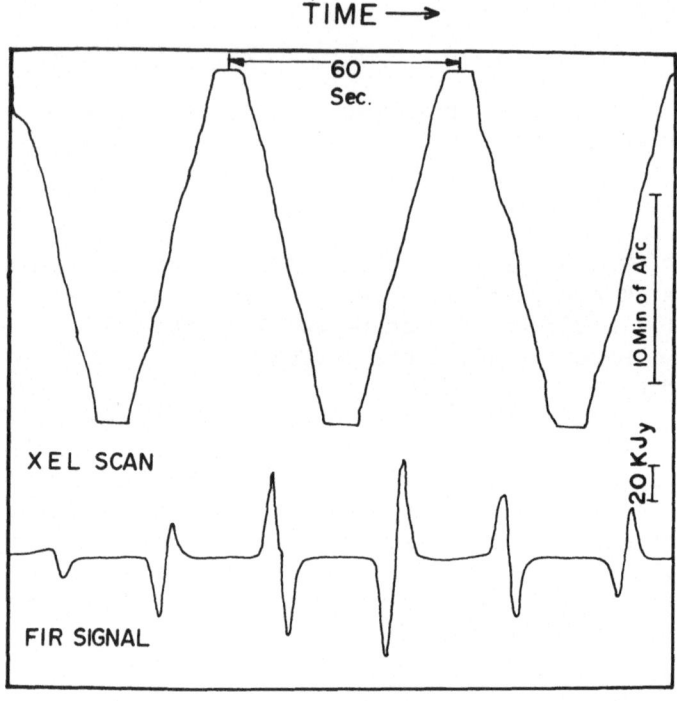

Figure 1. The signals recorded during the raster scan of Saturn are presented. The telescope beam is slowly (0.8 min of arc/sec) moved in cross elevation in a periodic fashion, and at the end of each half cycle the elevation is changed by ~0.7 min of arc. The upper trace shows the cross elevation signal of the star tracker indicating the cross elevation coordinate of the telescope whereas the lower curve is IR signal after phase sensitive detection.

TABLE I

Source	Peak Flux Density at 160μ from our flight Jy	IRAS Flux Density	
		100μ Jy	60μ Jy
RCW 57	17,600	>38,567	>10,908
Carina I	4,700	>17,086	< 4,648
RCW 108	10,500	18,735	<12,220
G 351.6-1.3	13,100	22,329	>12,750
RCW 122	16,400	>35,864	< 9,906
W 31 #4	3,800	11,766	6,031
W 31 #5	10,200	>27,355	> 8,797

YOUNG STARS AND DENSE CORES IN NEARBY DARK CLOUDS

P.C. Myers
Harvard-Smithsonian Center for Astrophysics
60 Garden Street, Cambridge, Massachusetts 02138 USA

ABSTRACT. We review recent progress in the study of young stars and dense cores in dark clouds, as revealed by gas emission line observations from NH_3 (1.3 cm), CO (2.6 mm); by dust and stellar continuum observations with the IRAS satellite (12–100 μm), and with ground-based telescopes (1–20 μm). The main results are: (1) Dense cores are actively forming low-mass stars, typically on time scales $\sim 10^5$ yr. (2) Stars in cores frequently have CO outflow, and are then more luminous than their T Tauri neighbors. They have enhanced NH_3 line broadening, of order 0.2 km s^{-1}; and very cold infrared spectra, peaking at $\lambda \gtrsim 100$ μm. They have at least ~ 30 mag of visual extinction between $\sim 10^{14}$ and $\sim 10^{17}$ cm from the star. (3) Cores without stars have velocity dispersions that are nearly thermal, suggesting that their nonthermal support has largely dissipated.

1. INTRODUCTION

Visually opaque condensations, both isolated and in complexes, have been known since the time of Barnard [1] and were first proposed to form stars by Bok and Reilly [2] over 30 years ago. In the last decade molecular line studies have established the size, temperature, density, velocity dispersion, and mass of such objects within a few hundred pc of the Sun [3-7]. Table 1 summarizes this information, based on observations of emission in the (J,K)=(1,1) line of NH_3 at 1.3 cm [6].

In this report we describe recent work, much of it based on IRAS observations and ground-based followup of IRAS results.

F. P. Israel (ed.), Light on Dark Matter, 307–312.

Table 1
Properties of Dense Cores

FWHM Size (pc)	Kinetic Temperature (K)	Number Density (cm^{-3})	FWHM Line Width (km s^{-1})	Mass ($M_\odot$)	Free-fall Time (yr)
0.1	10	3×10^4	0.3	1	2×10^5

2. CORES FORM STARS

The evidence that cores with properties in Table 1 form low-mass
stars has become stronger as core positions and sizes have been compared
with stellar data at optical, near-infrared, and far-infrared
wavelengths. In Taurus-Auriga, T Tauri stars were found to be clustered
in loose groupings 1-3 pc in size [8], and each of the seven groups
studied was found to contain one or two cores in projection [6]. At
2 μm, a survey of 25 cores revealed six to have an associated star
within one core map diameter of the map peak [9]. Four of these six
stars are optically invisible due to obscuration. Our most recent count
of IRAS results indicates that some 19 of 40 nearby cores with NH_3 maps
have a star within one core map diameter of the map peak [10-12]. Most
of these 19 stars were not detected in the original 2 μm searches [9]
but have since been detected in more sensitive 2 μm searches over
smaller areas [12]. Most of these stars were also detected by IRAS at
25, 60, and 100 μm and have very similar spectra. Furthermore, many of
them have evidence of interaction with the core gas in the molecular
line widths and maps, as described in Section 3. Thus, there is little
chance that they are field stars, unrelated to the cores on which they
are projected.

A good illustration of the coincidences among obscuration, NH_3
cores, and IRAS sources is present in the well-known Taurus dark cloud
"Heiles' Cloud 2," where four prominent patches of obscuration agree
closely with NH_3 emission in position, map size, and orientation. Of
these, three have optically invisible IRAS sources within one NH_3 map
diameter. The best known of these cores, TMC-1, is associated with an
IRAS source of estimated bolometric luminosity 0.7 $L_\odot$.

The detection of 19 stars in 40 cores allows a crude estimate of
the interval τ_{wait} between the time when a core becomes dense enough to
be detected in the NH_3 surveys of [6] and the time when the core begins
its star-forming collapse. If known cores differ only in their

evolutionary state, then $\tau_{wait} \lesssim 10^5$ yr [7]. This result implies that cores without stars generally are now forming, or will soon form, stars: there is no inactive period long compared to the free-fall time. Consequently cores without stars are good candidates for observations seeking evidence of infall or other motions associated with very early stellar evolution.

These results do not bear on the closely related questions of how, or how long, cores evolve before they are dense enough to be detected in surveys of emission lines such as the NH_3 line, requiring gas density $\gtrsim 10^4$ cm^{-3}. This period could conceivably be much longer than 10^5 yr.

3. PROPERTIES OF CORES WITH STARS

3.1 CO Outflows

The association of a young star with spatially displaced maps of CO emission in high- and low-velocity line wings is now a well-known signature of stellar outflow, a very early phase of stellar evolution. CO surveys for evidence of outflow have detected 3 outflows in 180 opaque condensations whose stellar content was unknown [14; detection rate 2%] and 3 outflows in 28 T Tauri stars whose degree of association with dense cores was unknown [15; 11%]. Recently a survey of dense cores known to have young associated stars gave a significantly higher detection rate: of 16 such nearby cores, five have CO outflows, giving 31% [16]. These outflows are similar in spatial and velocity extent to others seen in dark clouds [17].

This comparison of detection rates is consistent with the idea that stars in cores are younger than T Tauri stars, and that young stars clear away their parent cores, perhaps via outflows, before becoming visible as T Tauri stars. If so, the high detection rate of outflows from stars in cores suggests that the outflow process typically lasts a significant fraction of the T Tauri age -- a time closer to 10^5 yr than to the $\sim 10^4$ yr dynamical time deduced from outflow velocity and spatial extent.

Another difference involving outflows from stars in cores and T Tauri stars concerns the bolometric luminosity deduced from the combination of IRAS, near-infrared, and optical observations. In Taurus and Ophiuchus, stars in cores are about as luminous as visible T Tauri stars found in the same complexes, but those with outflows are distinctly more luminous than their non-outflow counterparts, by a typical factor ~ 6 [12,16]. The number of outflow sources of unknown spectral type in these complexes is too large to be consistent with the known proportion of hot, luminous stars of type A or earlier. Consequently the outflow sources are likely to have late spectral type,

as do T Tauri stars. The relatively high luminosities of outflow stars
may therefore represent an evolutionary difference between them and
"normal" T Tauri stars.

3.2   NH$_3$ Line Broadening

The presence or absence of embedded stars is now known for some
~ 40 cores, and it is therefore possible to compare the gas properties
of the two groups for differences that might reflect the presence or
absence of star-core interaction. Line widths of NH$_3$ and C^{18}O are
slightly but significantly broader, by 0.1 - 0.3 km s^{-1}, in cores with
stars than in cores without stars [7,11,12]. This increase is
approximately equal to that expected for free-fall in the typical NH$_3$
cores, and the corresponding increase in mechanical energy is a small
fraction of that indicated by CO outflows. Some combination of these
two types of motion may thus be responsible.

The "turbulent" part of the peak of the NH$_3$ core decreases from
~ 0.4 km s^{-1} when the associated star is centered in the core to
~ 0.2 km s^{-1} when the associated star is one core diameter away, for
stars with 0.3 - 3 L$_\odot$ luminosity. If this change in turbulent motions
is due primarily to the change in star-core separation, evidently
~ 0.1 pc separation is sufficient to significantly reduce the turbulent
coupling [12].

For stars that are well-centered in their cores, the NH$_3$ line width
increases dramatically with stellar luminosity when a sufficiently large
range in luminosity is considered. For low-mass cores with stars of
L ~ 1 L$_\odot$, Δv ~ 0.3 km s^{-1} while for massive cores with stars of
L ~ 10$^{4\mbox{-}5}$ L$_\odot$, Δv ~ 3 km s^{-1} [12]. Studies of relationships among core
line width, core-star separation, and stellar luminosity may provide a
useful complement to analysis of CO outflows in understanding
interaction between young stars and their associated cores.

3.3   Infrared Spectra

The IRAS spectra of stars associated with cores have been
supplemented by 1 - 20 μm observations for some 30 stars [12]. The
positions and fluxes of the ground-based measurements generally agree,
within expected uncertainties, with the IRAS data. The 1 - 100 μm
spectra of flux density generally increase monotonically toward long
wavelength and can be divided into two groups according to spectral
shape: "shallow" spectra increase toward long wavelength on a log-log
plot with slope $\lesssim$ 1 while "steep" spectra increase with slope ~ 4 from
1 to 5 μm, and then increase with slope $\lesssim$ 1.5 from 5 to 100 μm. Nearly
all shallow-spectrum sources are optically visible on the Palomar
prints, and some are well-known T Tauri stars such as CW and HK Tau.
Nearly all steep-spectrum sources are invisible on the Palomar prints,

and they tend to lie closer to the peaks of their associated NH_3 cores than do the shallow-spectrum sources.

Spectra typical of the shallow and steep groups, from sources at the same distance, intersect at $3 - 6$ μm, with the steep fluxes weaker than the shallow fluxes at short wavelengths and stronger at long wavelengths. The difference at short wavelengths is consistent with absorption by $A_v \sim 25$ mag according to the standard values of interstellar extinction in the near infrared. The difference at long wavelengths requires more detailed modeling. Preliminary results indicate that typical $20 - 100$ μm flux values can be modeled by a symmetric distribution of circumstellar dust with $A_v \sim 25$ mag located between $r \sim 5 \times 10^{14}$ cm and $\sim 2 \times 10^{17}$ cm from the star; with dust temperature 10 to 100 K and with gas density $n \propto r^{-1.5}$, such that the volume-average density is consistent with NH_3 observations. Predictions of this model agree well with those of Adams and Shu [20], who explicitly model an accreting protostar. But this model, and others with similar density and temperature laws, predict emission much weaker than observed for wavelengths $\lesssim 12$ μm. Evidently significant contributions must arise from stellar and circumstellar emission inward of $\sim 5 \times 10^{14}$ cm.

4.   PROPERTIES OF CORES WITHOUT STARS

As noted in Section 2, cores without stars have line widths significantly smaller than cores with stars, and the stellar detection statistics suggest that cores without stars are now forming, or will soon (within $\sim 10^5$ yr) form low-mass stars. In addition, the typical core without an IRAS source has FWHM nonthermal width ~ 0.20 km s^{-1}. In such a core, the energy in thermal motions is ~ 0.9 of the total internal kinetic energy, and nonthermal motions are unimportant for support [6,7].

The cores "without IRAS sources" could have stars with emission weaker than the IRAS detection limits, typically 0.5 Jy at 12, 25, and 60 μm and 1.5 Jy at 100 μm [11]. We model the spectrum of the weakest detectable source with a "steep" spectrum as $S_\nu = 0.5(\lambda/25 \text{ μm})^{1.5}$ Jy, giving limiting luminosities (including bolometric correction) of 0.12, 0.15, and 0.72 $L_\odot$ in Taurus, Ophiuchus, and Perseus, respectively. Thus, in Taurus and Ophiuchus, cores that lack IRAS sources are unlikely to harbor stars more luminous than ~ 0.1 $L_\odot$.

REFERENCES

[1] Barnard, E. 1927, <u>Atlas of Selected Regions of the Milky Way</u>,
     eds. E. Frost and M. Calvert (Washington: Carnegie Institution).
[2] Bok, B., and Reilly, E. 1947, 'Small Dark Nebulae.' <u>Ap. J., 105</u>,
     255-257.
[3] Churchwell, E., Winnewisser, G., and Walmsley, C. 1978, 'Molecular
     Observations of a Possible Proto-Solar Nebula in a Dark Cloud in
     Taurus.' <u>Astr. Ap., 67</u>, 139-147.
[4] Martin, R., and Barrett, A. 1978, 'Microwave Spectral Lines in
     Galactic Dust Globules.' <u>Ap. J. (Suppl.), 36</u>, 1-51.
[5] Snell, R. 1981, 'A Study of Nine Interstellar Dark Clouds.'
     <u>Ap. J. (Suppl.), 45</u>, 121-175.
[6] Myers, P., and Benson, P. 1983, 'Dense Cores in Dark Clouds.
     II. NH_3 Observations and Star Formation.' <u>Ap. J., 266</u>, 309-320.
[7] Myers, P. 1985, 'Dense Cores and Star Formation in Nearby Dark
     Clouds.' In <u>Nearby Molecular Clouds</u>, IAU Specialized Colloquium,
     ed. G. Serra (Springer-Verlag), in press.
[8] Cohen, M., and Kuhi, L. 1979, 'Observational Studies of Pre-Main-
     Sequence Evolution.' <u>Ap. J. (Suppl.), 41</u>, 743-843.
[9] Benson, P., Myers, P., and Wright, E. 1984, 'Dense Cores in Dark
     Clouds: Young Embedded Stars at 2 Micrometers.' <u>Ap. J. (Letters),
     279</u>, L27-L30.
[10] Beichman, C. 1984, 'IRAS Observations of Solar Type Stars.' Talk
     presented at <u>Protostars and Planets. II.</u>, Tucson.
[11] Beichman, C., <u>et al.</u> 1985, in preparation.
[12] Myers, P., Benson, P., Mathieu, R., Fuller, G., Fazio, G., and
     Beichman, C. 1985, in preparation.
[13] Stahler, S. 1983, 'The Birthline for Low-Mass Stars.' <u>Ap. J.,
     274</u>, 822-829.
[14] Frerking, M., and Langer, W. 1982, 'Detection of Pedestal Features
     in Dark Clouds: Evidence for Formation of Low-Mass Stars.'
     <u>Ap. J., 256</u>, 523-529.
[15] Edwards, S., and Snell, R. 1982, 'A Search for High-Velocity
     Molecular Gas Around T Tauri Stars.' <u>Ap. J., 261</u>, 151-160.
[16] Myers, P., Hemeon-Heyer, M., Snell, R., and Goldsmith, P. 1985, in
     preparation.
[17] Goldsmith, P., Snell, R., Hemeon-Heyer, M., and Langer, W. 1984,
     'Bipolar Outflows in Dark Clouds.' <u>Ap. J., 286</u>, 599-608.
[18] Myers, P. 1983, 'Dense Cores in Dark Clouds. III. Subsonic
     Turbulence.' <u>Ap. J., 270</u>, 105-118.
[19] Myers, P., Linke, R., and Benson, P. 1983, 'Dense Cores in Dark
     Clouds. I. CO Observations and Column Densities of High-Extinction
     Regions.' <u>Ap. J., 264</u>, 517-537.
[20] Adams, F., and Shu, F. 1985, 'Infrared Emission from Protostars,'
     in preparation.

WATER MASERS COINCIDENT WITH IRAS SOURCES

J.G.A. Wouterloot, C.M. Walmsley
Max-Planck-Institut für Radioastronomie,
Auf dem Hügel 69, 5300 Bonn 1, F.R.G.

The availability of the IRAS point-source catalogue makes more detailed studies of the correlation between H_2O masers and far-infrared sources situated in dark clouds possible. Previous studies have concentrated on sources with bolometric luminosities above $10^4 L_0$ although it is known (see Genzel and Downes 1977, 1979) that some H_2O masers are associated with far infrared sources as weak as $10^2 L_0$. With IRAS sensitivity, one can expect to and does find towards nearby molecular clouds a large number of sources in the luminosity range 10–$10^4 L_0$ and it is of considerable interest to investigate the frequency of occurence of H_2O maser emission towards such sources. Among other things, it is hoped that this will allow an estimate to be made of the relationship between counts of water masers in a molecular cloud and the star formation rate.

With this in mind, we have started a survey of H_2O masers towards IRAS sources apparently associated with two molecular cloud complexes in Orion and Cepheus. We selected infrared sources towards the clouds L1630 and L1641 in Orion (Kutner et al. (1977)) as well as the concentration of clouds and HII regions in the Perseus arm at $L=110^o$, which also contains the Cep OB3 complex (e.g. Wouterloot and Habing (1985)). We used the Effelsberg 100-m telescope (HPBW=40") in conjunction with a maser receiver or system temperature $\sim$70 K. Using position-switching with typically 3 minutes on integration-time per source, we were able to reach rms noise levels of 0.15 Jy. The bandwidth was normally 6.25 MHz which corresponds to a range of ±42 km s^{-1} centred upon the molecular cloud velocity. We searched 101 sources in the two Orion clouds and 84 sources in the Perseus arm region. The selection criterion for these sources was that the 25 micron flux (S_{25}) should be greater than the 12 micron flux (S_{12}). A region of $\sim$0.5 degrees around OMC1 was omitted; interference from the Orion maser in the side-lobes of the telescope could be seen up to 2° from this source.

From this sample, we have detected 3 new masers in the Orion clouds, 2 in the Cep OB3 cloud and 7 masers situated in the Perseus arm as well as confirming several known masers. (The S146 maser (Blair et al. 1978), and the masers near HH1 and HH19-27 (Haschick et al. 1983) were not detected however down to a limit of 0.30 Jy). One of the most remarkable of the

F. P. Israel (ed.), Light on Dark Matter, 313–314.

new detections is the source connected with IRAS 05413–0104 for which we show the spectrum in figure 1. This source has an integrated luminosity in the H_2O line of 2 10^{-6} L_Θ as compared with a bolometric luminosity of 20 L_Θ which implies an unusually high H_2O/IR luminosity ratio. Another interesting result is given in figure 2 where we show the positions of our H_2O detections and non-detections in an IRAS color-color plot. We see that the sources with detected water vapour emission have color temperatures in the range 50–80 K. Also, the H_2O sources cluster closer to the equal color-temperature line in figure 2 than the non-detections. The 60/25 vs. 25/12 color-color plot (not shown) shows a similar picture. However contrary to figure 2, the non-detections in this diagram have 25/12 color temperatures that are typically higher than those of the maser sources. The implication of this is not clear but it may be related to the dust distribution in the neighbourhood of the young star. In any case, it suggests that comparisons of the properties of the near infrared counterparts of the IRAS sources with and without H_2O masers may be fruitful. We note also that, in general, our success rate in detecting H_2O emission was considerably higher for high luminosity objects ($>10^4$ L_Θ) than for intrinsically weaker infrared sources. For low mass pre-main-sequence stars, the phase of water vapour emission may be much shorter than in their higher mass counterparts.

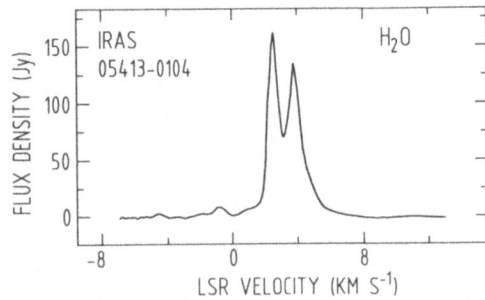

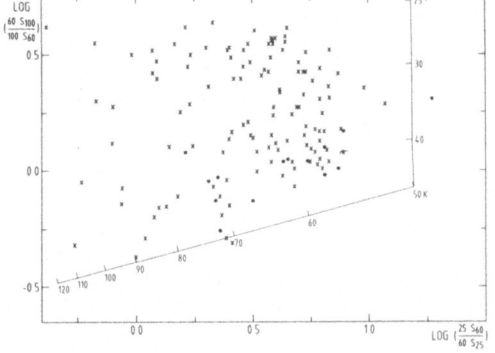

Fig. 1: H_2O spectrum towards the source IRAS 05413–0104 taken with the Effelsberg telescope in late March 1985.

Fig. 2: Color-color plot of IRAS sources surveyed for H_2O vapor emission. Filled circles are positive and crosses negative results. The full line shows the black body curve between 120 and 50 K. Lower temperatures are indicated for $T_{100/60}$.

REFERENCES

Blair, G.N., Davis, J.H., Dickinson, D.F. 1978, Astrophys. J. 226, 435
Genzel, R., Downes, D. 1977, Astron. Astrophys. Suppl. 30, 145
Genzel, R., Downes, D. 1979, Astron. Astrophys. 72, 234
Haschick, A.D., Moran, J.M., Rodriguez, L.F., Ho, P.T.P. 1978, Astrophys. 265, 281
Kutner, M.O., Tucker, K.D., Chin, G., Thaddeus, P. 1977, Astrophys. J. 215, 521
Wouterloot, J.G.A., Habing, H.J. 1985, Astron. Astrophys. (in press)

IR CCD IMAGING OF L1551-IRS 5: DIRECT OBSERVATIONS OF ITS
CIRCUMSTELLAR SHELL.

Andrea Moneti, Judith L. Pipher, William J. Forrest, and
Charles E. Woodward
Astronomy Dept. and C.E.K. Mees Observatory,
University of Rochester,
Rochester, NY 14627, USA

ABSTRACT. We present new, high spatial resolution images of L1551-
IRS 5 at 1.65, 2.2, and 3.8 μm which were obtained with the
University of Rochester's new Near-Infrared Array Camera mounted on
the KPNO 1.3-m telescope. We find that at these wavelengths IRS 5 is
about twice as large as a point source, and that it is elongated
along the direction of the polarization. These observations are
interpreted in terms of a flattened circumstellar shell that is
viewed from about 20° above its equatorial plane. In this model, the
central star is not seen directly, but only scattered light from one
of the polar regions, where the shell is thinnest, is observed. We
infer that the radius of the shell is about 1000 AU, and that its
density is more than 10^7 cm^{-3}. This model explains qualitatively the
large IR polarization. A comparison of our images with the VLA maps
of this source suggests that the radio jets are collimated by the
shell.

1. INTRODUCTION

IRS 5 is a highly obscured pre-main-sequence source which is
still interacting with its environment. IRS 5 is the source of a high
velocity molecular outflow (Snell et al. 1980), it is associated with
a double radio jet (Bieging et al. 1984) and with an optical jet
(Mundt and Fried 1983), it is highly polarized in the near infrared
(Nagata et al 1983), and it was suggested that the polarization is
due to scattering from a bipolar nebula. Near IR images of IRS 5 were
obtained to attempt to detect the bipolar nebula and to further study
the interaction between this source and its environment.

2. OBSERVATIONS

The observations were obtained with the University of
Rochester's near-IR array camera (Forrest et al. 1985) mounted on the
KPNO 1.3-m telescope. Fully processed images through the H(1.65 μm),

F. P. Israel (ed.), Light on Dark Matter, 315–318.

K(2.2 μm), and L'(3.8 μm) filters are presented in Figure 1.
Logarithmic scaling was used to emphasize the regions of low surface
brightness.

The sizes of IRS 5 at these wavelenghts, determined from cuts
through the images, are listed in Table 1 under "raw data"; the
"corrected" data has the beam subtracted in quadrature, and e is the
ellipticity of the image.

TABLE 1. WIDTH OF SOURCE IMAGES*

FILTER	RAW DATA			CORRECTED DATA		
	a	b	e	a	b	e
H	4.2	3.7	0.22	3.8	3.3	0.35
K	4.1	2.8	0.53	3.6	2.3	0.59
L'	4.6	3.3	0.49	4.2	2.9	0.52

*In arcsec. The beam has a = 1.9, b = 1.6, and e = 0.29.

3. DISCUSSION

The observed direction of elongation is inconsistent with the bipolar
nebula model (e.g., Elsässer and Staude 1978). Rather, both the shape
and the polarization of IRS 5 can be explained by a model in which a
central star is surrounded by a flattened circumstellar shell (see
sketch in Figure 2). In this configuration, which was studied
theoretically by Lefèvre et al. (1983), the observed image is made up
of light scattered by dust in the polar region that is toward the
observer. Direct light from the central star is not observed because
the optical depth through that part of the shell is too large. Based
on this model we find that: R_{shell} = 1000 AU, $A_v > 28^{mag}$, M_{shell} =
$4 \times 10^{-2} M_o$, and $M = 7 \times 10^{-7} M_o yr^{-1}$.

The radio jets are long and thin (their width is unresolved),
they are located within the shell, and they are along the minor axis
of the shell, which is probably the direction of steepest density
gradient. The observations are thus consistent with a model in which
the jets are collimated by the shell. The jets are also aligned with
the local magnetic field. That alignment may be a consequence of the
fact that the magnetic field caused the cloud core, out of which IRS
5 was born, to collapse farther along the field lines than
perpendicular to the field, and this then resulted in the steepest
density gradient being along the field direction.

REFERENCES

Biegling, J.H., et al. 1984, Ap. J. 282, 699.
Elsässer, H., Staude, H.J. 1978, A. A. 70, L3.
Forrest, W.J., et al. 1985, P. A. S. P. 97, 183
Lefèvre, J., et al., 1983, A. A. 121, 51.
Mundt, R., Fried, J. 1983, Ap. J. Lett. 274, L83.
Nagata, T., et al. 1983, A.A. 119, L1.
Snell R.L., et al. 1980, Ap. J. Lett. 239, 117.

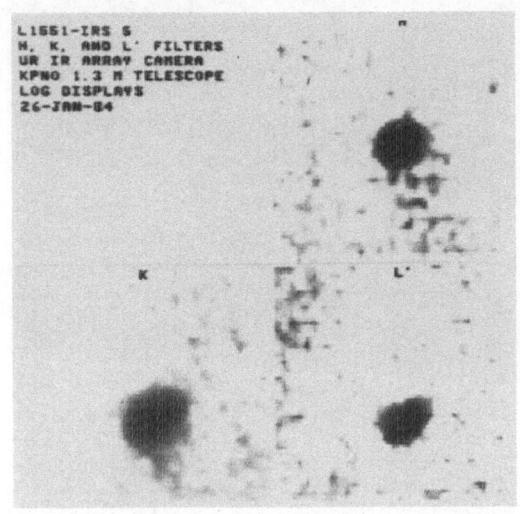
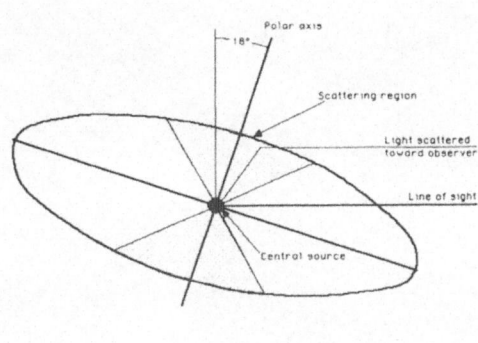

Figure 1   Figure 2

Figure 1.-IR images of IRS 5 at H (top right), K(bottom L'(bottom
right). The spatial resolution is 1'. North is up and east is to
left. Logarithmic scaling of the intensity levels was used to
emphasize the regions of low surface brightness. The lowest levels
displayed are: H - 3%, K - 0.4%, L' - 10%.

Figure 2. - A model for the circumstellar shell of IRAS 5. The
observed image is made up of light scattered from the upper polar
region. Direct light from the central star is not detected because
the optical depth through that part of the shell is too large.

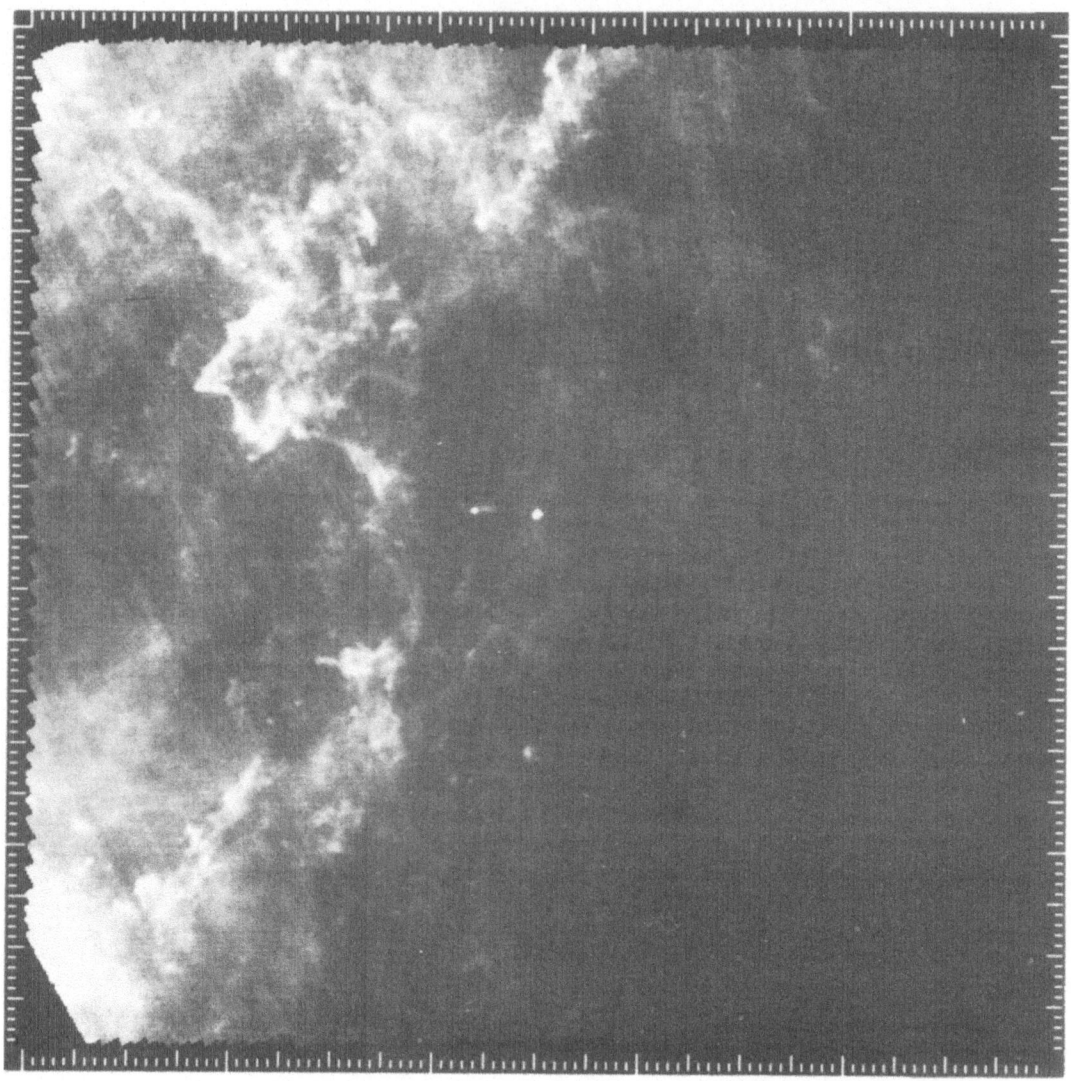

IRAS Field 173 $\alpha = 7^h 12^m$, $\delta = -45°$, HCON-3, 60 μm. Field in Puppis;
                  extensive cirrus.

A MODEL FOR BIPOLAR SOURCES IN MOLECULAR CLOUDS

Michael D. Smith
Department of Astronomy
University of Leicester
Leicester LE1 7RH
United Kingdom

ABSTRACT. A wind originating from a young star in a molecular cloud can be focussed into bipolar channels via shock deflection. The shocked material is here presumed to rapidly re-cool, flowing in a supersonic layer between the shock and the cloud. Stationary twin-jet flow patterns are possible even when the ambient pressure is isotropic about the central star. Single jet outflows can also occur when the stellar wind is interacting with the immediate protostellar environment.

1. THE MODEL

A large number of well-collimated outflows from young stars have now been discovered (Mundt et al 1985). Bipolar molecular outflows suggest that two-sided ejection occurs even when only one-sided optical jets and Herbig-Haro objects are directly observable. Canto(1980) and Canto and Rodriguez (1980) have developed a model to describe one-sided ejection, utilizing pressure gradients in the molecular cloud on the scale of order 10^{17}cm. However, jets are observed to be already collim-ated on the scale of 10^{15}cm (Mundt and Fried 1983). Nevertheless, I shall here demonstrate that the model can produce bipolar outflows on both large and small scales.

   The model assumes that the shocked stellar wind rapidly cools into a thin, dense layer (quantitatively expressed by Smith(1985)). The mat-erial then flows around between the shock and the molecular cloud at supersonic speeds. This provides a centrifugal pressure which must be added to the ram pressure of the wind to balance the ambient pressure. The ambient pressure is taken to be of the form $p \propto r^{-n}$. By fixing the 'impact angle' between the shock and wind directions Smith and Raine (1985) found a class of analytical solutions which require n(r). The flow patterns are reminiscent of double 'candle-flames'.

2. NUMERICAL SOLUTIONS

Figure 1 displays numerical solutions appropriate to constant pressure

319

F. P. Israel (ed.), Light on Dark Matter, 319–320.
© *1986 by D. Reidel Publishing Company.*

and spherically-symmetric protostellar environments. The assumption is
made that the shock and interface coincide. The oblique shock structure
in the symmetry plane can be supported by a circumstellar disc. Well-
collimated jets are found for $1 < n < 2$. Single jets, in which the
whole wind leaves through one exit, can also form in the range
$0.7 < n < 1.7$. Full details and consequences of these preliminary
results will be presented elsewhere (Smith 1985).

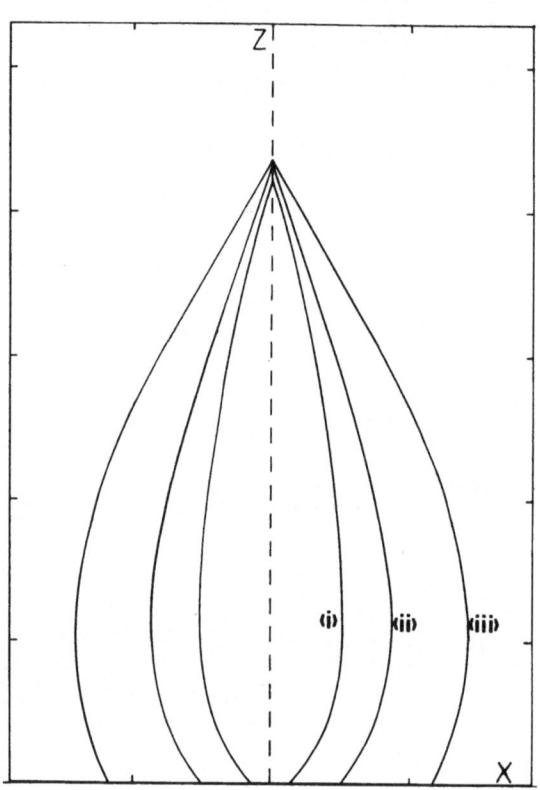

Figure 1. Jet boundary shapes on one side of the symmetry
plane for $n = 1.5$ and impact angles in the plane of (i) 30°,
(ii) 45° and (iii) 60°.

REFERENCES

Canto, J., 1980. Astr. Astrophys., 86, 327.
Canto, J. & Rodriguez, L.F., 1980. Astrophys. J., 239, 982.
Mundt, R. & Fried, J.F., 1983. Astrophys. J., 274, L83.
Mundt, R., Buhrke, T., Fried, J.F., Neckel, T. & Sarcander, M., 1985.
    Astr. Astrophys., 140, 17.
Smith, M.D., 1985. Submitted to M.N.R.A.S..
Smith, M.D. & Raine, D.J., 1985. M.N.R.A.S., 212, 425.

COMPARISON OF CO AND IR EMISSION OF IRAS UNIDENTIFIED SOURCES

F. Boulanger, F. Casoli, F. Combes, Ch. Dupraz & M. Gerin
Laboratoire de Radioastronomie de l'Ecole Normale Supérieure
24, rue Lhomond
75005 Paris
France

ABSTRACT. We study the correlation between IR and CO emission for a sample of 27 IRAS unidentified sources in the Galaxy. We have searched for ^{12}CO and ^{13}CO emission with the Bordeaux POM telescope. These sources seem to be associated with protostars or young stars still embedded in molecular clouds; in five cases, we detect high velocity flows of molecular gas associated with a forming star. There are fairly good correlations between the ^{12}CO and ^{13}CO luminosities and the total IR luminosity. Therefore, the luminosity of the central object seem to be related to the mass of the parent molecular clump.

1. OBSERVATIONS

IRAS has revealed in the galaxy a lot of point-like unidentified sources; in particular, they are not known HII regions, nor stars. We have undertaken mm-wave observations of a sample of 27 of such sources. The IR emission of the sources, selected from the shape of their spectra, is probably due to dust heated by stellar objects. However, the total IR fluxes of the sources are weaker than those of known star-forming regions, ranging from 10 to 600 10^{-13} W.m^{-2}.

Observations of ^{12}CO and ^{13}CO J=1-0 lines were made with the POM 2.5 m antenna in Bordeaux (beamsize = 4.4' ~ IRAS resolution). Typical system temperatures were 900 K for ^{13}CO and 1600 K for ^{12}CO. We used a frequency switching procedure. The frequency resolution is 0.26 km/s and spectra were integrated until a r.m.s. noise of 0.1 K is reached.

2. RESULTS

Among the 27 sources, only one has not been detected in ^{12}CO. We have obtained maps of fourteen: they reveal that the IR sources are always associated with the CO peaks, and thus indicate that the CO/IR association is real. Thus, we can use the CO velocity to determine the kinematic distance to the source, and to derive IR luminosities; our sample spans a range of distances of 0.3 to 3 kpc.

Five sources present high-velocity wings (Figure 1). Better

321

resolution observations of these sources will soon be done in order
to determine the extent and mass of the flows. One of these sources,
0537+238P, is 2' away from the Herbig-Haro object GGD4 (Gyulbudaghian
et al, 1978). We have observed this source with the Texas 5 m antenna
in the ^{12}CO J=2-1 line (beamsize 1.4') and detected a molecular flow
of extent ~ 2'. Such wings are generally interpreted as corresponding
to the ejection of molecular gas from the vicinity of young stellar
objects (Bally and Lada, 1983).

   In Figure 2, we compare the ^{12}CO luminosity in the central beam
to the total IR luminosity: there is a good correlation between the
two. We have found a similar correlation for the ^{13}CO emission, which
is generally optically thin and traces the column density of the
cloud. Therefore, the luminosity of the central object seem to be
related to the mass of the surrounding clump; we are now mapping the

^{13}CO emission around the sources, in order to determine the extent
and mass of the condensations where the stellar objects are born.

   Our preliminar study allows us to propose that denser clumps
tend to form more luminous protostars: this would constitute an
important result for the theories of star formation. To confirm this
conclusion, observations of new sources are needed.

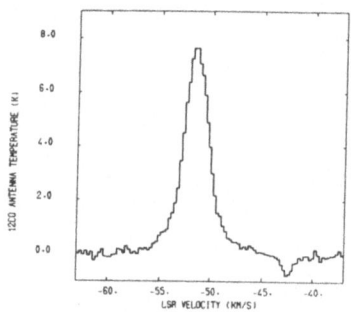

Figure 1. 12CO spectrum of IRAS
2250+597P. The negative feature at ~
43 km/s is due to the mesospheric CO
emission and to the frequency
switching procedure.

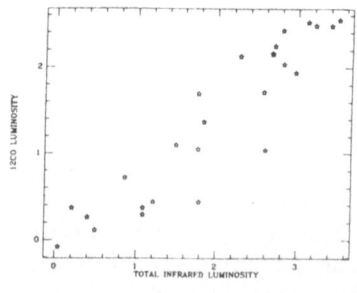

Figure 2. Plot of the 26 sources in a
diagram log (12CO luminosity) versus
log (total IR luminosity) in
arbitrary units.

REFERENCES

Bally, J., Lada, C.J.: 1983, Ap. J., 265, 778.
Gyulbudaghian, A.L., Glushkov, Y.I., Denisyuk, E.K.: 1978, Ap. J.,
      224, L137.

IRAS OBSERVATIONS OF SYMBIOTIC OBJECTS

Patricia A. Whitelock
S.A. Astronomical Observatory
P.O. Box 9
Observatory 7935   Cape
South Africa

ABSTRACT.   The three groups of symbiotics, defined according to their near IR colours (S: stellar, D: hot dust, D': cool dust), are shown to have distinctly different IRAS colours.   The D-type symbiotics divide into two groups, those with and without 10 μm silicate features.

1. INTRODUCTION

Symbiotic objects show the characteristics of both cool stars (usually molecular absorption) and high excitation nebulae (e.g., HeII emission lines).   Many of them (possibly all) are interacting binary systems, though the evidence for duplicity is often rather indirect.   It seems quite possible that the symbiotics do not in fact represent an homogeneous group.

Allen (e.g., 1979) has produced a useful classification scheme for these objects based upon their JHKL colours.   Thus S-type symbiotics have stellar colours, while D- and D'- (or yellow symbiotics) types have colours indicative of warm (1000K) and cool (500K) dust respectively. D-type systems involve long-period Mira variables and exhibit large temporal changes in the emission of and extinction by their associated dust (e.g., Whitelock et al. 1983).   The discussion here is limited to those objects detected in the 12,25 and 60 μm IRAS bands.   Full details will be published elsewhere.

2. DISCUSSION

The S, D and D' symbiotics are distinguished on the IRAS 2-colour diagram (Figure 1) where they fall in different regions.

2.1. Yellow Symbiotics

These are so called because they exhibit an absorption spectrum characteristic of an F, G or K star rather than an M star.   It is possible that this absorption spectrum actually originates from an accretion disk or thick shell.   They are all strong far IR sources. The IRAS colours indicate a dominant source of emission much cooler

F. P. Israel (ed.), Light on Dark Matter, 323–324.

($T_c \lessgtr 250K$) than the (K-L) measures suggest (T_c ~500K).   Even taking into account the effects of emission lines a range of dust temperatures is indicated.   Although classical planetary nebulae have much more extreme IRAS colours, typically $[25/12]$ $T_c < 150K$, one PN, BD $\pm$ 30° 3639, does fall near the symbiotics.   Yellow symbiotics may be related to PN.

2.2.   D-Type Symbiotics

All of the D-type objects show broad energy distributions indicative of dust covering a range of temperatures.   The 2 objects to the right of the blackbody line in Fig. 1 are distinctly different from the others.   They exhibit featureless 10 μm dust emission while the others have a silicate emission feature.   In fact the energy distributions of the objects on the right of Fig. 1 can be explained by a combination of blackbodies without the addition of any material showing normal small particle emissivities (i.e., $\varepsilon \propto \lambda^{-n}$, n ~ 1-2).   This would seem to imply the presence of very large particles in the environs of at least some symbiotics.

2.3.   S-Type Symbiotics

These are rather weak IRAS sources.   Their small far-IR excess is probably due to low temperature dust with a strongly λ-dependent emissivity.   Their far-IR characteristics are not significantly different from normal (non-symbiotic) late-type stars.

3.   REFERENCES

Allen, D.A., 1979.   IAU Coll. 46, p.125.
Roche, P.F. et al., 1983.   M.N.R.A.S. **204,** 1009.
Whitelock, P.A. et al., 1983.   M.N.R.A.S. **203,** 351.

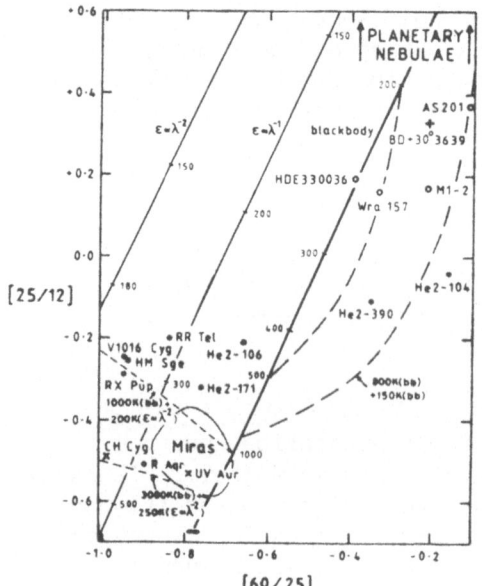

Figure 1.   IRAS 2-colour diagram. Symbiotics marked as follows. S-type: ×, D-type ●, D'-type o. Also marked are the blackbody locus and the loci for dust with emissivity $\propto \lambda^{-1}$, λ^{-2}.   Dashed lines mark the loci of combinations of dust shells with particular temperatures and emissivities.   The planetary nebula BD+30° 3639 and the area occupied by Mira variables and normal planetary nebula are also indicated.

A LARGE SCALE OH SURVEY IN ORION AND MONOCEROS

P.M.M. Jenniskens, H.J. Habing, J.G.A. Wouterloot,
P. te Lintel-Hekkert, A. Blaauw
Sterrewacht Leiden
P.O. Box 9513
2300 RA Leiden
Holland

ABSTRACT. Large scale mapping in both the 1665 and 1667 MHz molecular lines of OH shows overlap of two clouds near $l = 211^{\circ}$ and $b = -19^{\circ}$ in Orion South. Clouds associated with Monoceros R2 show a shell like structure. VB80 is associated with a large molecular cloud.

1. INTRODUCTION

A survey in both main lines of OH was conducted with the 25 metre Dwingeloo telescope, covering an area between RA = $5^{h}30^{m}$ until $7^{h}10^{m}$ and DEC = -3° until -13°. This area shows the molecular complexes Orion South, Monoceros R2, Van den Berg 80 and a small area north-west of Canis Major OB1. The first observational results are presented.

2. THE OBSERVATIONS

The survey contains 1500 gridpoints on a $0.^{\circ}3$ x $0.^{\circ}3$ grid. The half-power beam width of the telescope used is $0.^{\circ}5$. Each gridpoint is observed during 30 minutes, the main lines, 1665 and 1667 MHz, simultaniously by splitting the 256 channel autocorrelator into equal halves, each with a bandwidth of 625 kHz. Thus a one sigma noise level of 0.01 K in antenna temperature, and a typical signal to noise ratio of 1:4 is obtained. Velocity resolution is 0.9 km/s, velocity coverage in each line is 60 km/s centred on +10 km/s.

3. FIRST RESULTS

The overall velocity structure of the OH data is in good agreement with CO data. Typical linewidths are of order 1.8 km/s and distinctly larger (about 4 km/s) in a region up in Orion South, near $l = 211^{\circ}$ and $b = -19^{\circ}$. B-V plots show an overlap of two clouds here, one at 4-5 km/s, the other at 7-8 km/s. This shows there is no smooth velocity gradient in the region as is suggested by a "rotating cloud" model.

F. P. Israel (ed.), Light on Dark Matter, 325–326.
© *1986 by D. Reidel Publishing Company.*

One strong 1665 MHz maser was observed in the core of MonR2 near
position l = 213°42' and b = -12°38'. In this region a number of IRAS
point sources are found.

MonR2 shows up as a "shell" like structure with the OH maser near the
inner edge of the cloud at a velocity relative to the Local Standard of
Rest of +12 km/s. From velocity information and distance determinations
of the star associations vB80 and MonR2 (Herbst and Racine,1976), an
association of the corresponding clouds is probable, though a "bridge"
cloud between the complexes (Morris e.a.,1980) shows up at little higher
mean velocity (+10 km/s).

There is a good overall correlation between dust (IRAS spline maps) and
the molecular gas seen in this survey.

Extensive CO data of the area excist (Manddalena e.a,1985). A suggested
narrow fillament in the region (Morris e.a.,1980) connecting the Orion
Complex with gas in the galactic plane, is seen as a sequence of clouds
in our observations as well as in the IRAS sky flux maps. Though any
filamentary structure seen by Morris et. al., is just below our
detection limit.

REFERENCES

Baud B., Wouterloot J.G.A.; Aston. Astrophys. 90 (1980),297
Herbst W., Racine R.; Aston. Journ. 41 (1976), 840
Kutner M.L., Tucker K.D., Chin G.; Astroph. Journ. 215 (1977),52
Manddalena R.J., Morris M., Moscowitz J., Thaddeus P.; (1985)
                                        submitted to Ap.J.
Morris M., Montani J., Thaddeus P.; IAU-Symp. 87 (1980),197

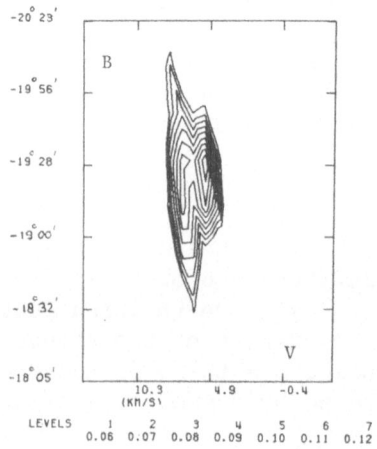

Figure 1: B-V plot of OH 1667 MHz line intensity for l = 211.°4.
          Contourlevels are in Kelvin antennatemperature.

SECTION 5.

GALACTIC BULGE AND GALACTIC STRUCTURE

THE GALACTIC DISTRIBUTION OF STELLAR SOURCES FOUND BY IRAS
(with some emphasis on the bulge of the Galaxy)

H.J. Habing
Sterrewacht, Huygens Laboratorium
Wassenaarseweg 78
Leiden (the Netherlands)

ABSTRACT. Most of the point sources detected by IRAS at its shortest wavelength, 12 μm, will be circumstellar shells around late type stars. Maps of the galactic distribution of these sources for different flux density intervals and different 25 to 12 μm colours show that sources are being found several kiloparsecs deep into the interior of the Galaxy. The faintest and reddest sources show the bulge of the Galaxy.

In the bulge the surface density distribution, $N(l,b)$, has an elliptical shape with an axial ratio of about 2:1; N varies with radial distance, r, according to the de Vaucouleurs law found in the visible light in elliptical galaxies, i.e. ln N is linearly proportional to $r^{1/4}$.

A large fraction of the sources are variable; the percentage variability decreases with increasing 25 to 12 μm colour, in the same way as for nearby long period variables (Mira's, variables from the 2.2 μm survey, and OH/IR stars). The sources in the bulge are therefore of the same nature, i.e. asymptotic giant branch stars with double shell burning. Estimates of the luminosities confirm this conclusion: lower limits on the luminosity are derived between 1000 and 10,000 L_O. There are more than 1000 stars in the bulge with luminosities exceeding those of the brightest stars in globular clusters. Possibly these bulge stars are not as old as the globular clusters.

1. OVERALL DISTRIBUTION

In this review I discuss only the sources from the IRAS point source catalog with a certain detection (i.e. a "good quality flux") at least at 12 μm and at 25 μm. I introduce two quantities: S_ν and R. S_ν equals the flux density at 12 μm as given in the catalog, i.e. without any colour correction. To calibrate the reader: most sources have a value of S_ν between 1 and 3 Jy; a 10 Jy source is "bright"; the catalog contains sources down to 0.5 Jy, but the completeness below ~ 1 Jy is not guaranteed. The quantity R is defined as the 25 to 12 μm flux density ratio, again, without colour correction. Hot sources, i.e. $T_{eff} \geq 1500$ k, are expected to have a Rayleigh-Jeans spectrum at 12 and 25 μm, and then R = 0.25. Cooler sources, specifically stars with circumstellar shells, have larger values of R: Mira variables, long

329

F. P. Israel (ed.), Light on Dark Matter, 329–338.

period variables from the IRC, and variable OH/IR stars form a narrow
sequence in a two-colour IRAS diagram (see figure 1, which is taken from
Olnon et al., 1984), with R between 0.2 and 3. [Point sources with R > 3
in figure 1 are planetary nebulae, others are probably transition cases
between long period variables (i.e. asymptotic giant branch stars) and
planetary nebulae - see below].

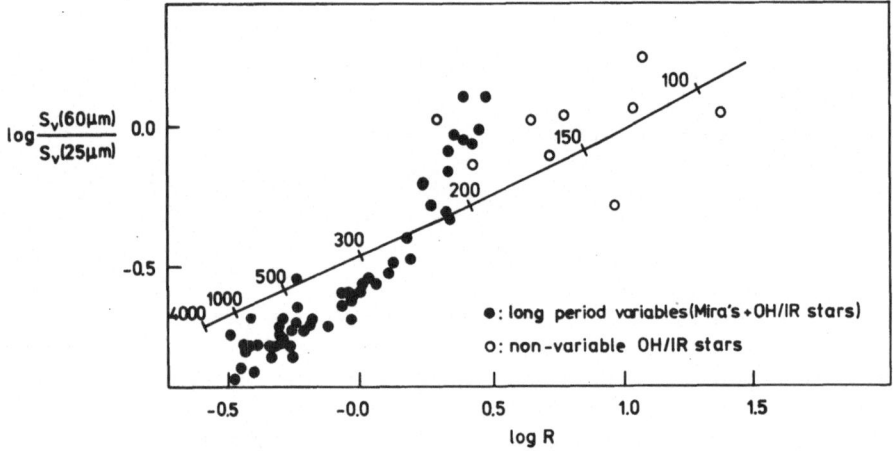

Figure 1. Two colour diagram for IRAS point sources identified
with known OH/IR stars and with known long period variables;
original diagram in Olnon et al. (1984).

    Figure 2 shows the distribution on the sky in galactic coordinates
of sources, separated in different intervals of S_ν and of R. Arranged
from the left to right are maps of successively weaker sources
(smaller S_ν); arranged from the top downwards are maps of successively
redder sources (increasing values of R).
    The maps contain two (minor) artefacts: (1) at $l \approx 90°$, at negative
latitudes, and at $l \approx 270°$, at positive latitudes, there is a lack of
sources due to gaps in sky coverage; (2) in the galactic plane, between
$l = 320°$ and $l = 40°$, there is also a conspicuous dearth of sources,
caused, in this case, by source confusion; for more details on both
artefacts the reader is referred to the Explanatory Supplement (IRAS,
1985). Outside these general areas the maps ought to be complete.
    All maps show very clearly the disk of the Galaxy; the disk gets
more pronounced when weaker sources and/or redder sources are sampled.
In the four maps in the lower righthand corner the bulge of our Galaxy
is prominent. Apparently in those maps we see sources quite far away,
beyond the galactic centre. Detailed analysis, through a comparison with
model distributions, will give us insight in the distribution of these
sources throughout the Galaxy, and will therefore be a useful population
study. Such a detailed analysis has not yet been made. In this section
we explore very superficially one of the diagrams (fig. 2a) and then
concentrate, in the next sections, on the galactic bulge. Figure 2a

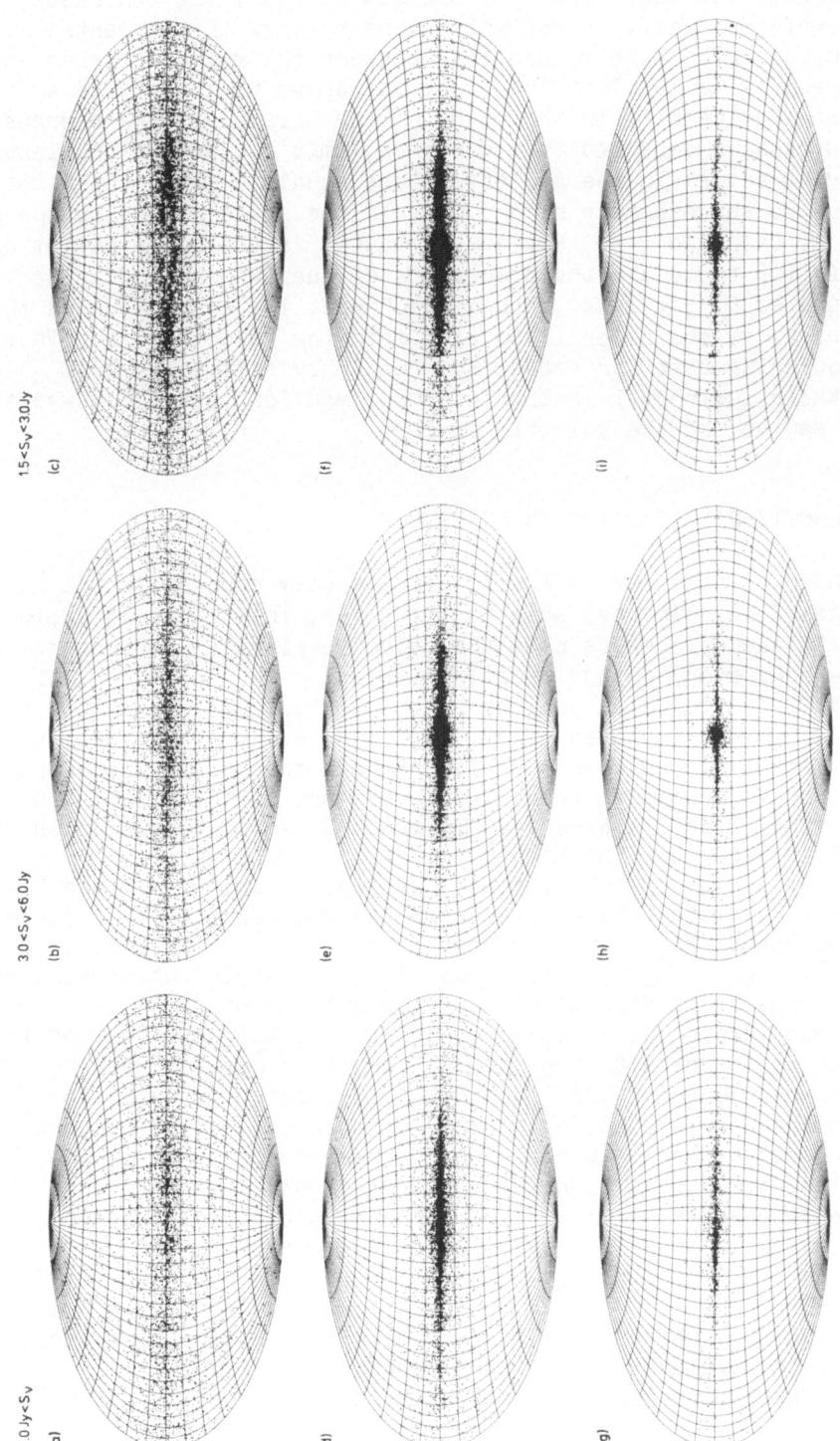

Figure 2. Sky maps of selected IRAS point sources. Each row contains maps in a different interval of R: top row $0.25 < R < 0.50$; middle row $0.50 < R < 1.0$; bottom row $1.0 < R < 2.0$. The maps in the left column contain objects with $S_\nu > 6.0$ Jy; in the middle column $6.0 > S_\nu > 3.0$ Jy; the right column $3.0 > S_\nu > 1.5$ Jy.

contains the brightest and bluest (i.e. least red) sources. The galactic
disk is prominent, but there are many sources outside. The contrast
between the surface density in the anticentre to that in the centre is
crudely estimated to be 1 to 2, and that between the galactic poles and
the anticentre as 1 to 10. The first contrast shows that there is a
gradient in stellar density in the plane of the Galaxy. If the sources
were standard candles detected to a distance limit r_o, and if the length
scale (e-folding scale) in the galactic plane equals α, then the first
contrast leads to the estimate $r_o/\alpha \simeq 0.7$. If the scale height in the z
direction is h, then $h/\alpha \simeq 30$. Now assume that h is the scale height of
F and G dwarfs (if these are the precursors of the IRAS sources) then
$h \simeq 150$ pc, and so $r_o \simeq 4.2$ kpc, and $\alpha \simeq 4.5$ kpc. The conclusion is that
even in figure 2a we see several kpc deep into the galactic disk. Thus
there is no surprise that for redder sources (they emit a larger
fraction of their luminosity in the infrared) and for apparently weaker
sources IRAS saw beyond the galactic centre.

2. SURFACE DENSITY DISTRIBUTION IN THE BULGE

In the remainder of this review I will discuss only sources in the
galactic bulge; I restrict the area to $|l| < 15°$, $|b| < 15°$. To avoid
problems due to incompleteness near the galactic plane (a consequence of
severe source confusion) an additional restriction is applied, namely
that $|b| > 2°$. Only source detections of the highest quality at
12 μm and at 25 μm are included (i.e. FSTAT = 3; see the Explanatory
Supplement). Finally I excluded sources brighter than 6 Jy. Therefore
the criteria for inclusion in the bulge sample are 6.0 Jy > S_ν > 0.6 Jy;
R > 0.7; the final list contains 2490 sources. Their distribution on the
sky is not shown here, but the reader will get an easy and fair
approximation by adding together, mentally, figures 2e, 2f, 2h and 2i.
    The resulting distribution of sources has two components: a bulge
and a disk component. The disk component is already much reduced by
excluding the sources brighter than 6 Jy, but some contribution remains.
I therefore subtracted a disk with a surface density of the
form $N_d = N_0 \exp(-|b/b_0|)$. By variation of N_0 and b_0 the effect of the
disk component on the results was studied; the best results were
obtained when $N_0 \simeq 10$ (sq. deg)$^{-2}$ and $b_0 = 2.5°$; but even assuming $N_0 =$
0, the further results are not critically affected; i.e. the properties
of the bulge donot depend critically on the assumed presence of the
disk. After the disk subtraction the numbers of objects were counted in
rings between successive ellipses around the galactic centre, each
ellipse with axial ratio 2:1. This results in an average surface density
of bulge sources $\langle N_r \rangle$ (number per square degree) as a function of radial
distance r, where $r = \sqrt{(l^2 + 4b^2)}$; the relation between $^e\log \langle N_r \rangle$ and $r^{1/4}$
is shown in figure 3. Obviously a linear relation can be fitted through
the points, if one excludes the 3 innermost points ($r^{1/4} < 1.7$
or $r < 8.4°$). The broken line, a fit by eye, corresponds to the relation
$\langle N_r \rangle = \exp(17.2 - 8.5\, r^{1/4})$, a type of relation found experimentally for
elliptical galaxies by de Vaucoulers (see Mihalas and Binney, 1981, page
311). Whether the deviation of the innermost points is real, is, in my

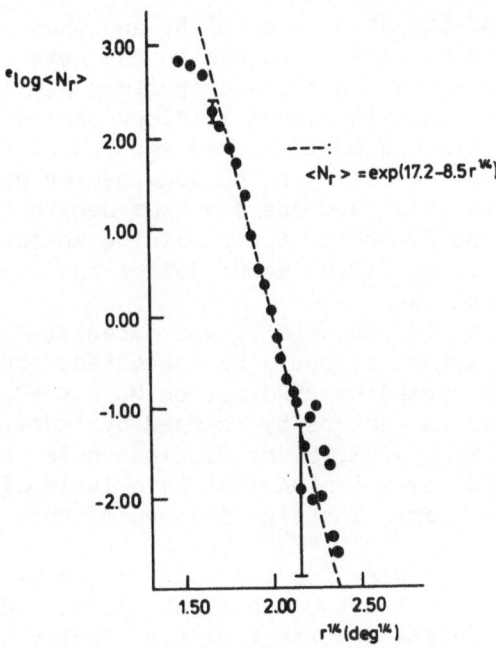

Figure 3. The decrease of the surface density of bulge sources with increasing distance from the galactic centre. See text.

opinion, undecided. According to chapter VIII-D of the Explanatory Supplement, source confusion sets in at $\langle N_r \rangle = 25$, or ^{e}log $\langle N_r \rangle = 3.2$, suggesting that source confusion has not affected $\langle N_r \rangle$. However, the analysis in the Explanatory Supplement may not be valid for the very complex area close to the galactic centre, source confusion may occur at lower values of $\langle N_r \rangle$ and some caution is needed.

A dynamical explanation for the existence of the de-Vaucouleurs relation is apparently lacking (see Mihalas and Binney, loc. cit.). However, the fact that the infrared sources have a distribution very similar to those of the visible light in elliptical galaxies suggests that the visible light of the bulge of our Galaxy will also display a de-Vaucouleurs relation, and that the distribution of the infrared sources therefore also represents the distribution of the visible stars.

3. LUMINOSITIES

For most of the approx. 2500 objects flux densities are available only at two wavelengths, 12 µm and 25 µm. To derive a total luminosity of the star one has to estimate the flux at both shorter and at longer wavelengths, and one has to assume a distance. For the distance I used 8.7 kpc, the distance to the galactic centre according to Oort and Plaut (1975), and I ignored the fact that some objects will be closer and others farther away. To estimate the flux at other wavelengths I define a dimensionless quantity $C_b \equiv$ total flux$/(\nu S_\nu)_{12\mu m}$. Obviously C_b will be a

function of the redness of the star, i.e. of R, and thus of the overall
spectrum. Supported by the results discussed in the next section, I
assume that the bulge stars have an overall spectrum similar to those of
the long period variables shown in figure 1 ("long period variables"
includes Mira's, IRC objects and OH/IR stars) and that $C_b(R)$ is
identical for the two sets. For most of the long period variables in
figure 1 the total flux is known and one can thus derive C_b (I am
indebted to J.H. Burger and J. Herman for providing me the data). A
lower limit is $C_b = 2.0$; if $C_b = 2.0$, about 30% of the stellar flux is
contained in the IRAS 12 μm band.

For a random sample of 141 objects C_b was calculated from the value
of R and thus the total luminosity could be estimated. About 80% of the
sources have luminosities exceeding 2500 L_o, or $M_{bol} < -3.77$. According
to Frogel (1981), the maximum luminosity reached by individual stars in
globular clusters equals $M_{bol} = -3.75$. My conclusion is therefore that a
large fraction of the 2500 stars in the list have luminosities exceeding
those of globular cluster stars. The significance of this conclusion
will be discussed below.

The luminosities derived here for the bulge stars overlap with
those of the OH/IR stars, but the two distributions may not coincide
completely. Among the OH/IR stars taken from e.g. Baud's survey (Baud,
1978; Herman and Habing, 1985) there is a significant fraction with
$L > 10,000$ L_o, whereas our sample does not contain stars with so large a
luminosity. The difference may be artificial, and due, e.g., to the fact
that I used an upper limit of 6 Jy in defining the bulge sample. If,
however, the difference is real, it may indicate that in Baud's sample
stars of higher luminosity, and therefore higher main sequence mass
occur; Baud's sample represents the galactic disk, and such a difference
between bulge and disk would not be surprising.

4. VARIABILITY

IRAS surveyed the area of the bulge three times. Each observation
consisted of four sightings within 36 hours in each wavelength band. The
second observation followed the first within a few weeks, typically 16
days. The third observation was made 6 months after the first. Between
the observations differences in the flux densities are often found that
exceeded the photometric uncertainty, suggesting that the source is
truly variable. In the IRAS catalog a variability index, p, is given
that measures flux density variations between different observations at
12 and at 25 μm, in so far as the variations are correlated between the
two wavelengthbands. The value of p is between 0% (no variability
detected) and 99% (definitely variable). Three observations of a
variable source may not be sufficient to detect variability, even when
the photometric accuracy is much better than the amplitudes of the source.
Thus, for individual objects p = 0% will not give much information; the
source could still be variable; however averaging p for a large sample
should give a good indication of the fraction of variable sources within
that sample. In table 1 I show <p>, average values of p for samples of
stars, obtained by ordering the objects according to redness, i.e. R.

TABLE 1
Distribution and variability according to redness (R)

redness	total number of sources	average variability
log R		$\langle p \rangle$
-0.15/0.00	1615	68%
0.00/0.15	555	76%
0.15/0.30	163	74%
0.30/0.45	53	53%
0.45/0.60	18	22%
0.60/0.75	24	15%
0.75/0.90	17	17%
0.90/1.05	22	9%
1.05/1.20	20	13%
1.20/ ∞	3	3%

The table shows two very clear effects: First, the numbers of stars drop rapidly with increasing R. Second, $\langle p \rangle$ is rather constant in the first three bins, but then drops rapidly; the turning point occurs in the fourth bin, where $2.0 < R < 3.0$. About 2/3 of the sources with $R < 2.0$ are variable, and about 1/6 of the sources with $R > 3.0$. This turning point is the same as that in Figure 1, and suggests strongly that the sources seen in the galactic bulge are similar.

I drew a sample of about 100 sources with $p > 60\%$, and checked the variability. In all cases the significant variation occurred between the second and the third observation (180 day interval), never between the first and second (16 day interval). This is additional support for the assumption that the variable sources in the bulge are long period variables.

5. DISCUSSION

Most sources in the bulge of the galaxy are too weak to have their 60 μm flux density measured; they are also all below the detection limit of the Low Resolution Spectrograph. The only complete information that we have on their nature is the parameter R and the variability parameter p. In my view the correlation found between these two gives substantial weight to the conclusion that the IRAS sources in the galactic bulge are long period variables with large mass loss; the hypothesis can be advanced that, once 60 μm measurements will be available, the sources will form a sequence like the objects with $R < 2.0$ in Figure 1.

Because the bulge of the Galaxy contains long period variables with large mass loss and strong at 12 μm, it will contain also red giants with much less mass loss and strong in the near infrared; most likely such giants will outnumber the 12 μm sources, because one expects the evolution to increase its speed when mass loss begins. The bulge of the Galaxy seen in large beam measurements at 2.4 μm by Maihara et al.

(1978) is caused very likely by such red giants - as Maihara et al. did point out.

Bedijn (see elsewhere in this volume) has calculated model spectra for a large number of stars embedded in a thick, outflowing dusty wind. His calculations show that the sequence in Figure 1 can be understood as one of increasing optical thickness of the outflowing material. Stars with different luminosities will lie on (approximately) the same sequence if the opacity law (i.e. the nature of the dust grains) is sufficiently similar. The suggestion is thus made that the sequence is in fact one of evolution: a star enters the sequence at the bluest side, and, when its mass loss gradually increases on a time scale of thousands of years, it moves along the sequence towards larger values of R. Optically detected long period variables, IRC sources and OH/IR stars are thus subsequent stages of evolution passed through by stars of very different luminosity. In this view the evolution stops when the material surrounding the nucleus of the star is exhausted. The fact that the end of the sequence also corresponds to the point where pulsation stops (when R ≈ 2.2) adds significant weight to Bedijn's proposal. Thereafter the circumstellar shell is rapidly ejected and a planetary nebulae emerges. In this context the numbers of sources in the different bins in log R, shown in table 1, become meaningful - these numbers should be proportional to the time spent in evolving through the bin; the small number with R > 2 shows how fast the evolution occurs after pulsation stops and how quickly the transition is made to a planetary nebula.

It is obvious in my mind that the asymptotic giant branch stars found by IRAS in the bulge of the Galaxy deserve further studies. I see two important lines of research. The first leads to a study of the structure of the bulge. Can the de-Vaucouleurs relation be extended inward? Can one measure radial velocities of the stars, and so establish whether the bulge rotates and what the distribution function is of the velocities? Probably quite a number of the bulge stars are OH masers, and then a very accurate radial velocity measurement is easy; OH/IR stars with large random velocities have already been found near the centre of the Galaxy (Habing et al. 1983).

The second line is, I think, more uncertain, but not less interesting. Whereas there is probably little dispute that the stars are asymptotic giant branch objects, evolved from red giants, there is much uncertainty on what main sequence mass they had originally and how old they are. I have drawn attention to the large luminosity of most of the bulge stars compared to the luminosity reached in globular clusters. Can one conclude that they are significantly younger than the globular clusters? If so, significant star formation has occurred throughout the bulge much after the era of globular cluster formation. Although I would like to draw the conclusion, there are several caveats. An important caveat (stressed by Frogel, 1981) is that stars with large metal abundance have been found in the central parts of the Galaxy (Whitford and Rich, 1983), and large metal abundance slows down the pre-AGB evolution quite drastically (see e.g. the model calculations by Mengel et al., 1979). Thus, although the large luminosity shows the large main sequence mass of the bulge stars, their age may be much larger than for similar stars with solar metal abundance. Clearly, the nature of the bulge stars has to be further investigated. Important information will come from a more accurate determination of the luminosities and from a determination of their periods. No doubt, that both pieces of information will become available in the near future.

ACKNOWLEDGEMENTS

Uncharacteristically, I have made most of the analysis presented in this review by myself. Wrong conclusions will be entirely my fault. But the analysis was made possible because data and access software had become available through the efforts of numerous workers, on both sides of the Atlantic. Also, in the last few years I have profited very much from many discussions with colleagues and students. I thank all of them.

REFERENCES

Baud, B. 1978, thesis, Leiden University

Frogel, J. 1981, in "Physical Processes in Red Giants", eds. I Iben and A. Renzini (Dordrecht, Reidel), p. 63

Habing, H.J., Olnon, F.M., Winnberg, A., Matthews, H.E., Baud, B. 1983, Astron Astrophys. **128**, 230

Herman, J., Habing, H.J. 1985, Physics Reports **124**, 257

Maihara, T., Oda, N., Sugiyama, T., Okuda, H. 1978, Publ. Astron. Soc. Japan **30**, 1

Mengel, J.G., Sweigart, A.V., Demarque, P., Gross, P.G. 1979, Ap. J. Suppl. Ser. **40**, 733

Mihalas, D., Binney, J. 1981, "Galactic Astronomy" (San Francisco, Freeman)

Olnon, F.M., Baud, B., Habing, H.J., de Jong, T., Harris, S., Pottasch, S.R. 1984, Ap.J. (Letters) **278**, L37

Oort, J.H., Plaut, L. 1975, Astron. Astrophys. **44**, 259

Whitford, A.E., Rich, R.M. 1983, Ap.J. **274**, 723

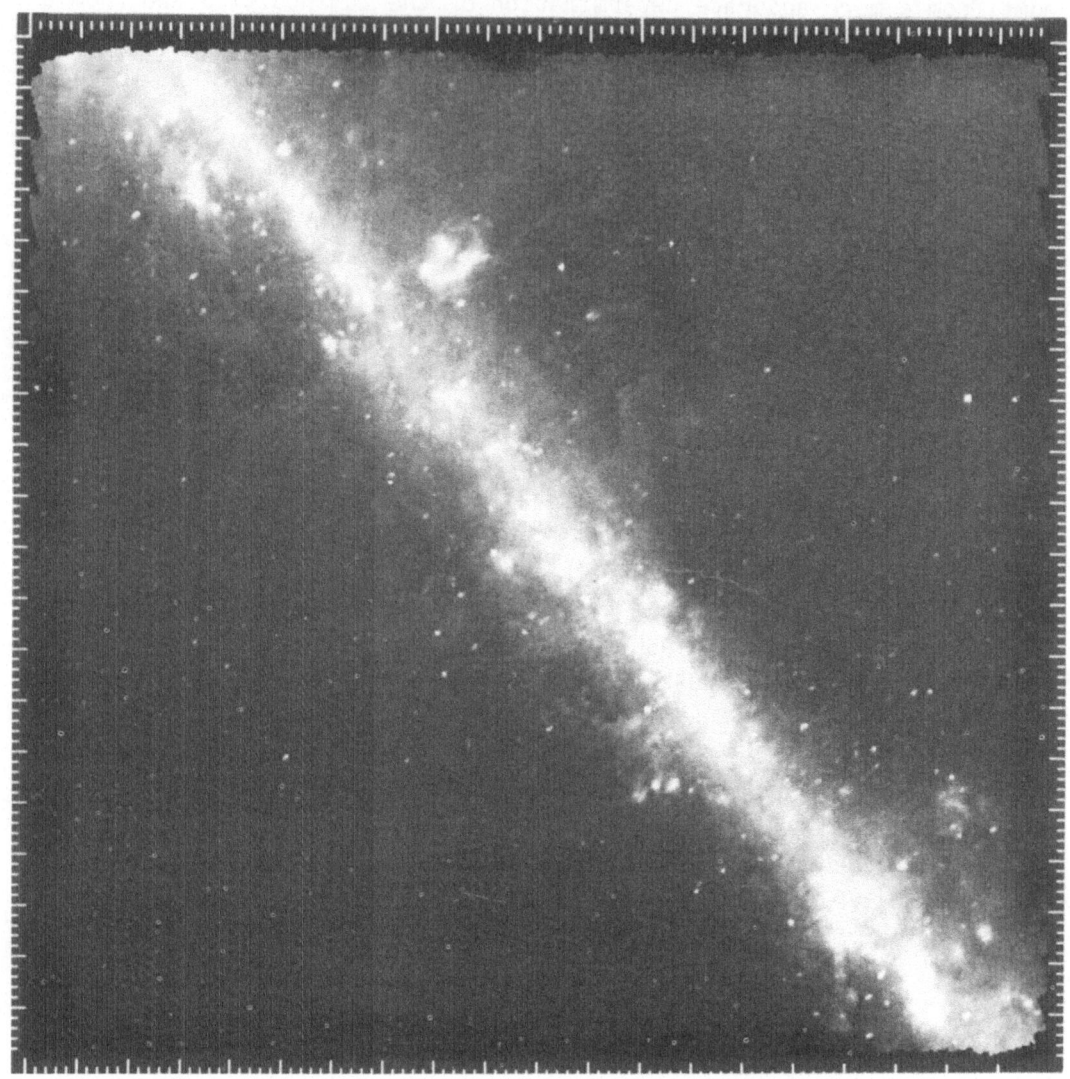

IRAS Field 181 $\alpha = 16^h48^m$, $\delta = -45°$, HCON-3, 12 μm. Field in Norma,
                    Ara and Scorpius; optically largely observed. RCW
                    106 and G333.6-0.2 is at bottom right; RWC108 is
                    just to the lower left of the center. The structure
                    above the plane is IC 4628.

VARIABLES, THE GALACTIC BULGE AND IRAS

M.W. Feast
S.A. Astronomical Observatory
P.O. Box 9, Observatory 7935,
South Africa

ABSTRACT. Evidence from the Baade windows suggests that the majority
of IRAS sources which delineate the bulge are Mira variables
(including probably their long period extention, the OH/IR
variables). No observational differences, at a given period, have yet
been firmly established between bulge Miras and local or globular
cluster Miras. Miras cover a range of ages from ~15 Gyr (short
periods) to ~4 Gyr (long periods) with main sequence progenitor
masses of ~1.3 M_O at periods of ~450 days and less at shorter
periods. Significantly higher masses (~2 M_O or more) might be
required if very luminous, long period (~2000 day) OH/IR Miras are
present in the bulge. In the solar neighbourhood the velocity
dispersion of Miras is a function of period. In the bulge it is
independent of period indicating a relaxation time of ≤5 Gyr for the
bulge. The velocity dispersion appears to be constant within the
central ~1 kpc but drops outside this. OH/IR sources near the Centre
show a rotational velocity increasing with distance out to ~150 pc.
Further out the rotation must drop since the mean velocity of stars
in the Baade windows (z ~ 500 pc and ~500 pc from the rotational
axis) is small. Ground based studies of the new (probable) Miras
found by IRAS should substantially improve our present understanding
of the kinematics and other properties of the bulge population.

1. INTRODUCTION

One of the most striking results obtained by IRAS is the picture of
the disk and bulge of our own Galaxy as delineated by point sources
whose 12 and 25 μm fluxes indicate circumstellar dust shells at a
temperature of ~400 K (Habing & Neugebauer 1984). The bulge revealed
by IRAS is highly flattened. Looking at it, two questions come most
obviously to mind: (1) What is the nature of these point sources: (2)
What are their kinematics? This paper discusses these two topics.

2. THE IRAS BULGE SOURCES

The most obvious way to discover the nature of the IRAS bulge sources

339

F. P. Israel (ed.), Light on Dark Matter, 339–348.
© *1986 by D. Reidel Publishing Company.*

is the look at those in the Baade windows of low extinction close to
the centre of the bulge. Table 1 lists the number of Mira variables
studied by Lloyd Evans (1976) in the three windows, the number of
IRAS point sources in the same areas (with a few doubtful exceptions
these all have typical IRAS "bulge object" 12 and 25 μm colours) and
the number of objects in common between the two surveys. In the NGC
6522 field which is perhaps the most completely studied field
optically, the identification of IRAS sources with known Miras is
essentially complete. In the other fields a very substantial number
of IRAS sources are known Miras. By contrast,of the 306 M giants
listed by Blanco, McCarthy and Blanco (1984) in a part of the NGC
6522 field, none are IRAS sources except 5 known Miras. Evidently a
major part of (perhaps all) of the IRAS sources in the bulge are Mira
variables of M spectral type. That is, cool giant variables, arguably
the last stage of stellar evolution before planetary nebula ejection.

TABLE 1. Numbers of Miras and IRAS Sources in the Baade Windows.

	NGC 6522	Sgr I	Sgr II	Total
Miras	38	57	12	107
IRAS Sources	12	15	9	36
Objects in Common	11	7	6	24

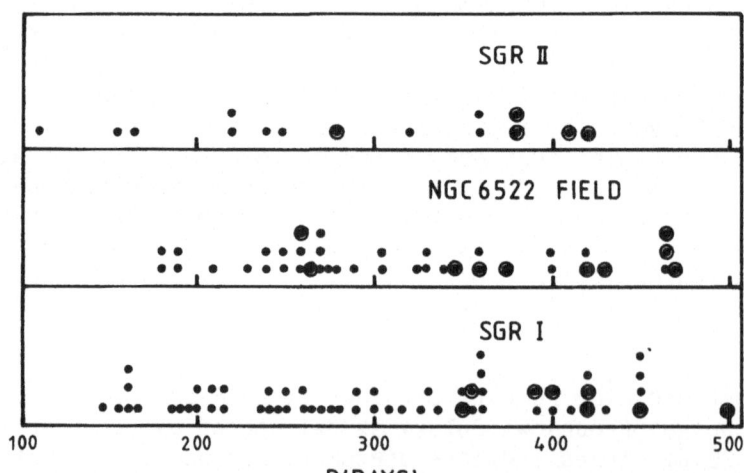

Figure 1. Period distribution of Miras in the Baade windows (Lloyd
Evans 1976). IRAS identifications are circled.

Dust shells are well known amongst Mira variables and tend to
become more obvious at longer periods. Figure 1 shows that this is
true for the bulge sources. As will be discussed, the type II OH/IR
sources form a natural extention of the Mira sequence and since such
stars are known to be present in the bulge (e.g. Baud et al. 1981)
they will be included in the discussion.

3. THE BULGE MIRAS

Wood and Bessell (1983) have suggested that the bulge Miras follow
different period-luminosity and period-(J-K) colour relations from
those elsewhere and are thus to be distinguished from other Miras.
Their discussion leads to the view that the longer period Miras in
the bulge have masses of 3 or 4 solar masses and ages of a few
times 10^8 years. If this were so it would be a remarkable result
since other evidence suggests that the bulge consists of old, low
mass stars (ages greater than about 4 Gyr and main sequence
progenitor masses less than about 1.3 M_O (cf. Whitford 1985). However,
recent work indicates that observational differences between bulge
Miras and others are less than suggested by Wood and Bessell.

Fig. 2 shows a plot of K magnitudes against log period for Miras in
the Baade windows (data from Glass & Feast 1982a and Wood & Bessell
1983). The line is a best fit for a period-luminosity relation of the
slope found in the LMC (Feast 1984) using data from Glass & Lloyd
Evans (1981), Glass & Feast (1982b) and Wood, Bessell & Paltoglou
1985. The LMC period-luminosity relation has been further
strengthened by new data of Glass & Reid (1985). Evidently the bulge
Miras show a period-luminosity relation (the scatter can be
attributed primarily to scatter in distance; cf. Glass & Feast 1982a)
and within the present uncertainties this does not differ from the
LMC slope. It will be possible to test this result much more
stringently by using the IRAS catalogue as a finding list for
detailed studies of additional bulge Miras.

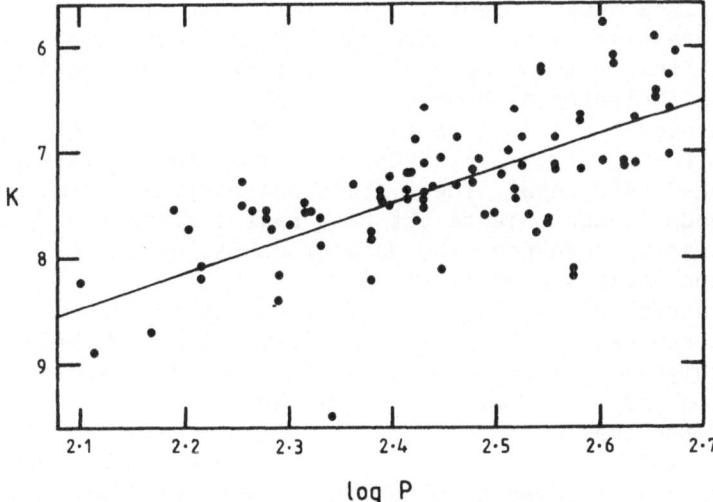

Figure 2. K Magnitude against log P for Baade window Miras. The line
has the slope of the LMC period-luminosity relation.

Possible differences between the period-(J-K) relations in
different systems is more difficult to discuss, partly because work

is still in progress to ensure that all observations are on the same system and partly because of interstellar adsorption. The Miras in the solar neighbourhood (data from Catchpole et al. 1977) show a rather good relation between J-K (corrected for reddening on a galactic model) and period (to be published). That the scatter in this relation is due primarily to the statistical nature of the reddening corrections is shown by the fact that the Miras in globular clusters (data from Menzies & Whitelock 1985) agree with the relation in the mean but have a smaller scatter about it. The bulk of the Miras in the Baade windows can be fitted to this relation if a mean reddening of E_{J-K} = 0.28 is adopted for them. This corresponds to A_V = 1.87 ± 0.1 which is greater than is often adopted for these fields. However Walker (1985) has recently obtained accurate BVI light curves of 11 RR Lyrae variables near NGC 6522 using a CCD system and finds A_V = 1.77 ± 0.1 in very good agreement with the Miras. He also finds a scale error in existing RR Lyrae magnitudes in the NGC 6522 field. Correcting this and adopting $\langle M_V \rangle$ = 0.6 for RR Lyraes he finds a distance to the Centre, R_0 = 8.2 kpc agreeing (exactly) with that obtained from Miras in the Baade windows using the data and discussion of Glass and Feast (1982a) but calibrating the Mira period luminosity relation from Miras in globular clusters (Menzies & Whitelock 1985) and adopting M_V = 0.6 for the cluster horizontal branches.

In addition to the stars discussed above, the Baade windows contain a number of Miras with periods in the 400-500 day range and very red colours (J-K up to about 2.8). This period range is the region of overlap between the "normal" Miras and the "classical" OH/IR variables with periods usually in the range 500-2000 days. We expect to see in the 400-500 day period range some variables whose circumstellar shells are sufficiently thick (at least along our line of sight) to produce significant reddening. For obvious reasons there is a bias against finding such stars in a solar neighbourhood sample selected by optical brightness. However the general solar neighbourhood contains stars such as IK Tau (($J-K)_0$ = 2.52, P = 460 days) and WX Ser (($J-K)_0$ = 2.91, P = 425 days) mean $(J-K)_0$ values based on unpublished SAAO data). These stars not only have similar colours to the Baade window objects but also have similar 1-20 μm spectra as can be seen from the data of Jones, Hyland & Robinson (1984) and Jones & Merrill (1976).

The above discussion suggests that, at least as a first approximation, we can assume that the Miras (including the IRAS sources) in the bulge are similar to those of the same period in the solar neighbourhood and in globular clusters. We shall thus assume that estimates of the ages and masses of these latter objects are also applicable to the bulge variables. Evidently the large number of (probable) bulge Miras discovered in the IRAS survey will allow this matter to be discussed in much more detail. It is at present not clear, for instance, whether a group of mildly redder Miras with periods near 340 days in the Sgr I field are abnormal or simply subject to somewhat higher obscuration.

4. THE AGES AND MASSES OF MIRAS

The shorter period Miras occur in globular clusters and are thus old low mass objects. There is a relation in globular clusters between Mira period and cluster metallicity on the Zinn (1980) scale (cf. Feast 1981). The implied range of abundances in the bulge is not surprising in view of the fact that it contains both mildly metal-poor RR Lyraes and super metal-rich K giants (cf. Whitford 1985). There has been recent discussion on whether or not all globular clusters have closely the same age or whether there is a marked dependence on metallicity. Table 2 gives two possible alternative views. Evidently Miras with periods less than ~300 days are older than ~ 5×10^9 years (at least) and less massive 1.2 M_0 (main sequence progenitor mass).
(Note: The masses are from Buzzoni's interpolation of the Yale isochrones (Ciardullo & Demargue 1977, cf. Iben & Renzini 1984)].

TABLE 2.   (a)   Ages and Masses of Mira and SRd variables
                       inferred from Globular Clusters.

P (days)	[Fe/H]	Carney (1980)		Sandage (1982)	
		Age (Gyr)	Mass (M_0)	Age (Gyr)	Mass (M_0)
60	-2.0	18.7	0.73	17	0.76
200	-0.7	10.5	0.93	17	0.79
270	0.0	6.1	1.18	17	0.87

(b)   Ages and Masses of Miras from Local Kinematics.

P days	P days	N	σ km$^{\tau 1}$	V kms^{-1}	Age (Gyr)	Mass (M_0)
149-200	178	41	150	-109±25	10	~0.9
300-350	324	80	61	- 22± 6	} 5	1.25
350-410	328	54	60	- 13± 8		
>410	457	33	68	- 6± 8	4	~1.33

σ_τ = Totale velocity dispersion; V = Deviation from circular
velocity   N = Number of stars

For the longer period Miras, ages and main sequence masses can be estimated from a comparison of the kinematics of solar neighbourhood Miras (Feast 1963) with the kinematics of nearby stars of known age (Wielen 1974), see Table 2(b). Evidently even out to ~400 days the Miras are low mass, old objects. It may be of relevance to galactic structure problems that (cf. Table 2(b)) the short period Miras have

kinematics similar to those of the "thick disk" proposed by Gilmore
and collaborators (cf. Gilmore 1985) which has σ_τ ~ 150 kms^{-1} V ~
100 kms^{-1}.

Pulsation masses consistent with the above discussion are
derived if we assume overtone pulsation and fix the effective
temperature scale using occultation (Glass & Feast 1982b). Table 3
gives revised pulsation masses based on a globular cluster distance
scale with M_v(HR) = 0.6 as discussed above and with two
representative values of the pulsation constant in the first overtone
(Fox & Wood 1982).

TABLE 3.   Pulsation Masses.

P (days)	Q=0.040 $M(M_O)$	Q=0.050 $M(M_O)$	P(days)	Q=0.040 $M(M_O)$	Q=0.050 $M(M_O)$
100	0.44	0.68	500	0.84	1.31
200	0.58	0.90	1000	1.11	1.74
300	0.68	1.06	2000	1.48	2.30
400	0.78	1.20			

5. AGES AND MASSES OF OH/IR VARIABLES

As noted above OH/IR variables are known in the galactic bulge.
Masses as high as 5 M_O have been suggested for OH/IR variables (cf.
Hermans & Habing 1985) which would be a remarkable result for bulge
objects. There are a number of reasons why such high masses may be
questioned for the OH/IR variables generally and for the bulge
sources in particular.

Engels et al. (1983) used kinematic distances to suggest that
the OH/IR variables extended the Mira P-L relation. The scatter about
this relation can be attributed to a velocity dispersion of 30 km^{-1}
(Feast 1985). Such a velocity dispersion for OH/IR sources was also
derived by Johansson et al. (1977) from quite different considerations and leads to an estimated age of 4×10^9 years and a
(main sequence) mass of ~1.3 M_O from local kinematics.

The much higher masses come from studies of the z distribution
of OH/IR sources or from modelling of the kinematics and
distribution. There are two points of uncertainty in this approach.
Firstly, some supergiant OH/IR sources (e.g. VX Sgr) are known and in
some samples their numbers might be significant. Secondly the
Goldreich-Scoville effect (Goldreich & Scoville 1976, Huggins &
Glassgold 1982) may significantly affect the apparent distribution of
these objects. In this mechanism a general ultraviolet field is
needed to dissociate H_2O into OH in the circumstellar shell. This
field is obviously strongest in the plane, and may vary in the plane.
That the effect may well be significant is suggested by the work of
Whitelock (1985). On the basis of groundbased observations of IRAS
sources outside the bulge she points out that there are OH/IR-like
objects at moderately high galactic latitudes and that these might be
OH-quiet. Some of the sources have not been shown by Engels et al.

(1984) and Lewis, Eder & Terzian (1985) to be weak OH emitters. Evidently the IRAS catalogue allows us to sort out a sample of OH/IR-like objects unbiased by the Goldreich-Scoville effect. The galactic bulge may well contain such objects.

Finally we still have to determine periods for the bulge OH/IR sources. We would expect a significant mass difference between 500 day variables (M_{Bol} ~-5) and 2000 day variables (M_{Bol} ~-6.3). If stars as bright as the latter exist in the bulge then their masses may be ~2 $M_\odot$ or more with ages of ~1×10^9 years or less. (cf. Iben & Renzini 1983 Fig. 7. Note that metallicity may affect this diagram.) However the preliminary results of Habing (this conference) suggest that variables with $M_{Bol} < -5.5$ may not be present in the bulge. This is equivalent to an upper limit to the period of ~840 days and the discussion of local kinematics (above) suggests that the bulge (main sequence) masses are not necessarily greater than ~1.3 $M_\odot$ and ages not necessarily younger than ~4×10^9 years. These tentative results can clearly be tested by ground based observations of IRAS bulge sources.

6. THE KINEMATICS OF THE BULGE POPULATION

In the solar neighbourhood the velocity dispersion of the Miras is high at the short period end (~200 days) but drops rapidly with increasing period (Feast 1963). Table 4 shows the results for Miras in the Baade windows (Feast, Robertson & Black 1980, Feast & Spencer

TABLE 4.   Velocity Dispersions in Bulge.

Baade Windows	$\bar{P}$	σ	N	σ_{SN}
Miras 147 < P ≤ 250 days	202	108±19	17	75
250 < P ≤ 465 days	320	138±24	17	40
all	261	122±15	34	
M giants		113±12	49	
K giants		120±11	62	
Mean		118± 7		
Mean (corrected for foreground)		126± 7		
Miras in outer bulge		98± 2	12	
RR Lyraes in outer bulge		70±10	26	
Mean Outer bulge		80±10		
OH/IR sources ≤ 2°, b ≤ 5°		134±22	20	
		145±19	30	
OH/IR sources within 1° of Centre				
		115±16	28	

σ = Observed velocity dispersion (kms^{-1}).
σ_{SN} = Observed velocity dispersion in the Galactic plane for Miras in the solar neighbourhood.

Jones 1985). It is evident that the velocity dispersion in this case
does not drop with increasing period. Within the scatter the
dispersion is independent of period. This together with the previous
discussion suggests that the relaxation time of the bulge is 5 Gyr or
less. Table 4 contains the dispersions for K giants in the NGC 6522
field (Rich & Whitford 1984) and the M giants in this field (Mould
1983). These objects and the Miras all have similar dispersions.
Table 4 also contains the velocity dispersions of faint Miras in the
outer part of the bulge (Feast 1966) and RR Lyraes in the outer bulge
(Groningen fields 2 and 3, Rodgers 1977). These objects at moderate
galactic latitudes (b ~ 10°, z ~ 1.5 kpc) have a somewhat lower
dispersion (80 ± 10 kms^{-1}). Correcting the Baade window results for a
few stars at distances of ~1 kpc or more from the tangential point
leads to a dispersion of 126 ± 7 kms^{-1} at ~500 pc from the Centre.
Habing et al. (1983) have determined velocities of OH/IR sources
within 1° of the Centre. The dispersion depends on how the data is
treated (Table 4) but is not significantly different from that in the
Baade windows though referring to a region within ~100 pc of the
Centre. These results therefore suggest that the velocity dispersion
remains constant out to perhaps 1 kpc after which it decreases.

It would be particularly interesting to study the rotation of
the bulge, especially in view of its marked flattening.
Unfortunately the data, summarized in Table 5, is still very limited.
Minkowski (1964) found evidence for a relatively small rotation
effect from planetary nebulae in the general direction of the bulge.
A rather similar effect was found later for Miras in the same
direction (Feast 1966, 1972). Unpublished data (Kinman, Lasker &
Feast) on some additional planetaries towards the Centre show a
similar result. Recently Habing et al. (1983) found evidence of
rotation for OH/IR sources within 1° of the Centre. The precise
gradient adopted depends on how the data is treated (See Table 5).
The results suggests that at 1° from the Centre (~140 pc) there is a
rotational velocity of 100-200 kms^{-1}. However the rotational gradient
must fall off very considerable outside this inner region. Table 5
shows the mean velocity of Miras and M giants in the Baade windows,
NGC 6522 and Sgr I (corrected to the local standard of rest and for
the circular velocity of rotation at the Sun) (data from references
above). This velocity is very small. Because these objects will not
all be at the tanential point but are distributed around it, the
observed velocities will in the mean be less than the rotational
velocity. Rough estimates suggest that the mean velocity must be
increased by a factor of about four in the NGC 6522 field and three in the
Sgr I field to get an estimate of the rotational velocity. Even so,
as Table 5 shows, the rotational velocity is quite small for these
objects at z ~ 500 pc and at ~500 pc from the axis of rotation.

TABLE 5.   Rotation of Bulge.

Planetaries with $|\ell| < 8°$ (all distances)		~+14 km/sec/degree
22 Miras $|\ell| < 5°$ (distant)		+26±8 km/sec/degree
30 OH/IR sources within 1°	least squares	+108±56 km/sec/degree
28 "	least squares	+160±45 km/sec/degree
28 "	model	+204 km/sec/degree

Mean velocity in Baade windows +5±13 km/sec (83 stars)
Mean rotational velocity of Baade window stars +22±46 km/sec (83 stars).

Many of the probable Miras discovered by IRAS in the bulge should be sufficiently unobscured that optical velocities can be obtained for them. This should enable us to radically improve on the first few hints at the kinematics of the bulge which are summarized above.

ACKNOWLEDGEMENTS

I am grateful to colleagues at SAAO for scientific discussions and for help in other ways. My thanks are particularly due to P.A. Whitelock, R.M. Catchpole, A.R. Walker, I.S. Glass, B.S. Carter, J. Spencer Jones and C.A. Black.

REFERENCES

Baud, B., Habing, H.J., Mathews, H.E. and Winnberg, A., 1981. Astron. Astrophys., 95, 171.
Blanco, V.M., McCarthy, M.F. and Blanco, B.M., 1984. Astron. J., 89, 636.
Carney, B.W., 1980. Astrophys. J. Suppl., 42, 481.
Catchpole, R.M., Robertson, B.S.C., Lloyd Evans, T.H.H., Feast, M.W., Glass, I.S. and Carter, B.S., 1977. S. Afr. Ast. Obs. Circ., no. 4, p.61.
Ciardullo, R.B. and Demarque, P., 1977. Trans. Yale Univ. Obs. Vol. 33,34,35.
Engels, D., Habing, H.J., Olnon, F.M., Schmid-Burgk, J. and Walmsley, C.M., 1984. Astron. Astrophys., 140, L9.
Engels, D., Kreysa, E., Schultz, G.V. and Sherwood, W.A., 1983. Astron. Astrophys. 124, 123.
Feast, M.W., 1963. Mon. Not. R. Astr. Soc., 125, 367.
Feast, M.W., 1966. Mon. Not. R. Astr. Soc., 132, 495.
Feast, M.W., 1972. Vistas Astr., 13, 207.
Feast, M.W., 1981. In Physical Processes in Red Giants ed. I. Iben. and A. Renzini (Reidel: Dordrecht) p.193.
Feast, M.W., 1984. Mon. Not. R. Astr. Soc., 211, 51p.
Feast, M.W., 1985. The Observatory, in press.

Feast, M.W., Robertson, B.S.C. and Black, C., 1980. Mon. Not. R. Astr. Soc., 190, 227.

Feast, M.W. and Spencer Jones, J., 1985. To be published.

Fox, M.W. and Wood, P.R., 1982. Astrophys. J., 259, 198.

Gilmore, G., 1985. Science from Measruing Machines (RGO meeting) in press.

Glass, I.S. and Feast, M.W., 1982a. Mon. Not. R. Astr. Soc., 198, 199.

Glass, I.S. and Feast, M.W., 1982b. Mon. Not. R. Astr. Soc., 199, 245.

Glass, I.S. and Lloyd Evans, T.H.H., 1981. Nature 291, 303.

Glass, I.S. and Reid, No., 1985. Mon. Not. R. Astr. Soc. In press.

Goldreich, P. and Scoville, N.Z., 1976. Astrophys. J., 205, 144.

Habing, H.J. and Neugebauer, G., 1984. Scientific American Vol.251, 42 (November).

Habing, H.J., Olnon, F.M., Winnberg, A., Mathews, H.E. and Baud, B., 1983. Astron. Astrophys. 128, 230.

Herman, J. and Habing, H.J., 1985. Rep. Prog. Phys. In press.

Huggins, P.J. and Glassgold, A.E., 1982. Astron. J., 87, 1828.

Iben, I. and Renzini, A., 1984. Physics Reports 105, 329.

Johansson, L.E.B., Anderssen, C., Goss, W.M. and Winnberg, A., 1977. Astron. Astrophys., 54, 323.

Jones, T.J., Hyland, A.R. and Robinson, G., 1984. Astron. J., 89, 999.

Jones, T.W., and Merrill, K.M., 1976. Astrophys. J., 209, 509.

Lewis, B.M., Eder, J. and Terzian, Y., 1985. Nature 323, 200.

Lloyd Evans, T., 1976. Mon. Not. R. Astr. Soc., 174, 169.

Menzies, J.W. and Whitelock, P.A., 1985. Mon Not. R. Astr. Soc., 212, 783.

Mould, J.R., 1983. Astrophys. J., 266, 255.

Rich, R.M. and Whitford, A.E., 1984. Pub. Ast. Soc. Pacif., 96, 794.

Rodgers, A.W., 1977. Astrophys. J., 212, 117.

Sandage, A.R., 1982. Astrophys. J., 252, 553.

Walker, A.R., 1985. To be published.

Whitelock, P.A., 1985. Mon. Not. R. Astr. Soc., 213, 51p.

Whitford, A.E., 1985. Pub. Ast. Soc. Pacif., 97, 205.

Wielen, R., 1974. In Highlights of Astronomyn 3 (ed. G. Contopoulos) (Reidel: Dordrecht) p. 395.

Wood, P.R. and Bessell, M.S., 1983. Astrophys. J., 265, 748.

Wood, P.R., Bessell, M.S. and Paltoglou, G., 1985. Astrophys. J., in press.

Zinn, R., 1980. Astrophys. J. Suppl., 42, 19.

GROUND BASED OBSERVATIONS OF NUCLEAR BULGE STARS

Jay A. Frogel
National Optical Astronomy Observatories
Cerro Tololo Inter-American Observatory
950 North Cherry Avenue
Tucson, AZ 85726

ABSTRACT Infrared photometry and CCD spectroscopy have been obtained
for about 400 M giants at several latitudes in the Galactic nuclear
bulge. Their colors and luminosities differ significantly from M
giants in the solar neighborhood which are used in stellar synthesis
models for ellipticals and the bulges of spirals. Initial examination
of the IRAS "disk and bulge" stars indicates that rather than representing a new class of objects, the disk stars, at least, are quite
similar to and extend to redder colors the latest and most luminous
group of spectroscopically identified M giants.

1. WHY STUDY THE NUCLEAR BULGE?

Observations of stars very close to the Galactic nucleus is extremely
difficult because of over 30 magnitudes of visual extinction. Baade
(1963), however, recognized that most of the obscuring material is
near the sun and closely confined to the galactic plane. He identified a number of "windows" with relatively low extinction through
which stars within a few degrees of the galactic center could be observed. Our line of sight to these windows, although as low as
b^{II}=3°, quickly passes out of the galactic plane and hits the bulge
after a passage through relatively clear space. There is little or no
dust mixed in with the stars of the bulge itself.

Pioneering studies of stars in Baade's Window at b^{II}=-3.9° were
done by Arp (1965) and van den Bergh (1971). Their BV color magnitude
diagrams of the brighter stars led them to two important conclusions.
First, the stars in the bulge are old - probably as old as the oldest
stars known - and that they are metal rich - certainly as metal rich
as the globular cluster 47 Tucanae and probably as metal rich as the
sun.

From photometric and spectroscopic studies carried out over the
past few decades it appears that the stars which are the main contributors to the integrated visual light of the nuclear bulges of all but
the latest type spiral galaxies and of elliptical galaxies are quite

349

similar to one another. Stars in the two types of systems have similar luminosity functions and ranges in age and chemical composition.

It is a well established fact that M giants make a significant
contribution to the bolometric luminosity of ellipticals and the
bulges of spirals: Stebbins and Whitford (1948) showed that both of
these types of systems have strong radiation at one micron. Johnson
(1966) interpreted the red K-L colors he measured for these galaxies
as indicative of a cool stellar population. The TiO and Ca II absorption lines measured by O'Connell (1976) were interpreted by him as
arising from late type giants. The VJHK colors and CO and H_2O indices
measured by Frogel, et al. (1978) and Aaronson, Frogel, and Persson
(1978) could most easily be explained as resulting from large numbers
of cool, luminous stars. Finally, Whitford's (1977) inability to
detect the Wing-Ford bands of FeH in these same systems argued strongly against the possibility that the red stars were dwarfs rather than
giants. M giants, while contributing less than 10% of the light at V,
dominate the radiation longward of one micron in ellipticals and spiral bulges.

Detailed interpretation of integrated light data for ellipticals
and spiral bulges has to rely on modelling procedures since direct
observation of their stars is generally impossible from the ground.
Such stellar synthesis models use globular cluster and solar neighborhood stars as building blocks (e.g. Aaronson, et al. [1978] and
Tinsley and Gunn [1976]). Astrophysically reasonable constraints are
imposed on how groups of stars can be combined and various mathematical techniques are employed to achieve the best fit between the models and the data subject to the constraints.

Two critical assumptions in the modelling are first that globular
cluster and solar neighborhood stars are the proper ones to be used in
the models, and second that they are a complete set of stars, i.e.
that spheroidal stellar systems do not contain a significant population of stars not present in clusters or in the solar neighborhood.
(In addition to uncertainties regarding the cool stellar component of
spheroidal stellar systems, there are some problems with understanding
what kinds of very hot stars make contributions to the integrated
light. Although these hot stars make only a minor contribution to the
bolometric luminosity, they may have important implications for
understanding the star formation histories of these systems.)

A comparison between infrared data and stellar synthesis models
revealed that a serious incompleteness existed in the latter (Frogel,
Persson, and Cohen 1980). The observed colors were too red and the CO
indices too strong to be accounted for by the stars included in the
models. Frogel, et al. concluded that real galaxies contained stars
of much higher metallicity and/or luminosity than any contained in the
models.

Morgan (1956) showed that photographic spectra of small patches
of sky in the Galactic nuclear bulge qualitatively resembled similar
spectra of the integrated light of ellipticals and spiral bulges.
Photoelectric scans of small areas in Baade's Window by Whitford
(1978) demonstrated quantitatively the similarity between the different systems. He showed that the continuous energy distribution of the

integrated light from Baade's window and the strengths of various atomic and molecular absorption features are virtually indistinguishable from what is seen in typical ellipticals and spiral bulges. It also appears that the stars in Baade's Window are, in the mean, extremely metal rich. Whitford and Rich (1983) found that 60% of a sample of K giants in the Window have metal abundances greater than solar with some having [Fe/H] values nearly 10 times solar.

Another key similarity between Baade's Window and spheroidal systems is the presence in both of large numbers of very cool giants. Evidence for their presence in external galaxies has been cited above. In the Galactic nuclear bulge their presence has been revealed by low resolution, large field spectroscopic surveys in the near infrared by Nassau and Blanco (1958) and Blanco, Blanco, and McCarthy (1978). A detailed description of the initial survey results is given by Blanco, McCarthy, and Blanco (1984).

It would seem, then, that the best technique for determining the cool stellar content of spheroidal systems lies in a study of the stars of the nuclear bulge of the Milky Way. The next section of this paper will briefly summarize some current research on this topic. The third section of this paper will discuss the relevance of the IRAS "disk and bulge" sources to these stars and describe preliminary results from ground-based observations of these sources by Don Terndrup (from Lick) and myself.

2. CURRENT RESEARCH IN THE GALACTIC NUCLEAR BULGE

Much of the work summarized here is being carried out in collaboration with V. M. Blanco of CTIO and A. E. Whitford and D. Terndrup of Lick.

The first step in our research on the Galactic nuclear bulge is the low resolution spectroscopic survey for M stars which Blanco carries out with a grism at the prime focus of the CTIO 4-meter or with an objective prism on the Curtiss-Schmidt. These surveys are nearly 100% complete for M6 and later stars. Incompleteness becomes a problem only for the earliest M's, not because of faintness but because of the difficulty of classifying the stars at the low dispersion used. Figure 5 of Blanco, et al. (1984) shows that the distribution of I magnitudes of M6 and later stars in the -3.9° window is sharply peaked. This illustrates two important points. First, the rapid fall off on the faint side one magnitude brighter than the survey limit shows that the survey is not magnitude limited. Second, the narrowness of the distribution demonstrates that the stars are clustered quite closely together in space. The B magnitudes of a sample of RR Lyrae variables in the same field show a very similar distribution (Blanco 1984).

The infrared data which we have obtained so far consists of JHK, CO and H_2O photometry for about 400 stars in windows at b^{II} = -2.9, -3.9, -6, -8, -10, and -12° and zero longitude. For a subset of these stars we have also measure magnitudes at L and 10 μm.

CCD spectra have been obtained in the 6000 - 9000 Å region for half of the stars in the IR sample. Most of the spectra are dominated

by absorption bands of TiO. While it will be impossible to get absolute abundances from these spectra, we expect to be able to examine
the distribution over metallicity in the various windows. For the
hottest stars observed the TiO bands are weak and an absolute metallicity determination should be possible. These same data will also
yield radial velocities with an accuracy of 20 km/sec or better. Thus
the velocity dispersion as a function of distance along the minor axis
can be studied. This will complement a dynamical study of stars in
the -3.9° window being carried out by Mike Rich at Caltech. Examination of metallicity distributions in bulge stars via optical photometric data is underway by Rich, Terndrup, and Whitford (cf. reviews by
Whitford 1985a,b).

A number of other groups are actively investigating stars in the
nuclear bulge. In a talk at this conference Michael Feast has described work going on in South Africa particularly concerning the long
period variables. Leonard Searle is using the delta S technique to
investigate RR Lyrae variables in a number of bulge fields. The
Australians are pursuing a program of infrared photometry and spectroscopy of bulge giants (cf. Wood and Bessell [1983] and Jones,
Hyland, and Robinson [1984]).

Two important early conclusions from the infrared studies of the
-3.9° window M giants are first that there are significant numbers of
stars with bolometric luminosities greater than that predicted for
core helium flash in first ascent giants, and that the mean luminosities of the M giants are up to one magnitude fainter than the values
used by Tinsley and Gunn (1976) in their stellar synthesis models
(Frogel 1981; Frogel and Whitford 1982). Data from the -8° window
indicate that its M giants are quite similar to those at lower latitudes (Frogel, Blanco, and Whitford 1984).

The first of these results has a number of significant implications for the age of the bulge stars. Frogel and Whitford (1982) and
Whitford (1985a,b) have argued that the combination of high metallicity with unknown mass loss rates can still allow these stars to be
luminous and old, particularly if a reduced galactic center distance
is used. Wood and Bessell (1983), on the other hand, claim that the
pulsational masses, which depend sensitively on effective temperature,
lead to an age of only a few Gyr. Initial results from main sequence
photometry in the less crowded -8° field (Rich 1984; Terndrup, Rich,
and Whitford 1984; Whitford 1985b) seems to rule out any significant
number of stars younger than 10 Gyr.

In addition to their luminosities, the infrared colors of the
bulge giants are quite different from their solar neighborhood counterparts. For some colors the trend is as would be expected for a
metal rich population based on knowledge of globular cluster and nearby field stars. An example are the JHK colors. In a J-H, H-K plot
bulge giants from all windows are displaced from the mean field giant
line by 0.05 mag in a direction opposite to the displacement of metal
poor globular cluster giants (Figure 2 of Frogel, Blanco and Whitford
1984). However, for a few colors the trend is just the opposite of
the expected one. For example in a color-magnitude diagram the J-K
and V-K colors of both K and M bulge giants are bluer than those of

field giants or of metal rich globular cluster giants (Frogel, Whitford, and Rich 1984; Frogel and Whitford 1985). Quantitative understanding of the mechanisms responsible for the color shifts in either case is lacking. Nevertheless these marked differences from solar neighborhood stars indicate another shortcoming in existing stellar synthesis models for galaxies.

A well defined subset of the reddest M7-9 giants in the -3.9 and -8° windows have a space density which falls off much more rapidly than an R 1/4 law (Frogel, Blanco and Whitford 1984). Blanco and Blanco (1984) find that at -12° the number of M5 and later giants has a surface density 100 times less than than expected from an extrapola- tion of either a Hubble or de Vaucouleurs density law from a latitude of -3.9°. One possible origin of these effects would be a steep de- cline in the number of very metal rich stars with distance from the galactic center since the Hayashi track shifts to warmer temperatures as the metallicity decreases. Another possibility is that these red, luminous stars belong to a population with a spatial distribution different from that of the majority of the bulge stars.

Most of the large amplitude variables that we (Frogel and Whitford 1985) have observed at 10 μm in the -3.9 and -8° windows have excess infrared emission while very few of the non-variables do. Such excess emission provides fairly unambiguous evidence for circumstellar dust and, by inference, for substantial mass loss rates. These data should be able to provide a significant clue to the variation of mass loss rate for giant stars as a function of metallicity. Knowledge of this relationship is crucial for a detailed understanding of the evo- lution of these stars. Preliminary analysis (Frogel 1984) of these data plus similar data for variables in relatively metal poor globular clusters (Elias and Frogel 1985) gives no indication for any substan- tial differences in mass loss rate for stars with widely differing heavy metal abundance.

3. GROUND BASED OBSERVATIONS OF IRAS "DISK AND BULGE" STARS

Earlier at this conference our attention was drawn to the fact that the distribution on the sky of the "disk and bulge" stars from the IRAS point source catalogue is closely confined to a disc with a well defined central bulge. This distribution is qualitatively similar to that found for the radiation at 2.4 μm from balloon observations of the central 120° of the galactic plane (Hayakawa, et al. 1981). Although this near infrared radiation most likely comes from M giants (Nogushi, et al. 1981; Mikami, et al. 1982), there is no reason to believe that these are the same stars which are being found by IRAS. The latter, in fact, are probably of much later spectral type.

The number of IRAS sources in the bulge fields surveyed by Blanco is less than 3% of the number of M6 and later stars in these windows. It is important to determine the nature of these sources: Do they represent a stellar population distinct from the spectroscopically identified M giants? What fraction of the IRAS sources are being found in the spectroscopic surveys? What is their relative contribution to the integrated light of the bulge? In the -3.9 and -8° windows there are 10 IRAS sources. On the basis of 10 µm photometry with the CTIO 4-meter Don Terndrup and I have identified all 10 with known M giants found spectroscopically. So the answer to the second question is "a high fraction".

In order to answer the other questions, Terndrup and I are obtaining infrared and spectroscopic data for samples of IRAS disk and bulge stars. We have nearly completed gathering data for a sample with the following characteristics: b^{II} between +2 and +15°; l^{II} between 304 and 306°; F(12) >1.0 with Q=3; F(12)/F(25) <1.5. Twenty stellar objects and one galaxy met these criteria. Unfortunately, we were able to obtain CCD spectra for only 9 of the sources. Four more have certain optical counterparts; three or four of the remaining 7 probably do.

From the CCD spectra 7 of the stars are late M type, one is a carbon star, and one is a peculiar bright blue (visually) emission line object. We have 1.2 - 20 µm magnitudes and CO and H20 indices for all 20 of the sources. These data agree with the spectroscopic classifications and show that 9 of the remaining 11 are definite M's while the last 2 are probable M's. Hence, 18 of the 20 sources are M giants.

The spectra obtained so far are dominated by very strong TiO and VO absorption bands. They are indistinguishable from those for the redder stars found in the bulge window surveys, although the IRAS sources extend the sequence of spectral types to cooler temperatures as judged by the relative strengths of the absorption bands. A number of objects have quite strong Hα emission; some have traces of stellar H_2O. On the whole, though, they do not appear to differ from previously known cool stars.

From 1.2 to 4.8 µm the colors can be easily fit by grey-body emission with color temperatures between 3000 and 7000° K. The hotter end of the distribution overlaps the regions in color-color plots where most of the large amplitude variables previously know to exist in the bulge lie (e.g. Frogel, Blanco, and Whitford 1984 and Frogel 1984). A few - the redder - of these latter stars lie considerably far along the sequences defined by the IRAS sources.

About 80% of the IRAS sources have a well defined silicate emission feature at 10 µm. Only one has a clear absorption feature. There may be a weak inverse correlation between the strength of the emission feature and the redness of the colors.

The main conclusion so far from the sample of 20 IRAS sources observed is that they are not qualitatively different from previously known M giants. That they extend previously determined color sequences to redder values, presumedly due to thicker dust shells and probably higher mass loss rates, is not surprising in view of IRAS's detection threshold and our criteria for selecting the sample. Only a small fraction of the IRAS disk and bulge stars would not be detected with Blanco's spectroscopic search technique.

We must determine if the sample of IRAS disk stars studied at l^{II}=305° has the same physical characteristics as those located in the bulge. If so, one would conclude that they are as old and metal rich as the bulge stars. The implication would be that some of the material that collapsed to form the disk did so at about the same time as the central spheroid and also underwent fairly rapid enrichment. A qualitatively similar picture is envisioned by Zinn (1985) to explain the disk like distribution of metal rich globular clusters: he suggests that the disk formed rapidly after the nearly free-fall collapse of the halo (cf. Eggen, Lynden-Bell, and Sandage 1962) since, he argues, the difference in age between the disk and the halo globular clusters does not appear to be more than 3.4×10^9 yr. Norris, Bessell, and Pickels (1985) also present evidence for a relatively metal rich component of field giants with kinematic characteristics similar to the disk globular clusters. It should be straight forward to determine the metallicity of the IRAS sources relative to the bulge M giants since there are a number of luminosity independent color criteria to use which are also be unaffected by reddening differences.

Finally, we must consider if the bulge visible in the distribution of the IRAS sources really corresponds to the old, metal rich optically delineated bulge of the galaxy. An alternative possibility is that what we see in the IRAS data is an increase in the scale height of the galactic disc with decreasing radial distance combined with a rapidly increasing stellar density function. Examination of these diverse possibilities is an objective of our current program.

REFERENCES

Aaronson, M., Cohen, J. G., Mould, J., Malkan, M. 1978, Ap. J., 223, 824.

Aaronson, M., Frogel, J. A., persson, S.E. 1978, Ap. J., 220, 442.

Arp, H. 1965, Ap. J., 141, 45.

Baade, W. 1963, in Evolution of Stars and Galaxies, ed. C. P. Gaposhkin (Cambridge: Harvard University Press), p. 279.

Blanco, B. M. 1984, Astr. J., 89, 1936.

Blanco, B. M., Blanco, V. M., McCarthy, M.F. 1978, Nature, 271, 638.

Blanco, V. M., McCarthy, M. F., Blanco, B.M. 1984, Astr. J., 89, 636.

Blanco, V. M., Blanco, B.M. 1984, Mem. Soc. Astr. Ital., in press.

Eggen, O. J., Lynden-Bell, D., Sandage, A.R. 1962, Ap. J., 136, 748.

Elias, J. H., Frogel, J.A. 1985, in preparation.

Frogel, J. A. 1981, in Physical Processes in Red Giants, ed. I. Iben, Jr. Renzini (Dordrecht: Reidel), p. 63.

Frogel, J. A. 1984, Mem. Soc. Astr. Ital., in press.

Frogel, J. A., Blanco, V. M., Whitford, A.E. 1984, in IAU Symp. No. 105, Observational Tests of Stellar Evolution Theory, ed. (Dordrecht: Reidel), p. 571.

Frogel, J. A., Persson, S. E., Aaronson, M., Matthews, K. 1978, Ap. J., 220, 75.

Frogel, J. A., Persson, S. E., Cohen, J.G. 1980, Ap. J., 240, 785.

Frogel, J. A., Whitford, A.E. 1982, Ap. J. (Letters), 259, L7.

Frogel, J. A., Whitford, A.E. 1985, in preparation.

Frogel, J. A., Whitford, A. E., Rich, R.M. 1984, Astr. J., 89, 1536.

Hayakawa, S., Matsumoto, T., Murakami, H., Uyama, K., Thomas, J. A., Yamagami, T. 1981, Astr. Ap., 100, 116.

Johnson, H. L. 1966, Ap. J., 143, 187.

Jones, T. J., Hyland, A. R., Robinson, G.R., 1984, Ap. J., 89, 999.

Morgan, W. W. 1956, Pub. A. S. P., 68, 509.

Nassau, J. J., Blanco, V.M. 1958, Ap. J., 128, 46.

Nogushi, K., Hayakawa, S., Matsumoto, T., Uyama, K. 1981, Pub. A. S. Jap., 33, 583. Mikami, T. Ishida, K., Hamajima, K., and Kawara, K. 1982, Pub. A. S. Jap., 34, 223.

Norris, J., Bessell, M. S., Pickels, A. 1985, Ap. J. Suppl., 58, 463.

O'Connell, R. W. 1976, Ap. J., 206, 370.

Rich, R. M. 1984, Mem. Soc. Astr. Ital., in press.

Stebbins, J., Whitford, A.E. 1948, Ap. J., 108, 413.

Terndrup, D. M., Rich, R. M., Whitford, A.E. 1984, Pub. A. S. P., 96, 796.

Tinsley, B. M., Gunn, J.G. 1976, Ap. J., 203, 52.

Van den Bergh, S. 1971, Astr. J., 76, 1082.

Whitford, A. E. 1977, Ap. J., 211, 527.

Whitford, A. E. 1978, Ap. J., 226, 777.

Whitford, A. E. 1985a, Pub. A. S. P., 97, 589.

Whitford, A. E., 1985b, Erice conference on Spectral Evolution of Stellar Populations, in press.

Whitford, A. E., Rich, R.M. 1983, Ap., J., 274, 723.

Wood, P. R., Bessell, M.S. 1983, Ap. J., 265, 748.

Zinn, R. 1985, Ap. J., 293, 424.

THE GALACTIC MORPHOLOGY OF THE INTERSTELLAR DUST DETECTED BY IRAS

W.B. Burton, E.R. Deul, H.J. Walker, and A.A.W. Jongeneelen
Sterrewacht Leiden
P.O. Box 9513
2300 RA Leiden
The Netherlands

Most of the gaseous component of the interstellar medium in our
Galaxy is in the form of cool or cold neutral gas at temperatures
ranging from a few K to a few hundred K. Most of the dust component
of the interstellar medium also resides in this temperature range. At
the lower end of the range, between a few K and about 40 K, the gas
is mostly molecular. The molecular material is found in compressed
clumps, confined to an annulus in the inner Galaxy. At the upper end
of the temperature range, between about 40 K and a few hundred K, the
gas is almost entirely in the form of atomic hydrogen; it is much
more pervasively distributed than the colder, clumped molecular gas.
The global distribution of the cool, atomic gas shows the same
relative deficiency in the inner few kpc of the Galaxy as the gas in
the molecular annulus, but its distribution extends to much larger
galactocentric distances than the colder gas. Where the molecular and
neutral gas distribution overlap, there is approximately the same
amount of mass residing in each distribution. The dust component of
the interstellar medium contributes a smaller, but not insignificant,
amount of interstellar mass. How the dust is distributed in our
Galaxy, and how its distribution compares to those of the gaseous
components, are questions which naturally arise. Is the IRAS-detected
dust confined, like the molecules, to dark regions of high density
and low volume-filling factor, or is it much more pervasively
distributed, like the atomic gas? Is the dust confined to a very thin
layer, like the molecules and the most recently formed stars? Does it
show the same sort of central deficiency as the gas? How far does the
dust distribution extend in radius? We address these questions in a
preliminary fashion here; we also discussed some of these questions
in our contribution to the Aspenäs Workshop on (Sub)-millimeter
Astronomy (being published by ESO).

1. GLOBAL DISTRIBUTION OF THE COLD AND COOL GAS COMPONENTS OF THE
   INTERSTELLAR MEDIUM

The interstellar gas distribution displays a wide variety of
structures whose physical characteristics are determined by the

F. P. Israel (ed.), Light on Dark Matter, 357–372.
© *1986 by D. Reidel Publishing Company.*

mechanical and radiation influences of their environment with which
they coexist in approximate pressure equilibrium. The condition of
pressure equilibrium allows study of the gas to proceed in terms of
temperature hierarchies, which are revealed, in general, more
directly than are density hierarchies.

1.1. Cold, clumped molecular gas

Hydrogen in the molecular form predominates over all other gaseous
material in cold regions of high enough density to be self shielding
against photodissociation. Having no dipole moment, H_2 does not emit
in the radio or optical windows. The H_2 Lyman absorption bands suffer
from extinction due to dust, and so reveal little from transgalactic
paths. The molecule CO, itself like H_2 stable at low temperatures, is
second in abundance to H_2. There are about 10^4 H_2 molecules for each
one of CO, but still there are 100 times more CO molecules than other
ones. The most important source of excitation of the CO rotational
transitions in the mm radio window involves collisions with H_2, and
so CO can serve as a surrogate tracer of H_2.

Observations gathered during the past decade (in the CO surveys
summarized by Israel, 1985) have shown how the cold molecular clouds
are distributed in the Galaxy, as well as their number, typical size
and separation, and relative speeds. The molecular cloud ensemble is
largely confined to an annulus extending over galactic radii 4 kpc to
8 kpc. (In these remarks the galactic scale is set by the solar-
vicinity distance and rotation velocity, R_o=10 kpc and θ_o=250 km s^{-1}).
The thickness of the cloud layer is characterized by a z-dispersion
of 60 pc. In this layer only about 1% of the volume is filled by the
clumps of compressed material. The clumps display a range of sizes,
and are often arranged in larger complexes. A substantial complex
might be 30 pc across - some are larger - and might be made up of 10
smaller clumps. The temperature of the compressed gas follows from
the high opacity of the CO gas. In clouds without special sources of
heat, for example from just-formed stars, an excitation temperature
of 12 K is respresentative.

Estimating the density of the compressed gas involves critical
assumptions. Because of its high opacity, the principal isotope ^{12}CO
does not provide column densities. Lines from the much less abundant
isotope ^{13}CO are usually unsaturated and so can give CO column
densities under certain assumptions. The intervening assumptions
multiply and introduce important uncertainties when the CO
intensities are converted to H_2 densities. Aspects of this conversion
are reviewed by Lequeux (1981), Liszt (1984), and Israel (1985).
Sources of uncertainty include the values of the $^{12}CO/^{13}CO$ and C/H
abundance ratios, the fraction of C bound in CO, and the degree of
constancy of these quantities over the Galaxy. Galactic gradients in
metallicity, temperature conditions, and radiation field can
certainly not be ruled out. Molecular densities in the cold,
compressed clumps are typically 10^3 cm^{-3} or more. The larger
complexes of clumps tend to have H_2 masses of some 10^5 M_o, although
just as the sizes show a wide range so do the masses.

1.2. Cool, pervasive atomic gas

Hydrogen in the atomic form is found throughout the galactic disk, where no region on a pc or larger scale has ever been identified as empty of HI. A residual amount of atomic hydrogen coexists with molecules in dense clumps at the temperatures, of order 10 K, prevalent in these clumps. Diffused throughout the whole intercloud volume is a hierarchy of HI structures. The pervasiveness of the emission hampers isolation of individual low-latitude features for separate study (see Heiles, 1980). The blending problem is exacerbated by thermal and mass-motion broadening, but mostly by the accumulation of emission from unrelated material, possibly separated by long path lengths but emitting at overlapping velocities.

Descriptions of the HI morphology thus refer to the collective properties of a hierarchy of structures. As a group these structures are organized in a coherent layer whose extent, shape, thickness, kinematics, and gross properties of optical depth, density, and temperature can be studied rather directly. HI cooler than a few hundred K is confined to a layer extending over galactic radii 4 to 25 kpc with a scale thickness in the region interior to the solar distance of about 120 pc. At larger radii the HI layer is dramatically warped, and much thicker; but at these radii neither molecules nor dust are present, so discussion of the outer-Galaxy HI layer is not needed here. Neither is discussion needed of the hot HI at temperatures of several thousand degrees which contributes insignificantly to the gas density budget.

Determination of the amount of HI in the gas layer requires knowledge of the macroscopic optical depth characteristics of the gas; low optical depths are indicated in some directions (where the integrated intensities are proportional to the path lengths), but high opacities are indicated in others. Many of the overall characteristics of the HI emission spectra may be simulated by a model distribution of a ubiquitous gas with spin temperature 135 K and density 0.4 cm^{-3}. More detailed studies (limited by constraints of line blending mostly to higher latitudes, that is, to the solar neighbourhood) indicate a range of temperatures and densities, but centered near these values. It seems, moreover, that these values of temperature and density remain roughly constant over most of the galactic HI disk.

Thus the cool, rather diffusely distributed HI shows a galactic morphology different from that of the cold, clumped molecular gas. The hierarchy of cool HI structures fill much of the volume of the disk gas layer; the cold clumps fill only about 1% of it. The galactic disk as defined by HI is at least three times as large as that defined by the ionized and molecular states of hydrogen, and it is about twice as thick. Both the molecular and atomic constituents are depleted in the inner few kpc of the Galaxy, relative to the situation at larger radii.

2. GLOBAL DISTRIBUTION OF THE IRAS-DETECTED DUST COMPONENTS OF THE INTERSTELLAR MEDIUM

It is natural to ask how the global properties of the dust component

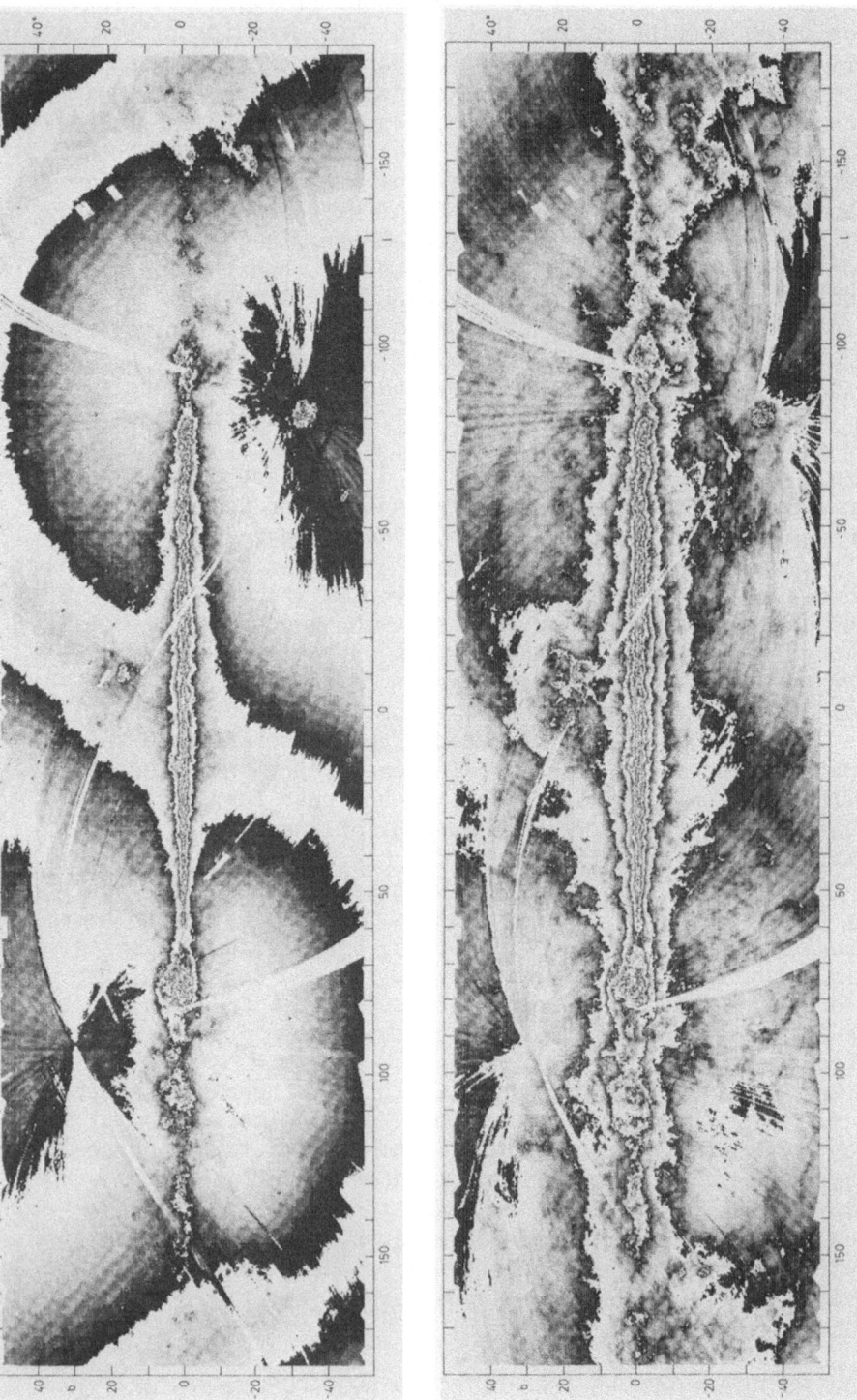

Figure 1. Maps in galactic coordinates of IRAS measurements of emission in the 60μm (upper) and 100μm (lower) bands. These SplineI data are uncorrected for contamination from zodiacal dust emission. The intensities, in units of 10^{-7} Wm^{-2}sr^{-1}, are plotted on a pseudo-logarithmic scale to enhance the low level emission.

of the interstellar medium compare to those of the gas components.
Access to dust properties on a galactic scale has been provided by
the IRAS mission; before this satellite was launched, balloon-mounted
telescopes provided important exploratory information but with much
more limited angular and spectral resolution, sky coverage, and
sensitivity than available with IRAS.

After launch in January 1983 the IRAS satellite surveyed the sky
at 12μm, 25μm, 60μm, and 100μm until November 1983. The Dutch
National Aerospace Laboratories and the Space Research Laboratory in
Groningen supplied software which allowed maps of the compressed (by
a fifth-order B-Spline curve fitting) 60μm and 100μm data to be
examined a few months after launch. The Spline I maps were originally
intended as a quick-look product, to be superseded by the Skyflux
maps from the Jet Propulsion Laboratories. They have, however, proved
to be of a quality sufficient for much interpretative work. The
dataset from Groningen consisted of overlapping 6°x6° maps in
equatorial coordinates. In Leiden, the data were converted to
galactic coordinates and added to make large maps covering all
galactic longitudes between latitudes -51°.2 to +51°.2 at 9 arcmin
angular resolution. Maps in galactic coordinates of the infrared
emission in the IRAS 60μm and 100μm bands are shown in Figure 1.

2.1. Accounting for contamination by emission from the zodiacal dust

The Figure 1 maps show clearly the dust emission confined to the
galactic equator, but they also show substantial emission from solar-
system dust confined to the zodiac. Because the zodiacal dust is
efficiently heated by the Sun to higher temperatures than the
temperatures found in the galactic dust, it predominates at the
shorter-wavelength band. Obviously subtraction of the contribution
from solar-system dust must precede study of the galactic
contribution.

An empirical model of the zodiacal dust emission was constructed
as follows. The Spline I data were regridded to ecliptic coordinates,
as appropriate for solar-system material. Being interested in the
large-scale characteristics of the zodiacal emission, we smoothed it,
using the Kappa-Sigma technique described by Herzog and Illingworth
(1977). This technique involves organizing the data in boxes and
determining the local background level by iteratively discarding data
pixels deviating more than two sigma from the mean. Each box was
subsequently represented by a single data pixel, with the resolution
consequently degraded, in our case to 0°.5. During this process
emission contributed by galactic, rather than zodiacal, dust was
identified and masked out by visual inspection of the maps on the
Leiden image processing equipment.

Having thus isolated and smoothed the zodiacal emission, we
determined the profiles of the emission in ecliptic latitude, at
constant values of ecliptic longitude. The latitude profile had
approximately the same shape at all longitudes, but the value of the
latitude centroid varied with longitude. The flanks of the latitude
profile were well fitted by an exponential function with scale
length ~20°. A standard latitude profile was derived and found to be
useful in bridging the gaps caused by masked out regions of galactic

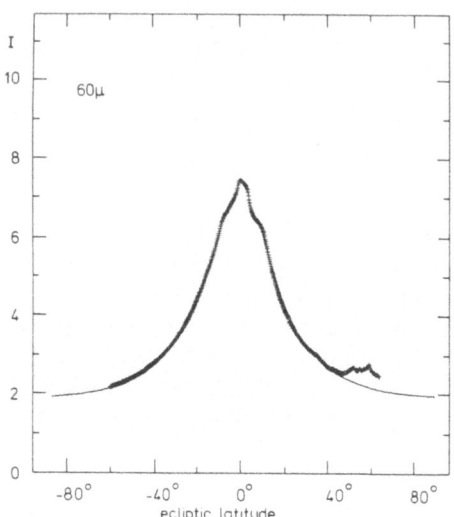 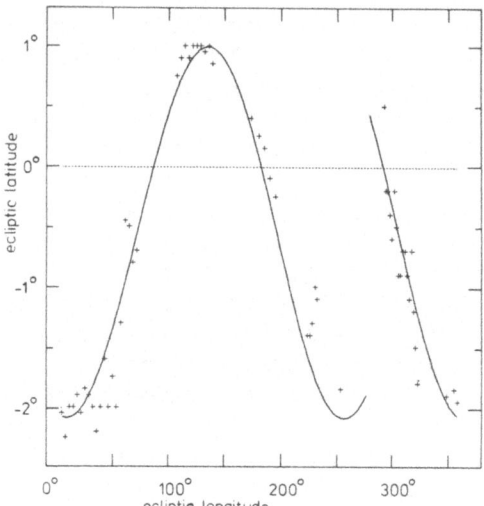

Figure 2. (left): Variation in ecliptic latitude of the zodiacal emission at 60μm. The form of the variation changes little with ecliptic longitude. (right): Variation of centroid latitude of the zodiacal emission with ecliptic longitude. The cosine fit to the variation does not have a period of 360° because of changes in the orientation of the satellite scanning direction during its flight.

emission; the profile is plotted on the left side of Figure 2. The points plotted on the right side of Figure 2 show the longitude variation of the emission centroid, which is offset from ecliptic latitude 0°. A cosine function was fitted through these points. The period of this function is not 360° because of the changes in the orientation of the satellite scanning direction during its flight; the IRAS survey commenced at ecliptic longitude 60°, but when it finished the location of the satellite within the zodiacal band had of course changed, resulting in changed emission characteristics. The approximately biweekly changes in the orientation of the orbit of the satellite with respect to the direction to the Sun caused a saw-tooth pattern to be superimposed on the zodiacal emission; although we do not include for this effect here, it is discussed in the paper by Jongeneelen and Deul (1985) giving details of the empirical model.

The upper part of Figure 3 shows the intensities of this empirical model after regridding back to galactic coordinates. The lower part of Figure 3 shows the observed 60μm emission after subtraction of the model. Most of the contamination by zodiacal dust is removed. Being quite hot, the zodiacal dust emits much less strongly at 100μm than at 60μm; we are currently using the 100μm data uncorrected in this regard.

The maps of the galactic dust emission resemble, at first glance, maps of integrated HI emission (see for example such maps in Burton and te Lintel Hekkert, 1985). Lines of sight through the inner Galaxy lying closer than, say, 10° to the galactic equator are

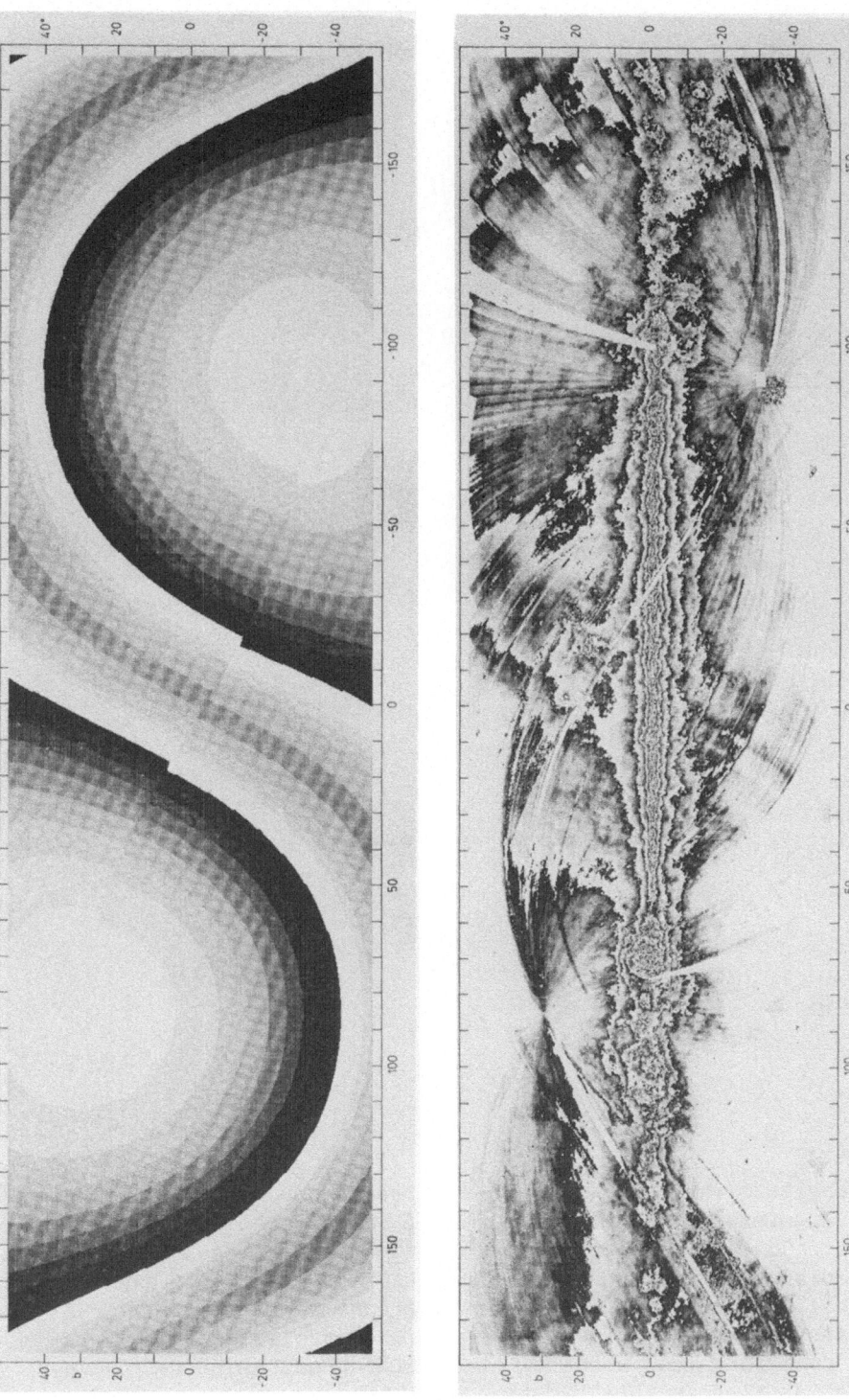

Figure 3. Model (upper) of the zodiacal dust emission at 60μm derived empirically from the IRAS measurements in that band. Galactic dust emission (lower) at 60μm after subtraction of the modelled zodiacal dust contamination. The representation of intensities is the same as in Figure 1.

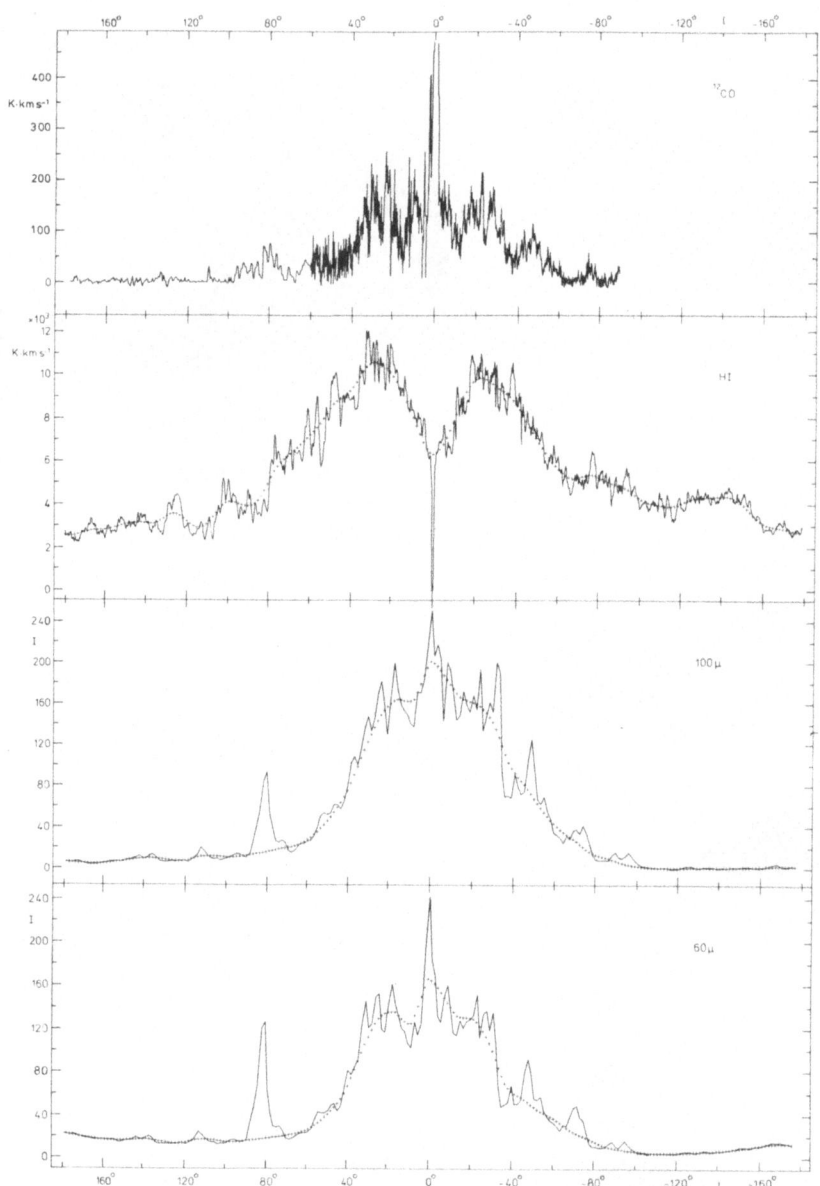

Figure 4. Longitude variation along the galactic equator of ^{12}CO and
HI integrated emission, and of the 60μm and 100μm intensities. The
units of the ^{12}CO data are K(of T_A^* x km s^{-1}; of the HI data, K(of T_b)
x km s^{-1}; and of the infrared intensities, 10^{-7} Wm^{-2}sr^{-1} (of the
Spline I data). The crosses follow the longitude variation after
removal of the emission (or in the case of HI, also possibly
absorption) from localized sources; the radial geometrical unfolding
algorithm was applied to this smoothed variation.

evidently sampling the collective properties of the dust. Like the HI, the dust is distributed so ubiquitously that separate features can not easily be isolated at low latitudes. In this regard the dust emission sampled at 60μm and 100μm does not resemble the cold clumps sampled in CO lines.

2.2. Longitudinal and radial distribution of the dust

Figures 1 and 3 also show that like the CO, but in this regard unlike the HI, the dust occurs sparsely in the 2nd and 3rd longitude quadrants representing R > 10 kpc. Figure 4 allows comparison of the longitude distributions along the galactic equator of CO (from Cohen and Bronfman, private communication), of HI (from Kerr et al., 1981 and Westerhout's 100-m data, private communication), and of dust. The distributions are consistent with the conclusion which holds generally: all tracers accessible on a galactic scale except HI show a morphological confinement to the inner Galaxy.

Conversion of the longitudinal distribution to the radial distribution of emissivities shown in Figure 5 was done using the kinematic information inherent in the spectral-line data, but using for the dust (for which such information is lacking) a geometrical unfolding process as follows. First the contribution from point sources was removed using the Kappa-Sigma clipping technique, resulting in maps of the diffuse infrared emission in both the 60μm and 100μm bands at angular resolution degraded to 1°5. The longitudinal variation of the emission at b=0° is plotted in the lower two panels of Figure 4. Emission regions localized in space, of which the local Cygnus complex near ℓ=80° is the most obvious example, would distort the radially unfolded distribution if not removed. The emission from identifiably local features was approximately accounted for by interpolation in the longitude profile, given the dotted-line variation shown in Figure 4.

Using the unfolding algorithm appropriate for continuum data described by Strong (1975), we determined the emission measure in a given galactocentric ring. The procedure started by setting the emission measure to zero in rings at R > 10 kpc. Moving from ℓ=90° to ℓ=0° in steps of a half degree, we solved for the emission measure, incorporating all previously derived values. Consequently the emissivity errors increase cumulatively with decreasing radius. The unfolding procedure was tested by carrying it out also on the velocity-integrated HI and CO data, to verify that the unfolded radial abundance distributions agreed with the ones derived using the kinematic information.

The radial emissivity plots of Figure 5 show that the dust emissivity at 60μm and 100μm is enhanced over the range 3 < R < 6 kpc, corresponding to the inner part of the molecular cloud annulus; it is relatively weaker near R ~ 7 kpc, where molecular clumps are still relatively plentiful. It is in the inner few kpc of the Galaxy that the dust emissivity differs most strongly with the gas emissivity. In this region, at least outside of the innermost few hundred pc, molecular clouds are scarce and HI emission is relatively

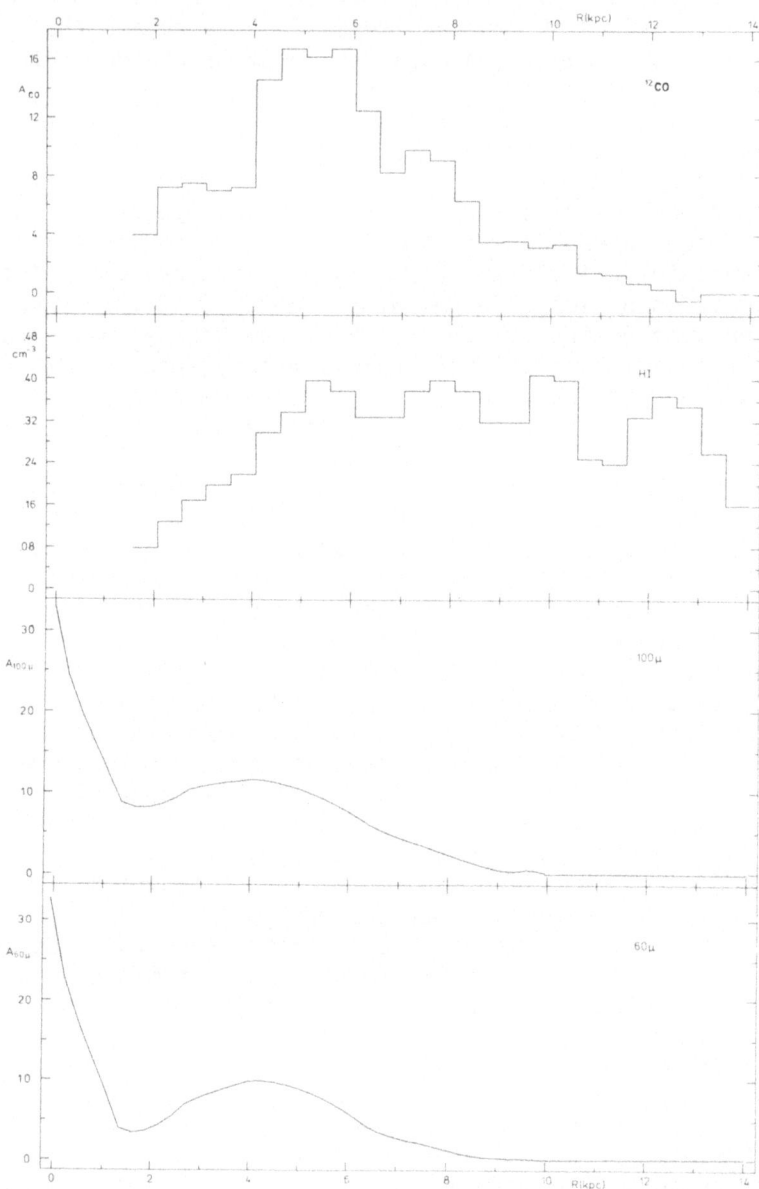

Figure 5. Radial variation of emissivities in the galactic equator
from ^{12}CO representing the cold, clumped component of the
interstellar gas; from HI, representing the cool, more diffuse,
component of the gas; and from dust in the 60µm and 100µm bands. The
^{12}CO and HI emissivities (from Burton and Gordon, 1978) incorporate
the kinematic information available for spectral-line data; the 60µm
and 100µm emissivities are based on an unfolding algorithm
appropriate to continuum data.

weak. There is generally a dearth of interstellar tracers in this region: hardly any HII regions are found there, the diffuse ionized hydrogen emission is very weak, and the gamma-ray emissivity is anomalously low (Blitz et al., 1985).

In each case it is important to decide which physical parameters are responsible for a given emissivity and for its variation.

Thus for the case of the CO data, uncertainties are introduced to the apparent radial abundance by uncertainties in what we know about possible radial gradients in metallicities, in the chemical composition of the interstellar gas, and in what we know about temperatures and densities in the molecular clouds. There is indeed some evidence that the cold-gas cloud column density is higher in the inner part of the molecular annulus. Liszt et al. (1984) showed that ^{13}CO, which effectively traces density, is more intense relative to ^{12}CO at R ~ 4 kpc than at ~ 8 kpc. Solomon et al. (1985) showed furthermore that the inner annulus contains more effective heat sources than the outer annulus. The narrowly confined radial distribution of diffuse HII emission (Lockman, 1976), which traces these heat sources, confirms this conclusion.

For the case of the HI data, uncertainties in the apparent radial abundance distribution are introduced primarily by optical depth effects. Toward the galactic center material on long path length tends to emit over a short velocity span; enough is known, however, about galactic kinematics so that a correction can be attempted. Localized self-absorption from the colder-than-typical residual HI in compressed, primarily molecular clumps, would diminish perceived intensities and would do so in a manner which reflects the clump morphology. We note that galaxies similar to ours show similar confined molecular annuli and extensive HI disks with a central hole.

For the case of the IRAS data, interpretation of the radial distribution of emissivities in terms of the distribution in the abundance of the dust is not so straightforward. Temperature is a particularly important parameter, and especially so in view of the limited spectral coverage of the IRAS detectors. Higher emissivities can indicate higher dust densities, but they can also indicate higher temperatures. A dust-to-gas ratio of the emissivities which are plotted in Figure 5 shows a smooth rise with decreasing distance, continuing into the center. Whether temperature or dust dominates this behaviour is an important distinction to make. Walterbos and Schwering (1985) stress that the enhanced infrared emissivity observed in the bulge region of M31 entails a relative minimum of dust density, if the enhancement is attributed to temperature effects alone. Dust on the other hand which might be well mixed with the cold gas in molecular clouds (at ~ 12 K) would emit quite weakly in the IRAS detector bands. Such dust, which certainly would play a crucial role in star-forming processes is not the dust to which the IRAS detectors are most sensitive.

2.3. Temperature of the dust dominating the 60μm and 100μm bands

A map representing the effective colour temperature of the dust dominating the 60μm and 100μm bands is given in Figure 6. The

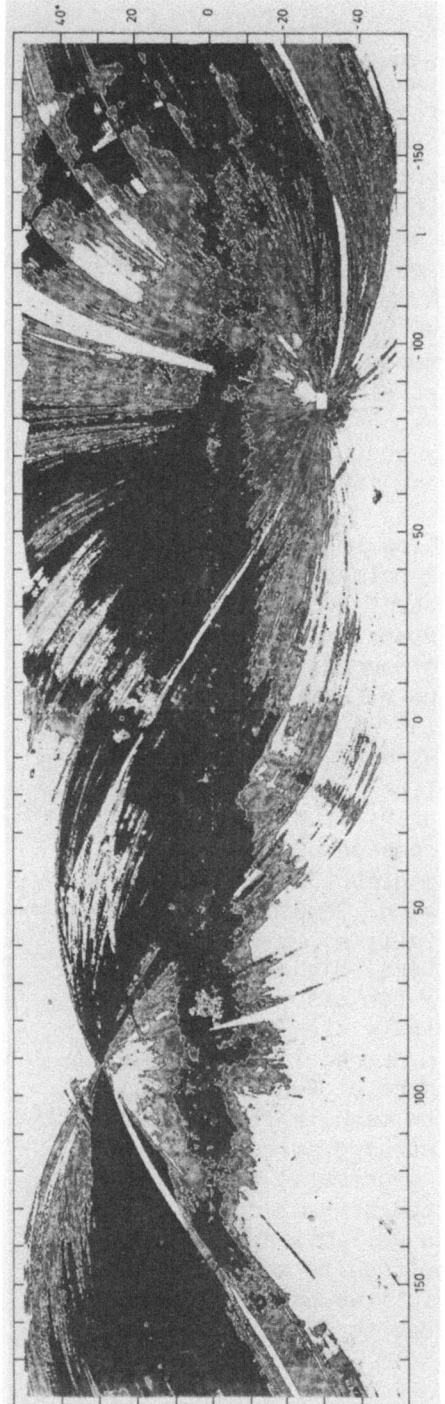

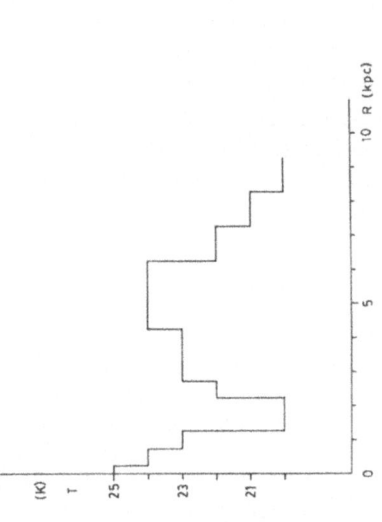

Figure 6. Distribution in galactic coordinates of the dust color temperature, derived from the ratio of IRAS 60μm and 100μm data using a λ⁻² dependence for the dust absorption efficiency. Gray scale levels are drawn at 1 K intervals starting at 15 K. Contour levels are at 5 K intervals, starting with a contour at 17 K, then at 20 K.

Figure 7. Galactic radial dependence at b=0° of the dust temperature from the 60μm/100μm ratio of intensities. There is uncertainty in the fluxes at the galactic center due to saturation effects.

temperatures follow from the ratio of intensities, assuming a λ^{-2} dependence for the dust absorption efficiency. For the temperatures characterizing the galactic dust emitting at these wavelengths, the exact form of the absorption efficiency does not sensitively affect the resulting temperature. Figure 6 shows a number of discrete sources of hot dust, strongly confined to the galactic equator and distributed like galactic HII regions; their distribution and number are quite like those of the subset of molecular clouds which are observed to have embedded heat sources. But most of the dust emission is characterized by a smooth temperature distribution. Evidently the dust is so ubiquitously spread that, like the HI, the hierarchy of structures - which no doubt exists - is so blended that a single, approximately harmonic mean, temperature suffices to describe the collective behaviour. This is quite a different situation than prevails for the cold discrete molecular clumps.

The radial distribution of the effective 60µm/100µm dust temperature plotted in Figure 7 results from a radial unfolding of the data in Figure 6 at b=0°, after removal of the heat-source and local spiral arm contribution. The temperature has a mean value of ~ 22 K, but it is somewhat higher where the warmer CO cloud cores are found, and in the central bulge region, where the illuminating radiation field is plausibly different from that in the Galaxy at large. The temperature within ~ 1 kpc from the center is uncertain because of saturation of the 100µm band. The mean value of the temperature is not very different from the mean value of 23 K found by Hauser et al. (1984), characterizing the diffuse dust emission at 250µm, nor from the value of 25 K found by Ryter and Puget (1977) from the far infrared emission.

2.4. Thickness of the dust layer

The thickness of the dust layer contributing the diffuse infrared emission has been determined in the following manner. We described the emission measure as decreasing exponentially with increasing z-height. Using the radial emission distributions as determined by the radial unfolding technique and plotted in Figure 5, emissivity profiles at constant longitude were calculated. By means of successive approximations we iteratively determined the scale length of the exponential z-height function. Examples of the fits of the calculated profiles to those observed are shown in Figure 8 for three representative longitudes. The situation observed at 60µm and 100µm can be simulated with dust in a layer of normal-distribution dispersion 120 pc. This thickness is equivalent to that of the pervasive HI gas but is substantially larger than that of the molecular layer. Figure 9 shows results of work indicating that the 60 pc thickness of the cold, compressed gas is about half that of the 120 pc thickness of the cool, more diffuse layer.

The large thickness of the dust layer is, to us, an unexpected result. It is important to be sure that no instrumental characteristic could influence this determination. Effects of hysteresis in the responses of the bolometers were considered. The

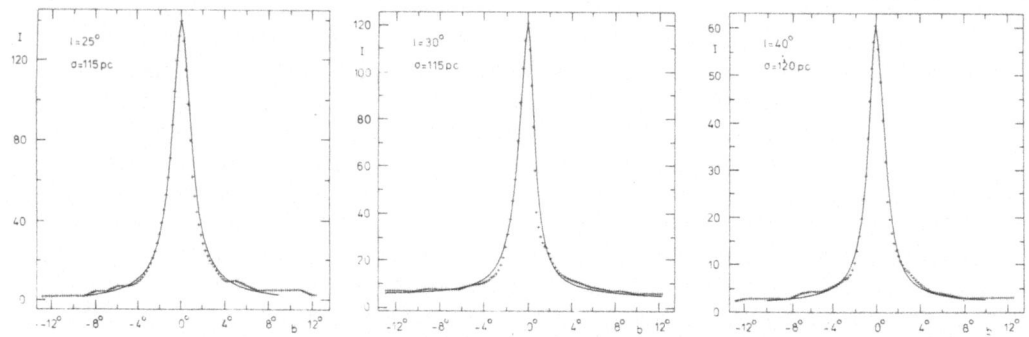

Figure 8. Sample latitude profiles of the dust emission at 100μm, at
the indicated longitudes. The lines represent curves of exponential
form, with the indicated scale lengths. The dust layer intensity vs.
latitude profiles can be well fitted, except at the low levels, by
such curves with scale heights 120 ± 5 pc. Evidently the dust
emitting at 100μm is distributed in a thicker layer than the cold
molecular cloud material.

IRAS Explanatory Supplement explains that effects of photon-induced
responsivity enhancement were tested in flight and later were also
studied using fluxes of intense point sources located near the
galactic equator. These tests showed flux deviations of order 15%,
with the effects predominating in the 12μm and 25μm bands. This
predominance can be clearly see by inspection of the photographic
prints of the Skyflux maps; on the 60μm and 100μm maps hysteresis
effects are much less obvious by such inspection, even for intense
point sources. We numerically estimated these effects by making use
of the Zodiacal Observation History File, which contains fluxes for
the four passbands recorded at 30 arcminute intervals and stored as
scans, with the information on direction of scanning across the sky
retained. We were thus able to compare fluxes from scans made a half
year apart so that the scan direction across the galactic dust layer
differs by approximately 180°; hysteresis effects should be
efficiently revealed by such a comparison. (We note that a problem
with the algorithm creating the Zodiacal Observatory History File
evidently introduced a lag of 30 arcminutes in the recorded scan
direction. We corrected for this lag before testing for hysteresis).
Hysteresis effects were detected in the intense 60μm and 100μm
emission from zodiacal dust. The flux differences from galactic
diffuse emission amount to only a few percent in these bands. These
differences result in errors in the z-height determination which are
less than the estimated statistical error of ~ 5 pc.
     We performed similar tests for hysteresis using the available
Hours Confirmation (HCON) Skyflux data files. Of these files HCON3 is
a first release and is less reliable than HCON1. Assuming, however,
that both HCON files are calibrated in a sufficiently similar way, we
compared the HCON1 and HCON3 Skyflux maps, which were made in
approximately opposite directions of satellite scanning. This test

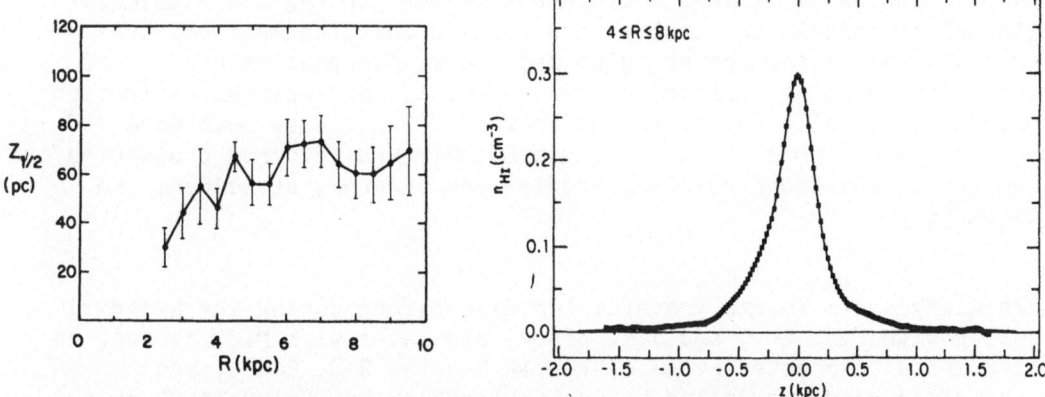

Figure 9. (left): Thickness (half-width at half-maximum intensity) of
the layer of CO emission as a function of galactocentric radius
(Sanders et al., 1984). (right): z-distribution of HI densities
characterizing the region 4 < R < 8 kpc (Lockman, 1984). The cold,
compressed gas has a layer thickness of about 60 pc; the cool, more
diffuse gas layer is twice as thick.

also showed flux differences limited to a few percent. We conclude
that the actual hysteresis effects are of negligible importance to
the thickness determination compared to the uncertainties in the
unfolding techniques. The verification process will be fully
described elsewhere (Deul and Walker, 1985).

3. SUMMARY REMARKS

We conclude that dust sampled at 60µm and 100µm is distributed more
like HI than like CO (except regarding the uniquely large radial
extent of HI). The 100µm dust fills a larger fraction of interstellar
space than the cold molecular clouds, and fills a thicker layer than
the cold cloud ensemble. As a consequence the derived physical
properties, most importantly dust density and temperature, refer to
collective properties along heavily blended lines of sight. At higher
latitudes individual structures may be seen. We believe that the
features called infrared cirrus are local examples of the features
heavily blended along transgalactic paths and dominating the 60µm and
100µm emission at lower latitudes, just as the filamentary HI
structures seen easily at higher latitudes (see e.g. Heiles and
Jenkins, 1976) are the structures dominating heavily blended HI
profiles at lower latitudes.
        Dust observed at wavelengths different than those of the IRAS
detectors may give different global properties, caused by a different
component of the dust. Hauser et al. (1984) concluded from data
at ~ 250µm gathered by a balloon-borne telescope that the emission in
the far infrared arises predominantly from dust that resides in
molecular clouds. Presumably the difference between their conclusion

and ours can be found in the different wavelength regimes sampled, although if this is the case we are puzzled that the mean colour temperatures are similar at 100μm and 250μm. Comparative studies of the regions seen as infrared cirrus in CO, HI, and optical extinction would be especially useful in this regard, as would be much more detailed comparisons of low-latitude CO, HI, and infrared properties than those reported here. Such studies are underway at Leiden, and elsewhere.

Acknowledgement: We are grateful for discussions during the Noordwijk meeting with F.J. Low, and J.R. Houck, and later with M.G. Hauser, on instrumental characteristics of the IRAS data. R.S. Cohen, and L. Bronfman kindly supplied before publication the Columbia CO data plotted in Figure 4.

References

Blitz, L., Bloemen, J.B.G.M., Hermsen, W., Bania, T.M.: 1985, Astron. Astrophys., 143, 267.

Burton, W.B., Gordon, M.A.: 1978, Astron. Astrophys., 63, 7.

Burton, W.B., te Lintel Hekkert, P.: 1985, Astron. Astrophys. Suppl., in press.

Dame, T.M., and Thaddeus, P.: 1985, Astrophys. J., submitted.

Deul, E.R., Walker, H.J.: 1985, Astron. Astrophys., in preparation.

Hauser, M.G., Silverberg, R.F., Stier, M.T., Kelsall, T., Gezari, D.Y., Dwek, E., Walser, D., Mather, J.C., Cheung, L.H.: 1984, Astrophys. J., 285, 74.

Heiles, C.: 1980, Astrophys. J., 235, 833.

Heiles, C., Jenkins, E.B.: 1976, Astron. Astrophys. Suppl., 46, 333.

Herzog, A.D., Illingworth, G.: 1977, Astrophys. J. Suppl., 33, 55.

Israel, F.P.: 1985, in 'New Aspects of Galaxy Photometry', J.-L. Nieto (ed.), Springer Verlag, p. 101.

Jongeneelen, A.A.W., Deul, E.R.: 1985, Astron. Astrophys., submitted.

Kerr, F.J., Bowers, P.F., Henderson, A.P.: 1981, Astron. Astrophys. Suppl., 44, 63.

Lequeux, J.: 1981, Comments on Astrophysics, 9, 117.

Liszt, H.S.: 1984, Comments on Astrophysics, 10, 137.

Liszt, H.S., Burton, W.B., Xiang, D.-L.: 1984, Astron. Astrophys., 140, 303.

Lockman, F.J.: 1976, Astrophys. J., 209, 429.

Lockman, F.J.: 1984, Astrophys. J., 283, 90.

Ryter, C.E., Puget, J.L.: 1977, Astrophys. J., 215, 775.

Sanders, D.B., Solomon, D.M., Scoville, N.Z.: 1984, Astrophys. J., 276, 182.

Solomon, P.M., Sanders, D.B., Scoville, N.Z.: 1985, Astrophys. J. Letters, 292, L19.

Strong, A.W.: 1975, J. Phys. A., 8, 617.

Walterbos, R.A.M., Schwering, P.: 1986, Proceedings First IRAS Symposium, F.P. Israel (ed.), this volume.

THE ASSOCIATION OF CLOUDS IN THE CARINA ARM WITH IRAS SPLINE MAPS

H.J. Walker, E.R. Deul, H.M. Butner, W.B. Burton
Sterrewacht, Postbus 9513, 2300 RA Leiden, The Netherlands

1. INTRODUCTION

The position of the emission peaks at 100 μm were compared with
molecular clouds, identified using CO and HII regions.Data used were
1. IRAS spline data. The 60 μm and 100μm spline data were used (see
Burton et al., 1986, for a description of the dataset) from
$280° \leq 1 \leq 330°$, $-15° \leq b \leq +15°$. The positions of the peaks in
the 100μm map were measured by hand. A 'temperature' map was produced
by subtracting a simple zodiacal light model (Burton et al., 1986)
from the 60μm map and dividing it by the 100μm map.
2. CO clouds Cohen et al. (1985), using the CO line at 2.6mm,
identified 37 molecular clouds, in the region $282° \leq 1 \leq 336°$,
$-5° \leq b \leq +5°$ belonging to the Carina spiral arm.
3. HII regions Positions were found from the optical surveys by
Shaver & Goss (1970) and Rodgers et al. (1960), from the 11cm survey
(covering $228° \leq 1 \leq 330°$, $-2° \leq b \leq +2°$) and the H109α survey at 6cm
from Wilson et al. (1970) (covering $280° \leq 1 \leq 345°$, $-4° \leq b \leq +4°$).
4. IRAS point sources Potential Pre-Main Sequence sources close to
the CO cloud positions were extracted from the IRAS catalog.

2. THE ANALYSIS

Following the technique of Myers et al. (1985) we compared the
positions of the sources common to the various datasets. Walker et
al. (1985a) shows the resulting histograms, with the conclusions
that, although the 100μm extended peaks are associated with the CO
peaks, the CO clouds associate more closely with the IRAS point
sources. The extended emission correlates better with the HII region
positions. In figure 1 the CO positions are plotted with
the 100μm spline data (units watts. m^{-2}. sr^{-1}). It shows that the CO
clouds are in the vicinity of the most intense 100μm emission.
        Figure 2 shows the 'temperature' map, using $\lambda^{-2}.B(\lambda)$,
where $B(\lambda)$ is the blackbody energy distribution. The gray scale
increases in intervals of 1°K from 17K (=0.19), and the counter
levels at 5°K intervals from 20K (=0.35). The scale is not
significantly changed if a model with the graphite and silicate
mixture from Draine & Lee (1984) is used. The galactic plane region

F. P. Israel (ed.), Light on Dark Matter, 373–374.
© *1986 by D. Reidel Publishing Company.*

has a very uniform temperature, except for isolated peaks. The
position of the H109α sources and some optical HII regions are
marked, to show they are the cause of the peaks.

ASSOCIATION OF CLOUDS IN THE CARINA ARM WITH IRAS DATA

REFERENCES
Burton, W.B., Deul, E.R., Walker, H.J. and Jongeneelen, 1986, This vol.
Cohen, R.S., Grabelsky, D.A., Alvarez, H., Bronfman, L., May, J.,
    Thaddeus, P., 1985. Astrophys. J. 290, L15.
Day, G.A., Thomas, B.M., Goss, W.M., 1969. Austr. J. Phys. Astrophys.
    Suppl. 11, 11.
Draine, B.T., Lee, H.M., 1984. Astrophys. J. 285, 89.
Myers, P.C., Dame, T.M., Thaddeus, P., Cohen, R.S., Siverberg, R.F.,
    Dwek, W., Hauser, M.G., 1985. Submitted to Astrophys. J.
Rodgers, A.W., Campbell, C.T., Whiteoak, J.B., 1960. Mon. Not. R.
    Astr. Soc. 121, 103.
Shaver, P.A., Goss, W.M., 1970. Austr. J. Phys. Astrophys. Suppl. 14,
    133.
Thomas, B.M., Day, G.A., 1969, Austr. J. Phys. Astrophys. Suppl. 11,
    3.
Walker, H.J., Deul, E.R., Butner, H.M., Burton W.B., 1985(a). Mitt.
    Astron. Ges. 63, 160.
Wilson, T.L., Mezger, P.G., Gardner, F.F., Milne, D.K., 1970. Astron.
    Astrophys. 6, 364.

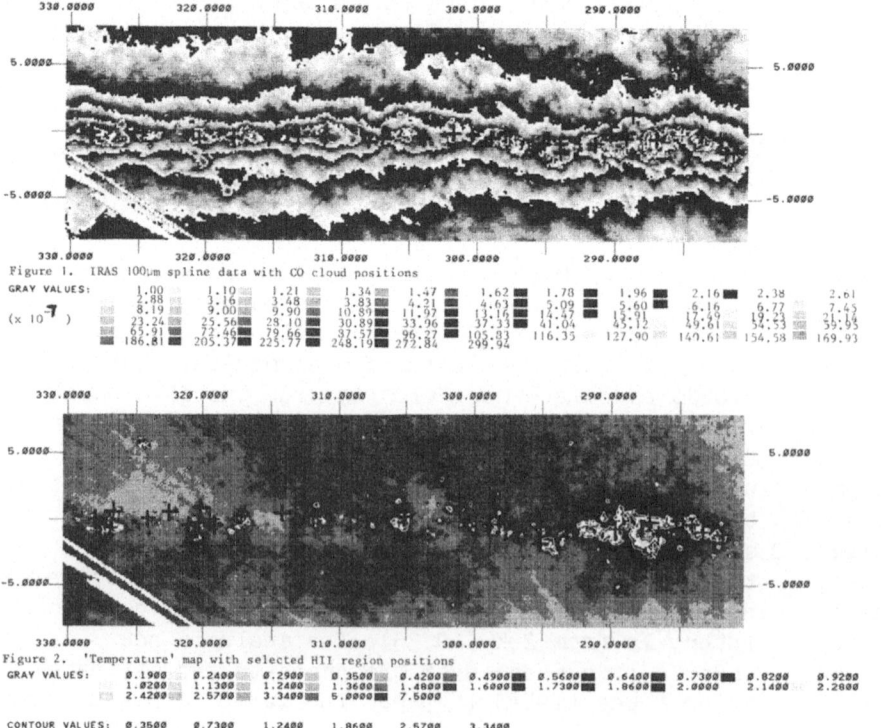

Figure 1. IRAS 100μm spline data with CO cloud positions

Figure 2. 'Temperature' map with selected HII region positions

DARK CLOUD STATISTICS

J.V. FEITZINGER, J.A. STÜWE
Astronomical Institute, Ruhr-University
P.O. Box 102148
D-4630 Bochum 1
F.R.G.

ABSTRACT. The dark cloud catalogues of Lynds (1962) and Feitzinger and Stüwe (1984) are combined. We deduce for the Northern (N), Anticenter (A) and Southern (S) sectors the obscuration percentages and the cloud distribution in galactic latitude.

STATISTICS

The average cloud area in the three sectors ($\Delta l=120^{\circ}$) is $F_{N,A,S} = 1.05$, 0.76, 0.58 sqdeg, respectively. The clouds of the Northern sky are twice as large than those of the Southern sky. Taking into account the distribution with respect to the opacity classes (Fig. 1), the above results are mainly influenced by the low opacity ($\leqslant 4$) clouds. These are 2.5 times smaller in the Southern than in the Northern sky. For the high opacity clouds (> 5) it is just the opposite. They are 1.6 times larger in the South. Generally the dark clouds of the Southern sky show more similarity to the clouds of the Anticenter region than to the clouds of the Northern part. The apparent size frequency distribution of the dark clouds for the three parts of the sky is presented in Fig.2. In Fig. 3 we show the percentage of obscured area as a function of the apparent distance from the galatic plane. The latitude distribution at the 5 % obscuration level in the Northern part is much broader ($\Delta|b_N| = 12^{\circ}$) than in the South ($\Delta|b_S| = 6^{\circ}$). The anticenter region shows for the opacity classes 2,3,4 no central peak ($\Delta|b_A| = 9^{\circ}$). The mean latitude width of the dark clouds for all longitudes is $\Delta|b_{tot}| = 10^{\circ}$. The different dark cloud distributions in the three sectors reflect the location of the sun at the outkirsts of the local spiral arm. (An extended version of this investigation will be published elsewhere.)

REFERENCES

Lynds B.T., 1962, Astrophys. Journal Suppl. 7, 1
Feitzinger J.V., Stüwe J.A., 1984, Astron. Astrophys. Suppl. 58, 365

F. P. Israel (ed.), Light on Dark Matter, 375–376.

Fig. 1   Area distribution
of the opacity
classes for the N,
A and S sector.

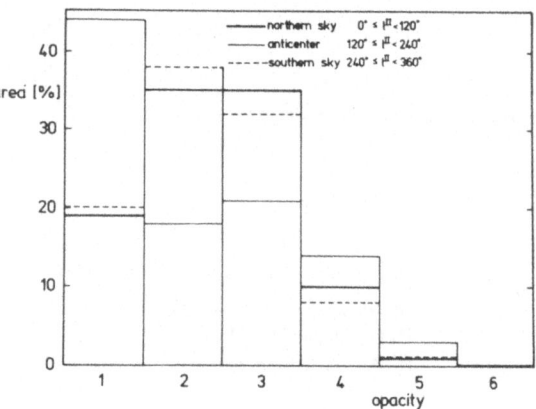

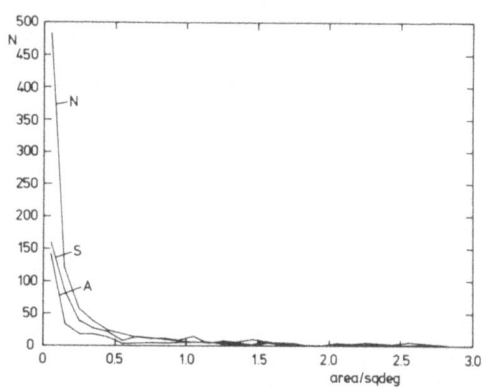

Fig. 2   Frequency function
of the dark cloud
areas for the N, A
and S sector.

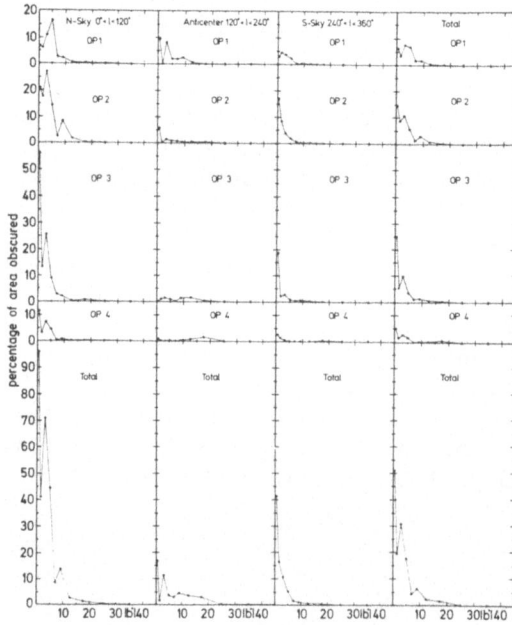

Fig. 3   Percentage of obscured area as function of galactic latitude.
"Total" means in the horizontal lines N+A+S and in the vertical
lines that all opacity classes are taken together.

HIGH LATITUDE MOLECULAR CLOUDS: COMPLETENESS OF THE SURVEY AND IMPLICATIONS FOR MOLECULAR SURVEYS

Loris Magnani and Leo Blitz
Astronomy Program, University of Maryland
Elizabeth Lada
Astronomy Department, University of Texas

ABSTRACT. The fractional completeness of the Blitz, Magnani, and Mundy (1984; hereafter Paper I) high latitude molecular cloud catalog is obtained and used to determine the contribution of the clouds to the local interstellar medium and to the Galaxy as a whole.

Initial results of a survey to search for high galactic latitude ($|b| > 20°$) molecular gas are presented in Paper I. Because the regions surveyed were chosen on the basis of extinction visible on the Palomar Sky Survey prints, the search catalog is incomplete. The results quoted in Paper I (number of clouds, surface filling fraction, and mass surface density) are expressed in terms of a quantity ε which denotes the fractional completeness of the survey. An unbiased survey was undertaken to determine ε and the ensuing high latitude molecular cloud properties.

The observations were carried out at the 5 meter telescope of the Millimeter Wave Observatory[1] near Fort Davis, Texas in 1984 June and November using the CO(1-0) line. A description of the observing procedure is included in Magnani, Blitz, and Mundy (1985). Two thousand-five hundred points were observed by stepping the telescope through 10x10cos(b) square degree grids in one degree intervals of ℓ and b. The positions of the grids were chosen to cover as much of the sky as possible and their locations are shown in Figure 1 of Magnani, Lada, and Blitz (1985).

Of the 1750 points surveyed in the northern galactic hemisphere, four lines of sight showed CO emission. In the southern galactic hemisphere CO was detected from five out of 750 lines of sight. The

---

1 The Millimeter Wave Observatory is operated by the Electrical Engineering Research Laboratory of the University of Texas at Austin with support from the National Science Foundation and McDonald Observatory.

F. P. Israel (ed.), Light on Dark Matter, 377–378.

distribution of the detections confirms the north-south asymmetry reported in Magnani, Blitz, and Mundy (1985) which was interpreted as a displacement of the Sun of 30 pc above the galactic midplane.

The results of the unbiased survey coupled with the calculations presented in Paper I lead to the following results:
1) The total number of individual clouds is 120^{+40}_{-30} (an individual cloud is defined as an individual entity with a closed contour of 0.5 K), and the surface filling fraction is 4.5×10^{-3}.
2) Within 100 pc of the Sun the high latitude molecular clouds should contribute $\sim 5 \times 10^3$ $M_\odot$ of material which results in a mass surface density of ~ 0.2 $M_\odot$ / pc^2. This value is $\sim 40\%$ of the surface density of GMCs in the local neighborhood (the local surface density of GMCs is calculated with the assumption that there is one GMC per OB association) but is only 10% of the total surface density of molecular gas at 10 kpc (Bloemen et al. 1985).

Since there is a clear relation between some of the IRAS 100μm high latitude cirrus (Low et al. 1984) and the high latitude molecular clouds (Hauser et al. 1985), the above results should be consistent with estimates based on the IRAS cirrus which contain molecular material.

If the high latitude molecular clouds are distributed uniformly throughout the Galaxy and if their distribution mimics the CO emissivity described by Scoville and Solomon (1975) and Burton and Gordon (1976), then the mean free path for the small clouds wil be $\sim 3-5$ kpc at the peak of the molecular ring and the integrated CO antenna temperature contribution from these objects will be ~ 7 K km-s^{-1}. The total number of clouds within 10 kpc of the Galactic center is 2×10^6 and the total mass (based on the average mass per cloud derived in Magnani, Blitz, and Mundy (1985) is $\sim 8 \times 10^7$ $M_\odot$. This value is $\sim 10\%$ of the value of 1.1×10^9 $M_\odot$ obtained by Bloemen et al. (1985) from gamma ray data. A small but substantial fraction of molecular material resides in these clouds.

REFERENCES

Blitz,L.,Magnani,L.,and Mundy,L. 1984, Ap.J.Letters,282,L9 (Paper I).
Bloemen,J.B.G.M. et al. 1985, Astron. and Ap.,submitted.
Burton,W.B. and Gordon,M.A. 1976,Ap.J.Letters,207,189.
Hauser,M. et al. 1985,in preparation.
Low,F.J. et al. 1984,Ap.J.Letters,278,L19.
Magnani,L.,Blitz,L.,and Mundy,L. 1985, Ap.J.,in press.
Magnani,L.,Lada,E.,and Blitz,L. 1985, Ap.J.,submitted.
Scoville,N.Z. and Solomon,P.M. 1975, Ap.J.Letters,199,L105.

DUST IN HIGH VELOCITY CLOUDS

B.P. Wakker
Kapteyn Laboratorium
Postbus 800
9700 AV Groningen
Netherlands

ABSTRACT. The facilities available in Groningen to produce IRAS maps were used to make comparatively high-quality maps of two fields where a high-velocity cloud (HVC) is observed in the 21-cm line. The HVC's are not discernable in these maps although one would expect to see them if the relation between 100 micron emission and HI column density found by Boulanger, Baud & van Albada (1985) is used. The implications of this non-detection are discussed in this contribution.

1. MAPS OF HVC-REGIONS

The IRAS satellite gives for the first time the opportunity to say something about the dust content of HVC's. Using the programs of the so-called Spline-I project that were developed in Groningen allows one to make high-quality maps of IRAS observations that have a signal to noise ratio sufficiently good to be able to detect a HVC as is shown by the following argument.

Empirically a relation has been found between the 100 micron emission and the HI column density (Boulanger, Baud & van Albada 1985). They found that the 100 micron emission scales linearly with the HI-emission, were the constant is 1.4E-8 if the 100 micron in-band flux is expressed in $W/m^2/sr$ and the HI column density in units E20 $/cm^2$.

The constant in this relation is found to vary across the sky, however. De Vries found a factor that was 3 times lower. The reason for this is probably that he used a field of high extinction were A_V was higher than 0.5 mag, while Boulanger et al. used a field with A_V less than 0.2 mag. The two HVC fields are in regions of low extinction, so the relation of Boulanger et al. is applicable.

Maps were made of the regions around the two HVC's called MI and AIV (see Giovanelli, Verschuur & Cram 1973). As the IRAS maps were made at a resolution of 8', they are directly comparable to the 10' maps of Giovanelli, Verschuur & Cram. The predicted peak intensities are of the order of 0.8 E-8 $W/m^2/sr$, which should be just visible as a small extended structure.

Because of the low level of the signal being looked for, the

F. P. Israel (ed.), Light on Dark Matter, 379–380.

problems with the calibration are important. Therefore a flat-fielding method developed by Boulanger & van Albada was used to improve the calibration and diminish the effect of "stripes". The zodiacal light was subtracted by fitting a cosec(beta)+constant law to the data.

2. RESULTS

Both HVC's are not detected on the maps made, although they should have been visible. There are three possible explanations for this:
1) the dust is very cool (T below 20 K);
2) there is less dust than in local HI clouds;
3) the content of large grains is relatively low, while that of small grains is high.
The first explanation seems less likely because Westerbork observations show that the neutral hydrogen is at a temperature of at least 25 K. This argument is only valid however if there are enough collisions between HI atoms and dust grains to establish an equilibrium.

The other two arguments have a higher probability of being true, but more study is needed to confirm them. In the case of possibility 2, there are two ways to get less dust: either the dust never formed or the dust grains have been destroyed by some means. The first of these is less likely because in two other HVC's one has detected Ca II absorption lines in a quasar spectrum (West et al. 1985 and Schwarz et al. 1985). Assuming that these results are representative for other HVC's, it means that the HVC material is not primordial. In that case there is no reason that dust has not been formed in a normal way. If then the dust was destroyed by some means, the non-detection of the two HVC's by IRAS argues in favor of an ejection model.

References

Boulanger F., Baud B., van Albada G.D., Astron. & Astroph., 144, L9, 1985
Giovanelli R., Verschuur G.L., Cram T.R., Astron. & Astroph. Suppl, 12, 209, 1973
Schwarz et al., in preparation, 1985
West K.A., Pettini M., Penston M.V., Blades J.C., Morton D.C., submitted to M.N.R.A.S., 1985

SECTION 6.

GALAXIES

IRAS OBSERVATIONS OF THE MAGELLANIC CLOUDS

F.P. Israel, P.B. Schwering
P.O. Box 9513
2300 RA Leiden

ABSTRACT: First results of the IRAS coverage of the Magellanic Clouds is briefly discussed in terms of the overall infrared properties of the Clouds.

1. INTRODUCTION

The Magellanic Clouds were extensively observed by IRAS, both as part of the all-sky survey, and as part of the additional observations (AO) program. AO's include observations made with the survey instrument and the the Dutch additional instrument (chopped photometric channel CPC and low resolution spectrograph LRS). The location of the Magellanic Clouds close to the southern ecliptic pole made it possible to obtain observations with different scan directions, including scan directions at right angles. As a consequence, maximum spectral resolution in two orthogonal directions could be obtained, unlike the situation for most other objects observed by IRAS.

Table 1 summarizes IRAS coverage of both the LMC and the SMC. For the meaning of abbreviations and a description of individual products we refer to the Explanatory Supplement (IRAS, 1985c) and the AO User's Guide (Young and Neugebauer, 1985).

Because of the large number of data, and because some questions of absolute calibration are still unresolved, we are still in the early stages of interpretation. For this reason, much of the following will be based upon a first analysis of the AO-DPM and SplineI maps of the LMC; it anticipates more detailed publications in preparation.

2. THE MAGELLANIC CLOUDS AND THE GALACTIC FOREGROUND

In Figure 1 we show an area of 50x50 degrees, starting at the Galactic Plane and extending southwards to include both Magellanic Clouds, constructed from SplineI maps. Figure 1 represents the

F. P. Israel (ed.), Light on Dark Matter, 383–396.

Table 1.

Product	#Observations LMC	SMC	Best Resolution	Remarks
Survey-skyflux	2(191,205)	2(202,203)	~6'	Field Numbers
SplineI	11(32,33,34,55, 56,57,58,84,85, 86,87)	6(30,28,29, 50,52)	9'	Field numbers
AO's-DPM*	68	41	1!2-4!4	
DPS	114	10		
DPS05	41	12		scans of 5° length
DSD	18	29	1!0-4!4	
DMS,DPP,DNS	56	20	-	
CPC	10	-	1!5	
LRS	~ 400	~ 80	-	poor quality
Point sources	~ 1200	~ 200	-	IRAS PS catalog

* Several overlapping fields were combined into full-coverage maps.

distribution of 100μm emission; the intensity coding is non-linear and is meant to highlight weak structures in the map.
At the top, the southern half of the Milky Way is visible; dust clouds associated with the Carina Arm are seen at top right. The LMC is just below the center, and only in the 30 Doradus region 100 micron intensities rise to the level seen in the Galactic Plane. At bottom left the SMC is seen at rather low intensity levels; its 100 micron surface brightness is generally lower even than that of the LMC. Note that the SMC is depicted upside down in Figure 1 as compared to conventional representations. Much of the upper right of the map is filled by weak emission structures associated with the foreground Milky Way, i.e. 'cirrus'. The radial striped pattern vaguely visible is due to imperfect removal of 'detector-striping'; it originates in the south ecliptic pole at the top right corner of the LMC. Even at the galactic latitude of the LMC (b $\approx$ -35°) long, thin streamers or filaments can be seen, notably two filaments originating from the lower left corner of the LMC. These features have weak optical counterparts, and have been interpreted as tidal arms having their true origin in the LMC (De Vaucouleurs, 1955, see Figure 2). However, careful comparison with southern hemisphere HI observations (Heiles and Cleary, 1979, Colomb, Pöppel, and Heiles, 1980) shows that they have precise HI counterparts at velocities corresponding to local Galactic material. In particular the striking feature extending downwards from the LML has a detailed HI counterpart at V=+13 km s^{-1} (see Cleary and Heiles, 1979). Likewise, the 'bridge' between the between the upper right corner of the LMC

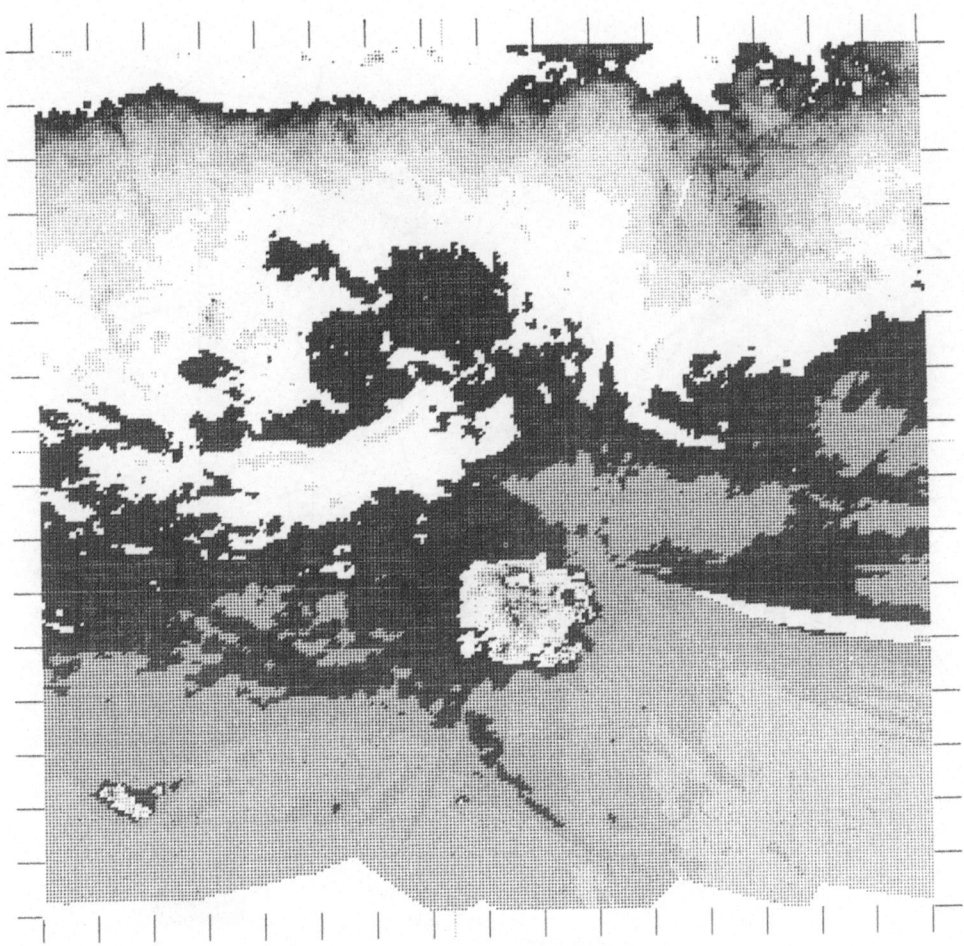

Figure 1. 50°x50° 100µm image showing the Galactic plane (top), LMC
          and SMC (center and bottom left). Virtually all extended
          emission is due to the Galactic Foreground, including the
          filement below the LMC.

and Galactic cirrus (NE of the LMC) is actually associated with
foreground material and therefore does not have anything to do with
the LMC. The pattern of small-scale structures down to the Spline-
IRAS resolution limit of 9 arcmin indicates considerable and rapid
variation of the galactic foreground extinction across the LMC; thus
use of an 'average' galactic foreground extinction of E(B-V)=0.07 mag
as is commonly used may lead to very misleading results when applied
to specific localities in the LMC. The foreground extinction may
actually vary from E(B-V)=0 to E(B-V)=0.15 on scales as small as 9

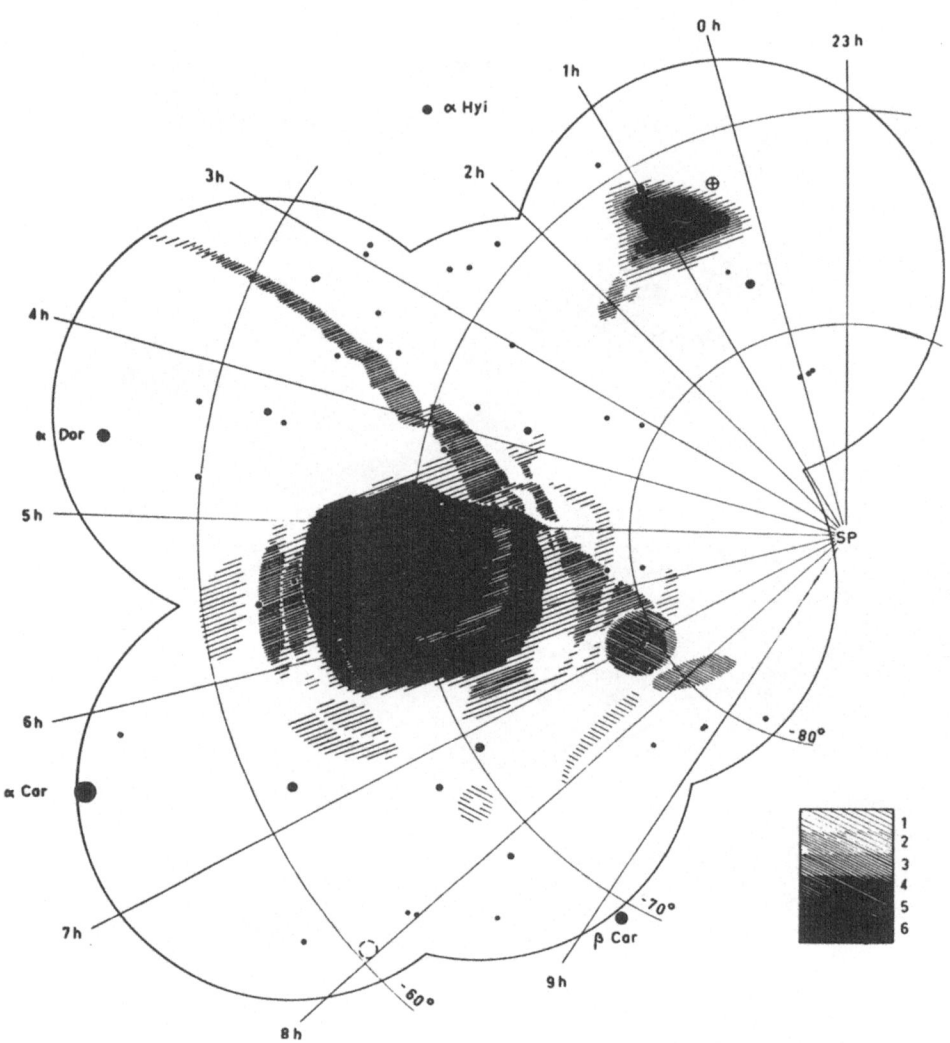

Figure 2. Sketch of optical emission in the LMC/SMC region (De
         Vaucouleurs, 1955).

arcmin, as compared to a peak extinction E(B-V)=0.40 (30 Doradus
region) in the LMC itself! This problem appears to be less severe in
the direction of the SMC (where usually an average extinction E(B-
V)=0.02 mag is assumed).

     Finally, we note that down to the sensitivity level in Figure 1,
there is no trace of an infrared counterpart of either the LMC-SMC HI
bridge or of the Magellanic Stream.

3. INFRARED IMAGES OF THE SMC.

In Figures 3 and 4 we show 25 micron and 100 micron (DMP) images of

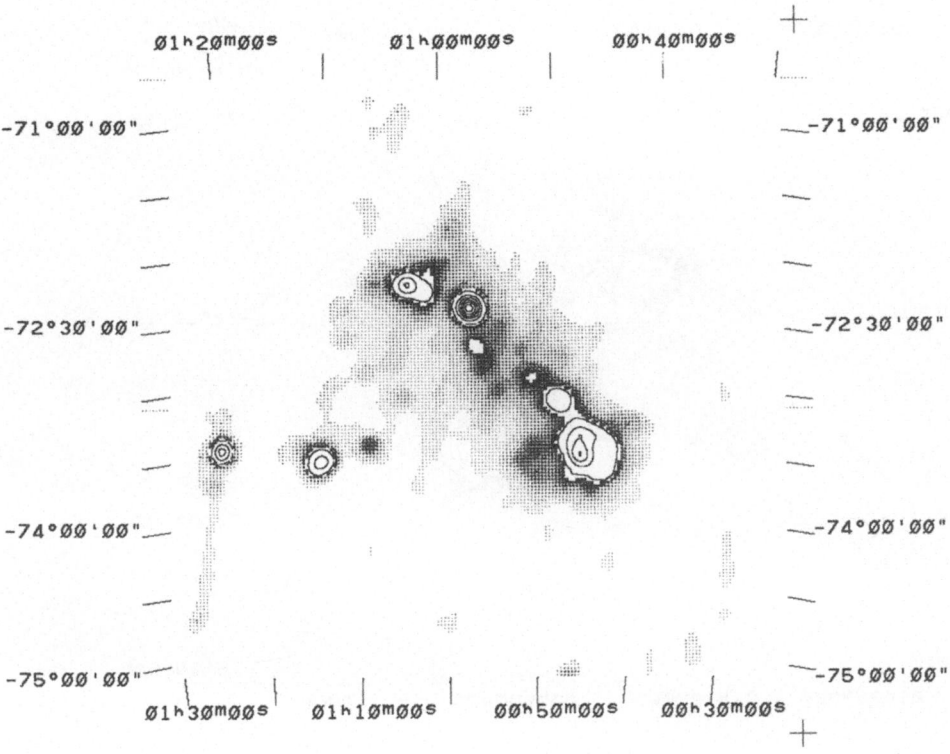

Figure 3. 25 micron map of SMC, resolution 8 arcmin.

the SMC. Some very weak galactic foreground emission is present,
mainly south of the SMC. The infrared images follow the optical
outline of the SMC closely. Most-intense emission (but still quite
weak by galactic standards) is found in the southwest bar; here
individual sources are confused, even at the highest (12μm) IRAS
resolutions. This is also the region where most dark clouds were
identified by Van de Bergh (1974) and Hodge (1974). Several authors
(Mathewson and Ford, 1984; Caldwell and Coulson, 1985) have presented
evidence that in this part of the SMC the line of sight is unusually
long, which would be a good explanation for the profileration of
sources seen in the infrared images. On the whole, the far IR images
closely resemble radio continuum maps of the SMC (Klein, priv.
communication). Major peaks in the remainder of the main body of the
SMC are associated with the bright HII region complexes N66 and N76
(Henize, 1956). The Shapley Wing is characterized by weak diffuse
emission, with a number of discrete sources most of which correspond
to HII region complexes (e.g. N81, N83, N88). This indicates the
presence of localized dust clouds not seen optically, most likely
because the stellar density in the wing is insufficient to show dust
cloud silhouettes. At this point we note that only weak CO emission

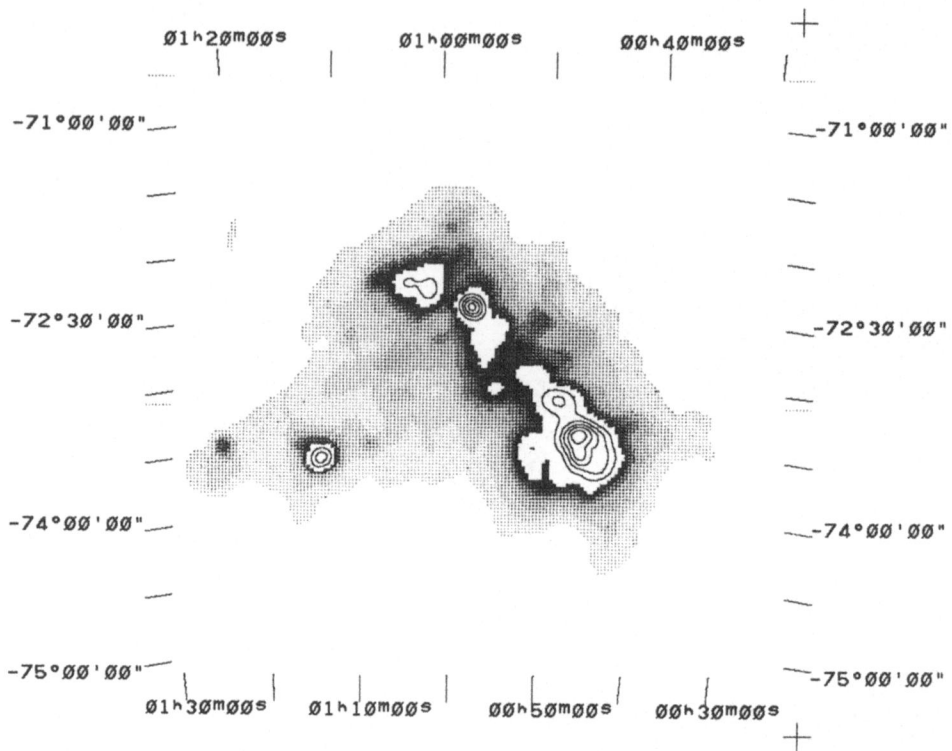

Figure 4. 100 micron map of SMC, resolution 8 arcmin

has been seen towards N81 (Israel et al., 1986), but that both N81
and N88 are associated with relatively strong shocked H_2 gas
(Koornneef and Israel, 1985; Israel and Koornneef, 1986; this
volume). It can thus be concluded that in the SMC, as in the Galaxy
and the LMC, star formation regions are associated with clouds of
dust and molecular gas. A more quantitative analysis awaits further
analysis (see Israel, 1984; Israel et al., 1986).

4. Infrared Images of the LMC.

In Figures 5 and 6 we show 25 micron and 100 micron (DPM) images of
the LMC. As in the case of the SMC, the 100 micron image in
particular strongly resembles optical images: virtually all infrared
sources correspond to HII region/dark cloud complexes (c.f. Henize,
1956; Hodge 1972; Van de Bergh 1974) The infrared peak in all
infrared maps coincides with 30 Doradus. The 30 Doradus region at all
wavelengths contains a significant fraction of the total flux. Other
notable infrared peaks coincide with the HII region complexes N11,

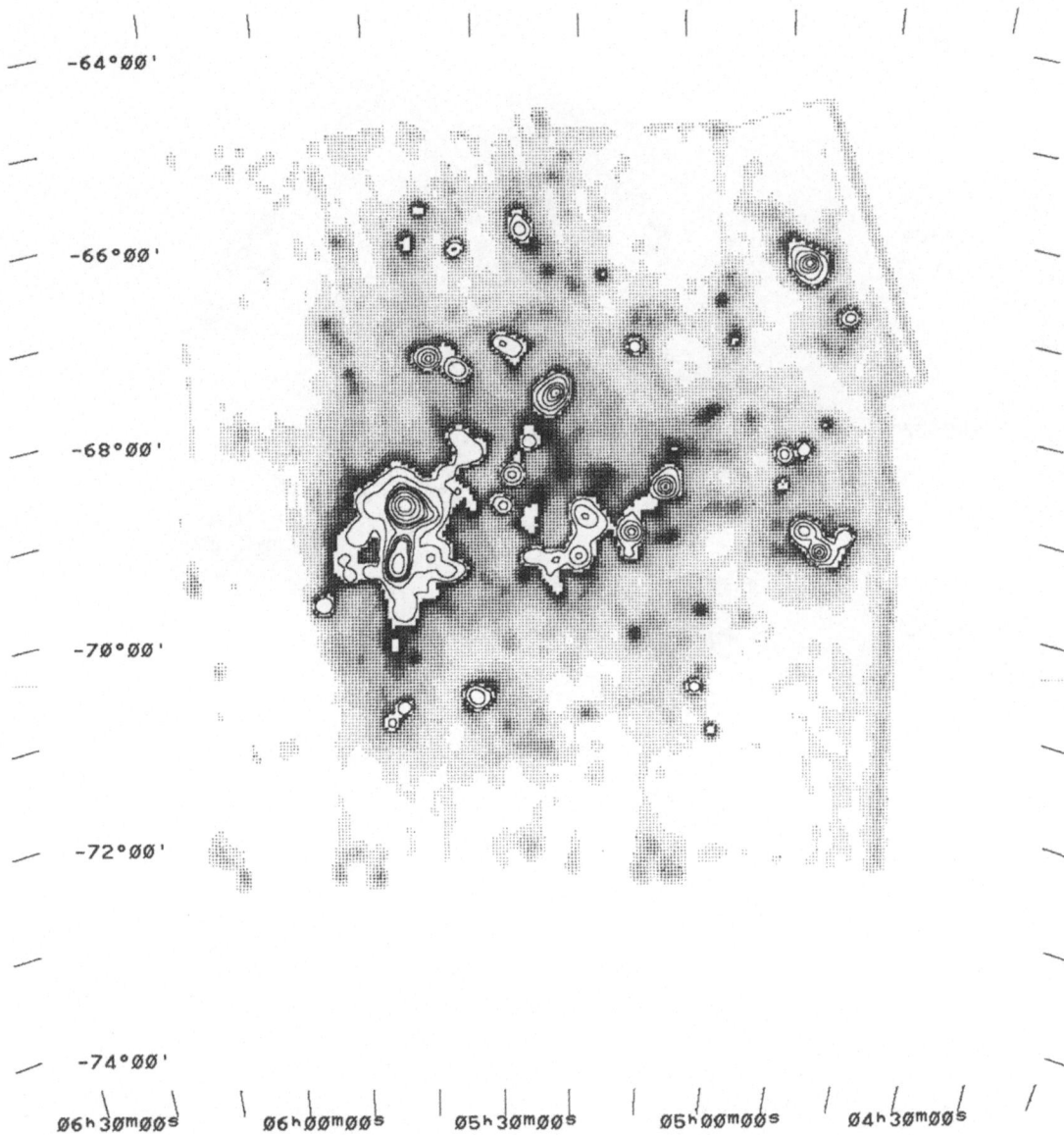

Figure 5. 25 micron map of LMC, resolution 8 arcmin.

N79 and N44. The LMC bar is less prominently visible, and mainly so
by the presence of discrete IR sources associated with known HII
regions or dust clouds. Again we note that CO and shocked H_2 has been
detected in the direction of several of the far-IR sources seen in
Figure 5 and 6 (Israel et al 1986; Israel and Koornneef, 1986; this
volume); again the far-infrared images strongly resemble radio
continuum maps of the LMC (Klein, private communication). Quite
remarkable is the 'hole' at the northern edge of the LMC, associated
with HII region complexes N48, N51, N59 and N55. This hole has

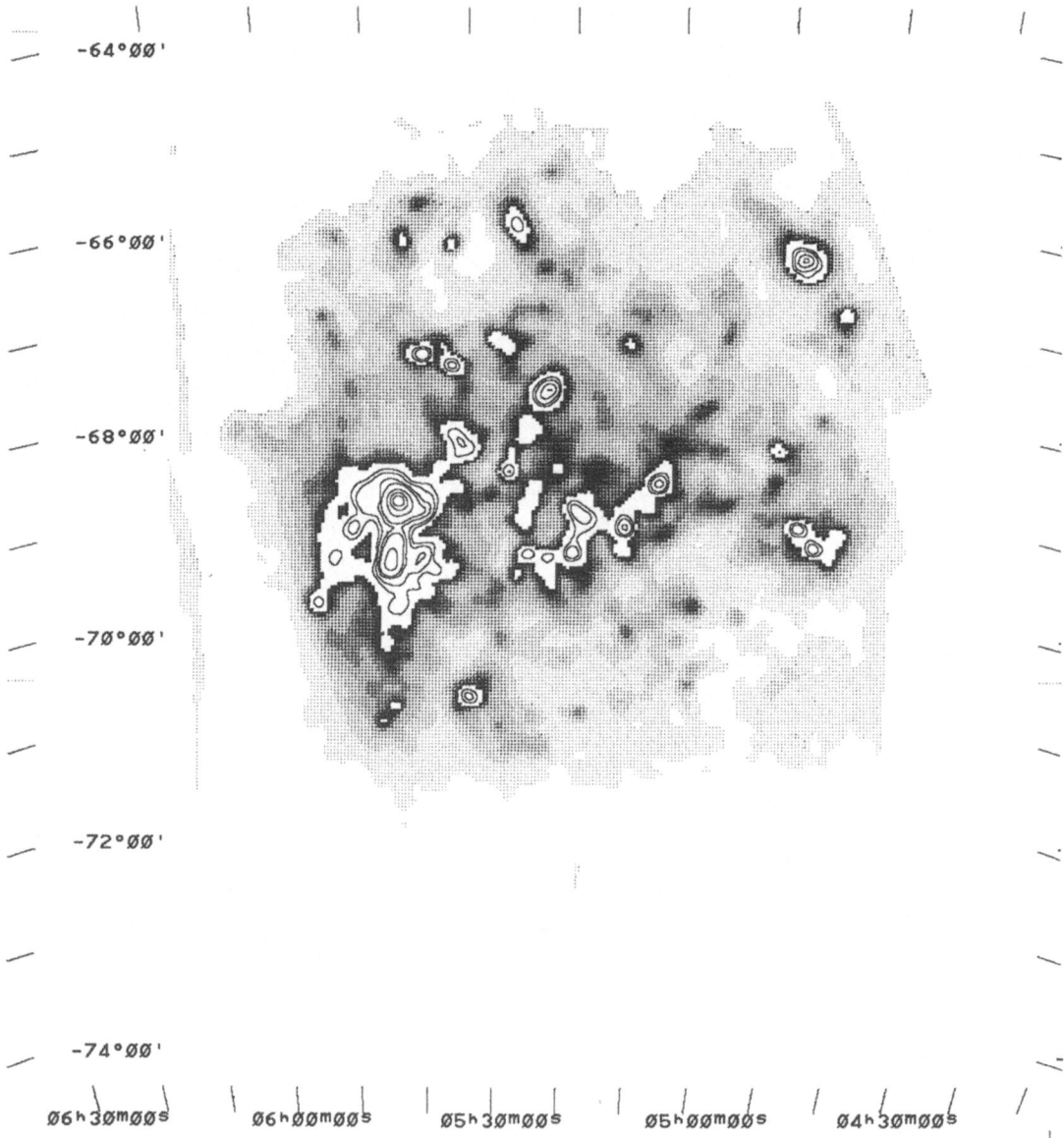

Figure 6. 100 micron map of LMC, resolution 8 arcmin.

already been noted in radio continuum and HI maps; it has been
suggested that it is a 'bubble' blown by a super-supernova or strong
stellar winds (see e.g. Westerlund and Mathewson, 1966, Meaburn,
1980). The most remarkable feature is the infrared source associated
with N55 inside the bubble. It has no HI counterpart. The presence of
hot dust and the absence of HI suggests that the neutral interstellar

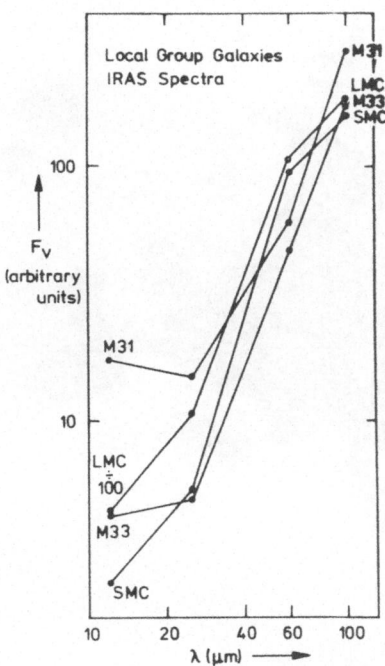

Figure 7. Integrated IRAS Spectra of LMC, SMC, M33 and M31

gas associated with this HII region is largely molecular. Thus, the N55 complex appears to represent star formation in an originally present dense clump passed and possibly compressed by the passing blast wave that formed the bubble. Two other features are noteworthy: the relatively sharp edge of the IR distribution east of the 30 Doradus region, and the faint mission extending southwards from the 30 Doradus/N159 region. It has been suggested that the 30 Doradus region represents a star formation sequence on a relatively large scale. (Israel, 1984; Jones et al, 1986) The sequence would have started with the now evolved complex of 30 Doradus; at present active star formation takes place south of it in N159. The presence of considerable amounts of dust and molecular material (Figures 5 and 6; Cohen et al, 1984) south of N159 idicates that the sequence has not yet reached its end. The location of the sharp edge (see als Figure 1) might be due to compression of material at the leading edge of the LMC, moving through e.g. Galactic halo material (c.f. Mathewson and Ford, 1984). This compression may have triggered the sequence of star formation we appear to be presently witnessing. The IR maps suggest that the dense material involved extends over two degrees, corresponding to about 2 kpc. If it could be shown conclusively that we are dealing with a coherent star formation sequence, this would be

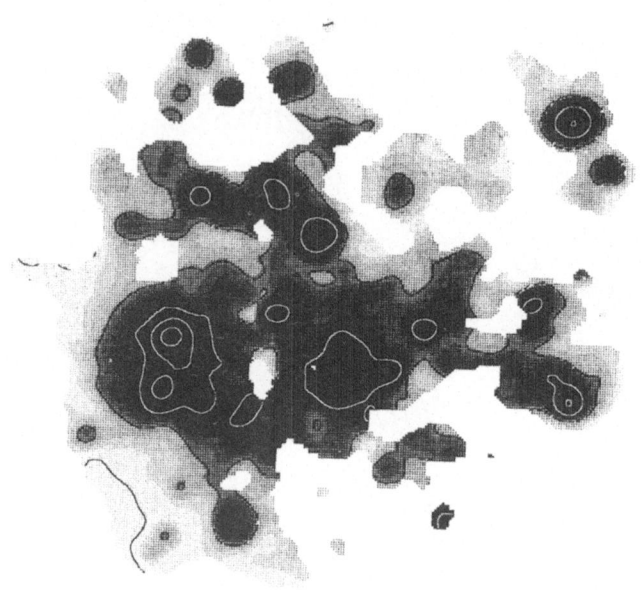

Figure 8. Temperature distribution over the LMC, hottest regions are
          depicted darkest. Resolution 15 arcmin

of great importance, since such large scale lengths are an essential
physical basis for otherwise mathematical theories of stochastic star
formation (see e.g. Matteucci and Chiosi, 1983; also Feitzinger,
1980).

5. LUMINOSITY, TEMPERATURE AND COLUMN DENSITY.

We have used the DPM maps of LMC and SMC to obtain more quantitative
information on dust conditions in the Magellanic Clouds. Figure 7
shows the integrated IR spectrum of LMC and SMC, together with these
of M31 and M33 (Walterbos and Schwering, this volume; Rice, private
communication). Because some problems of relative band calibration
still exist, we give intensities in arbitrary units only. The LMC and
SMC spectra are remarkably similar and characteristic of warm dust
clouds associated with HII regions. The M31 and M33 spectra are
cooler (i.e. they peak at longer wavelengths) and they show an excess
at 12μm, unlike LMC and SMC. In reality, the contrast at 12μm is
greater than depicted, because the 12μm fluxes of LMC and SMC are
contaminated by Galactic foreground stars for which no correction has
been applied. This effect is negligible for M31 and M33 because of

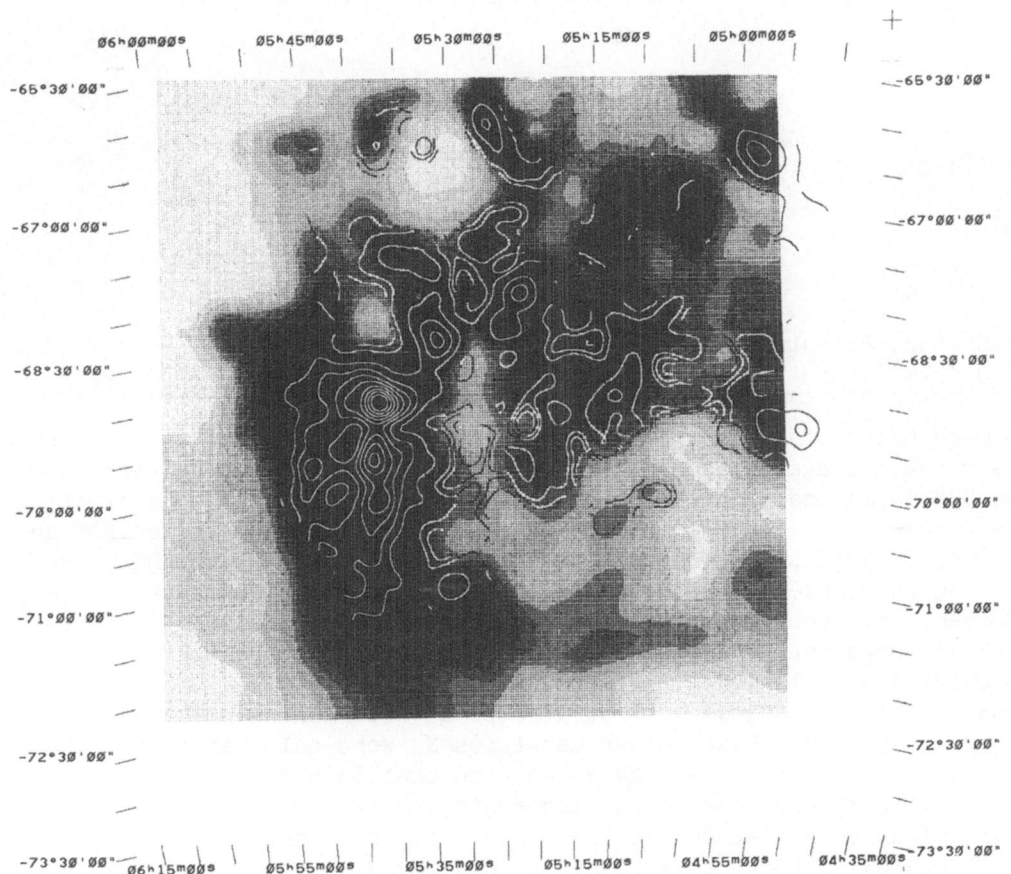

Figure 9. 'Dust Column Density' estimated map of the LMC. Resolution: 15 arcmin. Contours depict dust column density; grey scales integrated HI emission.

the much smaller angular extent of these galaxies.

Global physical parameters for the four galaxies are shown in Table 2. Dust temperatures were calculated for $Q(\lambda) \propto \lambda^{-1}$; luminosities were detemined from the observed 60 and 100 micron flux densities following the procedure outlined in chapter IV of Cataloged Galaxies and Quasars (IRAS, 1985b).

The higher dust temperatures for the Magellanic Clouds reflect the greater proportion of dust closely associated with HII regions.

Remarkable are the nearly identical IR luminosities of in particular the LMC and M33. Translating these luminosities into surface brightness, the earlier type spiral M31 is characterized by a three times lower value than the other galaxies.

This is, however, not so surprising in view of the fact that most FIR emission from M31 originates in only a fraction of the galactic disk (see Walterbos and Schwering, this volume) whereas the filling factor

Table 2   IR Parameters

	SMC	LMC	M31	M33
D (kpc)	63	53	720	720
F_{60}/F_{100}	0.58	0.60	0.21	0.28
$T_D(K)$	33	38	26	28
L (10^8 $L_\odot$)	0.6	5	9	6
IR Surface Brightness (10^{-7} $Wm^{-2}ster^{-1}$)	1.7	2.5	0.7	2.2

is much closer to unity for the other three galaxies. The temperature
distribution, again based on the 60 and 100 micron fluxdensities (in
the DPM maps) and again assuming $Q(\lambda) \alpha \lambda^{-1}$ is shown in more detail
(resolution 15') for the LMC in Figure 8. Temperature peaks of 40K up
to 50K (30 Doradus) are found for emission associated with the major
HII region complexes. Although locally dust temperatures peak on the
LMC-Bar, the structure of the Bar is not easily recognized. Thus, the
stellar component represented by the Bar does not appear to be
associated with large amounts of warm dust.
Finally, Figure 9 shows a first attempt at determining dust-to-gas
ratios in the LMC. Dust column densities N_d were calculated from the
100 micron DPM map, under the assumption that in a given line of
sight all dust radiates at the temperature shown in Figure 8. Thus,
dust column densities derived are lower limits, because the emission
from possibly present cooler dust is not included (i.e. actual dust
emissivities may be overestimated resulting in a column density
underestimate). Since Figures 8 and 9 show a reasonably good
correlation between temperature and density distributions, the major
effect of neglecting possible cool dust will be that the contrast
between high T_d/high N_d and low T_d/low N_d regions is underestimated.
For comparison, the integrated HI distribution from Rohlfs et al
(1984) is also shown in Figure 9 as a gray scale image. On the whole
the HI and dust distributions are in good agreement at least
indicating that dust enhancements and HI enhancements coincide. Quite
remarkable is the extent of relatively cold dust (cf Figure 8)
southwards of 30 Doradus/N159 (lower left of Figure 9). This dust
extension has a clear HI counterpart. Presumably, this is relatively
dense, cold material not yet engaged in star formation. It is
tempting to speculate (as mentioned earlier) that this material
represents the site of future star formation in a sequence extending
from 30 Doradus through N159 (present star formation site). In any
case, Figure 9 shows that, as far as dust content goes, the greater
30 Doradus region occupies a central place in the LMC. To put this
last statement in further perspective, we note that the greater 30
Doradus complex by itself has a FIR surface brightness three times
higher than e.g. the Bar, that about a third of the total radio flux

from the LMC originates in this complex, and that the FIR luminosity of the complex is more than half of the total LMC luminosity. Thus, the global characteristics of the LMC are significantly influenced by a phenomenon that has a typical lifetime of order 10^8 years.

6. CONCLUSION

The above represents only a first attempt at interpreting the richness of the IRAS database on the Magellanic Clouds. Further analysis will teach us more about the SMC - largely untouched in this paper - and detailed processes and structures associated with these dwarf companions of the Galaxy. The interpretation will be enriched by close comparizon with the wealth of e.g. optical data available on the Clouds. In particular, detailed properties of star forming complexes in the LMC and SMC are possible with the full resolution IRAS data and will be among the first studies carried out by us.

References:

Caldwell, J.A.R., Coulson, I.M., 1985, M.N.R.A.S. preprint.
Cohen, R., Montani, J., Rubio, M., 1984 in: 'Structure and Evolution
    of the Magellanic Clouds', IAU Symp. 108, Eds. S. van den Bergh
    and K.S. de Boer, Reidel, Dordrecht, p. 401.
Colomb, F.R., Pöppel, W.G.L., Heiles, C., 1980, Astr. Ap. Suppl. 40,
    47.
De Vaucouleurs, G., 1955, A. J. 60, 126.
Feitzinger, J.V., 1980, Sp. Sc. Rev. 27, 35.
Heiles, C., Cleary, M.N., 1979, Astr. J. Phys. Supp. No. 47.
Henize, K.G., 1956, Ap. J. Suppl. 2.
Hodge, P.W., 1972, P.A.S.P. 84, 365.
Hodge, P.W., 1974, Ap. J. 192, 21.
IRAS, 1985a, IRAS Explanatory Supplement, JISWG.
IRAS, 1985b, Cataloged Galaxies and Quasars observed in the IRAS
    Survey, JISWG.
Israel, F.P., 1984 in: 'Structure and Evolution of the Magellanic
    Clouds', IAU Symp. 108, Eds. S. van den Bergh and K.S. de Boer,
    Reidel, Dordrecht, p. 319.
Israel, F.P., De Graauw, Th., Van de Stadt, H., De Vries, C.P., 1986,
    Ap. J. April 1, in press.
Israel, F.P., Koornneef, J., 1986 preprint.
Jones, T.J., Hyland, A.R., Straw, S., Harvey, P.M., Wilking, B.A.,
    Joy, M., Gatley, I., Thomas, J.A., 1986 in press.
Koornneef, J., Israel, F.P., 1985, Ap. J. 291, 156.
Mathewson, D.S., Ford, V.L. in: 'Structure and Evolution of the
    Magellanic Clouds', IAU Symp. 108, Eds. S. van den Bergh and K.S.
    de Boer, Reidel, Dordrecht, p. 125.
Matteucci, F., Chiosi, C., 1983, Astr. Ap. 123, 121.
Meaburn, J., 1980,M.N.R.A.S. 192, 365.

Rohlfs, K., Kreitschmann, J., Siegman, B., Feitzinger, J.V., 1984,
    Astr. Ap. 137, 343.
Van den Bergh, S., 1974, Ap. J. 193, 63
Westerkind, B.E., Mathewson, D.S., 1966, M.N.R.A.S. 131, 371.
Young, E.T., Neugebauer, G., 1985 'A User's Guide to IRAS Additional
    Observations', IPAC, CIT (Pasadena).

COLLISIONALLY HEATED DUST IN LMC SUPERNOVA REMNANTS

J.R. Graham[1], W.P.S. Meikle[1], A. Evans[2], M.F. Bode[3], J.S. Albinson[2]

(1) Astronomy Group, Imperial College, London, SW7 2BZ.
(2) Dept. of Physics, University of Keele, Staffs., ST5 5BG.
(3) Dept. of Astronomy, University of Manchester, M13 9PL.

ABSTRACT. Preliminary IRAS additional observations (AOs) are presented, which show that luminous ($\sim 10^5$ $L_\odot$) infrared sources are associated with supernova remnants. Comparison of the infrared and X-ray data show that the infrared emission of these sources is due to collisionally heated dust.

1. INTRODUCTION

It has been predicted for some time that dust grains may be an important coolant of astrophysical plasmas because dust grains embedded in hot X-ray emitting gas are heated by collisions with electrons and ions. A supernova remnant (SNR) should constitute an ideal laboratory where grain cooling by this mechanism can be investigated[1]. LMC SNRs are particularly suitable for such an investigation with IRAS as they are (i) at a known distance (46 kpc); (ii) less likely to suffer confusion than Galactic SNRs; (iii) essentially point sources; (iv) well studied, those in our survey being triply confirmed at radio, optical and X-ray wavelengths. This note gives a preliminary report of our survey of LMC SNRs.

2. RESULTS

The observations consisted of co-added survey-like scans. Of the seven SNRs for which data have been received, four show unambiguous (> 10σ) detections in at least two bands. In each case the position of the infrared emission coincides with the X-ray centroid. From the source density in our AOs we estimate that the probability of a chance coincidence of a source being associated with a SNR is ~ 0.002.

Infrared luminosities were derived by fitting a blackbody to the fluxes in bands III and IV and integrating over all wavelengths; the results are given in Table 1.

F. P. Israel (ed.), Light on Dark Matter, 397–398.
© *1986 by D. Reidel Publishing Company.*

3. DISCUSSION

From the infrared luminosities of Table 1, and an assumed dust opacity, we can compute the mass M_d of radiating dust; see Table 1. The mass M_x

TABLE 1

SNR	T(K)	$L_d(10^5 L_\odot)$	$M_d(M_\odot)$	M_d/M_x	L_d/L_x
N49	60	0.80	0.28	0.007	23.7
N49B	28	0.17	0.98	0.005	10.9
N63A	40	1.55	2.2	0.02	16.2
N186D	33	0.81	2.23	0.16	2320

of hot ($\sim 10^6$ K) gas is calculated from known X-ray luminosities[2]. If the dust is collisionally heated by the X-ray emitting gas, the ratio $M_d/M_x \sim 0.01$ and it is predicted[3] that $L_d/L_x \sim 30$. We see from Table 1 that, with the exception of SNR N186D, this is indeed the case. Clearly the infrared emission of collisionally heated dust in SNRs is a significant drain on their energy budget. Hitherto treatments[4] of SNR evolution have neglected this crucial component; the present work demonstrates that this assumption cannot be justified.

The case of N186D is clearly different. It's infrared properties seem to imply that a large fraction of its infrared luminosity arises from dust embedded in cool gas that was not detected by the Einstein observatory. Obviously an additional source of energy is indicated and the most obvious mechanism is radiative heating. Catalogues of LMC early-type stars[5] contain several possible candidates in and around N186D. In particular we have identified a star that is coincident with an IRAS band I source which falls within the optical/X-ray limits of N186D (no similar stars were identified within the confines of the other remnants).

4. CONCLUSIONS

These data confirm the prediction[1] that collisional heating of dust is ~ 10 times more efficient a coolant than atomic processes and must have a considerable effect on the evolution of SNRs. A more detailed discussion of this work will be presented elsewhere.

References

1. Draine, B.T., Ap.J., 245, 880 (1981).
2. Long, K.S., Helfand, D.J., Grabelsky, D.A., Ap.J., 248, 925, (1981).
3. Graham, J.R., PhD thesis, University of London (1985).
4. Raymond, J.C., Ann. Rev. Astr. Astrophys., 22, 75 (1984).
5. Sanduleak, N., Contributions Cerro Tololo InterAmerican
   Observatory # 89 (1970).

GROUNDBASED INFRARED OBSERVATIONS OF MAGELLANIC CLOUD HII REGIONS

J. Koornneef, ESA, Space Telescope Science Institute,
Baltimore, USA
F.P. Israel, Sterrewacht Leiden, The Netherlands

We have obtained near-infrared JHKL'(M) photmetry of some fifteen
small ($\leq$ 10") HII regions and HII regions knots in both the LMC and
the SMC. K magnitudes of the observed objects typically range from 11
to 14 mag. For several of the brightest sources (K $\leq$ 12 mag) we also
obtained CVF spectrophotometry in the 2.2μm (and in a few cases the
3.7μm) window. All observations were made with the ESO 3.6 m
telescope and an InSb detector at La Silla, Chile. Results on the SMC
HII region N81, have been published (Koornneef and Israel, 1985). For
this source, and for the others (Israel and Koornneef, 1986) the K
window CVF spectra are similar to those obtained by other authors for
Galactic HII regions such as G333.6-0.2 and Orion A.

A major result of the present study is the detection of
significant amounts of hot molecular hydrogen through its (1-0)Q
branch emission at ~ λ2.4 μm. Observed Q branch luminosities range
from 5 to 25 L_O. The high Q branch/S1 line ratios observed indicate
typical H_2 extinctions of order $A_{2.4μm}$ = 0.5-6 mag, well in excess of
extinctions derived for the associated ionized gas (typically A_V = 0-
4 mag). This indicates that the molecular material is located mostly
behind the associated HII regions, in agreement with their optical
appearance. The best estimates for the ratio of extinction corrected
total hydrogen luminosities to total available stellar luminosity is
of order $L(H_2)$/L star = 10^{-4} - 10^{-3}. Although these numbers suffer
from considerable uncercainty they are consistent with the notion
that the observed H_2 is shock-excited by the interaction of stellar
winds with an ambient cool molecular cloud complex: L(wind)/L(star)
= 2.5 10^{-3} generally for OB stars. Taking into account that only a
small fraction of all (cool) H_2 in each complex is expected to be
(shock)excited, the high H_2 detection statistics show that both the
LMC should contain considerable amounts of molecular hydrogen.

The observed near-infrared broad band flux densities can be
understood as originating from a combination of
- a (reddened) hot stellar contribution (Rayleigh-Jeans tail) mainly
  in J and H
- nebular continuum (free-free and 2 mg)

F. P. Israel (ed.), Light on Dark Matter, 399–400.
© *1986 by D. Reidel Publishing Company.*

- nebular recombination lines (e.g. Pα in J, BY in K and Bα in L')
- relatively hot dust (T_{dust} up to 1000 K), mainly in L' and M.

The IRAS observations indicate the presence of large amounts of additional cool dust with color temperatures decreasing to longer wavelengths (typically T_c(60-100μm) ≈ 50 K). By way of example we show the results obtained on the SMC HII region N88 in Figures 1 and 2.
A full analysis of both the spectroscopic and photometric samples is in preparation.

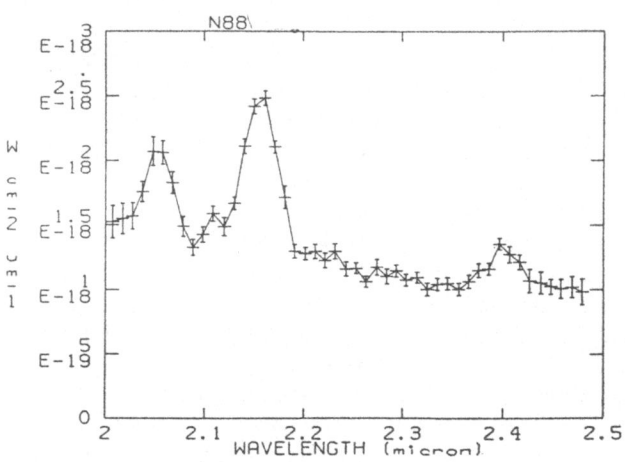

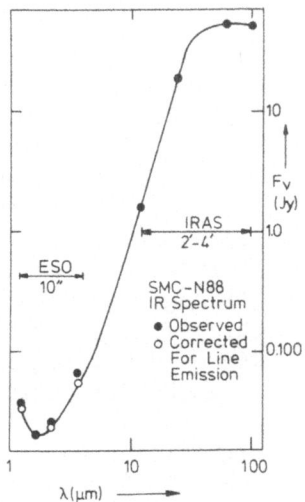

Figure 1. K-band CVF Spectrum of N88 in a 10" aperture, showing emission due to HeI (2.06μn), B_γ (2,17μm) and shocked H_2 (2.12, 2.4 μm). Spectral resolution $\lambda/\Delta\lambda = 50$.

Figure 2. IR broadband spectrum of N88; composite of 10" groundbased observations and spaceborne IRAS measurements with a larger aperture (point-source flux densities, colour-corrected).

References

Koornneef, J., Israel, F.P., 1985, Ap. J., 192, 156.
Israel, F.P., Koornneef, J., 1986, submitted.

DUST IN M31 - OBSERVATIONS IN THE IR WITH IRAS

R.A.M. Walterbos and P.B.W. Schwering
Sterrewacht, Leiden, The Netherlands

We have analysed IRAS observations of M31 at 12, 25, 60, and 100 micron. The data were obtained as part of the 'Additional Observations' Program. Two large fields together covering M31 were mapped with the survey instrument in all four wavelength bands. Preliminary results of these data have been presented by Habing et al. (1984). Additional higher resolution data have been obtained at 50 and 100 micron with the CPC for several small fields in the nuclear region and the NE arm.

To be able to compare the emission in the four bands, we have used special software developed by R. Braun (1985) to obtain maps at the same angular resolution with a well defined, Gaussian shaped beam. Figures 1a and 1b show the 12 and 60 micron maps obtained this way. The emission at 12 micron is remarkably similar to the 60 micron map. The same is true for the maps at 25 and 100 micron. The IR emission closely follows the optical dust lanes, HI, and other Pop I material. This means that even at 12 and 25 micron the main contribution comes from extended dust and not stars with dust shells, unless these stars would be distributed exactly as the dust lanes which is very unlikely.

We have tried to separate the various components contributing to the IR emission, using the ratios of the surface brightnesses in the various bands as a diagnostic. The bulk of the IR emission from the disk (>80%) can be described by a single spectral component with a 60/100 micron ratio indicating a colour temperature of about 21-23 K (lambda -2 emissivity). The relatively low temperature indicates that most of the IR emission originates from dust associated with diffuse HI (and perhaps molecular) gas, but not from HII regions. This is in agreement with radio continuum observations (Walterbos et al., 1985) which have shown that the HII regions in M31 are weak, implying a mild star formation activity. Model calculations by Draine and Anderson (1985) show that heating by a general interstellar radiation field of comparable intensity as in the solar neighborhood can account for the dust temperature. The spectrum of the main component is indicated in figure 2. There is a large excess at 12 and 25 micron compared to the emission expected from standard models. The excess is similar to that observed in

F. P. Israel (ed.), Light on Dark Matter, 401–404.
© *1986 by D. Reidel Publishing Company.*

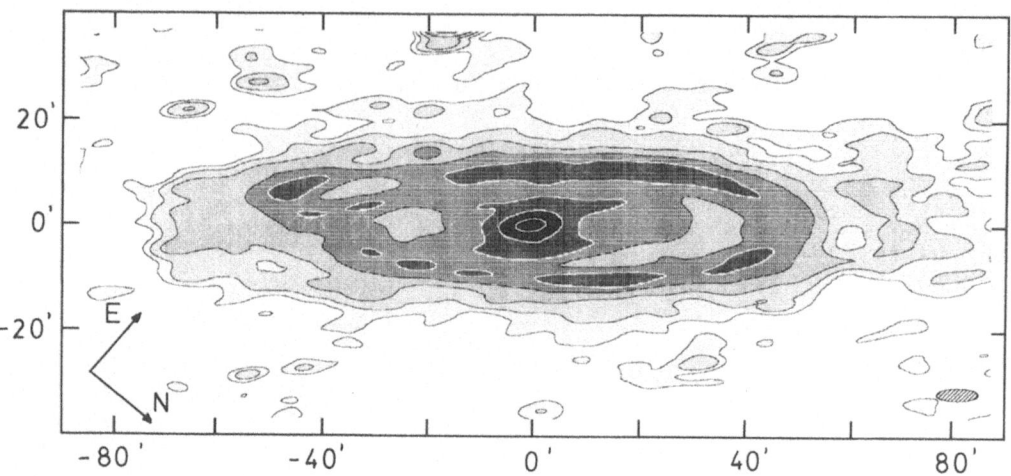

Figure 1a. 12 Micron map of M31 at a resolution of 2!5 x 7'. Contourlevels are at 0.65, 1.3, 2.6, 5.2, 10.4, 20.8, and 42 units, one unit being 10^{-8} $Wm^{-2}sterad^{-1}$ (no color corrections applied).

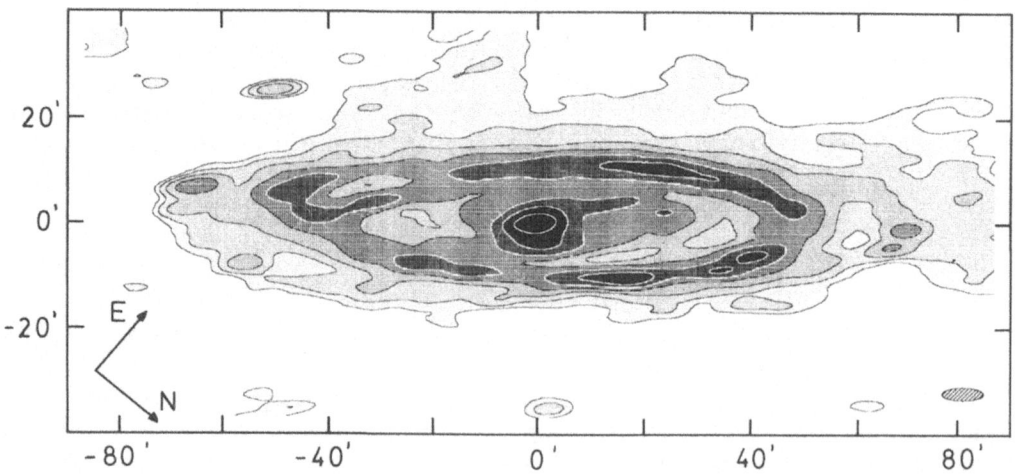

Figure 1b. 60 Micron map of M31 with the same resolution as the 12 micron map. Contourlevels are at 0.42, 0.85, 1.7, 3.4, 6.8, 13.2, 26.4, and 52.8 units, one unit is 10^{-8} $Wm^{-2}sterad^{-1}$.
Note the striking similarity with figure 1a.

some isolated clouds in our Galaxy (Boulanger et al., 1985; De Vries, 1985). It can be interpreted as evidence for very small dust grains transiently heated to high temperatures (e.g. Leger and Puget, 1984). The nucleus and some regions in the main ring of emission which coincide with complexes of HII regions are warmer. A detailed analysis shows that

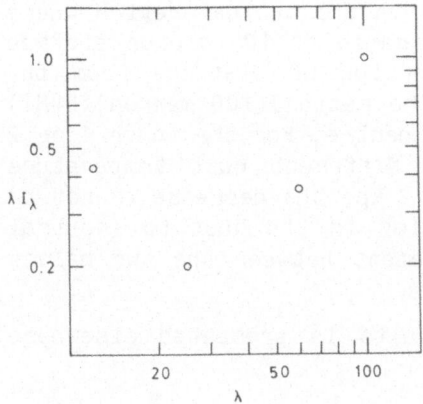

Figure 2. Spectrum of the main component of the infrared emission from M31 (wavelength in microns, brightness in arbitrary units).

in the warmer regions in the ring the excess at 12 and 25 microns is smaller than in the cooler dust (see also figure 1). The small grains are probably destroyed in the warmer regions by collisions with thermal atoms and ions. In the center the situation is more complicated, because there we perhaps do see a contribution from stellar photospheres and dust shells at the short wavelengths; the ratio 12 to 100 micron emission is even larger here than for the cool dust.

We have studied the relation between the infrared emission and the HI gas using the HI survey by Brinks and Shane (1984). There is a tight

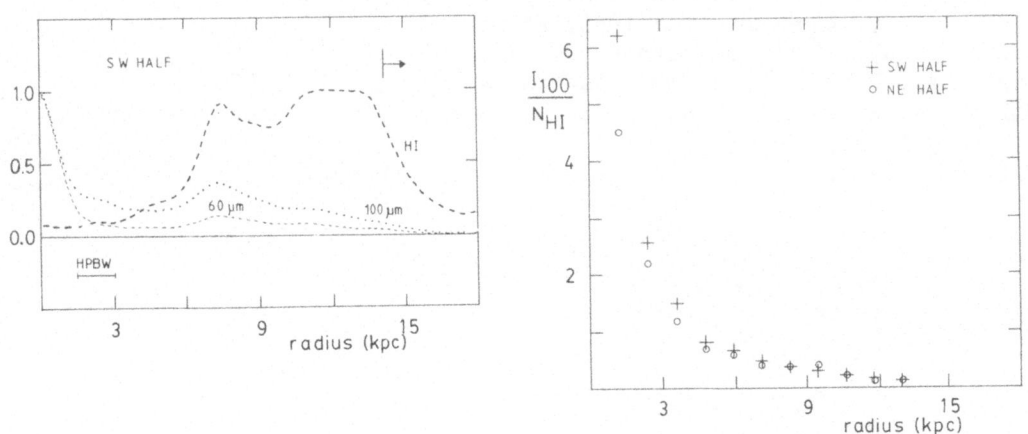

Figure 3a. Radial distribution of HI column density and 60 and 100 micron surface brightness for the SW half of M31 in arbitrary units. The arrow indicates the beginning of the region where the HI coverage is incomplete.
Figure 3b. Ratio of 100 micron surface brightness and HI column density as a function of distance from the center for the two halves of the galaxy. HI data taken from Brinks and Shane (1984).

spatial correlation between the two, except in the nuclear region where there is a lack of HI. Figure 3 shows the ratio of 100 micron surface brightness to HI column density as a function of distance from the centre for the two halves of the galaxy. The ratio I(100 micron)/N(HI) decreases with increasing distance from the centre. For the inner 1 or 2 kpc this is, at least partly, due to a different dust temperature (nuclear region is warmer). However, beyond 2 kpc the decrease cannot be explained this way. It indicates a variation in the dust to (neutral hydrogen) gas ratio. Note the overall agreement between the two halves of the galaxy.

A more detailed report of these results is presented elsewhere (Walterbos and Schwering, 1985).

ACKNOWLEDGEMENTS

R. Walterbos acknowledges support by the Netherlands Foundation for Astronomical Research (ASTRON) with financial aid from the Netherlands Foundation for the Advancement of Pure Research (ZWO).

REFERENCES

Boulanger, F., Baud, B., van Albada, G.D.: 1985, Astron. Astrophys. 144,L9
Braun, R.: 1985, Astron. Astrophys. (in prep.), and this volume
Brinks, E., Shane, W.W.: 1984, Astron. Astrophys. Suppl. 55, 179
De Vries, C.: 1985, Mitt. Astron. Gesells., 63, 158
Draine, B.T., Anderson, N.: 1985, Astrophys. J. 292, 494
Habing et al.: 1984, Astrophys. J. 278, L59-L62
Leger, A., Puget, J.L.: 1984, Astron. Astrophys. 137, L5
Walterbos, R.A.M., Brinks, E., Shane, W.W.: 1985, Astron. Astrophys. Suppl. 61, 451
Walterbos, R.A.M., Schwering, P.B.W.: 1985, in prep.

STATISTICAL PROPERTIES OF IRAS GALAXIES

George Helou
Infrared Processing and Analysis Center, 100-22
California Institute of Technology
Pasadena, California 91125
U. S. A.

ABSTRACT. Statistical approaches to the study of galaxies seen by IRAS are reviewed, stressing the contrast between infrared complete and optically complete samples. A statistical overview is given of the Cataloged Galaxies and Quasars Observed in the IRAS Survey. The bi-variate distribution of ESO/U galaxies as a function of 60 μm flux and optical size is used to predict the size of an infrared complete sample from the optically complete sample. To within the uncertainties, both samples are found to belong to the same population, leaving little room for a fundamentally new class of extragalactic objects, besides infrared bright galaxies like Arp220.

1. BACKGROUND

As soon as the first IRAS data became available, it was clear that the optical luminosity of a galaxy could not be used to predict the infrared luminosity. The wide range explored by the ratio of infrared to blue luminosity (IR/B) was well illustrated by the differences between two samples, one chosen to be complete to a limiting optical magnitude (de Jong et al. 1984), and the other complete to a limiting flux density at 60 μm (Soifer et al. 1984). The IR/B distribution for the two samples barely overlapped: whereas 83 of 86 IRAS selected galaxies had IR/B>1, only 12 of the 88 optically selected sample had IR/B>1. One implication of course is that selection by infrared completeness will net objects too faint to appear in optical catalogs. Indeed, less than half of the Soifer sample (38 out of 86) were galaxies previously cataloged optically, whereas de Jong's sample consisted entirely of the Shapley-Ames galaxies (Sandage and Tammann 1981) in the same area of the sky. Further statistical comparison between IRAS selected and optically selected samples of galaxies is given in Section 2.

405

F. P. Israel (ed.), Light on Dark Matter, 405–414.

More dramatic yet was the early discovery by Houck et al. (1984) of IRAS point sources with only faint or no visible counterparts on the Palomar Observatory Sky Survey prints. These sources may be thought of as the extreme cases in the Soifer sample, but do constitute the decisive step in "the discovery of fundamentally new types of objects, one of the potentials of every new instrument having unique capabilities" (Houck et al. 1984). When followed up, these Houck objects were in fact identified as visually faint galaxies with relatively normal colors in the visible, total luminosities on the order of $10^{12}L_{\odot}$, IR/B on the order of 100, and redshifts in the range 0.1 to 0.2 (Houck et al. 1985).

The hypothesis that the Houck objects belonged to a fundamentally new class could be re-examined with new evidence when Soifer et al. (1984) pointed out a relatively nearby extragalactic system, Arp220 = IC4553, with a remarkable IR/B~100 at a redshift of only 0.02. If Arp220 is indeed a prototype of the Houck objects, the latter can be called a new class since they are the manifestation of a physical process, probably accelerated star formation, carried to the extreme of dominating the energy budget. The search for new objects takes then a somewhat different direction, with two specific questions to ask: Are the Houck objects uniformly distributed in the volume sampled by IRAS? Is there yet another new class of objects among the IRAS sources? A statistical and unbiased approach to these questions would be to extrapolate from optically cataloged galaxy counts to the total number of galaxies that IRAS should have detected; new objects are presumably the difference between this prediction and the total number of point source catalog entries with galaxy-like colors. This procedure is carried out in Section 3, and the results are discussed in Section 4.

2.   CATALOGED GALAXIES AND QUASARS OBSERVED IN THE IRAS SURVEY

As discussed by T. Chester at the beginning of this conference, the IRAS point source catalog contains between 20,000 and 25,000 extragalactic sources with galaxy-like colors, i.e., more flux density at 60 than at 25 μm. But IRAS color alone cannot distinguish between this population and Galactic sources: this is illustrated by a plot of all IRAS point sources with galaxy-like colors and a high quality flux at 60 μm. Figure 1 is an Aitoff projection of an all-sky map in Supergalactic coordinates (de Vaucouleurs et al. 1976); each source is plotted as a symbol whose size scales with the log of its flux. The Galactic Plane is quite obvious running along the periphery of the map and up the

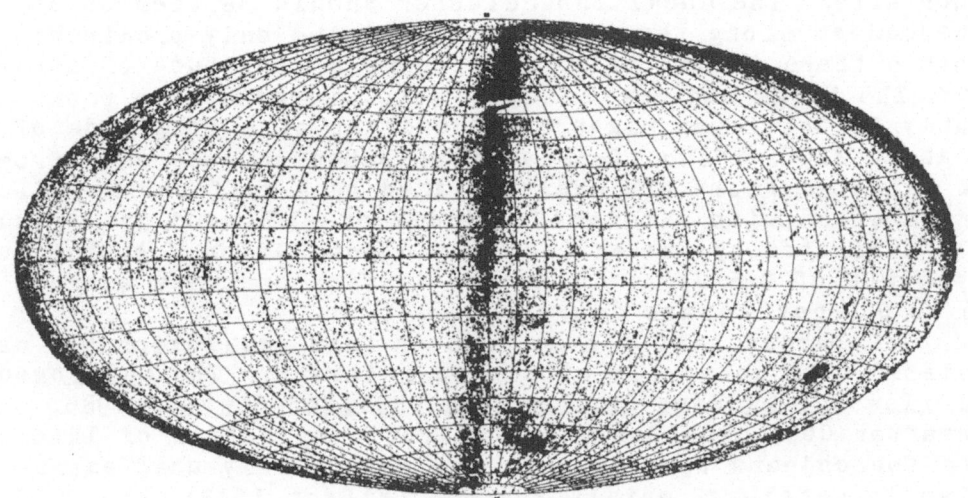

Figure 1. Equal area projection of an all-sky map in Supergalactic coordinates showing all IRAS point sources with galaxy-like colors.

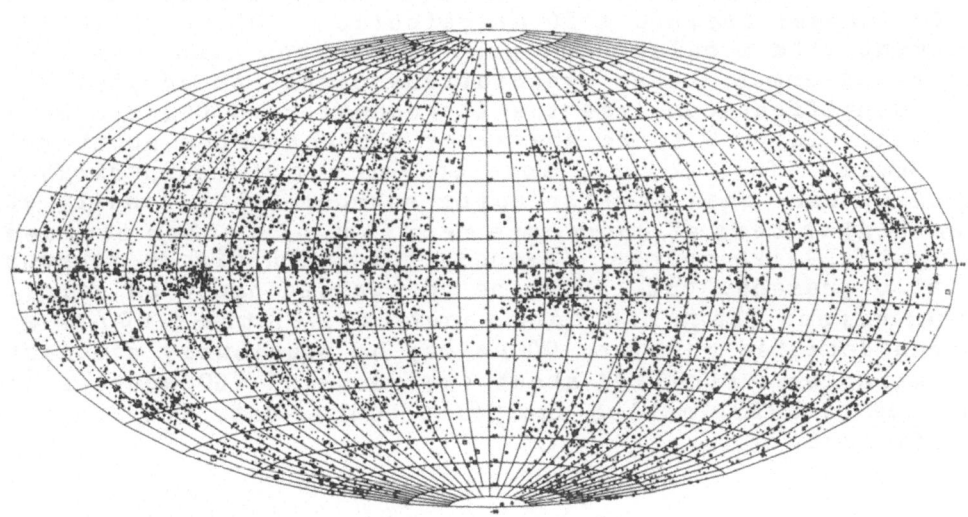

Figure 2. Equal area projection of an all-sky map in Supergalactic coordinates showing all IRAS point sources in the CGQ.

minor axis. The Local Supercluster should be seen as an
enhancement along the major axis, but the only prominent
feature there is the Virgo CLuster at a longitude of about
100°. The Large and Small Magellanic Clouds appear prominently in the lower right hand quadrant at a longitude of
of about 220°. The uniformly distributed sources away from
the Galactic Plane are characterized by a differential flux
distribution with a power law index of -1.5, and are presumably dominated by the extragalactic population. At the survey limit of 0.5 Jy at 60 μm the density of these sources is
0.6 to 0.7 per square degree.

Only about half the sources in this population are associated with optically cataloged galaxies. The Cataloged
Galaxies and Quasars Observed in the IRAS Survey (1985,
hereafter CGQ) provides a convenient compilation of IRAS
data for objects appearing in the most widely used extragalactic catalogs, mainly the UGC (Nilson 1973), the CGCG
(Zwicky et al. 1963-1968), the ESO/U (Lauberts 1982), and
the MCG (Vorontsov-Velyaminov et al. 1962-1974). The CGQ
contains 11,444 IRAS point sources that can be positionally
associated with optical galaxies and quasars; they are plotted in Figure 2, using the same coordinate system as in Figure 1. The Local Supercluster is somewhat more pronounced
here, and is still high-lighted by the Virgo Cluster.

There are about 650 galaxies detected in all four
bands; about 1,300 in three bands, at 25, 60 and 100 μm; and
about 6,800 detected in 2 bands, at 60 and 100 μm. About
1,000 sources have only a 60 μm detection, whereas more than
10,000 sources include a 60 μm detection. Since positional
agreement with a relatively generous search window is the
only requirement, these associations cannot be adopted as
true identifications (see the Introduction to CGQ for more
details). In fact there are among these sources about 300
with stellar IRAS colors, most of which must be spurious
associations. When stellar color sources (flux density at
12 μm higher than at 25 μm) are removed, all other sources
show galaxy-like colors. Associations with cirrus sources
are also present, but cannot be recognized based on color
alone; cirrus flags indicate about half of the 1,000 or so
sources detected only at 100 μm may be spuriously associated
with galaxies. Finally, about one thousand CGQ sources are
resolved by IRAS, making the point source data inadequate or
insufficient.

3.   METHOD AND RESULTS

The method used to extrapolate from an optically complete to
an IRAS complete sample describes each galaxy by two parameters, its optical diameter and its 60 μm flux. The only
assumption in the method is that the relation between sizes

and fluxes is invariant with distance, which is equivalent
to absence of evolution out to the redshift reached by the
IRAS survey. Since IRAS galaxies are rarely found at redshifts beyond 0.3 the assumption is quite reasonable. This
assumed invariance allows the distribution of sizes at each
flux level to be scaled to another flux level by an appropriate change in distance to the galaxies.

The first step is to establish the size distribution at
the highest flux levels where all galaxies detected are
cataloged in the optical. Then using the shape of that distribution the number of all galaxies detected at a lower
flux level is deduced from the number of optically cataloged
galaxies detected at that lower level. In order to mini-mize
the dependence of the results on the details of the shape
adopted for the size distribution, the shape is characterized in the simplest possible way, by the median and quartile points.

The method was implemented on the ESO/Uppsala catalog
(Lauberts 1982), in the region South of declination -17.5°
and at |b|>17.5°, where it is reasonable to assume the
cataloge uniformly complete to a galaxy diameter of about
1'. Circular areas 9° and 5° in diameter centered respectively on the Large and Small Magellanic Clouds were removed
from consideration. In this region, the IRAS point source
catalog contains a total of 4,958 entries (hereafter called
the infrared complete sample) with a high quality 60 μm
flux density above 0.55 Jy and with galaxy-like colors,
defined to be a larger flux density at 60 than at 25 μm.
Among these are included 2,058 entries (the ESO/U sample)
associated in the CGQ with ESO/U galaxies. The bi-variate
distribution of this population as a function of 60 μm flux
density and optical diameter is shown in Figure 3, where
each curve is labeled by the flux bin center, and the ab-
scissa gives the size bin center. The curve labeled "A"
combines entries from two flux bins (log f_ν =0.64 and 0.84).
The curves on Figure 3 show the observed counts. If the
invariance assumption stated at the beginning of this Section is exactly true, the curves would be similar in shape
and obtainable each from the one below it by shifting by one
bin towards smaller sizes and doubling the amplitude; with
each shift the ESO/U completeness limit at about 1' cuts
more deeply into the curve.

The first step in the implementation is to establish
the median point of the size distribution at high flux
levels. But even above 3.5 Jy, 74 out of 271 sources in the
infrared complete sample are not contained in CGQ. When examined individually on the ESO(B) Survey prints, they all
associate with optical counterparts with the following morphological interpretation: (i) 40 sources turn out to be of
Galactic origin, including 24 in the Ophiuchus region, 10 in
reflection nebulae and cirrus; (ii) seven are due to large

galaxies with multiple IRAS sources; (iii) seventeen associate with galaxies below the ESO/U completeness limit;
(iv) six more appear to do the same but are uncertain; and
(v) four remain undecided. When curve "A" in Figure 3 is
adjusted for sources in categories (ii) and (iii) only, it
acquires 16 new entries at the low end and 2 new ones at the
high end of the size distribution. The median is then found
to be about 100" at 4.4 Jy. It is clear that with this
value the expected position for the median shifts below the
ESO/U completeness limit before the lowest flux level is
reached. Using the three lowest curves on Figure 3, the
upper quartile point (on the side of larger sizes) is
therefore derived in an intermediate step, and is found to
be about 174" at 4.4 Jy. This size and flux combination
can then be scaled with distance to find the expected position of the upper quartile point on each of the curves.
The last step in the procedure is simply to sum under each

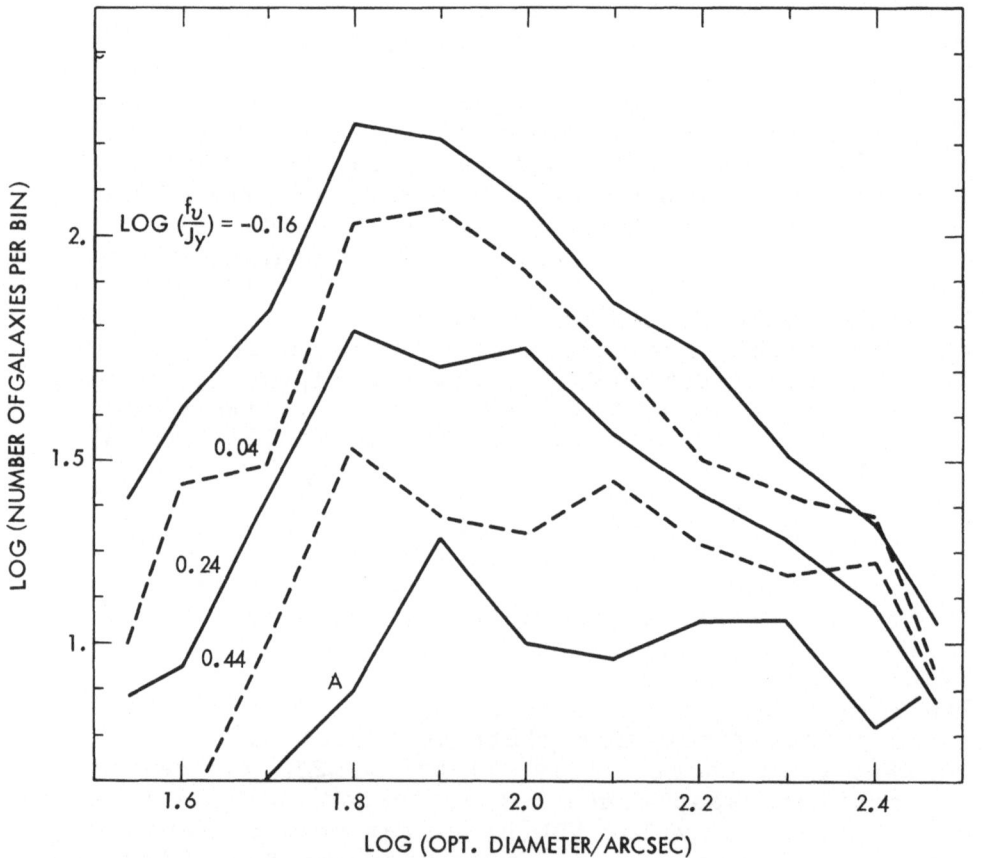

Figure 3. The distribution of ESO/U galaxies (in the CGQ) as a function of optical
diameter and 60 μm flux density.

curve from the upper end down to the expected quartile
point, then multiply this sum by four to obtain the prediction for the total number of galaxies that should have been
detected at that flux level.

   The first result is that the projected number of detections reproduces a power law flux distribution with an index
of -1.5, just like the infrared complete sample. Second,
the 2,058 ESO/U galaxies project to a total of 3,940 detections expected, out of 4,958 sources found in the infrared
complete sample. But as indicated above some of these 4958
infrared sources are of Galactic origin. To check on that
possibility, the ratio of source count in the infrared complete sample to source count in the ESO/U sample is plotted
as a function of Galactic latitude in Figure 4. Each hori-
zontal bar gives that ratio in the |b| interval it covers;
the over-all ratio for the total samples is 2.41. The number just below the bar is the number of galaxy-like point
sources in that interval. The higher ratio at lower galactic latitudes is a result of contamination of the infrared
complete sample by galactic sources. Given the clean linear

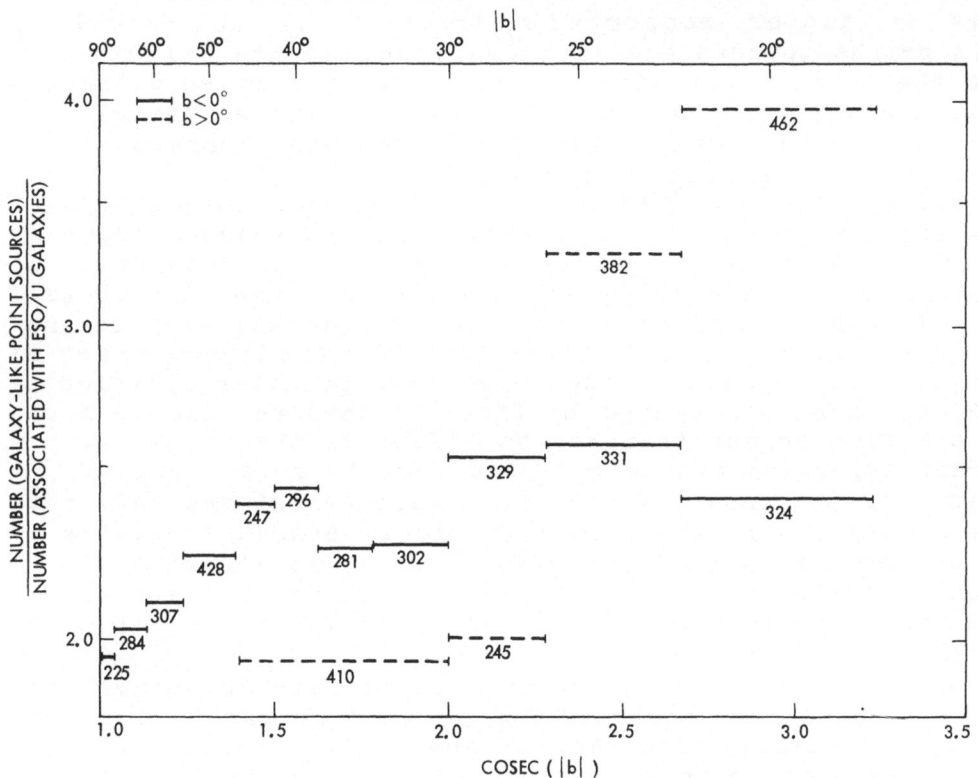

Figure 4. The ratio of infrared complete
source counts to detected ESO/U galaxy
counts as a function of Galactic latitude.

drop in the ratio above $|b|=40°$ in the Southern Galactic hemisphere, and the value above $b=30°$ in the Northern hemisphere, a value of 1.95 is adopted for the ratio of truly extragalactic point sources to ESO/U galaxies detected. This places the number of extragalactic sources at 4,013 out of 4,958 with galaxy-like colors. 4,013 is remarkably close to the 3,940 projected; in fact, the two numbers are equal in view of statistical uncertainties alone in the method.

4.   DISCUSSION

Besides statistical fluctutations, the main source of error in the method is that probably a few hundred of the galaxies in ESO/U are resolved by IRAS at 60 μm, so their fluxes are underestimated systematically in the point source catalog. This flux error may critically affect the results because it occurs more often for galaxies with large sizes and changes asymmetrically the shape of curve "A" in Figure 3. Smaller uncertainties are involved in determining the fraction of the infrared complete sample which is in fact extragalactic, for Figure 4 offers excellent circumstantial evidence for the value adopted, supported by the fact that at least 40 of the 74 bright sources examined individually are Galactic. While the total uncertainties cannot be quantified without detailed analysis, they could not be as small as the 2% difference found above between projected and observed numbers of extragalactic sources.

   The conclusion is thus reached that to within the uncertainties in the method all extragalactic sources above 0.55 Jy at 60 μm can be accounted for as an extrapolation of the optically cataloged galaxies detected above that threshold. Infrared complete and optically complete samples are drawn from the same population. Thus if a truly new class of extragalactic objects besides high IR/B galaxies typified by Arp220 has been discovered by IRAS, it does not number more than the few percent that can be hidden in the margin of uncertainty, which can be narrowed down by more detailed analysis of the data. While the result cannot yet rule out a new population it predicts a difficult search, becoming probably harder as the margin of uncertainty shrinks. Unfortunately the typical redshifts reached by IRAS are too low to make the expected limits apply to a substantial fraction of the universe.

   Because Arp220 is in the UGC, it is safe to assume that Houck's systems are accounted for by this extrapolation from cataloged galaxies. The results above imply then that the space density of these systems does not increase appreciably out to a redshift of 0.2 or 0.3, which is not surprising since the look-back time sampled is only about 10^9 years. Having identified these systems as a new class in the sense of being taken over by a "new" physical process (probably star formation) IRAS finds this same process pervades normal galaxies at more moderate levels. The IRAS data thus demonstrate that infrared bright and normal quiescent galaxies cannot be studied separately. While these new systems cannot be called normal galaxies given their tremendous infrared luminosity, they should be viewed as part of a continuous distribution that encompasses normal galaxies, i.e. as spectacular extremes rather than radically different objects. The continuity of the distribution then leads to the interpretation of these extremes as short lived phenomena initiated in otherwise normal galaxies.

ACKNOWLEDGEMENTS

I would like to thank B. T. Soifer for interesting and enlightening discussions. This work was supported through the Jet Propulsion Laboratory (JPL), California Institute of Technology, under contract with the National Aeronautics and Space Administration.

REFERENCES

Cataloged Galaxies and Quasars Observed in the IRAS Survey, 1985, prepared by Lonsdale, C.J., Helou, G., Good, J.C., and Rice, W. Preprint D-1932, Jet Propulsion Laboratory (CGQ).

de Jong, T., et al. 1984, Ap. J. (Letters), 278, L67.

de Vaucouleurs, G., de Vaucouleurs, A., and Corwin, Jr., H.G. 1976, Second Reference Catalogue of Bright Galaxies (Austin: University of Texas Press).

Houck, J.R., et al. 1985, Ap. J. (Letters), 290, L5.

Houck, J.R., et al. 1984, Ap. J. (Letters), 278, L63.

Lauberts, A. 1982, The ESO/Uppsala Survey of the ESO(B) Atlas (Munich: European Southern Observatory)(ESO/U).

Nilson, P. 1973, Uppsala General Catalogue of Galaxies (Uppsala: Royal Scientific Society)(UGC).

Sandage, A., and Tammann, G. A 1981, A Revised Shapley-Ames Catalog of Bright Galaxies (Washington: Carnegie Institution of Washington).

Soifer, B.T., et al. 1984, Ap. J. (Letters), 278, L71.

Soifer, B.T., et al. 1984, Ap. J. (Letters), 283, L1.

Vorontsov-Velyaminov, B.A., Arhipova, V.P., and Krasnogorskaja, A.A. 1962-1974, Morphological Catalogue of Galaxies, I-V (Moscow: Moscow State University) (MCG).

Zwicky, F., et al. 1961-68, Catalogue of Galaxies and of Clusters of Galaxies, I-VI (Pasadena: California Institute of Technology) (CGCG).

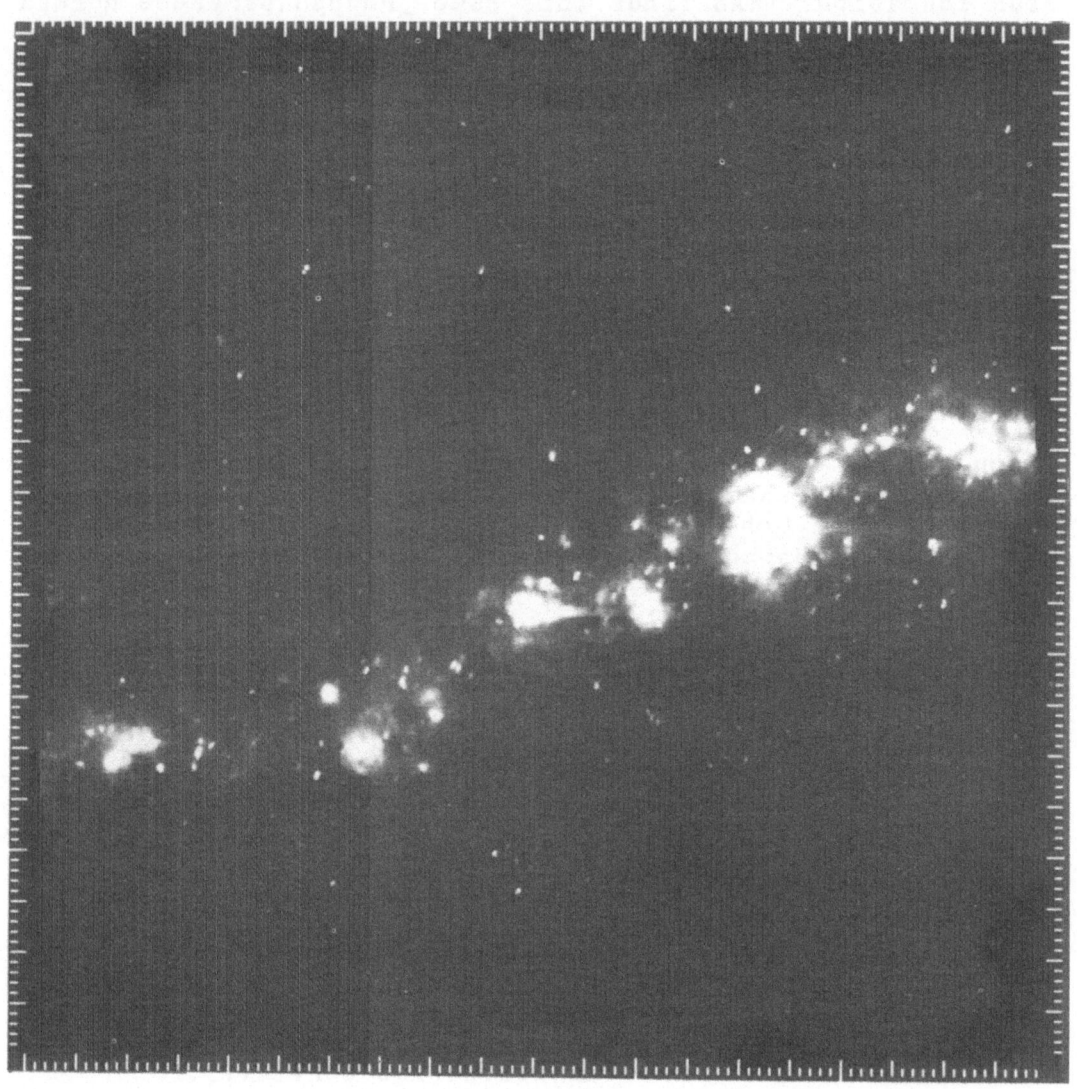

IRAS Field 194 $\alpha = 11^{h}12^{m}$, $\delta = -60°$, HCON-3, 24 µm. Field in Vela,
                 Carina and Centaurus; showing η car nebula (RCW 57),
                 brightest object at right.

INFRARED RADIATION FROM NORMAL GALAXIES

E.E. Becklin
Institute for Astronomy
University of Hawaii
2680 Woodlawn Drive
Honolulu, Hawaii.

ABSTRACT. We review the possible sources of thermal infrared radiation in a complete sample of optically bright normal galaxies. The
sample comes primarily from the Virgo cluster but is augmented by
field galaxies at the same distance. Ground based 10 μm measurements
of the nuclear regions and the IRAS survey are both reviewed. We find
that elliptical galaxies are very weak at thermal infrared wavelengths
and that much of their emission comes from photospheres and dust
shells around late-type M stars. S0 galaxies, as a sample, are somewhat brighter than elliptical galaxies and in a small subset of the
sample the infrared emission appears to come primarily from nuclear
star-formation. Spiral galaxies are bright at thermal infrared wavelengths. On average, twenty-five per cent of the infrared radiation
from spiral galaxies comes from a nuclear bursts of star formation and
the remainder comes from the disk. The nuclear emission strength is
independent of morphological type whereas disks of Sc galaxies are
much stronger than the disks of Sa galaxies. Infrared emission from
peculiar objects such as Seyfert type activity appears relatively weak
in this sample.

1. INTRODUCTION

In this study we attempt to categorize the types of infrared radiation
seen longward of 5μm in a sample of nearby galaxies selected optically.
As a basic sample we use all the bright galaxies in the Virgo cluster
as given in the revised Shapley-Ames catalog (Sandage and Tammann,
1981) with a few additional field galaxies at approximately the same
distance. For the Virgo galaxies we take a distance of 17 Mpc and for
the field galaxies we adopt the distances given by Tully (1985). In
this review, the ground based observations at 10 μm are the work of
Devereux, Becklin and Scoville (1986; hereafter DBS); Impey, Wynn-
Williams and Becklin (1986; hereafter IWWB) and Scoville et al (1983;
hereafter SBYC). The 60 and 100 μm observations are IRAS results given
in the "Cataloged Galaxies and Quasars Observed in the IRAS Survey"
assimilated by Lonsdale and Helou. All the ground based observations
were made with a 5.5 arc-sec aperture (500 pc) on the IRTF. The IRAS

F. P. Israel (ed.), Light on Dark Matter, 415–420.

results at 60 and 100 μm have a spatial resolution of approximately
3'(18 kpc). The ground based observations therefore sample the nuclear
emission and the IRAS measurements the whole galaxy. Luminosities
were calculated using the methods given in DBS. The sample includes
15 elliptical, 34 S0 and 51 spiral galaxies in the Virgo cluster and
39 spiral field galaxies in the field at approximately the same
distance as Virgo.

The types of emission process considered are: 1) photospheric
emission from stars; 2) circumstellar emission from dust formed in
the outer atmospheres of M stars; 3) thermal emission from interstellar dust in the disk; 4) thermal emission from regions of star-
formation in the disk; 5) a burst of star formation in the nucleus
and, 6) a peculiar object in the nucleus.

2. ELLIPTICAL GALAXIES

It is already known that IRAS detected very few elliptical galaxies at
any wavelength (De Jong et al. 1984). This immediately implies that
there is little infrared radiation above that expected from stellar
photospheres. IWWB measured the nuclei of 65 nearby ellipticals at
10 μm and detected 1/3 of the sample with a signal to noise ratio
greater two. By comparing the results with 2 μm measurements they
conclude that about 30% of the observed flux is from the photospheres
of M stars. In the nucleus of M32 IWWB show that the spatial distribution of the excess flux above the photospheric emission follows the
stellar distribution. This implies that the excess emission is also
directly related to the old stellar population. By modelling Frogel's
(1985) 10 μm measurements made in Baade's window, IWWB conclude that
30% of the 10 μm flux measured in the nuclei of elliptical galaxies
could come from circumstellar dust shells around M8-M9 stars on the
asymptotic giant branch. The dust shells are produced by mass loss
from the giants. A portion of the 10 μm emission in IWWB's sample of
ellipticals is related to a peculiar nuclear sources such as that seen
in NGC 1052, NGC 838 and M87. Only a few elliptical galaxies have
been detected at 60 or 100 μm in the IRAS survey; for the 15 bright
elliptical galaxies in the Virgo cluster the IRAS measurements show
the average infrared luminosity to be $L(IR) \underset{\sim}{<} 5 \times 10^5$ Lo per galaxy.

3. S0 GALAXIES

As with elliptical galaxies, very few S0 galaxies were seen by IRAS.
For the 34 S0 galaxies studied by DBS there was significant average
10 μm emission in the nucleus (d = 500 pc). Approximately one half of
the emission is attributable to the photospheres and circumstellar
dust shells of M stars, but the remainder is of unknown origin. Of
the 34 Virgo S0 galaxies, 7 were detected by IRAS at 60 or 100 μm.
These 7 are in fact, among the brightest S0 galaxies seen by DBS at
10 μm and contributed most of the 10 μm flux not attributable to M
stars. Following the empirical relationships developed in SBYC we

find that the average nuclear luminosity for these seven S0 galaxies to be 9 x 10^8 Lo which may be compared with their mean IRAS luminosity of 2.2 x 10^9 Lo. These results are consistent with the idea that 40% of the IRAS 60 and 100 μm flux is coming from a nuclear burst of star-formation in the sub-sample of 7 galaxies. The IRAS results tell us that the average S0 far infrared luminosity is L(IR) = 5 x 10^8 Lo, even though a few S0 galaxies have L(IR) ∿ 5 x 10^9 Lo.

4.   SPIRALS

Over 90% of the 51 brightest spiral galaxies in the Virgo cluster (those listed in the Shapley Ames catalog) were detected at 60 and 100 μm in the IRAS survey. A check in the "Cataloged Galaxies and Quasars Observed in the IRAS Survey" shows many more spirals to have been detected in the Virgo cluster, but that the 51 brightest contribute more than 60 per cent of the total far infrared luminosity of the cluster. Further, SBYC found that the nuclear 10 μm fluxes from the 51 galaxies imply an average luminosity in the nucleus of 10^9 Lo per galaxy. In the same sample IRAS measured a mean luminosity of 4 x 10^9 Lo per galaxy. We conclude therefore that in the brightest 51 spirals in the Virgo cluster, 25% of the total infrared luminosity comes from the central 500 pc. Similar trends were seen in the 39 bright field spirals at the same distance as Virgo (DBS).

How is the nuclear 10 μm emission in spiral galaxies explained? Following SBYC we suppose that most of the emission comes from nuclear regions of star formation. The reasons for this are: 1) in well-studied nearby galaxies such as NGC 253 (Rieke et al. 1980) and NGC 2903 (Wynn Williams & Becklin 1985) the radiation appears to be from newly formed stars; 2) in the center of the Milky Way the observed 10 μm emission, over a 500 pc region, appears associated with regions of star formation (SBYC, Gautier et al. 1984), and 3) there is at most only a weak correlation between 10 μm flux and the 2.2 μm emission in the DBS sample of spiral galaxies.

In addition, SBYC and DBS find the following properties of the nuclear infrared emission from nearby spiral galaxies based on 10 μm observations. i) The nuclei show a continuous luminosity function with a slope of one in a log N vs log S plot. ii) The thermal infrared flux from the central 500 pc does not correlate with morphological class (i.e. at 10 μm L(Sa) = L(Sb) = L(Sc)). iii) The nuclear luminosity does not depend on whether the morphology is barred iv) B type stars probably contribute much of the luminosity based on the ratio of observed infrared flux to the number of ionizing photons in the nucleus of the Milky Way and the nucleus of NGC 2903 (Boisse et al 1981 and Wynn-Williams and Becklin 1985). v) Small, temporarily-heated grains (Sellgren 1984) may be important in producing the 10 μm emission (Wynn-Williams and Becklin 1985). vi) The 10 μm flux implies a high star-formation rate with 0.1 Mo per year to be converted into OBA stars in the central 500 pc (SBYC). vii) There is not enough stellar mass loss occurring in the nuclear region to provide the necessary interstellar material; it must come from outside.

If we assume that all the emission from a spiral galaxy that is not nuclear is coming from the disk, then it appears that on average the disk of a large spiral galaxy produces about 75% of the total infrared luminosity. Unlike the nuclear emission, the disk infrared luminosity does depend on the type of galaxy. The disks of Sc galaxies are much brighter than those of Sa.

$$L_{disk}(Sa) = 1 \times 10^9 Lo; \quad L_{disk}(Sb) = 3 \times 10^9 Lo; \quad L_{disk}(Sc) = 4 \times 10^9 Lo$$

As with the nuclear emission, the strength of the disk emission does not appear to depend on whether the galaxy has a bar or not. It should be noted that the relative strengths of the emission from interstellar dust and emission from molecular star-forming regions are unknown in the disks of spirals.

How important are peculiar objects (i.e. black holes) at the very centers of the galaxies in this sample? Although this question is very difficult to determine it appears for this sample of spirals that the average infrared luminosity from such objects is given by: L(peculiar) < 0.2 L(nuclear) = 0.05 L(total). There are two arguments which suggest this limit in a 500 pc region. Within the sample there are two Seyfert galaxies, NGC 3227 and NGC 4388 both of which are the two brightest galaxies at 10 μm. If all of the infrared radiation seen by IRAS in these two galaxies is the result of a peculiar object in their center, then the contribution by these two galaxies is still only 20% of the total nuclear infrared flux from the sample. Furthermore, in the Milky Way, whose infrared radiation has been studied in detail, L (peculiar) < 0.03 L(500 pc) (Becklin et al 1983).

5. CONCLUSIONS

We find that in the nuclei of elliptical galaxies approximately 1/3 of the 10 μm radiation comes from photospheres of M stars, 1/3 from mass loss dust shells around late M stars and 1/3 from peculiar nuclear objects. From the results of IRAS we find that for ellipticals the infrared luminosity beyond 5μm is at most 1% of the visible light from stars.

From a sample of 34 S0 galaxies in the Virgo cluster we find that about half the nuclear 10 μm flux comes from photospheres and circumstellar dust shells of late M stars. The remainder of the 10μm emission and the IRAS 60 and 100 μm emission appears to come from nuclear star formation in seven galaxies in the sample. For this sample L(IR) ∿ 0.05 L (visible).

Spiral galaxies are very bright in the infrared. About 25% of the emission comes from the nuclear (500 pc) region where the average luminosity in our full sample of galaxies is: L (IR) ∿ 10^9 Lo

The strength of the nuclear emission is relatively independent of
morphological type. Most of the luminosity comes from star formation
and less than 20% on average comes from peculiar nuclear objects.
The disks of optically bright spiral galaxies contribute approximately
75% of their infrared luminosity. The luminosity of the disk depends
on the type of spiral with Sa's having the faintest disks and Sc having
the brightest. For our complete sample of bright spiral galaxies the
average infrared luminosity is: LIR ∿ 4 x 10^9 Lo ∿ 0.5 L(visible).

This review would not have been possible without the dedicated
work of N. Devereux and collaborations with G.G. Wynn Williams, N.
Scoville, C. Impey and N. Devereux. I thank A. Webster for a critical
reading of this manuscript and M. McLean for typing of the manuscript.
This work was supported by NSF grant AST 84-18197. The review
was written while I was on sabatical leave at the Royal Observatory,
Edinburgh, through an NSF Co-operative Science Program with Western
Europe and a Guggenheim Fellowship.

REFERENCES

Becklin, E.E., Gatley, I., Werner, M.W. 1982. Ap.J. 258, 135.
Boisse, P., Gispert R., Coron, N., Wijnbergen, J.J., Serra, G.,
     Ryter, C., Puget, J.L. 1981. Astro.Ap. 94, 265.
DeJong, T., Clegg, P.E., Soifer, B.T., Rowan-Robinson, M.,
     Habing, H.J., Houck, J.R., Aumann, H.H., Raimond, E.
     Ap.J.Letters, 278, L67.
Devereux, N.A., Becklin, E.E., Scoville, N.Z. 1986. (DBS), Ap.J.
     (submitted).
Frogel, J.A., 1985. private communication.
Gautier, T.N., Hauser, M.G., Beichman, C.A., Low, F.J., Neugebauer, G.,
     Rowan-Robinson, M., Aumann, H.H., Boggess, N., Emerson, J.P.,
     Harris, S., Houck, J.R. Jennings, R.E., Marsden, P.L. 1984.
     Ap.J.Letters, 279, L57.
Impey, C.H., Wynn-Williams, G.G., and Becklin, E.E. 1986. (IWWB),
     Ap.J. (submitted).
Rieke, G.H., Lebofsky, M.J., Thompson, R.I., Low, F.S., and Tokunaga,
     A.T. 1980. Ap.J. 238, 24.
Sandage, A., Tammann, G.A., 1981. A Revised Shapley-Ames Catalog
     of Bright Galaxies, (Washington, D.C. - Carnegie Institute of
     Washington Publication).
Scoville, N.Z., Becklin, E.E., Young, J.S., Capps, R.W., 1983.
     (SBYC) Ap.J. 271, 512.
Sellgren, K., 1984. Ap.J. 277, 623.
Tully, R.B., 1985. Private communication.
Wynn-Williams, G.G., Becklin, E.E., 1985. Ap.J., 290, 108.

IRAS Field 196 $\alpha = 14^{h}24^{m}$, $\delta = -60°$, HCON-3, 12 μm. Field in
                  Centaurus and Circinus, showing the Galactic Plane.

MODELS FOR IRAS GALAXIES

M. Rowan-Robinson and J. Crawford
Theoretical Astronomy Init
Queen Mary College
London E.I.

ABSTRACT. We have modelled IRAS 12-100 μ observations of galaxies as a mixture of 3 components: a cool "disc" component, a warmer "starburst" component and a "Seyfert" component peaking at 25 μ.

1. THE SAMPLE OF GALAXIES STUDIED

We have selected from the IRAS Point Source Catalog all sources (i) identified with catalogued galaxies, (ii) with high quality fluxes in all 4 IRAS bands, (iii) with no associations with months-confirmed small extended sources, i.e. SES2 = 0 in all 4 bands, (iv) the 100 μm flux has good signal-to-noise relative to the cirrus, i.e. CIRR2 < 6. There are 209 such fully resolved IRAS galaxies with high quality data.

2. IRAS COLOUR-COLOUR DIAGRAMS

Figure 1a, b show the IRAS colour-colour diagrams for these 209 galaxies, with different symbols for Seyferts, starburst galaxies and others. A number of the latter have not had spectra taken, so may turn out to be Seyferts or starburst galaxies. Two features of these diagrams are immediately apparent: (a) the Seyferts spread out to the left, indicating the presence of a component peaking at 25 μm, (b) most of the starburst and other galaxies populate a well-defined band in each diagram, characteristic of a mixture line.

3. DECONVOLUTION INTO 3 COMPONENTS

We have therefore deconvolved these spectra into 3 components, whose colours are indicated by the symbol x: (1) a "disc" component (D), (2) a "starburst" component (B), (3) a "Seyfert" component (S). Figure 2 shows the spectra of the 3 components together with simple models for them. The "disc" component is compared with cirrus spectrum predicted by Draine and Anderson (1985). Although this may explain the

421

F. P. Israel (ed.), Light on Dark Matter, 421–424.
© *1986 by D. Reidel Publishing Company.*

100 μ emission, warmer dust, perhaps associated with star-forming regions in spiral arms, must also be present.

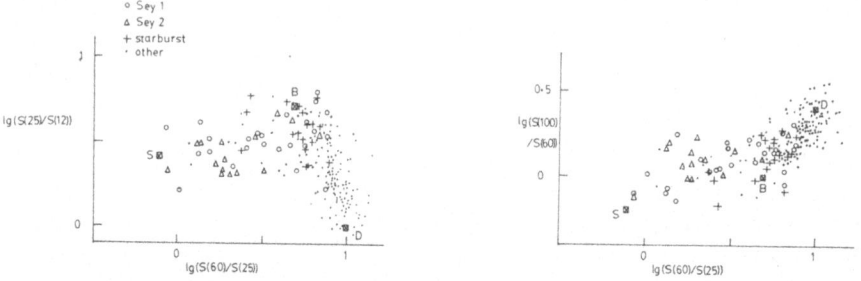

Fig. 1: IRAS colour-colour diagrams for fully resolved galaxies with high quality data, with different symbols for Seyferts, Starburst and other galaxies. The squared x's labelled D, B, S are the assumed ingredients in the 3-component model fitted here.

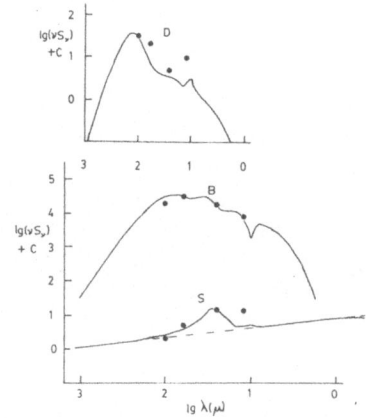

Fig. 2: Models for the "disc" (D) "starburst" (B) and "Seyfert" (S) components.

Circumstellar dust shells around late-type stars may also make a significant contribution at 12 μm. The "starburst" component is well fitted by a simple HII region model (Crawford and Rowan-Robinson 1986). The "Seyfert" component can be modelled as a mini-quasar embedded in a dust cloud with $A_v \simeq 0.25$: presumably this dust is associated with the narrow-line region of the quasar.

   The deconvolution is performed by matrix methods, requiring a positive (or zero) contribution from each component. Because the components are so dissimilar, the deconvolution is robust. The quasar 3C273 required an additional power-law component to achieve an acceptable fit.

4. CORRELATIONS

Figures 3a, b show the correlation of the luminosities in the "disc" and "starburst" components with the optical luminosity of the galaxy. Apart from early type galaxies, the "disc" component is well correlated with L_{opt}, showing that this component is associated with normal star formation and evolution. The "starburst" component shows no such correlation, as might be expected from a transient and anomaleous event like a starburst. The difference between these 2 diagrams shows that the

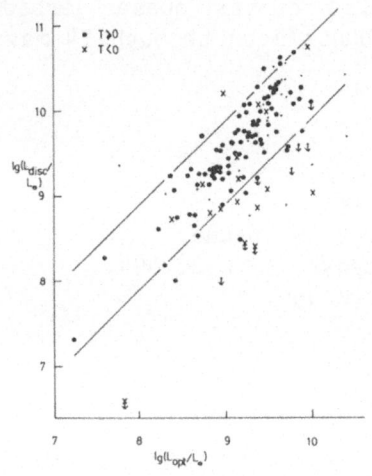
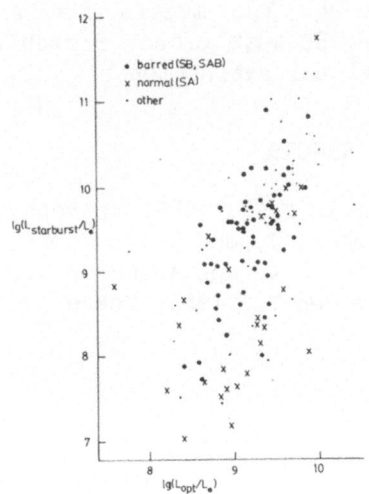

Fig. 3a: Correlation of luminosity in "disc" component with optical luminosity of galaxy.

Fig.3b: Correlation of luminosity in "starburst" component with optical luminosity of galaxy.

mixture-line interpretation of Fig. 1 (rather than, say, a one-parameter sequence) is correct. Note that barred spirals tend to have higher "starburst" luminosities than unbarred spirals (Hawarden 1986).

Almost all Seyferts require a "starburst" component. Most also require a "Seyfert" component, though 7 type 2's and 1 Type 1 do not. The luminosity in the "Seyfert" component is well correlated with the X-ray luminosity, consistent with the model of Fig. 2.

Finally, Figure 4 shows 2 possible models for Arp 220, a quasar embedded in a τ_{uv} = 80 dust cloud and a starburst seen through 74 magnitudes of visual extinction. The quasar model is a better overall fit and is supported by the $\leq$ 1" size measured by Becklin (1985) at 20 μm.

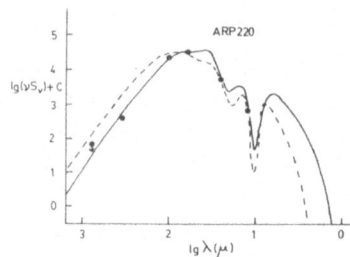

Fig. 4: Two models for Arp 220. Solid curve: quasar embedded in $\tau_{\mu v}$ = 80 dust cloud. Broken curve: starburst seen through 74 magnitudes of visual extinction.

REFERENCES

Becklin E.E., 1985, Astrophys. J. (in press)
Crawford J. and Rowan-Robinson M., 1986, this volume.
Draine B.T. and Anderson N., 1985, Astrophys. J. 292, 494.
Hawarden T., 1986, these proceedings, page 455.

THE IDENTIFICATION OF IRAS GALAXIES

R.D. Wolstencroft, R.G. Clowes, M. Kalafi, S.K. Leggett,
H.T. MacGillivray and A. Savage.
Royal Observatory,
Edinburgh, EH9 3HJ, Scotland

ABSTRACT
We report on the first results of a large-scale program to identify
sources in the IRAS Point Source (IRPS) Catalogue. The positions
and magnitudes of candidate identifications are obtained from UK
Schmidt Telescope J plates, which are measured by the automatic
high-speed measuring machine COSMOS: this program is ideally suited
to the facilities at ROE. There are 312 IRPS sources in a 300 deg^2
field centred on the South Galactic Pole (SGP) which are either
identified with stars (149) and galaxies (155), or are unidentified
(8 'empty fields') down to B_J = 21. The ratio R of far infrared to
blue luminosity for our sample of galaxies ranges from about 0.1 to
250. The statistical properties of this sample are discussed here
and compared with the infrared and optically selected samples of
Soifer et al (1984) and de Jong et al (1984) which both cover a
narrower range of R. Plans for follow-up observations and the
extension to other fields are briefly described.

1.   INTRODUCTION

It is clear from the publication of the first scientific results
from IRAS (Neugebauer et al, 1984; and following papers in Ap J
278) that an abundance of new important and exciting information is
contained in the IRAS Catalogues and Atlases which have recently or
will shortly be released. Probably the most valuable resource is
the IRAS Point Source (IRPS) Catalogue, which contains information
on about 250,000 sources. Only 28% of these sources are associated
with objects in astronomical catalogues (Beichman et al, 1984),
mostly objects with V < 16. It is clearly of great importance to
pursue the optical identification of the remaining 72% of the IRPS
Catalogue sources.

Since the machine readable version of the IRAS Point Source (IRPS)
Catalogue became available we have begun a large-scale program to
identify the optical counterparts of these sources. Our approach is
to search for candidate identifications using the United Kingdom
Schmidt Telescope (UKST) IIIa-J survey plates, each of which covers

425

F. P. Israel (ed.), Light on Dark Matter, 425–434.

a 6.4 x 6.4 deg^2 area of sky. Each plate is measured by the
automatic high-speed plate measuring machine COSMOS (MacGillivray
and Stobie, 1985): coordinates, magnitudes, angular diameters and
other parameters useful in the identification procedure are measured
for all objects in the central 5.4 x 5.4 deg^2 area of the plate.
The positional uncertainties in the IRPS positions are described
approximately by error ellipses whose parameters depend on
brightness, infrared colour, the path of the source across the
detector array and the number of sightings: thus these parameters
vary considerably from source to source. In searching for
associations of IRPS sources with objects in catalogues with small
positional errors, an error ellipse of semi-major and semi-minor
axes of 45 and 8 arc sec was used by the IRAS team (Beichman et al.,
1985, VII-35): this indicates an approximate upper bound on the
error ellipse. As a first stage in the identification procedure we
have developed software to extract the COSMOS data for all objects
within a larger area than this, namely a circle of one arc minute
radius centred on each IRPS position. The probability that the true
optical identification lies outside this area should be very small
(<0.1%) for stars and galaxies with accurate positions, even near
the plate limit. As a further aid in the initial stages of the
identification program we produce an overlay for each Schmidt plate
on which are plotted the positions of IRPS Catalogue sources and a
selection of positional reference stars.

2.   IDENTIFICATION : METHODS AND RESULTS
The methods used in the identification program are given in detail
in an accompanying paper (Savage et al., 1986) with the emphasis
here being on the astronomical results emerging from the program.
The fields we have studied most intensively so far have been at high
galactic latitude where we expect a minimum both in the probability
of spurious identification due to chance coincidence and in the
density of sources seen only at 100µm (Cirrus) (Rowan Robinson et
al, 1985). The results in this paper will be confined to those
obtained in 10 UKST fields centred on the South Galactic Pole (SGP):
312 sources of all classes are present in an area of 300 deg^2. We
have found that 149 of the sources are identified with stars, 155
with galaxies and 8 have no identification ('empty field' objects).
Other fields being studied are listed in Table 1.

TABLE 1

Field Name	Central RA(1950)	Coordinates DEC (1950)	Area (deg^2)	Source Number	b^o	Source density (deg^{-2})
SGP	00^h 45^m	-17° 30'	300	312	-90	1.04
Virgo	12 25	+12° 20'	130	212	+74	1.63
Centaurus	13 25	-30 00	41	125	+32	3.05
Scorpius	15 24	-25 00	41	264	+28	6.44
Orion	05 32	-04 00	41	678	-19	16.54

The parameters of the average error ellipse for sources in the SGP
field were determined from the distribution of positional
differences between COSMOS (mean accuracy = 0.5 arc sec) and IRPS
Catalogue coordinates for those cases (127) where the identification
could be made unambiguously with a bright SAO star. Although the
distribution of positional errors is not strictly Gaussian (Beichman
et al., 1984) it is convenient to describe the error ellipse in
terms of the best fitting Gaussian (equal area criterion): we
obtain semi-minor and semi-major axes of σ_{MIN} = 3 arc sec and σ_{MAX} =
15 arc sec at 65^{0} position angle. For our sample 95% of the errors
lie within ±30 arc sec and ±6 arc sec corresponding to an error
ellipse of area 565 (arc sec)^2 which we adopt as our search area.
The centre of the error ellipse is offset by 2 arc sec north
relative to the coordinates in the IRPS Catalogue.

We have used positional coincidence as the criterion for
identification except in a very few cases for which the wavelength
dependence of IRAS flux density was used to distinguish between
neighbouring candidates. Some examples of identifications are shown
in Plate 1. The plus sign marks the position of the IRAS source
where the identification is not clear cut; the error ellipse has
been drawn as a parallelogram for convenience. With a bright star
or galaxy within the error ellipse the identification is rarely
ambiguous although such cases do occur: an example is 01367-3010,
which is discussed by Savage et al.(1986). The identification
becomes less straightforward at fainter magnitudes when there is a
significant probability of finding one or more object within the
error ellipse. Down to the limit used in our survey (B_J = 21.0) the
expected surface density of all stars and galaxies is about 2.3 x
10^{-4} (arc sec)$^{-2}$ (Savage et al, 1986), and thus we expect only 0.14
objects on average in the adopted error ellipse. From the histogram
of B_J for unambiguous identifications we infer that chance
coincidences need be considered only for B_J>18 (about 8% of our
sample). In practice where only faint candidates are present in and
near the error ellipse it is important to consider each case
carefully to minimize the likelihood of chance coincidence. For
about 5% of the sample of 312 sources the identifications can be
fairly described as ambiguous and requiring further study. Of these
8 have no object within the error ellipse: we refer to these
sources as 'empty field' objects.

The identified stars and galaxies have been classified into 10
classes (Table 2). The first two stellar classes are based on
spectral types from catalogues. The classification scheme of de
Vaucouleurs (1959) has been used for the galaxies. Fainter galaxies
are more difficult to classify and in some cases these fall either
into class 8 (faint, unclassified) or class 10 (ambiguous
classification). Class 10 also includes 'peculiar' galaxies. The
table illustrates the clear difference between IRAS spectral
classification for stars and galaxies. The only spectral class of
sources which are not either all stars or all galaxies is (1,1,1,1).

The spectral class of the 8 empty field objects (class 11) tends to suggest that most of them are likely to be galaxies rather than stars. The statistics for interacting galaxies, which are difficult to classify objectively, represent a conservative lower limit especially for the newly identified galaxies (see also Table 3).

TABLE 2
SPECTRAL CLASSIFICATION OF SGP SOURCES

FLUX QUALITY[+] (λ, μm)				STARS (149)				GALAXIES (155)						EMPTY FIELDS	TOTAL
10	25	60	100	1	2	3	4	5	6	7	8	9	10	11	
1	0	0	0	7	83	0	6	0	0	0	0	0	0	0	96
1	1	0	0	0	40	1	2	0	0	0	0	0	0	0	43
1	1	1	0	0	3	3	0	0	0	0	0	0	0	0	6
0	1	0	0	0	0	0	0	0	0	0	0	0	0	1	1
0	1	1	0	0	0	0	0	0	0	0	1	0	0	0	1
1	1	1	1	0	0	4	0	0	0	3	0	0	0	1	8
1	0	1	1	0	0	0	0	0	0	1	0	0	0	1	2
0	1	1	1	0	0	0	0	1	0	10	0	1	0	1	13
0	0	1	0	0	0	0	0	1	6	7	11	3	5	1	34
0	0	1	1	0	0	0	0	7	10	52	6	5	15	2	97
0	0	0	1	0	0	0	0	0	2	5	2	0	1	1	11
TOTAL				7	126	8	8	9	18	78	20	9	21	8	312

+ 1=Moderate or high   0=Upper Limit

STARS   GALAXIES
1   Early type (O to F)   5   Elliptical
2   Late type   (G to M)   6   Lenticular
3   Very late type (eg carbon stars)   7   Spiral
4   Faint, unknown type   8   Faint, unclassified
                                                                   9   Interacting
                                                                   10 Peculiar or ambiguous
11 EMPTY FIELDS   classification

3.   STATISTICAL PROPERTIES OF THE IDENTIFIED IRAS GALAXIES
Almost exactly half of our sample, 155 objects, have been identified as galaxies. Classification of the 82 galaxies of our sample which are in the IRAS association catalogues (Table 3) shows the well established result (de Jong et al, 1984) that spirals (76%) predominate, with ellipticals being very rare (1%). For the 73 fainter, newly identified galaxies the distribution is quite different : this difference may just reflect the difficulty in classifying these faint galaxies which have $B_J > 16$. A program of short exposure CCD imaging at about 10 arc sec mm^{-1} should clarify the morphological classification of the newly identified galaxies. The 155 galaxies were found in a total area of 300 deg^2 corresponding to a surface density of 0.52 deg^{-2} for all galaxies in the IRPS Catalogue. Soifer et al. (1984) determined the surface density of galaxies in the IRAS minisurvey with 60μm flux density > 0.5Jy and m_{pg} < 18.0: they obtained 0.25 deg^{-2} with an uncertainty of approximately 50%. Applying the same limits to the SGP sample

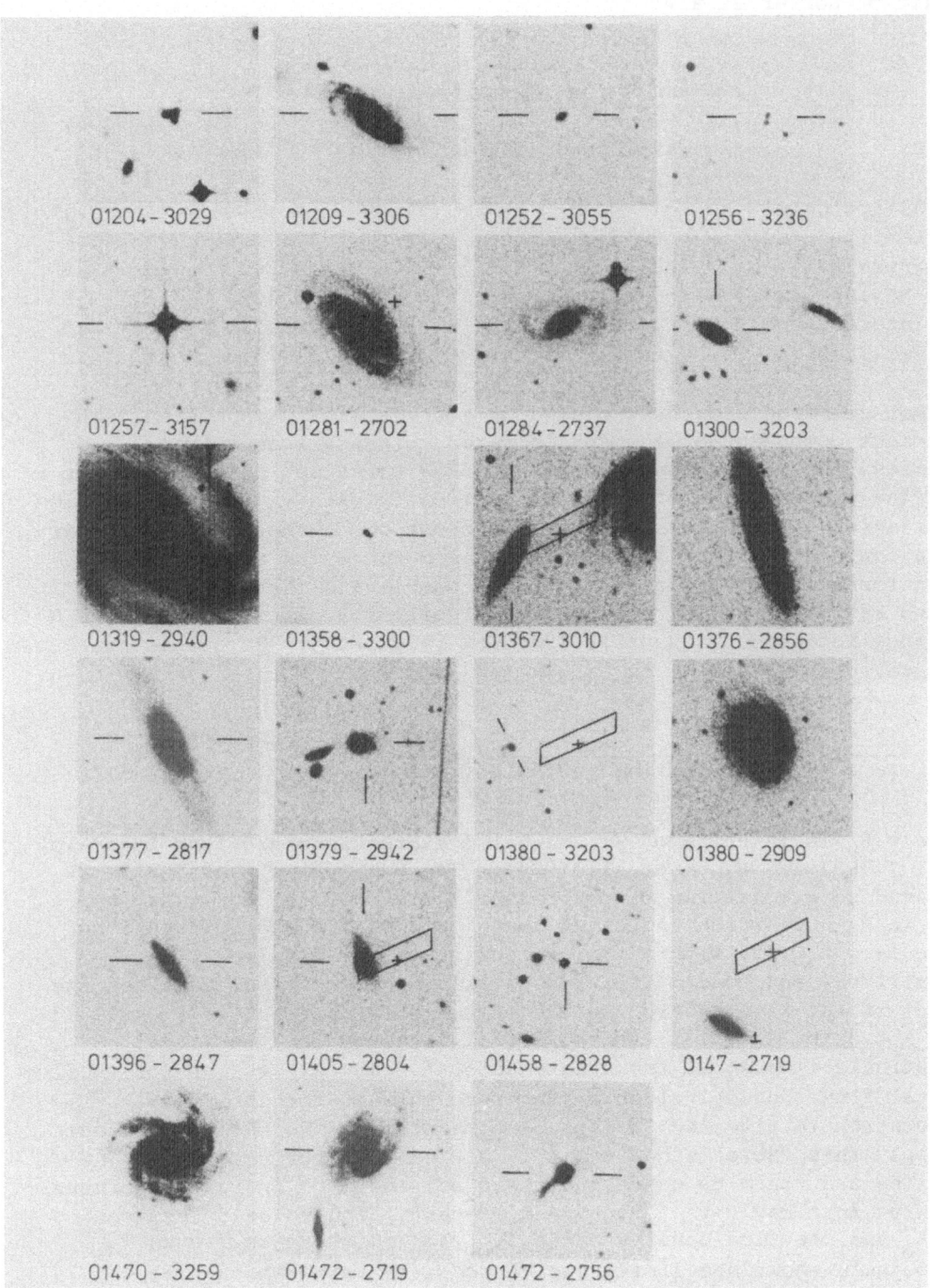

Plate 1
Examples of identified objects in the SGP field. A 2 x 2 (arc min)^2
field is shown with N up and E to the right. The IRAS position is
marked + and the error ellipse is indicated for convenience as a
parallelogram.

(using $B_J < 18.0$) yields 0.37 deg^{-2} which is compatible with the result of Soifer et al.

TABLE 3
Morphology of Identified Galaxies

Class	Catalogued		New Identification		All		$F_{60} > 0.5Jy$ m$_{pg}$ <18	
5 (Ellipticals)	1	1.2%	8	11.0%	9	5.8%	8	7.1%
6 (Lenticulars)	6	7.3	12	16.4	18	11.6	13	11.6
7 (Spirals)	62	75.6	16	21.9	78	50.3	65	58.0
8 (Fnt.,unclass.)	3	3.7	17	23.2	20	12.9	5	4.5
9 (Interacting)	4	4.9	5	6.9	9	5.8	5	4.5
10 (Pec./ambig.)	6	7.3	15	20.6	21	13.6	16	14.3

Thermal emission from dust heated by young stars is undoubtedly an important component of the far infrared emission from spiral galaxies, the most abundant class of IRAS galaxies, and it is clear that the distribution of infrared luminosity, L_{IR}, in a large sample of galaxies can provide important information on the process of star formation. At present the velocities and hence the distances are known for only about one third of our sample and therefore we turn to the ratio of far infrared to blue luminosity, L_{IR}/L_B, which is a distance independent quantity (for low redshifts). We assume that this ratio is identical to

$$R = \frac{(\nu F_\nu)_{FIR}}{(\nu F_\nu)_{BLUE}} \equiv \frac{(3.25 \, F_{60} + 1.26 \, F_{100}) \times 10^{-14}}{\text{Antilog} \, (-7.54 - 0.4 \, B_J)}$$

where νF_ν is in units of wm^{-2} and the expressions are from Lonsdale et al (1985) and Houck et al (1984). The assumption that $R = L_{IR}/L_B$ is based on a modelling of the redshift distribution for galaxies down to $B_J = 21$ which indicates that the most frequently occurring value is $z \simeq 0.15$, with a tail extending to $z \simeq 0.6$ (see MacGillivray and Dodd, 1982, for details) : for spiral galaxies the effect of the K correction on the value of L_B is relatively small ($\Delta m_K \approx 0.5$ for $z = 0.15$ (Pence 1976)), although for the (rare) ellipticals it must be taken into account. The expression for the optical flux should include a factor to correct for internal obscuration to give the face on blue magnitude. We have chosen not to apply this factor since it is tabulated only for catalogued galaxies and could be quite uncertain for the more infrared luminous galaxies that may have a high dust content. The relation between L_{IR}/L_B and the flux density ratio F_{100}/F_{60} is shown in figure 1. The objects shown are limited to those 94 galaxies which have moderate or high quality flux measurements at both 60 and 100μm and for which accurate magnitudes B_J have been determined by COSMOS (magnitude error <0.2 for B_J <21). The luminosity ratio correlates well with flux ratio, or temperature, with the hotter galaxies showing the largest values of L_{IR}/L_B. This is in agreement with the

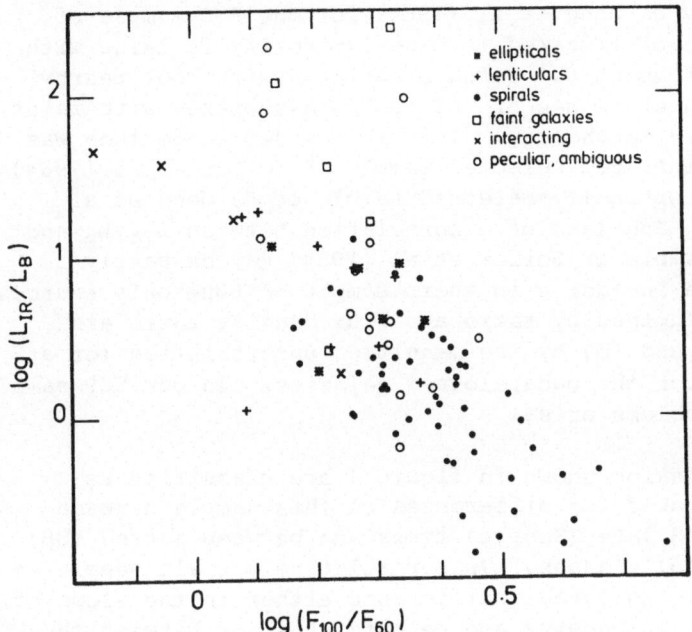

Figure 1
The infrared to blue luminosity ratio versus the 100μm to 60μm flux
density ratio for all (94) identified SGP galaxies with (i) moderate
or high quality 60 and 100μm flux densities, and (ii) accurate
magnitudes determined using COSMOS.

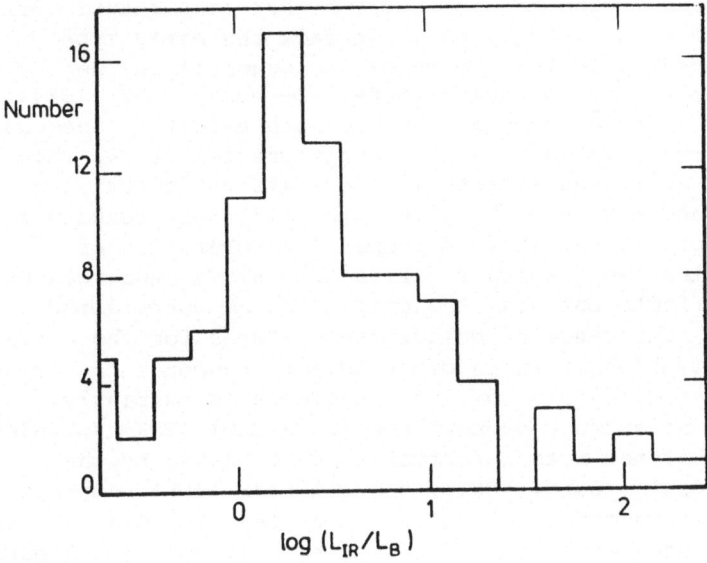

Figure 2
Histogram showing the distribution of the infrared to blue
luminosity ratio for the same sample as plotted in figure 1.

correlation found by de Jong et al (1984) for the RSA sample of
galaxies. The range of infrared to blue luminosity is large with
the lowest values of about 0.1 being associated with cool nearby
spirals and the highest values of 100 to 250 associated with faint
galaxies of uncertain morphology. This is a wider range than was
seen either in the infrared selected sample of Soifer et al (1984)
of 0.5 to 50 or the optically selected sample of de Jong et al
(1984) of 0.1 to 5. The lack of a correlation between L_{IR}/L_B and
flux ratio in the sample of Soifer et al (1984) may be partly
explained (a) by the inclusion in their sample of 60µm only sources
for which both the luminosity ratio and flux density ratio are
somewhat uncertain, and (b) by the magnitude uncertainties (of at
least 1 magnitude) for the uncatalogued galaxies. In our SGP sample
neither of these problems arise.

About 55% of the galaxies shown in figure 1 are classified as
spirals. We have looked for differences in this sample between
early (Sa, ab, b) and late (Sbc, c) types and between barred (SB)
and unbarred (SA, SAB) classes. In our relatively small sample we
find that there is no detectable difference either in the slope of
the relation between luminosity and colour ratio, or between the
average values of these quantities for barred and unbarred spirals.
The only noticeable difference is between early and late type
spirals, for both barred and unbarred classes. The infrared to blue
luminosity ratio for the early type spirals is about 3 times that
for the late type spirals. The early type spirals have a lower flux
density ratio, corresponding to a slightly higher temperature (35K
as opposed to 32K for the late types). In fact the early type
spirals are displaced along the 'correlation sequence' in the
direction of higher L_{IR}/L_B and temperature. De Jong et al (1984),
in their optically selected sample, find no such effect : instead
they find that barred spirals have a slightly greater L_{IR}/L_B than
unbarred spirals. Selection effects will operate differently on
these two samples and may in part explain the different results for
the infrared and optical samples. A proper interpretation of the
above results for the two samples requires not only an understanding
of the selection effects but also the origin of the correlation
shown in figure 1. The range of colour temperatures for the coolest
(25K) and warmest (42K) spirals in our sample corresponds to a range
of L_{IR}/L_B from about 0.15 to 12.5. This sequence is currently
discussed in terms of a two component model: a cool (25K) extended
disk component associated with interstellar dust heated by the
general radiation field (Jura, 1982), and a warmer (50K) component
associated with dust heated by obscured young stars associated with
molecular clouds. The latter component may be dominant in starburst
galaxies, is possibly centrally concentrated, and L_{IR}/L_B, which can
be viewed as a measure of the ratio of obscured to unobscured young
stars (for a given dust content), should be high for such galaxies.

In figure 2 we show the distribution of log (L_{IR}/L_B) for the sample
of 94 galaxies shown in figure 1. The range of log (L_{IR}/L_B) is -0.8

to + 2.4 with a median of 0.36 corresponding to L_{IR}/L_B = 2.3. The distribution differs significantly from the minisurvey sample of Soifer et al (median of 0.67 in the log), which contains a much narrower range of $\log(L_{IR}/L_B)$ from essentially 0.0 to 1.6. Our distribution indicates the presence of a small fraction (about 8%) of objects with very high values of L_{IR}/L_B between 1.6 and 2.4. There are 8 objects not yet identified in our sample which we have termed 'empty field' objects. The nature of these sources cannot be established until careful follow-up observations have been carried out, which will include imaging and photometry in optical, infrared and radio bands. However it is interesting to note, on the basis of the IRAS spectral classification (see Table 2), that the spectral classification of 6 out of the 8 sources suggests the identification will ultimately be with a distant galaxy. Using B_J = 21 to calculate a lower limit to L_{IR}/L_B for these 6 empty field sources, we obtain lower limits to $\log(L_{IR}/L_B)$ in the range 2.42 to 2.91, i.e. L_{IR}/L_B = 260 to 810.

4.   FUTURE PLANS

We have carried out the identification program for the SGP field in a semi-automated manner which has led to an enormous reduction in the manual effort required for position and magnitude measurements. However it has made us aware of other factors that need to be considered in a more automated program of identifications: (i) we need initial unambiguous identifications based on bright SAO stars, to determine the distribution of positional errors (Optical-IRAS) for any small area - thus the initial phase of the program may need to be iterative; ii) the chance coincidence rates in terms of both position and magnitude of candidates need to be expressed as probabilities so that the program draws our attention to the minority of ambiguous identifications for which visual inspection is essential: the percentage of ambiguous identifications depends on candidate and IRPS source density and is a minimum (~5%) in the SGP field; iii) the software should incorporate a search ellipse rather than a circle at an early stage in the listing of candidate identifications; and iv) the visual classification of morphological types (for the uncatalogued identifications) cannot be readily automated: however visual classification can be done fairly quickly at the rate of about 15 per hour.

The fields that we are currently surveying are listed in Table 1. The work on the Virgo field is well advanced and will be prepared for publication following submission of the SGP paper. The identification problem related to ambiguous and chance coincidences plus cirrus contamination increase towards the galactic plane, and it is not yet clear at what galactic latitude these problems will become severe using our approach: preliminary work on identifying sources in the Centaurus field indicates that our method will be successful at least for fields with <3 IRPS sources deg^{-2}. We plan to extend our program very soon to the following areas: a) 10 fields at the North Galactic Pole (300 deg^2) and (b) an equatorial

band of 20 UKST fields (820 deg^2) distributed uniformly in right
ascension but avoiding galactic latitudes $|b|<40^\circ$. This will allow
follow-up observations to be carried out from both hemispheres and
at most times of the year. Planned observations include imaging
with optical CCD's and with the UKIRT infrared camera (presently
under development), as well as radio continuum measurements at 20cm.

REFERENCES
Beichman, C., et al 1984, 'IRAS Explanatory Supplement'.
De Jong, T. et al 1984. Ap J 278, L67.
De Vaucouleurs, G., 1959, Handbook of Physics, Astrophysics IV:
  Stellar Systems, 8, 275.
Houck, J.R., et al 1984, Ap J 278, L63.
Jura, M. 1982, Ap J 254, 70.
Lonsdale, C.J. et al 1985, 'Catalogued Galaxies and Quasars Observed
  in the IRAS Survey'.
MacGillivray, H.T., and Dodd, R.J. 1982, Astr. Sp.Sc. 86, 437.
MacGillivray, H.T., and Stobie, R.S. 1985, Vistas in Astronomy 27,
433.
Neugebauer, G., et al 1984, Ap J 278, L1.
Pence, W. 1976, Ap J 203, 39.
Rowan-Robinson, M. et al 1985, preprint.
Savage, A., et al 1986, these proceedings, page 23.
Soifer, B.T. et al 1984, Ap J 278, L71.

Aitken, D.K.[1]   Roche, P.F.[2]   & Smith, C.[1]

1. Physics Department (RAAF Academy)
University of Melbourne
Parkville
Victoria 3052
Australia

2. Anglo-Australian Observatory
PO Box 296
Epping
NSW 2121
Australia

We have made spectroscopic observations between 8-13µm, using the AAT in September 1984 and the UKIRT in May 1985, of a small sample (6) of galaxies found to be bright in the IRAS bands but without apparent sign of nuclear activity. On closer inspection three of these show the presence of nuclear HII regions and their spectra display the familiar 11.3µm feature and [NeII] emission which are almost always seen in such sources (Roche & Aitken, 1985). The remaining three, NGC 7479, Zw 958+16 and NGC 4418 (whose 8-13µm and 17-23µm spectrum is shown here)

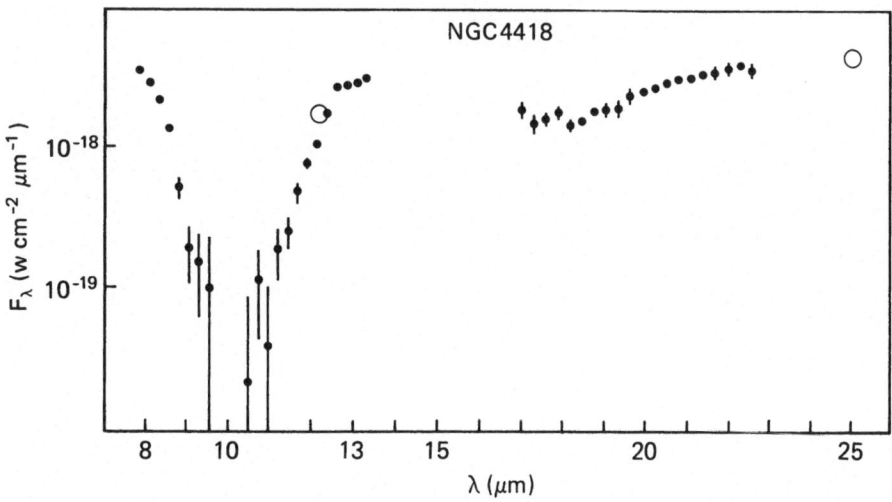

Fig 1. The 8-13µm and 17-23µm spectrum of NGC 4418. The IRAS points at 12 and 25µm are shown as open circles.

F. P. Israel (ed.), Light on Dark Matter, 435–436.

all show very deep silicate absorption. In NGC 4418 the feature has an optical depth of $\cong$ 7 near 10μm, which is as deep as the most heavily obscured galactic source known, W33A (Capps, Gillett & Knacke, 1978; Roche & Aitken, 1984). Its visible and near infrared spectrum, obtained at the AAT, is typical of a normal galaxy with no indication of nuclear activity. With a redshift of 0.009 its infrared luminosity is $\sim 10^{11}$ L_0 and the infrared colours suggest a Seyfert nucleus which in this case is completely obscured in the visible. NGC 7479 also has much deeper silicate absorption than previously observed in galaxy nuclei, and Zw 958+16 has an unusually deep feature as well. It seems that IRAS is detecting a class of active nuclei too heavily obscured to be observed in the visible.

References

Capps, R.W., Gillett, F.C. & Knacke, R.F., 1978. Astrophys. J., 226, 863.
Roche, P.F. & Aitken, D.K., 1984. Mon. Not. R. astr. Soc., 209, 33p.
Roche, P.F. & Aitken, D.K., 1985. Mon. Not. R. astr, Soc., 213, 789.

THE RELATIONSHIP BETWEEN BLUE AND FIR LUMINOSITIES OF SPIRAL GALAXIES

K.V.K. Iyengar, T.N. Rengarajan and R.P. Verma
Tata Institute of Fundamental Research
Homi Bhabha Road, Bombay 400005
India

We have studied optical and infrared properties of 86 galaxies from IRAS circulars, identified in RC-2 and UGC. For these galaxies face-on integrated blue magnitude (B_T°) and the integrated FIR flux have been estimated and plotted in Fig. 1. With this set of homogeneous blue magnitude ($B_T^{\circ} \leq 14.5$) it is seen that the FIR flux is proportional to the blue band flux. The mean value of L_{FIR}/L_B is 6.8.

In Fig. 2a we plot $L_{FIR}/L_{\odot}$ vs $R=F_{100}/F_{60}$ where F_{100} and F_{60} are flux densities at 100 and 60 μm. R is an indicator of the temperature of the dust responsible for FIR emission. The temperature scale corresponding to emissivity $\propto \lambda^{-1}$ is indicated at the top of the figure. In Fig. 2b, L_B is plotted against R. It is seen that there is no correlation for L_B while there is good correlation for L_{FIR}. If the spatial distribution of the dust surrounding young OB stars is, on the average, same for different galaxies, we expect the dust to be heated to higher temperature when the source luminosity is higher. Thus the correlation between L_{FIR} and R can be ascribed to rate of formation of young stars. The lack of correlation between L_B and R shows that the observed blue luminosity is not associated with young star formation activity.

Using initial mass function given by Miller and Scalo (1979) and assuming a constant rate of star formation, we calculate $L_{Bol}/L_B = 12$ for OB stars in the mass range of 75 to 5 $M_{\odot}$ (life time $\leq 10^8$ yrs). The observed value of L_{FIR}/L_B is 6.8 which is an upper limit to L_{Bol}/L_B because of large extinction for L_B. This also shows that L_B is unlikely to be associated with the formation of young stars.

CO observations are available for 45 galaxies, 25 of these in Virgo cluster, from Young et al. (1984) and Young (1985). Using FIR data from IRAS we have plotted in Fig. 3, L_{FIR} vs M_{H2} (within central 50") derived from their data using $N_{H2} = 2 \times 10^{20} I_{CO}$. Again, in general, FIR luminosity is proportional to the mass of the molecular hydrogen. The mean value of $L_{FIR}/M_{H2}(50")$ for Virgo galaxies is 54, after multiplying L_{FIR} by a factor of 2 for out of band emission. Estimating the amount of gas covered by IRAS beam, by assuming an exponential distribution with scale length of 4.7 kpc, we get $L_{FIR}/M_{H2} = 8.5 L_{\odot}/M_{\odot}$. This compares well with value of 7.6 $L_{\odot}/M_{\odot}$ obtained by Rengarajan (1984) for a sample of molecular clouds in the Galaxy. Using the value of total M_{H1}

F. P. Israel (ed.), Light on Dark Matter, 437–438.
© *1986 by D. Reidel Publishing Company.*

for galaxies common to our list and that of Davis and Seaquist (1983), we also find L_{FIR}/M_{H1} = 9 $L_\odot/M_\odot$.

Though L_B*is not associated with young star formation activity, we do find a correlation between L_{FIR} and L_B. L_B, the integrated luminosity of old stars, is a measure of the mass of the galaxy whereas L_{FIR} is a measure of its young star formation activity. Thus the observed correlation implies a relationship between the mass of a galaxy and its young star formation activity. One also notices from Fig. 2a and 3 that peculiar galaxies (filled symbols) have higher values for L_{FIR} and R. This may be due to star burst activity in these galaxies.

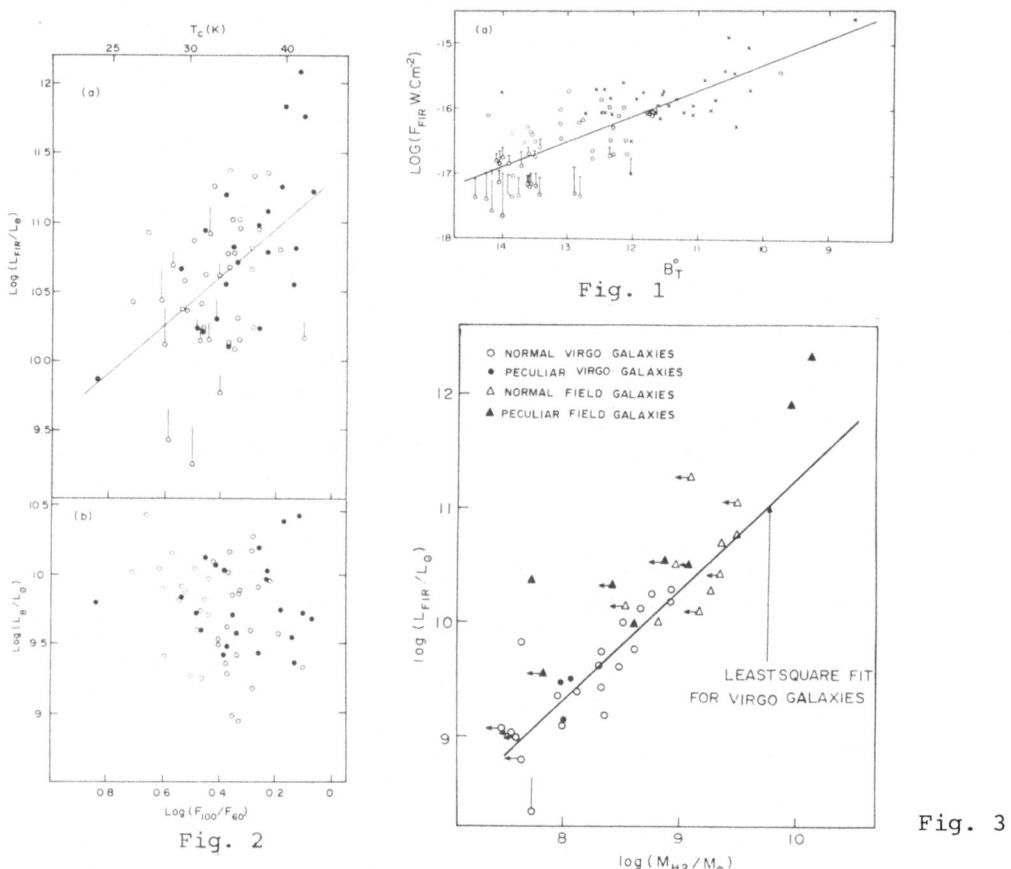

Fig. 1

Fig. 2

Fig. 3

References
Davis L.E., Seaquist, E.R., Ap.J. Suppl. 53, 269 (1983)
Miller G.E., Scalo J.M., Ap.J. Suppl. 41, 513 (1979)
Rengarajan T.N., Ap.J. 287, 671 (1984)
Young J.S., Ap.J. 288, 487 (1985)
Young J.S., Kenney J., Lord S.D., Schloerb F.P., Ap.J. 287, L65 (1984).

* In this paper L_B is in units of $L_\odot$ and not in units of the blue luminosity of the sun.

PRELIMINARY RESULTS OF AN HI SURVEY OF A SAMPLE OF IRAS GALAXIES

J. Thomas Armstrong
Alwyn Wootten
National Radio Astronomy Observatory
Edgemont Road
Charlottesville, Virginia   22903   U.S.A.

We report HI detections in 46 of a sample of 110 galaxies selected from the IRAS circulars and from the point source catalog.   This sample forms a subset of the 350 galaxies we have observed so far with the NRAO 91 m telescope at Green Bank, West Virginia.   A typical galaxy in our sample lies at a velocity of ~3000 km/s, is infrared bright (log (LIR)/L(opt)) ~0.3 to 0.5) with a luminosity of log (L(IR)/L(sol)) ~10.2, and contains a hydrogen mass of log (M(HI)/M(sol)) ~9.5.

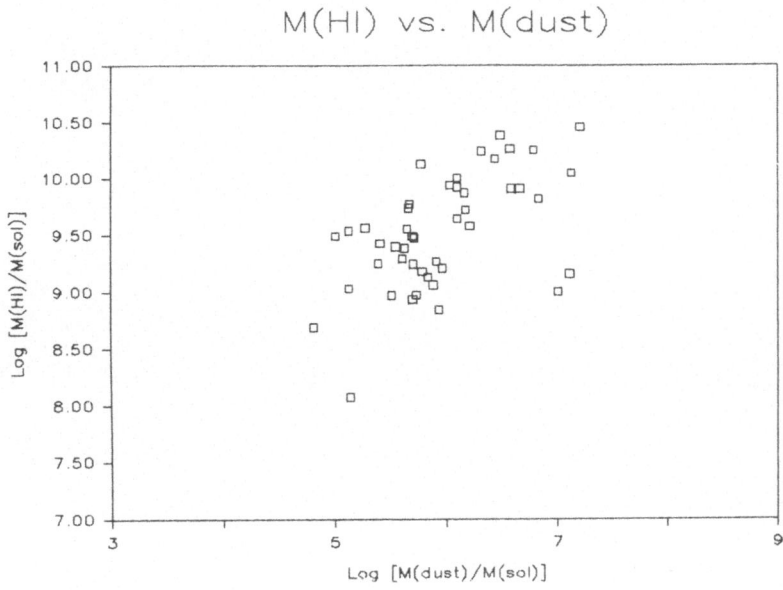

439

F. P. Israel (ed.), Light on Dark Matter, 439–440.

The hydrogen masses in the sample galaxies correlate roughly with the infrared luminosities calculated from the observed fluxes listed in the IRAS point source catalog. We infer that the infrared luminosity is dominated by heated dust. The 60 μm and 100 μm fluxes from the catalog were employed to calculate temperatures typical of the warm galaxian dust. Derived dust masses correlate more tightly with the HI masses than do the infrared luminosities. The mean mass ratio of HI to dust displays a scatter of a factor of two about the mean. If the warm dust lies primarily in regions of star formation, this result is somewhat surprising, since atomic gas does not generally correlate well with star forming activity. We conclude that the warm dust is well mixed with atomic gas in these galaxies.

FAR-INFRARED PROPERTIES OF MULTIPLE NUCLEUS GALAXIES

K.J. Fricke and W. Kollatschny
Universitäts-Sternwarte
Geismarlandstr. 11
3400 Göttingen
F.R.G.

ABSTRACT. The double nucleus Markarian galaxies have properties as expected from advanced stages of galaxy mergers. From their far-infrared, radio-continuum, and HI fluxes we conclude that these systems are sites of violent starburst activity due to massive star formation and supernova explosions.

A small number of Markarian galaxies shows double nuclear structure with component separation $\lesssim$ 10 arcsec ($\lesssim$ 5 kpc). These systems are intrinsically bright ($M_v \sim$ -21 to -22), show strong emission lines in their nuclei (starburst, Liner, Seyfert types), peculiar velocity fields in the ionized gas, and disturbed morphologies often with extended tidal arms. From a comparison of the optical emission line ratios the double nucleus galaxies (DNG's) are found to be on average more active than the nuclei of interactive Arp systems (Kollatschny and Fricke 1984; Kollatschny et al. 1986).

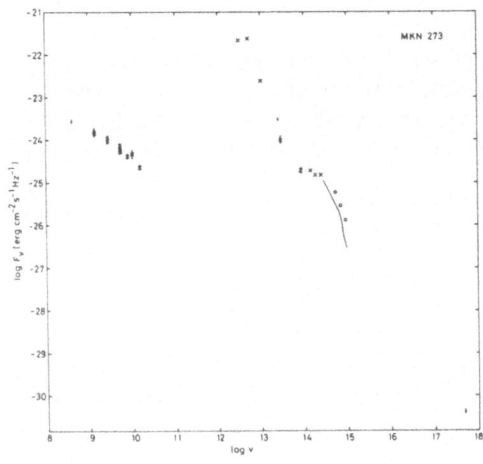

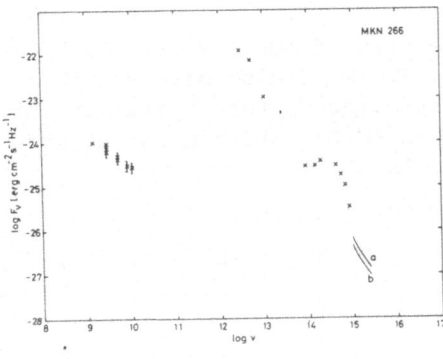

Fig.2: Overall continuum spectra of
2 double nucleus galaxies.

F. P. Israel (ed.), Light on Dark Matter, 441–442.
© *1986 by D. Reidel Publishing Company.*

Table 1: Star burst parameters of double nucleus galaxies

Object	$S_{100\mu}$ Jy	$S_{100\mu}/S_{60\mu}$	L_{IR}/L_B	$\dfrac{S_{6cm} \, (20 \text{ Mpc})}{L_B + L_{IR}}$ mJy/$L_\odot$	steady star formation rate $M_\odot$/yr
Mkn 544	1.42	1.53	4.3	1.19	32
Mkn 1027	8.18	1.55	12.0	2.04	592
Mkn 739	2.33	1.65	4.3	1.52	160
Mkn 788	2.37	1.52	4.5	3.06	110
Mkn 789	5.36	1.41	16.6	1.35	440
Mkn 266	11.80	1.64	9.2	7.43	728
Mkn 273	23.40	0.92	80.0	3.61	2716
Mkn 463	1.79	0.83	16.5	26.98	360
Mkn 673	5.35	1.88	13.4	3.55	568
Mkn 480	3.03	1.68	5.2	6.67	80
Mkn 296	1.36	2.23	6.5	2.96	26
Mkn 306	2.34	2.00	4.1	2.10	68
Mkn 314	1.55	1.20	2.1	14.55	6
Mkn 930	1.32L	1.12	4.0	3.68	36

All these objects have been detected in the far-infrared by the IRAS point-source survey with fluxes implying luminosities up to $L_{IR} \sim 10^{12} L_\odot$ and beyond. Taken together these properties indicate that the DNG's are advanced stages of galaxy mergers connected with violent star formation.

Fig.1 shows two examples (Mkn 266 and 273) of the overall spectra (radio through X-ray). We have analysed the far-infrared and continuum radio fluxes of 14 DNG's together with their optical colours and blue luminosities in terms of the starburst models of Biermann and Fricke (1977). The data are consistent with violent star formation and supernova activity triggered by the merging process. Star formation rates up to 10^3 $M_\odot$/yr (cf. Table 1) are found.

Together with the observed HI masses the corresponding gas consumption rates imply that mostly massive stars (50-100 $M_\odot$) are formed in violent bursts.

The details of our investigation on DNG's will be reported shortly elsewhere. There we also discuss the key importance of these objects for an understanding of the Seyfert and quasar activity which seems to be closely linked to interaction and merging processes of galaxies (Hutchings et al. 1984).

References

Biermann, P. and Fricke, K.J.: 1977, *Astron. Astrophys.* 54, 461
Hutchings, J.B. et al.: 1984, *Astrophys. J. Suppl.* 55, 319
Kollatschny, W. and Fricke, K.J.: 1984, *Astron. Astrophys.* 135, 171
Kollatschny, W., Fricke, K.J., Hellwig, J.: 1985, Proc. of the Workshop
    "Structure and Evolution of Active Galactic Nuclei", Trieste 1986,
    Reidel Publishing Co., Dordrecht.

Acknowledgements
This work has in part been supported by the Deutsche Forschungsgemein-schaft through grants Fr 325/21-1 and Fr 325/15-2.

YOUNG SUPERNOVAE IN THE STARBURST GALAXY M82

S.W. Unger, A. Pedlar, D.J. Axon, P.N. Wilkinson
Nuffield Radio Astronomy Laboratories, Jodrell Bank,
Macclesfield, Cheshire SK11 9DL
P.N. Appleton, Department of Astronomy, The University,
Manchester M13 9PL

ABSTRACT. MERLIN observations at 1666 MHz detect 18 compact (< 0.25 arcsec) radio sources within the central few hundred parsecs of M82. M82 is well known as a 'starburst' galaxy, and it is probable that these objects are very young supernova remnants. The distribution of radio sources appears asymmetric with respect to the centre of the galaxy, and we suggest that this is due to the current region of star formation lying within a molecular ring.

1. INTRODUCTION

The irregular galaxy M82 has been intensively studied in all wavebands since the discovery by Lynds & Sandage (1963) of a system of Ha filaments extending along the galactic minor axis out to 3 kpc from the galactic plane. The central few hundred parsecs of M82 appears optically as a complicated maze of star clusters, giant HII regions, and dust lanes (O'Connell & Mangano 1978) and is also an extended radio (Kronberg & Wilkinson 1975), infrared (Rieke et al. 1980) and X-ray (Watson et al. 1984) source. This nuclear activity is probably powered by a massive burst of star formation, and the proximity of M82 (about 3 Mpc) makes it an excellent prototype for 'starburst galaxies.

2. RADIO SUPERNOVAE IN M82

High resolution radio observations using MERLIN (Ungeretal. 1984) and the VLA (Kronberg & Sramek 1985) detect a large number of compact radio sources embedded in the diffuse radio emission. The 1666-MHz MERLIN observations detect 18 radio sources with flux densities greater than 2 mJy and angular sizes less than 0.25 arcsec.

It seems likely that these compact radio sources are very young supernova remnants. Although a high supernova rate is predicted for M82 by Rieke et al. (1980) on the basis of infrared observations, the large optical extinction (Av ~ 26: Rieke et al. 1980) to the centre of the galaxy means that these supernovae would not be observed optically. Radio counterparts to optically observed supernovae have however been detected in a few nearby galaxies (Weiler et al. 1983), and for M82 radio observations may provide the only direct means of observing such supernova events (c.f. NGC 4258: van der Hulst et al. (1983)).

443

F. P. Israel (ed.), Light on Dark Matter, 443–444.

3. STAR FORMATION IN A MOLECULAR RING,

The low resolution VLA observations of Kronberg et al. (1981) show
the extended radio source to have a central plateau, away from which
the radio emissivity falls steeply. The distribution of compact radio
sources is asymmetric with respect to the extended radio source, with
many of the compact radio sources (and all of the strongest ones)
lying close to the steep gradient in radio emissivity to the south of
the central plateau, whilst there are none in the north of the
plateau.

A number of observations lead us to suggest a model in which the
asymmetric distribution of the compact radio sources is due to them
lying within that part of a nearly edge-on ring which is tilted to
the south. Rieke et al. (1980) noted that the infrared emission at
2 μm, which probably arises from a central cluster of old stars, is
more compact than the emission at 10 μm, which is probably due to hot
dust associated with a more recent region of star formation. This
prompted Rieke et al. to suggest that the region of star formation is
moving outwards from the centre of M82.

Further evidence in favour of this view has been provided by
high resolution VLA observations of HI and OH absorption against
thenuclear continuum radio source, which show the presence of a
toroidal ring of cold gas (Weliachew et al. 1984). This ring has
strong parallels with the molecular ring in our own galaxy (Scoville
1972), and is likely to represent the region of current star
formation in M82. This hypothesis is given strong support by the fact
that the compact radio sources seen in the MERLIN observations all
appear to lie within this ring.

The observations and model described in this paper are discussed
in rather greater detail by Unger et al. (1984).

REFERENCES

Kronberg, P.P. & Wilkinson, P.N.: 1975, Astrophys. J., 200, 430.
Kronberg, P.P., Biermann, P. & Schwab, F.: 1981, Astrophys. J., 246,
    751.
Kronberg, P.P. & Sramek, R.A.: 1985, Science, 227, 28.
Lynds, C..R. & Sandage, A.R.: 1963, Astrophys. J., 137, 1005.
O'Connell, R.W. & Mangano, J.J.: 1978, Astrophys. J., 221, 62.
Rieke, G.H., Lebofsky, M.J., Thompson, R.I., Low, F.J. & Tokunaga,
    A.T.: 1980, Astrophys. J., 238, 24.
Scoville, N.Z.: 1972, Astrophys. J., 175, L127.
van der Hulst, J.M., Hummel, E., Davies, R.D., Pedlar, A. & van
    Albada, G.D.: 1983, Nature, 306, 566.
Unger, S.W., Pedlar, A., Axon, D.J., Wilkinson, P.N. & Appleton,
    P.N.: 1984, Mon. Not. R. astr. Soc., 211, 783.
Watson, M.G., Stonger, V. & Griffiths, R.E.: 1984, Astrophys. J.,
    286, 144.
Weiler, K.W., Sramek, R.A., van der Hulst, J.M. & Panagia, N.: 1983,
    In IAU Symposium 101, p. 171.
Weliachew, L., Fomalont, E.B. & Greisen, E.W.: 1984, Astron.
    Astrophys., 137, 335.

THE AMAZING TAIL OF NGC 2146

H. C. M. Caspers, W. W. Shane
Astronomical Institute, Catholic University
Toernooiveld
6525 ED Nijmegen
Holland

NGC 2146 is a disturbed looking SbII pec galaxy, distance 14.5 Mpc. Optically we see a nucleus, two spiral arms and a prominent dust lane almost crossing the nucleus. Apart from being a strong radio continuum source, IRAS data now show that it is a very strong infrared source, suggesting a recent burst of star formation. The galaxy was observed in 1981 with the WSRT in the 21-cm line of HI. The spectral resolution was 32 km/s and the map shown here has been convolved to a 1' beam.

The HI distribution in the optical galaxy is strongly bounded on the north side. It coincides well with the NW spiral arm and the dust lane. On the NE-side it is displaced southward from the optical arm, leaving this part almost empty. To the south the HI distribution becomes more extended and diffuse. Well beyond the optical image we find two long extensions to the south-east and south, without known optical counterparts.

Most remarkable is the gently curved southern tail (see figure). This is seen best at 870 km/s where it extends from 5' to 22' (95 kpc) southward from the nucleus. A weaker feature appears to split off toward the east at 870 km/s. There is practically no velocity gradient over the tail, but at 850 km/s a concentration is seen 3' east of the south extremity. The second - SE - tail, which is not shown in the figure, is first seen at about 700 km/s at 3' from the nucleus and extends outward, more or less along the major axis, to 6' (25 kpc) at 770 km/s. At 830 km/s there is an even more extended parallel feature displaced 5' SW, which is more irregular and shows little velocity gradient.

The presence of HI in the tails is in agreement with the single dish observations of Fisher and Tully (1976, hereafter FT). Because of primary beam attenuation we cannot as yet confirm the northward extension. We do see a very weak diffuse extension 10' toward the north (compare FT). Optically NGC 2146 has some features in common with NGC 520. As many mergers like NGC 520 are bright infrared sources (Joseph and Wright, 1985) we may well be looking at a merger, in which case the tails can be formed by tidal interaction and one could hope to see faint optical counterparts. Burbidge et al. (1959) reported double emission lines but later spectroscopic observations (Benvenuti et al. 1975) did not confirm this.

F. P. Israel (ed.), Light on Dark Matter, 445–446.
© 1986 by D. Reidel Publishing Company.

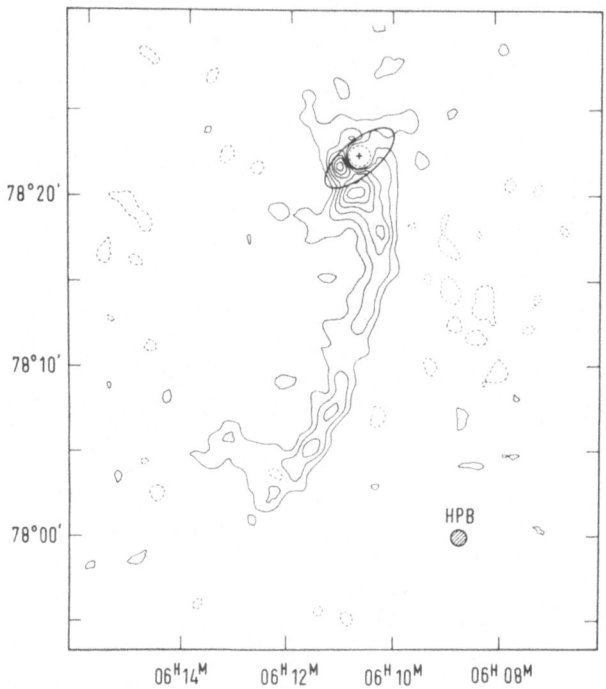

HI in NGC 2146 for 800 < V < 900 km/s. These channels include all HI in
the south tail. The contour interval is 1.75 mJy, zero contour omitted.
The heavy ellipse shows the optical outline of the galaxy.

    Another possibility is the stripping of hydrogen from the galaxy
by an intergalactic medium (G. D. van Albada, private communication).
This could also account for the removal of gas from the NE quadrant and
the low velocity gradient in the tails. In this case we might expect
X-ray emission from a halo around the galaxy. The distribution of the
hydrogen in the northern extension, seen by FT, is being investigated.

REFERENCES

Benvenuti, P., Capaccioli, M., D'Odorico, S. 1975, Astron. Astrophys.
    41, 91
Burbidge, E.M., Burbidge, G.R., Prendergast, K.H. 1959, Astrophys. J.
    130, 739
Fisher, J.R. and Tully, R.B. 1976, Astron. Astrophys. 53, 397
Joseph, R.D. and Wright, G.S. 1985, Mon. Not. roy. astron. Soc. 214, 87

STARBURSTS IN THE NUCLEI OF INTERACTING AND MERGING GALAXIES

R D Joseph
Blackett Laboratory, Imperial College, London SW7

ABSTRACT. We review the evidence for IR activity in interacting and merging galaxies. Interactions appear to induce bursts of star formation of exceptional IR luminosity with very high efficiency. Mergers produce 'super-starbursts' of even greater luminosity and spatial extent. As a result, interacting galaxies are exceptionally prominent among IRAS galaxies. Finally, we indicate some of the larger implications of interaction-induced starbursts for extra-galactic astronomy and observational cosmology.

1. INTRODUCTION

One of the recurring themes in the press releases which emerged during the IRAS mission was the 'surprising' prominence of interacting galaxies among the extra-galactic objects detected by IRAS. If this were true it would not be surprising to some of us who have been interested in interacting galaxies. But perhaps a bit of caution is in order: we have been here before. It would be ironic if, following the first high sensitivity all-sky IR survey, we followed in the footsteps of our radio brethren and announced, prematurely, that IR galaxies are commonly colliding galaxies. It is at least prudent to specify what we mean by interacting galaxies, and why we think that they are genuinely physical interactions.

I take interacting galaxies to be a pair of galaxies in close angular proximity, with disturbed morphologies accompanied by 'bridges' and 'tails' indicative of tidal disturbance. Moreover, the differences in redshift should be small, < 200 km/s. I should like to distinguish one subset of interactions: those which have resulted in a merger in which the two participating galaxies have lost their individual identities and appear as a single, coalesced object. The principal morphological evidence for a merger is two tidal tails emanating from a single disturbed galaxy (cf. Toomre & Toomre 1972).

2. 'INFRARED ACTIVITY' IN GALAXIES

An investigation of IR activity in galaxies should begin with a precise statement of what is meant by 'IR activity.' I will use the definition

447

given by Balick & Heckman (1983): '[emission which is] qualitatively
unusual, quantitatively energetic, compared to evolution of normal
stars.' A quantitative indicator of such activity is the mass-to-light
ratio, M/L_{bol}. If the luminosity is due to the thermonuclear energy
released in the 'evolution of normal stars' over the lifetime of a
galaxy, then one expects a mass-to-light ratio of $\sim 1\ M_\odot/L_\odot$ if $\sim 10\%$
of the galaxy's mass undergoes nuclear burning over a galaxy lifetime of
$\sim 10^{10}$ yr. For 'IR galaxies' such as those detected by IRAS, with a
steeply-rising continuum spectrum between 5 and 100 μm, $L_{bol} \sim L_{IR}$,
and so a primary indicator of IR activity is a ratio $M/L_{IR} \ll 1$.

3. EVIDENCE FOR NUCLEAR INFRARED ACTIVITY IN INTERACTING GALAXIES

3.1 JHKL Photometry

One of the most convincing ways to demonstrate a large IR luminosity and
therefore an indication of nuclear IR activity is through photometry at
10 μm. However, the rather severe sensitivity limitation due to photon
noise from warm, ground-based telescopes, makes it difficult to survey a
large number of galaxies at 10 μm. As an alternative strategy we
(Joseph et al. 1984a) chose to do JHKL photometry, and to use a K-L
colour index significantly redder than normal galaxy colours to infer a
steeply-rising far-IR continuum spectrum. We observed a representative
selection of 50 interacting galaxies from the Arp (1966) Atlas, composed
of various morphological types. The striking result which emerged was
that there was a K-L excess, i.e. K-L > 0.5 (cf. K-L = 0.3 for
early-type galaxies and the bulges of spirals), for one member of an
interacting pair for ~ 85% of the sample. In no case did we find more
than one member of a pair to exhibit a K-L excess. This suggests that
the K-L colour excesses are not merely distributed randomly among
galaxies generally, but are causally related to each specific
interaction. Thus, if a K-L colour excess is a good indicator of a
large IR luminosity and a small M/L_{IR}, it appears that interactions are
extremely efficient in triggering nuclear activity in galaxies.

3.2 Photometry at 10 μm

We have followed up the JHKL study with 10 and 20 μm photometry
(Joseph et al. 1984b, Wright et al. 1985a), and similar studies have
been carried out by Lonsdale et al. (1984) and Cutri & McAlary (1985).
In total there is 10 μm photometry now available in small apertures (5-8
arcsec) for the nuclei of about 30 interacting galaxies. The continuum
spectra all exhibit a steep rise from 3-20 μm. The 10 μm luminosities,
computed from $L_{10} = 4\pi D^2 \nu S_\nu$, have an average $\langle L_{10} \rangle = 3\times10^9\ L_\odot$, and
a range $L_{10} \sim (1\ \text{to}\ 100)\times10^8\ L_\odot$. By comparison, the starburst galaxies
M82 and NGC253 have luminosities $L_{10} = 10^9\ L_\odot$ and $6\times10^8\ L_\odot$ respectively.
While the sample of interacting galaxies is not exactly unbiased, this
comparison with the most luminous of non-interacting starburst galaxies
provides tantalising evidence that interactions induce unusually intense
nuclear activity.

3.3 IRAS measurements of interacting galaxies

It is not immediately clear that IRAS measurements, with arcminute
apertures, are relevant to a discussion of nuclear activity in galaxies.
Since we have shown above that there is good reason to expect that the
nuclear IR emission for the galaxies under consideration will be
unusually luminous, I have assumed that the far-IR IRAS flux densities
are dominated by emission from the nuclei.

There are three samples of interacting galaxies for which IRAS
detections have been investigated. In the sample studied by Joseph et
al. (1984a), 19 of their 22 pairs were detected by IRAS. The average
integrated IR luminosity of this sample is $\sim 5 \times 10^{10}$ $L_\odot$.

Cutri & McAlary (1985) defined a complete sample of 39 pairs of
galaxies brighter than mpg = 14 from the Karachentsev Catalogue of
Isolated Pairs of Galaxies (1972). They report IRAS detections for
three-fourths of the pairs in this sample. The average integrated IR
luminosity of the detected galaxies is $\sim 9 \times 10^{10}$ $L_\odot$, a result which is
probably heavily Malmquist-biased.

Lonsdale et al. (1985) have defined a "clean" subset of the entire
Arp (1966) Atlas with ~ 90 IRAS detections not confused by nearby
sources, "cirrus," or galactic emission. This sample has an average
integrated IR luminosity $\sim 4 \times 10^{10}$ $L_\odot$ (where I have removed the merging
systems as defined above).

Lonsdale et al. have also identified a control sample of spiral
galaxies, for which the average IRAS luminosity is 1.5×10^{10} $L_\odot$. Thus
the interacting galaxies have total IR luminosities 3 to 6 times that of
'normal' spirals. Clearly interactions produce luminous IR galaxies.

4. ENERGY SOURCE AND EMISSION MECHANISM

There are two popular candidates for the underlying energy sources
driving the nuclear activity in galaxies, 'starbursts,' and 'monsters'
(cf. Heckman et al. 1983). A first clue in distinguishing between these
two sources in interacting galaxies comes from the 3-20 μm continuum
spectra. The interacting galaxies have a quasi-thermal spectrum (a slow
rise compared to a Planck function) indicative of thermal emission over
a range of temperatures, rather than the power law spectrum typical of
quasars. Thermal emission suggests that starbursts are the underlying
energy source, with dust absorbing the hard stellar radiation and
re-radiating it in the IR, as is seen in galactic HII regions.

However, there could be an underlying 'monster' with an optically
thick dust cloud thermalising its emission. One way to distinguish
between this possibility and a starburst is to measure extended emission
from the (non-stellar) source of the IR activity, say at 10 μm. If a
central source of luminosity 10^{10} $L_\odot$ heats a dust grain of radius 0.1 μm
to a temperature of 300 K, the grain can be no more than a few parsecs
from the source. Thus if the nucleus of any galaxy has angular size
> 1 arcsec at 10 μm, the emission cannot arise from a uniform dust cloud
heated by a single central source, and it is most plausible that a
starburst is present.

In our 10 µm studies of the interacting galaxies with K-L excesses
(Wright et al. 1985a) we have found spatial extent on a 5 arcsec scale
for four bright galaxies, and Cutri & McAlary (1985) report spatial
extent for a fifth. Thus, the available evidence points strongly toward
starbursts, rather than monsters, in these interacting galaxies.

5. INFRARED ACTIVITY IN MERGING GALAXIES

We turn now to what I regard as the most interesting subset of
interacting galaxies: those interactions in which two disc galaxies have
lost their individual identities and have coalesced into a single
object. This excludes pairs of galaxies which may merge in future
(e.g. NGC3690 + IC694) as well as ellipticals such as cD galaxies and
elliptical-disc systems like NGC5128. Two of the much-discussed IRAS
extra-galactic sources, NGC6240 (Wright et al. 1984) and IC4553, alias
Arp220, (Soifer et al. 1985) are mergers under this definition.

5.1 Photometry of Mergers at 10 µm

We (Joseph & Wright 1985) began our investigation of IR activity in
merging galaxies by identifying all those mergers of which we were aware
for which 10 µm (pre-IRAS) photometry is available. There are 9 such
mergers, and their average 10 µm luminosity is $\langle L_{10} \rangle \sim 2 \times 10^{10}$ $L_\odot$, with
range $4 \times 10^9 - 5 \times 10^{10}$ $L_\odot$. Thus the average 10 µm luminosity of
these mergers is about an order of magnitude larger than that of the
non-merging interacting galaxies discussed above.
        We have measured extended 10 µm emission for 3 of these mergers,
and similar measurements are in the literature for 3 more. The extent
of the 10 µm emission in these galaxies is phenomenal, ranging from 1 -
5 kpc. [The situation with IC4553 is ambiguous, however. Becklin
(private communication) finds only a point source at the position of the
radio peak at 20 µm, whereas Rieke et al. (1985) found the source
extended over at least 15 arcsec at this position at 10 µm.] It appears
that, in general, these merging galaxies exhibit exceptionally extended
10 µm emission, and it is difficult to avoid the inference that the
similarly exceptional 10 µm luminosities of these galaxies are due
to recent bursts of star formation. This interpretation is supported by
optical and radio data. Thus it seems that, for this small sample at
least, disc-disc mergers result in super-starbursts of exceptional
luminosity and spatial extent.

5.2 IRAS Observations of Mergers

The publication of the IRAS catalogues has enabled us to extend our
investigation of IR activity in merging galaxies to a larger sample. We
(Joseph et al. 1985b) have combed the literature for all examples of
mergers, culling out those which do not conform to the criteria listed
above. There are 31 such objects for which redshifts are available, and
there are IRAS detections for 29 of them. This sample has an average
integrated IR (10 - 100 µm) luminosity $\sim 5 \times 10^{11}$ $L_\odot$ and a range

~ (2 to 200)x10^{10} L$_\odot$. In Fig. 1 we show the integral luminosity
distributions, based on IRAS data, of these mergers, the Cutri & McAlary
(1985) interacting galaxies, and the Miley et al. (1985) Seyferts,
compared with that for the Hercules cluster (Young et al. 1984). The
luminosity distributions of the (non-merging) interacting galaxies and
Seyferts are essentially equivalent, whereas the mergers are
significantly more luminous. This is admittedly a very rough
comparison, since there is undoubtedly significant Malmquist-biasing in
all four samples, but it suggests that the super-starbursts induced in
mergers produce the most luminous class of IR galaxies yet discovered.

Space does not permit discussion of recent IR spectroscopy of
interacting and merging galaxies (Fischer et al. 1983, Joseph et al.
1984b, Rieke et al. 1985). We have detected strong quadrupole emission
from H$_2$, and the 1.644 μm line of [FeII], in virtually every
interacting galaxy we have observed. These results provide additional
support for the starburst interpretation, as well as physical
diagnostics for the associated physical processes (Wright et al. 1985b).

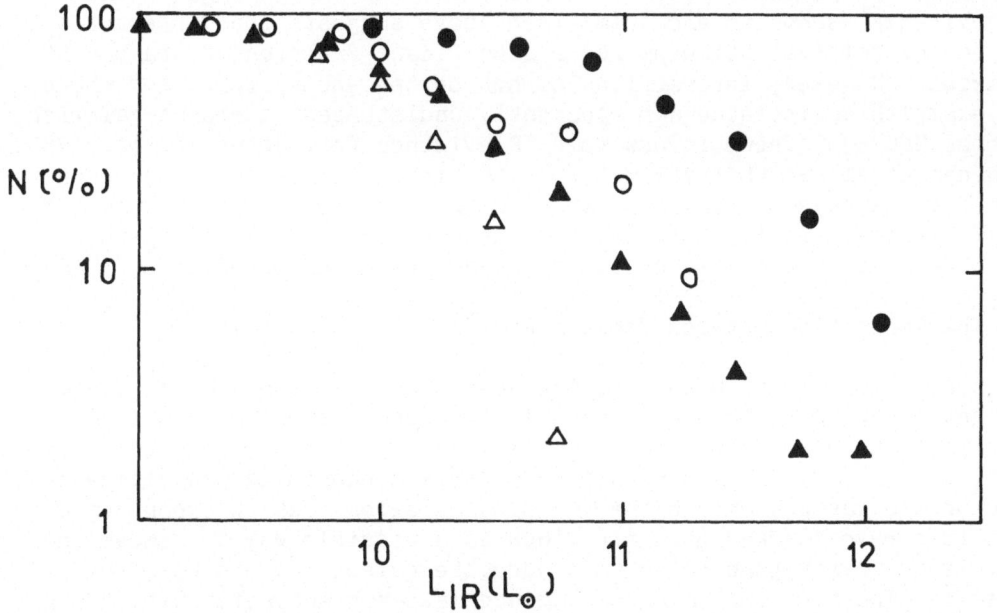

Figure 1. Integral luminosity distributions of merging galaxies (●),
non-merging interactions (O), Seyferts (▲), & the Hercules cluster (Δ).

6. ARE INTERACTIONS AND MERGERS OVER-REPRESENTED AMONG IRAS GALAXIES?

The data on IR luminosities of interacting and merging galaxies adduced
above would suggest that such galaxies must be over-represented among
the galaxies detected by IRAS.   Probably the best evidence comes from
investigation of the IRAS 'Minisurvey' galaxies by Lonsdale et al.
(1984). They examined 5x enlargements of POSS prints for each of the 81
galaxies detected in the Minisurvey and they conclude that, at minimum,

34% of the IRAS galaxies are interacting or merging systems.

There is also anecdotal evidence from programmes to identify unidentified IRAS sources whose IRAS colours suggest that they may be galaxies. Allen, using deep CCD exposures at the Anglo-Australian Telescope, reports (private communication) that about half of the IRAS sources he identifies with galaxies are, in fact, interacting. In a similar endeavour, Wolstencroft et al. (in this volume) are using plates taken with the UK Schmidt Telescope to identify IRAS sources, and they also find a 'surprisingly' large fraction of the galaxies identified to be highly disturbed or interacting. Thus, all the indications are that interacting galaxies are especially prominent among IRAS galaxies.

7. IMPLICATIONS

7.1 What Fraction of Starburst Galaxies Are Due to Interactions?

One of the outstanding questions posed by the discovery of starbursts in the nuclei of galaxies (cf. Rieke 1981) is, what has triggered them? The evidence assembled and summarised above suggests that interactions may be the critical stimulus for a significant fraction of starburst galaxies. However, interactions cannot be the whole story, for there are examples of isolated and apparently undisturbed starburst galaxies, such as NGC253. There is now some IR evidence (cf. Joseph et al. 1985a, Hawarden et al. in this volume) that the instabilities which produce barred galaxies also fuel a burst of star formation in the centres of such galaxies. It is possible that interactions and barred systems together account for most of the starburst galaxies observed by IRAS.

7.2 The Connection Between Mergers and Elliptical Galaxies

The suggestion has been current for some time that the violent dynamical relaxation which follows a merger will produce a stellar velocity distribution which looks very much like an elliptical galaxy (cf. Toomre 1977, White 1979). One uncertainty of this scenario is the ultimate fate of the gas present in the two disc galaxies. While exponents of this idea have invoked galactic winds as a possible way to remove the gas, it has never been clear that adequate galactic winds would be present. The fact that mergers engender super-starbursts, with M/L_{IR} < 1 (cf. Joseph & Wright 1985) over a spatial extent of kiloparsecs, does apparently lay this problem to rest. Graham et al. (1984) show that for a starburst with an M/L < 1, the associated supernovae will produce galactic winds sufficiently energetic to sweep the corresponding volume free of gas. Therefore a super-starburst, with spatial extent of kiloparsecs, should leave the merger remnant as gas-depleted as any elliptical galaxy.

7.3 Spectral Evolution of Galaxies

The probability of galaxy-galaxy interactions should increase rather rapidly as one looks back at epochs of increasing redshift. The probability of interaction will be proportional to the galaxies' peculiar velocities with respect to the Hubble flow, divided by the interaction mean free path. The probability of interaction should therefore increase with redshift roughly as $(1 + z)^4$. Thus we expect to see the effects of interactions and mergers--starbursts and super-starbursts--in a much larger fraction of galaxies at higher redshifts. Indeed, Lilly & Longair (1984) note evidence for just such effects of interactions in their study of distant 3CR radio sources.

7.4 The Relation Between Interaction-Induced Starbursts and Other Types of Activity in Galaxies

Balick & Heckman (1982) have recently reviewed the extranuclear clues to the origin of activity in galactic nuclei. Some of the clues they cite are likely to be associated with interactions. For example, Seyfert galaxies commonly exhibit tidal distortions and have companions. In the case of low redshift quasars, when an extended image is found, it tends to be distorted, or a companion is near (e.g. Stockton 1982). MacKenty & Stockton (1984) find strong emission lines in the extended luminous material surrounding the quasar Mrk1014, which they plausibly attribute to an interaction-induced burst of star formation.

8. CONCLUSIONS

The evidence marshalled above suggests that there is exceptional nuclear IR activity in interacting galaxies. The source of this activity seems to be a recent burst of star formation. Merging galaxies show similar, but more pronounced features; they may be the most luminous IR objects in the heavens. These results of galaxy-galaxy interactions should have rather wide-ranging implications for a variety of topics in extra-galactic astronomy and observational cosmology.

However, there are many lacunae in the story presented above. One of the objectives of this review is to identify just where these these gaps are, and to indicate some of the directions in which future studies of interacting galaxies might be most rewarding. With more time to digest the embarras de richesses of IRAS survey data and the additional observations, and further studies with ground-based instruments, more firm and detailed astrophysical conclusions will most certainly be presented in future IRAS conferences.

ACKNOWLEDGEMENTS

It is a pleasure to acknowledge the contributions of Gillian Wright, Norna Robertson, Peter Meikle, Jack Abolins, Ian Gatley, James Graham, and Richard Wade on various aspects of our studies of interacting galaxies, and to thank Eric Becklin, Roc Cutri, and Carol Lonsdale Persson for sending material in advance of publication.

REFERENCES

Arp, H., 1966. Atlas of Peculiar Galaxies. California Institute of
     Technology, Pasadena.
Balick, B. & Heckman, T., 1982. Ann. Rev. Astr. Astrophys., **20**, 431.
Cutri, R. & McAlary, C., 1985. Preprint.
Fischer, J., Simon, M., Benson, J., & Solomon, P.M., 1983. Astrophys.
     J., **273**, L27.
Graham, J.R., Wright, G.S., Meikle, W.P.S., Joseph, R.D., & Bode, M.F.,
     1984. Nature, **310**, 213.
Heckman, T.M., van Breugel, W., Miley, G.K., & Butcher, H.R., 1983.
     Astron. J., **88**, 1077.
Joseph, R.D., Hawarden, T.G., & Gatley, I., 1985a. Proceedings of the
     Workshop on the Virgo Cluster of Galaxies, p. 127, eds. Richter,
     O.-G., & Binggeli, B., European Southern Obervatory, Garching.
Joseph, R.D., Meikle, W.P.S., Robertson, N.A., & Wright, G.S., 1984a.
     Mon. Not. R. astr. Soc., **209**, 111.
Joseph, R.D., Meikle, W.P.S., Robertson, N.A., & Wright, G.S., 1984b.
     Proceedings of the Workshop on Star Formation (1983: Edinburgh),
     p. 177, ed. Wolstencroft, R.D., Royal Observatory, Edinburgh.
Joseph, R.D., Wright, G.S., & Abolins, J.A., 1985b. In preparation.
Joseph, R.D. & Wright, G.S., 1985. Mon. Not. R. astr. Soc., **214**, 87.
Karachentsev, I., 1972. Comm. Special Astrophys. Obs. USSR, **7**, 1.
Lilly, S.J. & Longair, M.S., 1984. Mon. Not. R. astr. Soc., **211**, 833.
Lonsdale, C.J., Persson, S.E., & Matthews, K., 1984. Astrophys. J.,
     287, 95.
Lonsdale, C.J., Neugebauer, G., & Soifer, B.T., 1985. In preparation.
MacKenty, J.W. & Stockton A., 1984. Astrophys. J., **283**, 64.
Miley, G.K., Neugebauer, G., & Soifer, B.T., 1985. Preprint.
Rieke, G.H., 1981. Infrared Astronomy, p. 317, eds. Wynn-Williams, C.G.
     & Cruikshank, D.P., Reidel, Dordrecht, Holland.
Rieke, G.H., Cutri, R.M., Black, J.H., Kailey, W.F., McAlary, C.W.,
     Lebofsky, M.J., & Elston, R., 1985. Astrophys. J., **290**, 116.
Stockton, A., 1982. Astrophys. J., **257**, 33.
Soifer, B.T., Helou, G., Lonsdale, C.J., Neugebauer, G., Hacking, G.,
     Houck, J.R., Low, F.J., Rice, W., & Rowan-Robinson, M., 1984.
     Astrophys. J., **283**, L1.
Toomre, A. & Toomre, J., 1972. Astrophys. J., **178**, 623.
Toomre, A., 1977. Evolution of Galaxies and Stellar Populations, p. 401,
     eds Tinsley, B.M. & Larson, R.B., Yale University Observatory, New
     Haven, Connecticut.
Young, E., Soifer, B.T., Low, F.J., Neugebauer, G., Rowan-Robinson, M.,
     Miley, G., Clegg, P.E., de Jong, T., & Gautier, T.N., 1984.
     Astrophys. J., **278**, L75.
White, S.D.M., 1978. Mon. Not. R. Astr. Soc., **184**, 185.
Wright, G.S., Joseph, R.D., & Meikle, W.P.S., 1984. Nature, **309**, 430.
Wright, G.S., Joseph, R.D., Robertston, N.A., & Meikle, W.P.S., 1985a.
     In preparation.
Wright, G.S., Joseph, R.D., Wade, R., Gatley, I., & Graham, J.R., 1985b.
     Proceedings of the RAL Workshop on Extra-Galactic IR Astronomy, ed.
     P.M. Gondhalekar, Rutherford Appleton Laboratory, Chilton, UK.

STARBURSTS IN NON-INTERACTING GALAXIES

T.G. Hawarden[1], J.H. Fairclough[2], R.D. Joseph[3], S.K. Leggett[4],
C.M. Mountain[1]
[1]Royal Observatory, [4]Astronomy Dept., Blackford Hill, Edin.
[2]RAL, Chilton, Didcot, Oxon, OX1 QX11
[3]Imperial College, Prince Consort Road, London

ABSTRACT   Examination of IRAS results for a complete sample of nearby
galaxies with morphological class T between 0 and 6 (inclusive)
suggests that episodes of star formation vigorous enough to dominate
the IR colours of a system occur <u>only</u> in barred galaxies. This is
understandable if the location of these "starbursts" are the small
complexes of HII regions which commonly surround the nuclei of barred
galaxies of these types and which probably mark circumnuclear
stagnation zones in a bar-driven inward flow of gas. The luminosity of
these complexes will be regulated by the supply of gas to the bar and
this model appears capable of explaining most of the mid-IR properties
of bright spiral galaxies.

1.   INTRODUCTION

1.1 The centres of barred galaxies

Remarkably little attention has been paid to the region surrounding the
nuclei of barred galaxies. This is partly because of the established
interests of many workers in the potentially exotic physics of the
nucleus itself and those of most of the remainder in the problems
associated with the large-scale properties of the main body of the
galaxy; even more so because these central regions are usually burned
out on the majority of sky survey photographs. Consequently, most
astronomers are unaware that many barred spirals have nuclei surrounded
by luminous ring-like complexes of HII regions, between a few hundred
parsecs and a kiloparsec or so in radius, intimately associated with
dust lanes which run down the edges of the bar. Several examples are
illustrated and discussed in the Hubble Atlas (Sandage, 1961).
Well-studied specimens include NGC 1512 (Hawarden et al., 1979;
Lindblad & Jörsäter,1981) for which quite detailed kinematics are
available and NGC 1097 (Rickard, 1975; Schempp, 1980; Wolstencroft,
Tully & Perley, 1984). This galaxy is discussed further, below, and is
illustrated in Figure 1.

F. P. Israel (ed.), Light on Dark Matter, 455–462.
© *1986 by D. Reidel Publishing Company.*

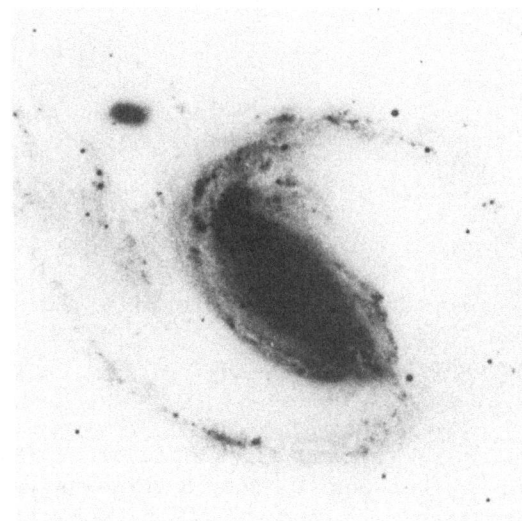 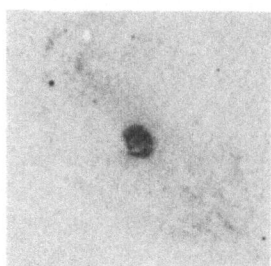

Figure 1: The luminous SBbI galaxy NGC 1097, showing its apperance on a deeply-exposed IIIaJ plate taken with the 1.2m UK Schmidt Telescope and, on the same scale, a detail of the central circumnuclear HII regions, taken from a short exposure in the U band by R Brent Tully. North is up, East is to the left. The smaller picture is 3.8 arcmins across.

The dust lanes have long been known to trace out shock fronts generated in the ISM by the rotating ellipsoidal potential of the bar, and have been studied in some detail by theoreticians. Thus Schwarz (1985) notes that these fronts curl around the nucleus in all his models and furthermore that they are responsible for very efficient movement of gas inwards along the bar. Indeed, his models indicate that almost all the gas within one bar radius is driven inwards to the vicinity of the nucleus on a timescale of about 10^8 y almost independent of the bar strength and pattern speed, an effect which Schwarz finds inconvenient for modelling the further evolution of the bar. The theoretical prediction of bar-driven inflow is supported by both HI (Sancisi et al., 1981) and optical (Blackman & Pence, 1984) observations of such flows in barred systems.

Any bar-driven inflow must be a function of the strength of the non-axisymmetric component of the potential. For any galaxy with a strong central concentration of mass this non-axisymmetric component must decline in importance very close to the centre and consequently there must be a radius at which the inflow will cease. This implies that there must exist a circumnuclear stagnation zone in which the gas driven by the bar accumulates. Such a zone is obviously a very likely location for vigorous star formaton and we contend that the ringlike circumnuclear HII-region complexes mark this zone.   Their association with the inner Lindblad resonance of these systems has been remarked on -obliquely - by Wolstencroft et al., (1984) and by Lindbland & Jörsäter (1981.).   If our arguments are correct, essentially all massive barred spirals should possess the dust lanes and HII-region complexes and we would expect a high fraction of these complexes to be the sites of substantial star-formation episodes.

We have systematically examined a large number of nearby galaxies on the most suitable available plates in the UKST Plate Library at ROE and on several hundred high-resolution, short exposure photographs of galaxies by R B Tully at the Institute of Astronomy of the University of Hawaii.   We find that the dust lanes in the bar are visible in most spirals of types SBO/a (or SABO/a) through SBc (or SABc), provided that the orientation of the galaxy is favourable and the plate material suitable.   The dust lanes, in fact, provide a powerful morphological diagnostic for the presence of any significant degree of "barredness", i.e. of a non-axisymmetric light (and mass) distribution in the inner parts of the galaxy.   Furthermore we believe that the complexes of HII regions are probably present whenever the dust lanes are there; as remarked above, however, these complexes can be hard to discern.

The especially beautiful ringlike central complex which occurs in NGC 1097 is shown in Fig. 1.   This ring has been mapped at 10 microns by Telesco & Gatley (1981) and is the site of a particularly vigorous current outburst of star formation, with total luminosity $\sim 10^{11}$ L$_\odot$. Almost all this energy originates in the ring; the actual nucleus contributes only a few percent.   We hypothesise that these rings in barred spirals may often be starbursts and have examined the IRAS Point Source Catalogue to explore this possibility.

2.   IRAS DATA

2.1 Data for nearby spirals.

Figure 2 shows IRAS fluxes from the Point Source Catalogue for two galactic HII regions of small angular diameter and for two galaxies, NGC 1097 and NGC 5033.   The latter object stands out from the others by the relative <u>steepness</u> of its spectrum between 25 and 100 microns and its <u>flatness</u> between 12 and 25 microns.   It is clear that NGC 5033 is dominated by two components, a relatively hot component, probably peaking shortwards of 25 microns and a very cool component peaking well to the red of 100 mincrons.   In sharp contrast the two HII regions and NGC 1097 - which, as we have seen, contains an exceptionally vigorous region of star formation in its inner parts - have spectra which appear to be dominated by a "warm" dust component.   These spectra rise strongly from 12 to 25 microns and less dramatically than that of NGC 5033 thereafter.

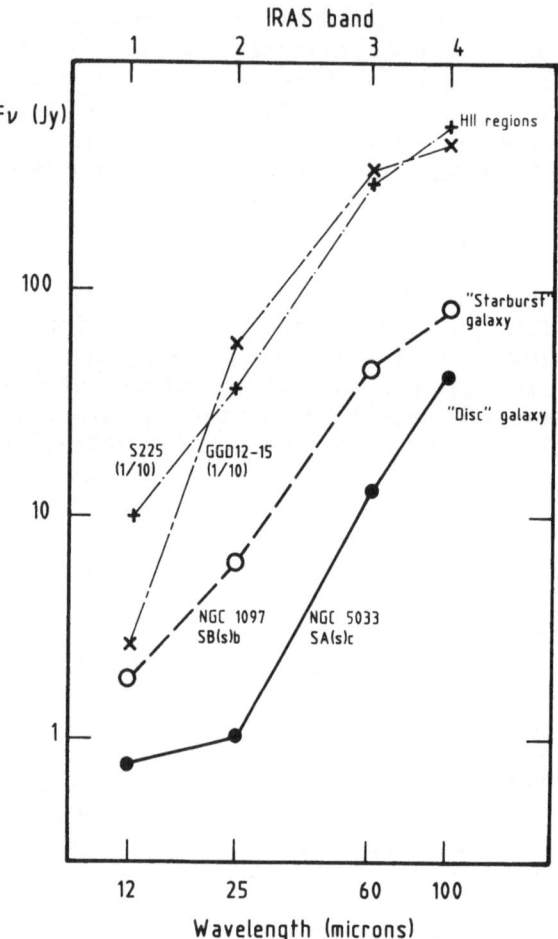

Figure 2: IRAS fluxes from the point source catalogue of two HII regions, S225 and GGD12-15 (LIR ~ 10^4 L$_\odot$) for comparison to the "starburst" galaxy NGC 1097 and "Disc" galaxy NGC 5033.

The form of these spectra strongly suggest that the degree to which regions of current star formation activity dominate the IR properties of a galaxy can readily be discerned by comparing the flux ratios measured by IRAS in bands 1 and 2, which will measure the dominance of the warm dust of the HII regions over the hot component, and in bands 2 and 4, which compare the HII region contribution to that of the cold dust component. In such a diagram "starburst" systems such as NGC 1097, NGC 253 and NGC 3310 should exhibit high values of the ratio R(2/1) between the flux in band 2 (25 microns) and that in band 1 (12 microns) and low values of the ratio R(4/2) between the fluxes in bands 4 (100 microns) and 2. (Comparisons involving the 60 micron band do not add much to this argument.)

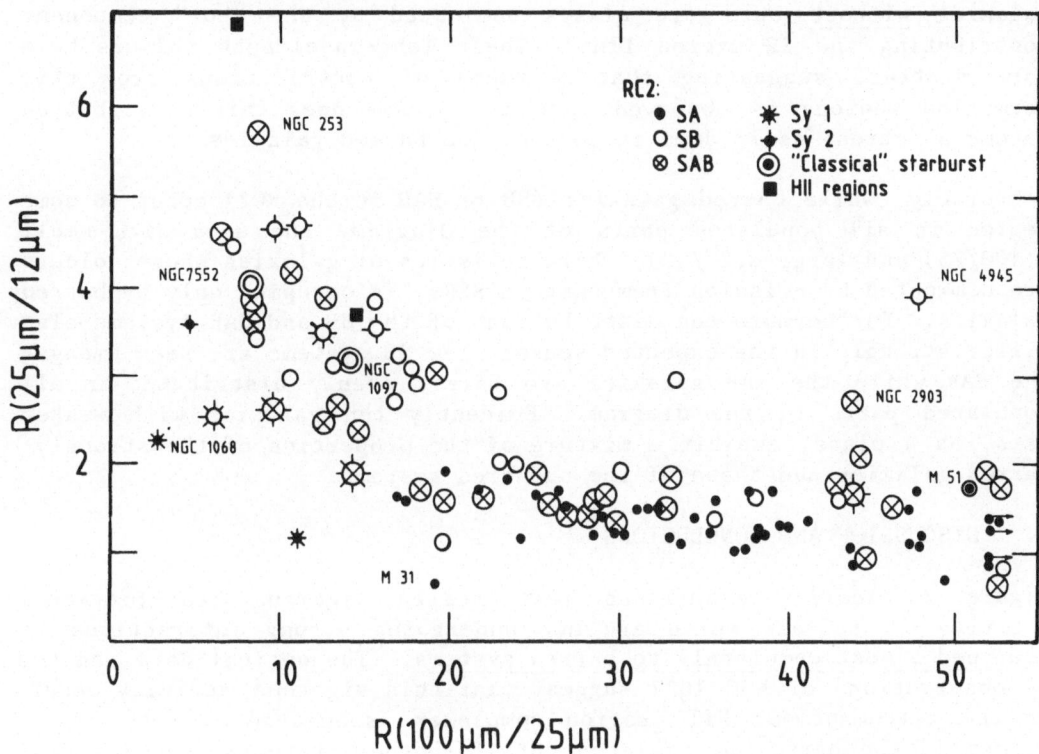

Figure 3: Plot of the IRAS flux ratios, R(2/1) against R(4/2), for a complete sample of nearby spiral galaxies. The HII regions S225 and GGD12-15 have also been included for comparison.

Figure 3 shows a plot of R(2/1) against R(4/2) for a complete sample of nearby spiral galaxies with moderate to high quality fluxes in all four IRAS bands. The IRAS point source catalogue has been compared (using a spatial tolerance of ±90 arcseconds) to all galaxies in the Shapley-Ames catalogue that have B_T ≤ 12.0m and that have been typed by de Vaucouleurs et al., (1976) as either SA, SAB(=SX) or SB with Hubble types 0<T<6.

No discrimination has been made on angular diameter. Nevertheless 95% of this sample have 100µm point source correlation coefficients of A(=100%) so size effects can in general be ignored. In only two galaxies (NGC 224 = M31 and NGC 3031) is the cirrus 2 flag > 3.

A remarkable degree of segregation occurs on this diagram. With the exception of one or two Seyfert galaxies, few or no SA systems (i.e. those classified in RC2 as exhibiting no signs of a bar) have R(25/12) > 2.0 or R(100/25) < 15 or so. The short-wavelength colours of galaxies <u>without</u> bars are always dominated by the "hot" component contributing the 12 micron flux. Their long-wavelength colours have more scatter, suggesting that a range of contributions from star formation regions is observed, but in no case does this contribution become as strong as it does in some of the barred galaxies.

Conversely, while barred galaxies (SB or SAB in the RC2) occur to some degree in all populated parts of the diagram, the area with small R(100/25) and large R(25/12), characteristics of galaxies whose colours are dominated by emission from dust in SFRs, is occupied <u>only</u> by barred galaxies. Furthermore the distributions of the SB and SAB systems also differ strongly in the expected sense: few SB systems are seen amongst the SAs while the SAB galaxies are fairly evenly distributed in all populated parts of the diagram. Evidently the galaxies with weaker bars, as a class, exhibit a mixture of the properties of the strongly-barred galaxies and those of the unbarred systems.

3.   DISCUSSION AND CONCLUSIONS

Figure 3 clearly establishes that really vigorous star-formation activity in spirals which are not undergoing strong interactions is <u>confined almost completely to barred systems</u>. The optical data and the IR observations of NGC 1097 suggest that this vigorous activity occurs in the circumnuclear HII region complexes associated with the shock fronts in the bars; accumulated data for other starburst systems such as NGC 253 (Rieke et al., 1980; Scoville et al., 1985) supports this identification of the starburst, if present, with a near-nuclear phenomenon such as the central complexes.

The immense IR luminosities of the ring in NGC 1097 and of the star bursts in, e.g., NGC 253 (Rieke et al., 1980; Scoville et al., 1985) cannot be sustained by star formation for more than a small fraction of a Hubble time without accumulating stellar masses greater than currently exist interior to these features. Consequently - as has long been realised - the full-luminosity starburst must be a transient phase in any galaxy. This may not be true of the bar, which must continue to pump any available material inward for as long as it can grasp fresh ISM. If bars are longer-lived than starbursts the majority of the central complexes in barred systems, therefore, must be starved of fuel and probably remain for long periods in a semi-quiescent state, subsisting on relatively meagre rations gleaned from the mass loss of stars in the general bulge. Examples of such galaxies are NGC 1512 and NGC 1300, both of which are strongly barred and contain circumnuclear HII rings but which are not especially luminous in the IR. Neither was detected by IRAS at 12 microns but they probably fall amongst the SA systems in Figure 3. However, the luminous phases <u>do</u> occur with some frequency, (17 out of 62 barred galaxies in our sample have $L_{TOT} > 5$ x $10^{10} L_{\odot}$ while relatively few SBs lie amongst the SAs) so extra rations of gas must be forthcoming in some circumstances. Perhaps the relatively minor interactions with satellite dwarfs - such as that currently occurring in NGC 1097 - which must be quite frequent over a

Hubble time, intermittently perturb sufficient new material within reach of the bar potential? We have not had time to examine these possibilities in any detail.

The infrared observational possibilities of our scenario are numerous. Spectroscopy - and mapping in the main emission lines - is needed to begin to understand the physical properties such as the ionising fluxes and molecular contents of the circumnuclear rings in their great variety of luminosities, and to disentangle extinction effects. Observations of the H_2 lines may uncover cases of systems where raw mechanical energy dumped into the gas enhances shock emission out of proportion to the total luminosity, a process which is probably happening (Joseph et al., 1984, see also this meeting) in the famous post-merger starburst galaxy NGC 6240, with its anomalously strong H_2 emission. At UKIRT we have detected H_2 emission from NGC 2903 and NGC 5236 of a strength roughly appropriate to the observed starburst luminosity in these galaxies.

Clearly, a lot of ground-based work badly needs doing to confirm and elaborate this; the Dutch "IRASFU" on UKIRT, Gatley's "8-Banger" and Telesco's "20-banger" bolometer arrays ought to see much service in this regard.

ACKNOWLEDGEMENTS

The IRAS team, of course, provided the core data for this work. We are also particularly grateful to Clive Davenhall for giving us HAGGIS, with which we explored the databases, and to Morag Brown for giving us access to the riches of the IRAS Point Source Catalog; our thanks are also due to Brent Tully for generating, and letting us loose with, his atlas of galaxy photographs, and, as nearly always, to the UK Schmidt Telescope Unit, for access to the finest suite of astronomical photographs in the world and to to the ROE Photolabs for producing the illustrations.

REFERENCES

Blackman, C.P., & Pence, W.D., 1984. MNRAS, 210, 547.
Hawarden, T.G., van Woerden, H., Melbold, U., Goss, W.M. & Peterson, B.A., 1979. Astron. and Astrophys., 76, 230.
Huntley, J.M., 1980. Ap.J., 238, 524.
Joseph, R.D., Wright, G.S. & Wade, R., 1984. Nature, 311, 132.
Jura, M., 1982. Ap.J., 254, 70.
Lindblad, P.O., & Jörsäter, S., 1981. Astron. and Astrophys., 97, 56.
Rickard, J.J. 1975. Astron. and Astrophys., 40, 339.
Rieke, G.H., Lebofsky, M.J., Thompson, R.I., Low, F.J., & Tokunaga, A.T., 1980. Ap.J., 238, 24.
Sandage, A., 1961. The Hubble Atlas of Galaxies.
Sancisi, R., Allen, R.J. & Sullivan, W.T., 1979. A. & A., 78, 217.
Schempp, W.V., 1980. PhD. thesis, University of Hawaii.
Schwarz, M.P., 1985. MNRAS, 212, 677.

Scoville, N.Z., Soifer, B.T., Neugebauer, G., Young, J.S., Matthews, K.
    & Yerha, J., 1985. Ap.J., 289, 129.
Telesco, C.M. & Gatley, I., 1981. Ap.J., 247, L11.
Vaucoulers, G. de, Vaucouleurs, A. de, Corwin, Jr. H.G., 1976. Second
    Reference Catalogue of Bright Galaxies.
Wolstencroft, R.D., Tully, R.B. & Perley, R.A., 1984. MNRAS, 207, 889.

NEW GROUND-BASED STUDIES OF TWO ACTIVE IRAS GALAXIES

C. G. Wynn-Williams, E. E. Becklin, D. L. DePoy,
J. N. Heasley, G. J. Hill, J. W. MacKenty
Institute for Astronomy, University of Hawaii
2680 Woodlawn Drive
Honolulu, Hawaii 96822, USA

C. A. Beichman
Jet Propulsion Laboratory
California Institute of Technology
4800 Oak Grove Drive
Pasadena, California 91109, USA

ABSTRACT. New data are presented for IRAS 0421+040P06 and NGC 6240. These two galaxies have properties that differ from both Seyfert and starburst galaxies.

1. INTRODUCTION

Although the majority of IRAS galaxies are explicable in terms of bursts of star formation, a small number have been found to show unusual types of behaviour not previously seen in active galaxies. This paper presents a progress report on our studies of two particular sources.

2. IRAS 0421+040P06

Some of the unusual properties of this galaxy have already been described in an earlier paper (Beichman et al. 1985). It is the only IRAS source in the "mini-survey" (Soifer et al. 1984) that has a double-lobe radio structure. Its nucleus has a spectrum similar to that of a Seyfert 2 or narrow-line radio galaxy, but it has the unusual property that ridges of luminous emission are seen connecting the galaxy to the radio lobes that lie well beyond the visible confines of the galaxy. In the earlier paper we speculated that these ridges were spiral arms. Subsequent long-slit spectroscopy and narrow-band imaging with the UH 2.2 m telescope at Mauna Kea (Figure 1) indicate that the light from these ridges has an essentially pure high-excitation emission-line spectrum. We therefore now believe that the ridges are more likely to be luminous curved jets produced as part of the radio-

F. P. Israel (ed.), Light on Dark Matter, 463–466.

galaxy phenomenon rather than regions of high stellar density in the
galaxy disk. More details of the radio lobes can be seen in Figure 2,
which is a new A-array VLA map of 0421+040P06. The radio lobes are now
seen to have the classic Fanaroff and Riley (1974) class I structure
characteristic of low-luminosity radio galaxies. No radio emission is
seen from the visible jets, which extend approximately as far as the
bright inner edges of the radio lobes, about 7" (7 kpc) from the
nucleus.

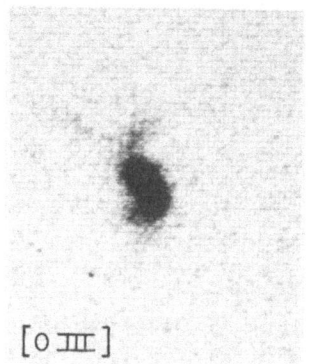 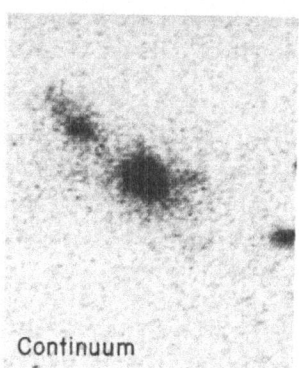

Figure 1. Narrow-band images of 0421+040P06 at the wavelength of the
[OIII] lines (left) and in an emission-line-free region of the
continuum near 6000 Å (right). The diffuse object to the northeast of
the main galaxy is a companion at a similar redshift. The images cover
exactly the same region of sky as shown in Figure 2.

3. NGC 6240

NGC 6240 (IRAS 1650+023P04), with a bolometric luminosity of 4 x $10^{11}L_o$,
is one of the most powerful infrared galaxies known. Its nuclear
region appears as a 1.6" (800 pc) double object at both radio and
visible wavelengths. It is one of the few galaxies from which
molecular hydrogen emission has been detected (Becklin et al. 1985;
Rieke et al. 1985; Joseph et al. 1984). We have now obtained spectra
of the galaxy with the 3.8 m UKIRT that span most of the 2.0 to 2.5
μm atmospheric window (Figure 3), plus the regions around Paschen-α at
1.9 μm and Bracket-α at 4.05 μm.
     Four molecular hydrogen lines have now been detected. The line
ratios are similar to those in Orion and NGC 1068, and correspond to
the values predicted for shock excitation. The S(1) line has a width
of 600 km/sec, similar to those of the optical lines, but is broader
than the H_2 lines seen in any other source. The power emitted in the
H_2 lines exceeds that from any other observed object, and comprises
0.02% of the total bolometric luminosity of NGC 6240.

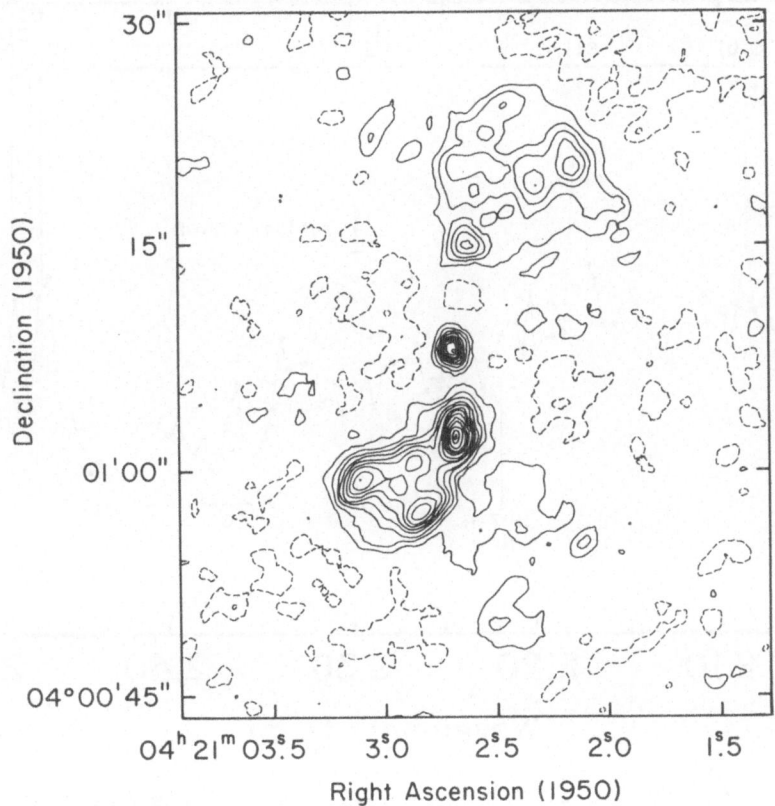

Figure 2. VLA map of 0421+040P06 at 6 cm. The beamsize is 1.5". The central source coincides with the infrared galaxy and, in higher resolution maps, appears as a 0.8" double radio source.

The only hydrogen recombination line that has been detected from NGC 6240 at infrared wavelengths is the Paschen-α line; it has a flux of $0.60 \pm 0.13 \times 10^{-16}$ W m^{-2}. The reddening deduced from a comparison of the Paschen-α to Hβ line ($A_v \sim 2.9$ mag) is very similar to that deduced from the ratio of the Hα and Hβ lines; we find no evidence for the large amounts of extinction proposed by Rieke et al. (1985). The weakness of the recombination lines is surprising; normalized to the bolometric luminosity, the ionization rate in NGC 6240 is some thirty times less than in M82 and in most other 'starburst' galaxies. We conclude that the infrared luminosity in NGC 6240 is probably <u>not</u> caused by a starburst unless, for some reason, the starburst is devoid of any significant number of O stars.

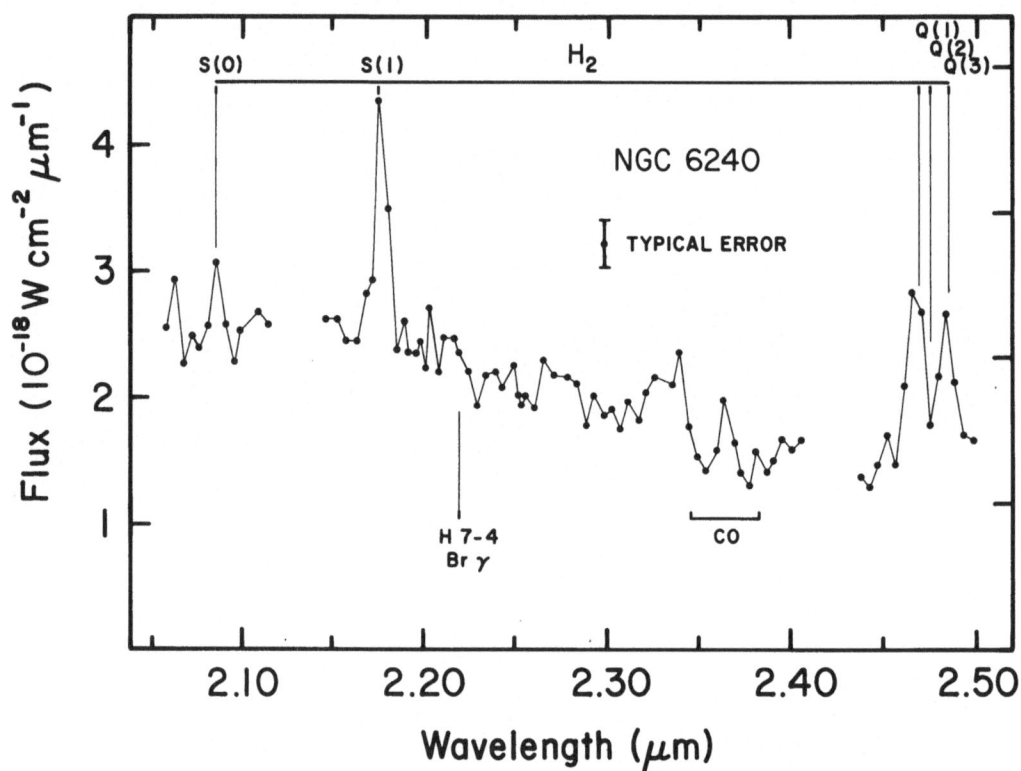

Figure 3. Spectrum of NGC 6240 obtained with a 5.5" diaphragm using
the Cooled Grating Spectrometer on UKIRT.

ACKNOWLEDGMENT

This work has been supported by NSF grants AST 82-17118 and
AST 84-18197.

REFERENCES

Becklin, E. E., DePoy, D. L., and Wynn-Williams, C. G.: 1985,
    Proceedings of the Infrared Detector Workshop, Laramie, Wyoming,
    May 15-16, 1984. In press.
Beichman, C., Wynn-Williams, C. G., Lonsdale, C. J., Persson, S. E.,
    Heasley, J. N., Miley, G. K., Soifer, B. T., Neugebauer, G.,
    Becklin, E. E., and Houck, J. R.: 1985, Astrophys. J. **293**, 148.
Fanaroff, B. L., and Riley, J. M.: 1974, M.N.R.A.S. **167**, 31P.
Joseph, R. D., Wright, G. S., and Wade, R.: 1984, Nature **311**, 132.
Rieke, G. H., Cutri, R. M., Black, J. H., Kailey, W. F., McAlary, C. W.,
    Lebofsky, M. J., and Elston, R.: 1985, Astrophys. J. **290**, 116.
Soifer, B. T., Rowan-Robinson, M., Houck, J. R., de Jong, T.,
    Neugebauer, G., Aumann, H. H., Beichman, C. A., Boggess, N.,
    Clegg, P. E., Emerson, J. P., Gillet, F. C., Habing, H. J.,
    Hauser, M. J., Low, F. J., Miley, G., and Young, E.: 1984,
    Astrophys. J. Letters **278**, L71.

SPECTROSCOPY OF ACTIVE AND STARBURST GALAXIES BETWEEN 8-13µm

P.F. Roche
Anglo-Australian Observatory
P.O. Box 296, Epping, NSW 2121 Australia.

We can divide most infrared-luminous galaxies into two groups;
1. those whose nuclear activity can be explained by photoionisation by hot stars - termed the 'starburst' galaxies.
2. galaxies showing evidence for a strong non-thermal energy source, such as QSOs and Seyferts - the 'active' galaxies.
Galaxies that show no evidence for nuclear activity generally show little excess infrared flux over the expected continuum from stars in the nuclear bulge; these are too faint for 10µm spectroscopy with existing instrumentation.

Spectra at 8-13µm of more than 50 galaxy nuclei have been secured and many have been published (e.g. Aitken & Roche 1985; Roche & Aitken 1985 and references therein). The division of the galaxies into two classes, active or starburst, is reflected in their 8-13µm spectra. The starburst galaxies generally have spectra dominated by the well known, but as yet unidentified, narrow dust emission features at 11.25, 8.65 and 7.7µm seen in a variety of Galactic sources together with strong ionic line emission from the 12.8µm [NeII] fine-structure line. By contrast, these features are seen only rarely in the active nuclei which instead have 8-13µm spectra well represented by a power law; in some cases there is evidence for silicate absorption.

Starburst galaxies.

The starburst galaxies are typified by the archetypes NGC 253 and M82, whose spectra show strong emission in the narrow unidentified emission features and the [NeII] line together with some silicate absorption (Gillett et al 1975). Indeed, 90% of the nuclear HII region galaxies show remarkably similar spectra (c.f. NGC 1808 in fig 1a), over a range in 10µm luminosity of 5 x 10^7 to 10^{10} $L_\odot$. The measured 10µm flux in these galaxies is usually aperture dependent, implying that the emitting regions are extended on a scale of a few hundred parsecs. The emission in the narrow bands can account for a few percent of the total luminosity of these galaxies. The equivalent widths of the

F. P. Israel (ed.), Light on Dark Matter, 467–470.

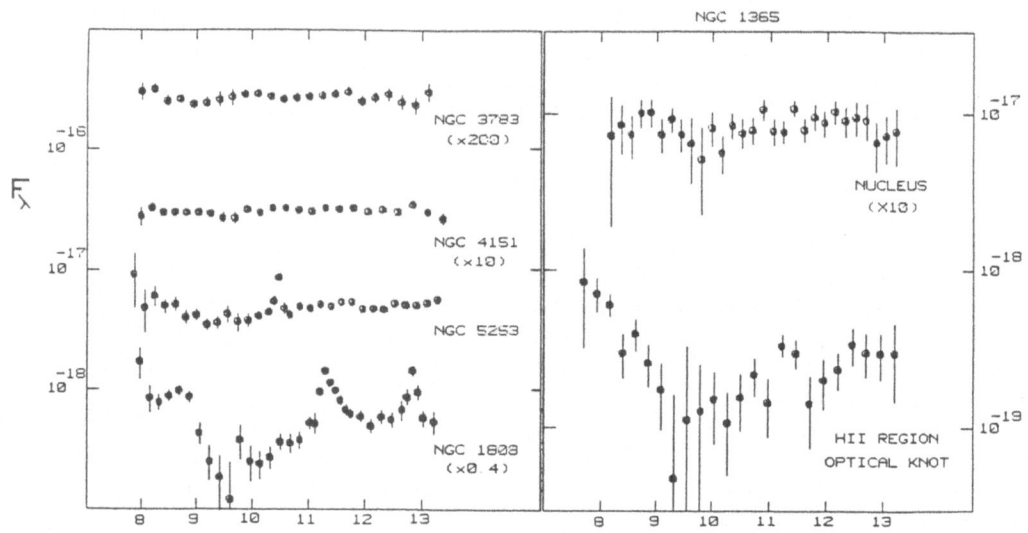

Fig 1a. 8-13μm spectra of two Seyfert and two Starburst nuclei.
1b. Spectra of the Seyfert nucleus and HII regions in NGC 1365.

emission features in the starburst galaxies are generally higher than in
Galactic sources and dominate the spectra to an extent that the silicate
emission signature seen in Galactic HII regions is swamped, if present
at all. The origin and excitation of the features are discussed at
length in these proceedings. By analogy with HII regions in our Galaxy
(Aitken et al 1979), it is likely that the grains are excited by blue
photons escaping from the ionised regions in the nuclear HII region
complexes. With the vigorous star formation in the nuclei of these
galaxies, it is probable that much of the central region is bathed in
radiation with $\lambda < 912$ Å. If non-equilibrium emission from very small
grains is invoked as the excitation mechanism (Sellgren 1983), these
non-ionising photons will be efficient in producing the unidentified
bands. The emission in these galaxies may be a more efficiently excited
version of the IR cirrus detected by IRAS in our Galaxy.

Two of the HII region galaxies, NGC 5253 (fig 1a) and II Zw 40 have
quite different spectra from the remainder. They display the 10.52μm
[SIV] line indicating higher excitation gas than the other starburst
nuclei, and the narrow emission features are not detected. The high
excitation, implying very hot and hence young stars, together with the
compact size at 10μm suggests that the IR emitting regions in these two
nuclei are very young. It may be that we are seeing a very early stage
in a starburst nucleus where the IR emission is dominated by one central
region, as opposed to the more extended emission in the other galaxies
with nuclear HII regions. We might then expect the dust emission to be
dominated by heating in the central core rather than diffuse emission
outside the ionised region.

Active galaxies.

The active galaxies encompass a huge range in luminosity from the quasars to the low luminosity Seyfert 2 galaxies, and generally have compact IR sources centred on the optical active nucleus. The 8–13μm spectra of almost all the active nuclei are broadly similar with little evidence of spectral structure that could be attributed to dust emission features (see NGC 3783 (S1)and NGC 4151 (S1.5) in fig 1a). The QSOs and Seyfert 1 galaxies in our sample mostly have very smooth 10μm spectra, and only 2 out of 13 show significant departures from a power law fit to their 8–13μm spectra. The two exceptions are NGC 7469 and Mkn 231, both of which are rather atypical Seyfert 1 galaxies. Several of the lower luminosity Seyfert 2 nuclei show evidence for silicate absorption.

However, with a low-luminosity active nucleus, contamination from HII regions in the central regions of the galaxy may be a problem. This can be seen clearly in the nucleus of the nearby barred spiral NGC 1365 where the 8–13μm spectrum of the Seyfert nucleus is smooth, and similar to other active nuclei, but the emission from two optical knots only $\approx$8 arcsecs away is strongly featured and typical of starburst galaxies (fig 1b). If this galaxy were located at a greater distance, these separate components would not be resolvable, and the contributions from the HII regions within the central few hundred parsecs would dominate the 10μm emission. Clearly, high spatial resolution is important, and the characteristic signature of a low- luminosity active nucleus may not be detectable in distant galaxies.

There is no evidence of the silicate emission feature which is almost ubiquitous in M giants and HII regions in our Galaxy. If the 10μm flux is produced by thermal emission from dust in the active nuclei, the heating mechanism or grain population must be such that the silicate grains do not emit strongly, although the approximately power law IR flux distribution requires emission from grains over a large temperature range. The remarkable absence of the narrow emission features in the active nuclei can be explained in terms of thermal spiking in small grains. The fact that the 11.25μm emission feature peaks outside the ionised region in Orion and NGC 7027 (Aitken & Roche, 1983) suggests that the grains producing it are destroyed by energetic photons. The active nuclei are characterised by strong non-thermal emission extending into the FUV and X-ray regions. This hard photon continuum will quickly destroy small grains by evaporation. However, because the grain absorption cross-section falls with increasing energy, larger grains, which require more energetic photons to heat them above the critical evaporation temperature, will survive. A consequence of this is that the regions around the active nuclei should have a cut off in the grain size distribution.

Overall, the 10μm spectra of active galaxies suggest two main components, namely the non-thermal energy source from the active nucleus and radiation from dust in the nuclear region. In the highest luminosity objects, the non-thermal source dominates with dust emission becoming more important with decreasing luminosity.

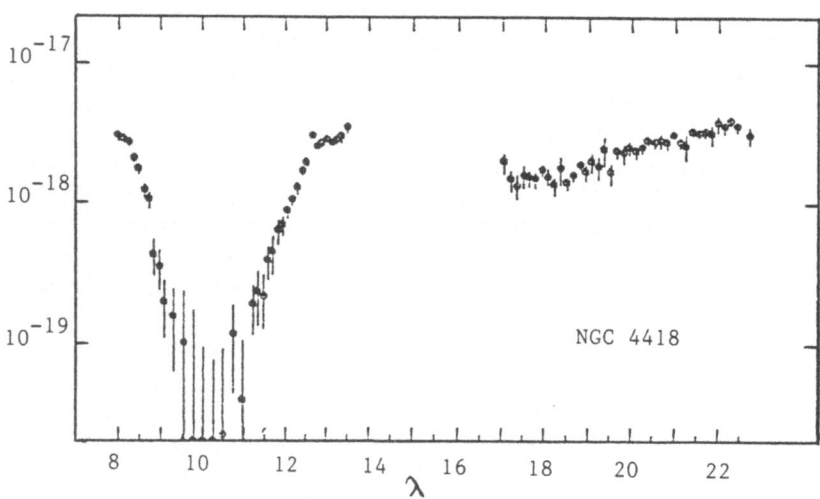

IRAS galaxies

The IRAS satellite has identified several thousand IR-bright galaxies with no *a priori* selection effects. A start has been made in investigating whether the IRAS-selected galaxies show up any differences in their 8-13μm spectral properties. From the small sample available (6 galaxies), it appears that a substantial number may have deeper silicate absorption than optically-selected galaxies, and indeed, some galaxies may be so heavily obscured that the activity giving rise to the powerful IR emission is not discernable in the visible. The 8-23μm spectrum of the most extreme of these galaxies, NGC 4418, is shown in figure 2 where the depth of the silicate absorption feature $\tau_{9.7} \sim 7$ corresponds to $A_V \geqslant$ 100. Optical spectra (3000-10000 Å) obtained at the AAT reveal little sign of the luminous IR source but convey the impression of a rather unremarkable galaxy. In this galaxy, the IR luminosity is $\approx 10^{11}$ $L_\odot$ and all this energy lies behind a very large column of obscuring dust. From the non-detection of ionic and dust emission structure in the 8-13μm spectrum and the relatively blue IRAS colours, it is likely that NGC 4418 harbours an extremely heavily obscured Seyfert nucleus.

References:
Aitken, D.K. & Roche, P.F., 1983. M.N.R.A.S., 202, 1233.
Aitken, D.K. & Roche, P.F., 1985. M.N.R.A.S., 213, 777.
Aitken, D.K., et al., 1979. Astr.Ap. 76, 60.
Cutri, R.M., et al., 1984. Ap.J., 280, 521.
Gillett, F.C., et al., 1975. Ap.J., 198, L65.
Roche, P.F. & Aitken, D.K., 1985. M.N.R.A.S., 213, 789.
Sellgren, K., 1984. Ap.J., 277, 623

IRAS OBSERVATIONS OF ACTIVE GALAXIES - A REVIEW

George Miley[1,2,3,4] and Rien de Grijp[1,3]
[1] Sterrewacht Leiden
[2] Space Telescope Science Institute, Baltimore, U.S.A.
[3] Affiliated to the Astrophysics Division, Space Science
  Dept. of the European Space Agency, Noordwijk
[4] On leave from Sterrewacht Leiden

ABSTRACT. IRAS observations of AGNs are reviewed. We consider both the
IRAS results on known AGNs and ground-based followup work on candidate-
AGNs from the IRAS Catalogue

1. INTRODUCTION[*]

In this talk we shall review IRAS results to date on active galaxies and
quasars. We shall deal mainly with conventional AGNs, i.e. galactic
nuclei which have high-ionization emission lines in their optical
spectra, or which have strong associated radio sources.

Although before IRAS ("BI") only a few bright AGNs had been studied
at wavelengths longer than 10 microns, there were already several
indicators to whet our appetites for further information about their
mid- and far-infrared properties (e.g. Neugebauer, 1978; Rieke and
Lebofsky, 1979; Smith 1983, Smith 1985). First, a few bright Seyferts
were known to emit their maximum energy in this part of the spectrum. In
the case of the nearby galaxy NGC1068, speckle work had shown a mid-IR
component of size about 60 pc which was shown to be thermal (Becklin et
al. 1973; Jones et al. 1977). Secondly, thermal IR emission is a unique
signature of both star formation and dust. The relation of these
phenomena to the various types of Seyfert activity is intriguing and had
received considerable attention (e.g. Smith 1985). Thirdly, the compact
nonthermal emission associated with many AGNs has a spectral "break"
between a few microns and 1 mm. It therefore seemed likely that
radiation emitted in this wavelength range must be of considerable
importance to the mechanisms responsible for the nonthermal nuclear
emission.
      Let us first remind you of the three sensitivity regimes within
which IRAS can study AGNs:

[*]Throughout this talk we shall assume a value of H_0 = 75 km/s/Mpc

 for the Hubble Constant.

F. P. Israel (ed.), Light on Dark Matter, 471–486.
© *1986 by D. Reidel Publishing Company.*

- low sensitivity. The published "strong source" survey covers about 96%
of the sky to flux levels about 0.5 Jy at 12, 25, and 60 microns and
about 1.5 Jy at 100 microns for point sources. The survey data is now
available and as we shall see has resulted in the detection of several
hundred AGNs.
- intermediate sensitivity. Coadding the various observations which
comprise the published survey will improve the sensitivity by a factor
of 2 to 3. Several programmes have commenced to obtain coadded data for
AGNs and we expect that these will result in the detection of up to a
few hundred more.
- high sensitivity. Directed or additional observations give improvement
of factors of from 5 to 15 over the strong source survey. About 500 AGNs
have been studied by the IRAS Science Team in this high-sensitivity mode
and the results are being analysed at present.

In using IRAS to study AGNs we have taken a two-pronged approach.
On the one hand we have studied the IR properties of previously known
AGNs (Sections 2, 3, 4, and 5). The other line of approach (which as we
shall see is proving just as fruitful) consists of ground-based followup
work on candidate-AGNs from the IRAS Catalogue (Sections 6, and 7).

2. KNOWN AGNs IN THE STRONG SOURCE SURVEY

Table 1 gives the detection rate of AGNs in the strong source survey for
different ranges of absolute magnitude. Using the listed identification
criteria, 222 objects from the compendium of Veron-Cetty and Veron
(1984) (V^2) were found to have IRAS counterparts. As is to be expected,
these detections are biased towards bright active galaxies. From the
histogram of apparent magnitudes in Fig. 1, the median magnitude of
detected AGNs is V ~ 14.4 with only 10% having V > 15.5.

TABLE 1
Matches Between Strong Source Survey and Veron-Cetty/Veron Catalogue

	Absolute Visual Mag. M_{ABS}	No. of[1] Matches	Est. No. of Spurious Matches	Est. Detection Percentage
QSOs	> -25	9	1	
	> -23	21	1	(1.5)%
	≤ -23	221	1	44%

[1]Match was accepted if positional coincidence was ≤ 55" for
M_{ABS} ≥ -23 (galaxies) and ≤ 30" for M_{ABS} < -23 (QSOs). All matches
with QSOs near (≤ 2' from) bright galaxies were rejected.

The sensitivity of the strong source survey is therefore
insufficient to enable more than a few QSOs to be detected. Using the V^2
classification of a QSO as an AGN with absolute visual magnitude

M_{ABS} < -23, the 21 QSOs listed in Table 2 satisfy our identification criteria. Most of these are rather piddling QSOs; only 9 have M_{ABS} < -25 and qualify for the major league. For these reasons the main statistical studies carried out so far on known AGNs have concentrated on the bright Seyferts.

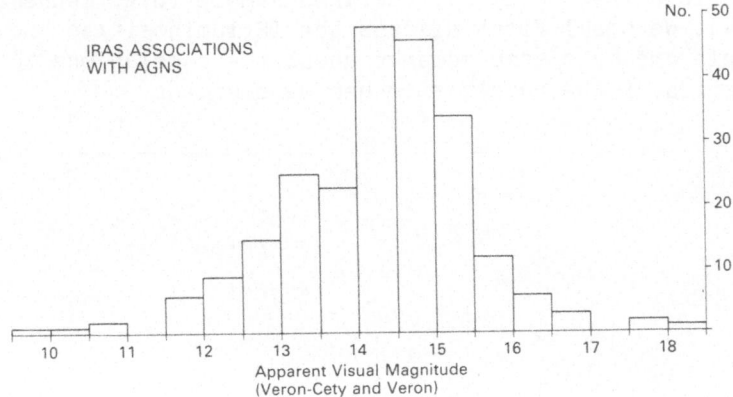

Figure 1. The distribution of apparent magnitudes for the objects in the compendium of Veron-Cetty and Veron (1984) identified with sources in the IRAS Catalogue.

TABLE 2

QSOs in Veron-Cetty and Veron (1983) detected by the IRAS Survey

Object	Name	Redshift	APP V-Mag	ABS V-Mag	Flux Densities (Jy)			
					12μm	25μm	60μm	100μm
0050+124	IZw1	0.061	14.03	-23.8	0.54	1.25	2.13	2.43
0134+329	3C48	0.367	16.2	-25.7	<0.25	<0.25	0.78	<1.0
0157+001	Mkn 1014	0.163	15.69	-24.2	<0.25	0.62	2.30	2.20
0248+430*	S4	1.361	15.5	-29.8	<0.25	<0.25	4.4	7.2
0420-014	PKS	0.915	17.7	-26.4	<0.4	(0.34)	0.58	<3.3
0537-441	PKS	0.894	16.48	-27.5	<0.30	0.25	0.52	<1.0
0710+457	Mkn 376	0.056	14.62	-23.0	<0.35	0.57	0.86	<1.3
1226+023	3C273	0.158	12.8	-27.0	0.54	0.93	2.17	2.80
1254+571	Mkn 231	0.041	13.84	-23.1	1.82	8.56	33.2	30.0
1351+640	PG	0.088	14.84	-23.8	<0.30	0.54	0.86	(1.01)
1353+186	Mkn 463	0.051	14.22	-23.2	0.60	1.62	2.16	1.79
1440+357	Mkn 478	0.079	14.58	-23.8	<0.25	<0.25	0.59	(1.01)
1613+658	Mkn 876	0.129	15.23	-24.1	<0.25	<0.25	0.63	(1.05)
1634+706*	PG	1.334	14.90	-30.4	<0.25	0.26	<0.4	<1.0
1641+399	3C345	0.594	15.96	-27.1	<0.35	0.29	0.68	(1.03)
1700+518	PG	0.292	15.43	-25.9	<0.25	0.26	(0.52)	<1.0
1807+698	3C371	0.051	14.22	-23.2	<0.25	0.25	(0.40)	<5.4
1916-587	ESO141-G55	0.037	13.64	-23.1	(0.31)	0.38	0.61	<4.5
2041-109	Mkn 509	0.035	13	-23.6	0.35	0.74	1.40	1.36
2130+099*	IIZw136	0.061	14.64	-23.2	0.3	(0.45)	0.52	<1.0
2223-052*	3C446	1.404	18.39	-27.1	<0.5	<0.4	0.66	<1.0

*Doubtful identifications.

3. IR PROPERTIES OF SEYFERTS

Recently an analysis has been made of the mid- and far-IR properties of Markarian and NGC galaxies having well-classified optical emission-line spectra in the compendium of AGNs by V^2 (Miley et al. 1985). This is a reasonably well-defined sample (e.g. Meurs and Wilson 1984). We shall first discuss the IR luminosities and colours of the sample and make some remarks about their relations to the characteristics of the sample at other wavelengths.

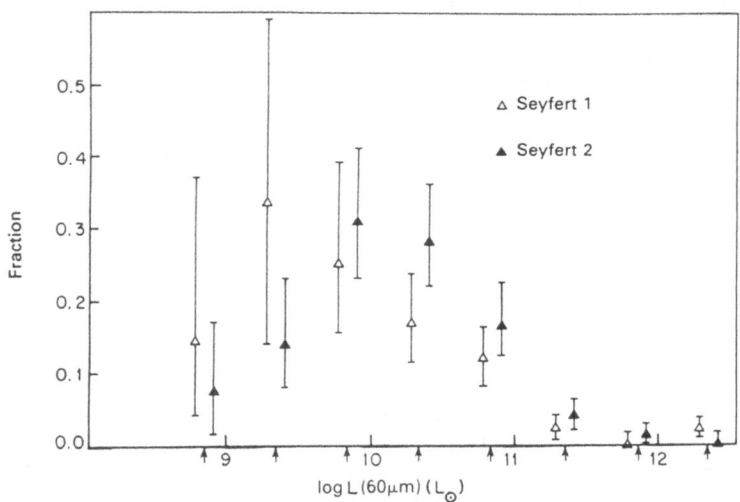

Figure 2. The fraction of Seyfert galaxies as a function of their 60 μm luminosity. This was derived by dividing the number of detected galaxies in a given luminosity bin by the number of objects in the sample that could have been detected in that luminosity bin. Although for ease of plotting each pair of points is plotted side by side the relevant luminosity for both points in the pair is the mean indicated by the arrows (from Miley et al. 1985).

3.1. IR Luminosities

66 of the 104 galaxies classified as Seyfert 1, and 66 of the 84 Seyfert 2s were detected in the strong source survey. Fig. 2 shows the fractional 60 micron luminosity distribution of this bright Seyfert sample taking into account the level at which each galaxy in the sample could have been detected.
- About 50% of the bright Seyferts have 60 micron luminosities in excess of 10^{24} WHz^{-1} (= $10^{36.7}$ Watt = $10^{10.1}$ L$_\odot$).
- There is no significant difference between the luminosity function of Seyfert 1 and Seyfert 2 galaxies.
- Seyfert 2 galaxies have both larger IR to optical-B and smaller IR to radio-1.4 GHz intensity ratios (Figures 3 and 4) than Seyfert galaxies.
        These data indicate that a large fraction of Seyferts (more than 60% of Seyferts 2s) have their maximum energy outputs (νF_ν) in the IR.

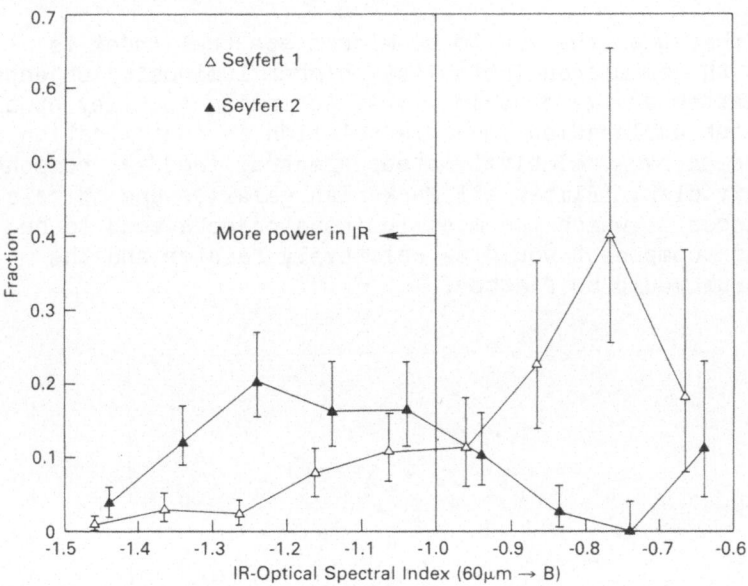

Figure 3. The fraction of Seyferts as a function of their IR (60 μm) to optical (B) luminosity ratio. This ratio is expressed in terms of a frequency spectral index α defined by $F_{(\nu)} \alpha \, \nu^{\alpha}$ Seyfert 1 galaxies are denoted by open triangles and Seyfert 2 galaxies by closed triangles.

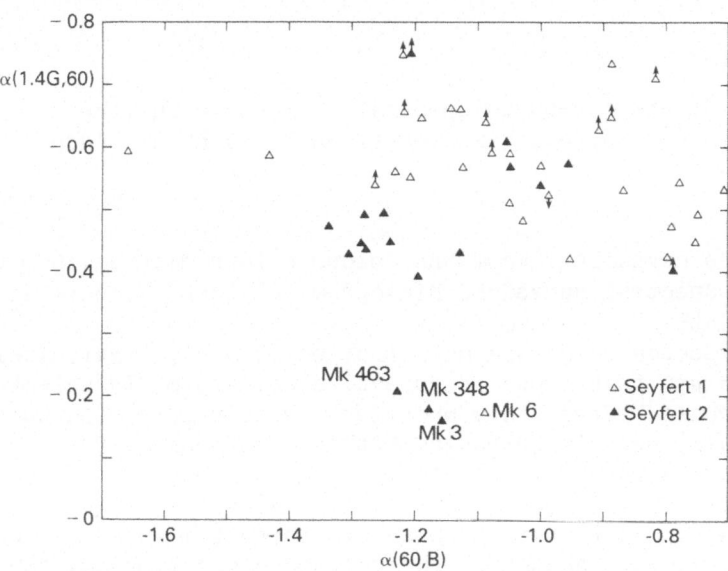

Figure 4. The ratio of radio (1.4 GHz) to IR (60 μm) luminosities plotted against the ratio of IR (60 μm) to optical (B). Both are expressed in terms of frequency spectral indices.

3.2. IR Luminosity-Colour Relation

Figure 5 shows that when the 100 to 60 micron spectral index is
plotted against the 60 micron luminosity, higher luminosity objects
tend to have flatter energy distributions. According to Miley et al.
1985, the likliest explanation for this relation is contamination of
the nuclear IR flux by a relatively steep spectrum (redder) component
due to the galaxy disk. (Almost all Markarian galaxies are spirals).
At larger distances at which the most luminous objects tend to be
located, the disk component would be relatevely fainter and the
composite spectrum would be flatter.

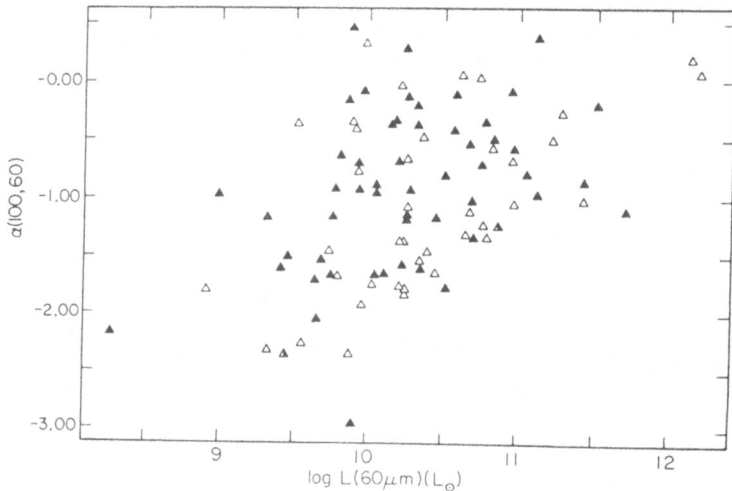

Figure 5. A plot of the 100-60 μm spectral indices versus the
60 μm luminosities for Seyfert galaxies (from Miley et al. 1985).

3.3. IR Colours.

The IR colours are revealing. From the colour-colour diagram in
Figure 6 and the spectral curvature histogram in Figure 7, several
points are apparent:
- Seyferts have spectra which are much flatter (bluer), particularly
between 60 and 25 μm, than those of nonactive spirals of IR selected
galaxies. As discussed later, this provides a colour criterion which
we have successfully used in selecting hitherto unknown Seyfert
galaxies from sources in the IRAS Catalogue.
- Different species of emission line galaxies have on average
different spectral characteristics. Although Seyfert 1 galaxies seem
to be distributed widely throughout the colour-colour diagram, the
lower right of the diagram are populated by objects with Seyfert 2
and H II region spectra. The colour differences between the three
classes of emission-line AGNs is most clearly seen in the histogram
of spectral curvature at 60 microns (Figure 7).

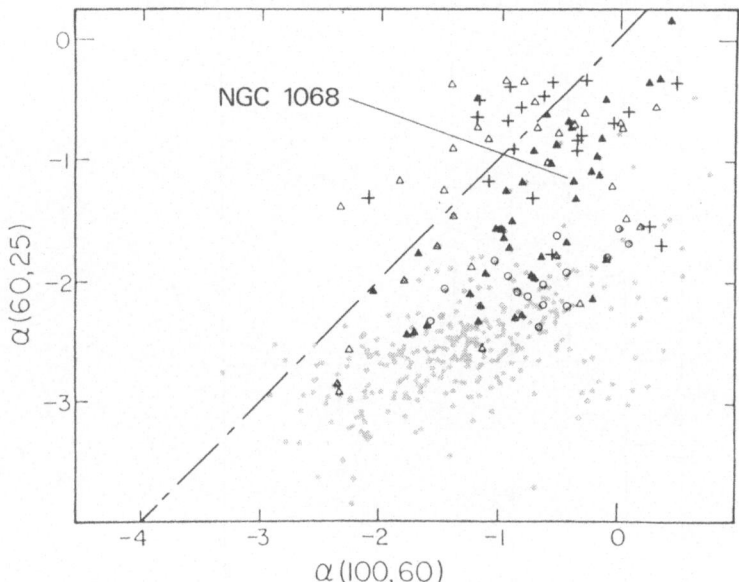

Figure 6. A mid- to far-infrared color-color plot for galaxies with
strong nuclear emission lines. The spectral indexes are defined in
the text. In all figures, Seyfert 1 galaxies are denoted by open
triangles, Seyfert 2 galaxies are denoted by filled triangles
galaxies with nuclear HII regions by circles and QSOs by crosses. The
shaded region denotes the colours of non-active spirals and of an IR-
defined sample of galaxies. The position of the archetypical Seyfert
2 galaxy NGC 1068 is indicated (adapted from Miley et al. 1985 and
Neugebauer and Rowan Robinson 1985).

These differences in colour between Seyfert 1s and 2s are consistent
with the conventional picture derived from near-IR observations, in
which the IR emission from Seyfert 1 nuclei is dominated by a
nonthermal component, while the IR emission from Seyfert 2s is mainly
thermal in origin (e.g. Rieke 1978; Ward et al. 1982). A further
argument for this viewpoint is provided by the fact that the mean IR
colours of the Seyfert 1 galaxies agree with those of the handful of
objects with strong flat-spectrum radio cores and known mid- and far-
IR colours. These nonthermal sources include NGC1275, 3C120, as well
as the QSOs 3C345, and 0537-441 (Neugebauer et al. 1984).

4. MID-IR THERMAL COMPONENTS

A clue to understanding the implications of the preferred 60 μm
curvature comes from the measurement of the archetypical Seyfert 2
galaxy, NGC1068. The location of this galaxy in Figure 6 is given
by α(100,60) = -0.49 and α(60,25) = -0.87, an undistinguished position.
For this object, speckle work at 10 microns have revealed an
0.9" (~ 60 pc) component (Becklin et al. 1973), while Jones et al.

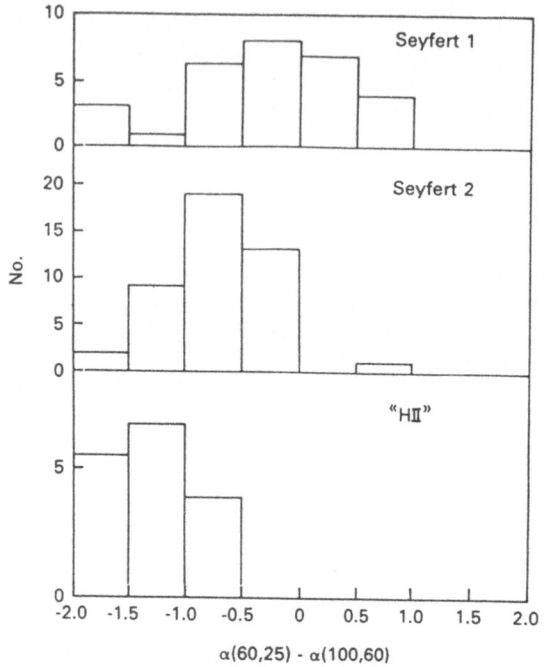

Figure 7. Histograms of the 60 μm spectral curvatures are shown for the three classes of emission line galaxies studied. The curvature is defined as the difference of the two frequency spectral indices α.

(1977) concluded that this component which probably peaks near 30 μm is thermal in origin. Indeed, for this galaxy the spectrum is known to be composite with much of the far infrared flux coming from star forming regions outside the nucleus. Given the quite ordinary location of NGC1068 in Figure 6, it seems reasonable to postulate the widespread existence of similar mid-IR thermal components in many Seyfert 2 galaxies as responsible for the observed spectral curvature. They are however usually swamped in the composite spectrum by the colder disk emission.

    Further evidence for the existence of these mid-IR components comes from measurements in the sensitive pointing mode made by IRAS on the radio galaxy 3C390.3 (Miley et al. 1984). This object shows a maximum in its spectrum near 25 microns, corresponding to a temperature of ~ 180 K. Since 3C390.3 is not in a spiral galaxy, in this case we may be seeing a "naked- mid-IR component, divorced from the cold disk. Several Seyferts in the survey also show clear peaks in their spectra within the IRAS wavelength range. These include Mkn 3, 335, 463 and 1210, ESO 350-IG58, 103-635, IC4329A and PKS 2048-573.

    Assuming that 30 micron thermal components with luminosities ~ 10^9 to 10^{11} L_o are widespread in AGNs (at least in Seyfert 2s) what can we say about the intrinsic parameters of the emitting regions? If the reradiating dust is optically thin and located at a distance R from a single nuclear source of heating, then the 30 micron components would imply (via Stefan's Law) sizes of R ~ 1.5 to 15 ($\epsilon^{-1/2}$ pc) where ε is the grain emissivity. The corresponding mass of dust derived using Planck's Law ranges from $M_d \sim 10^4 \, \kappa^{-1} M_\odot$ to $2.5 \, 10^6 \, \kappa^{-1} M_\odot$ where κ ($cm^2 \, kg^{-1}$) is the mass absorption coefficient at 30 microns.

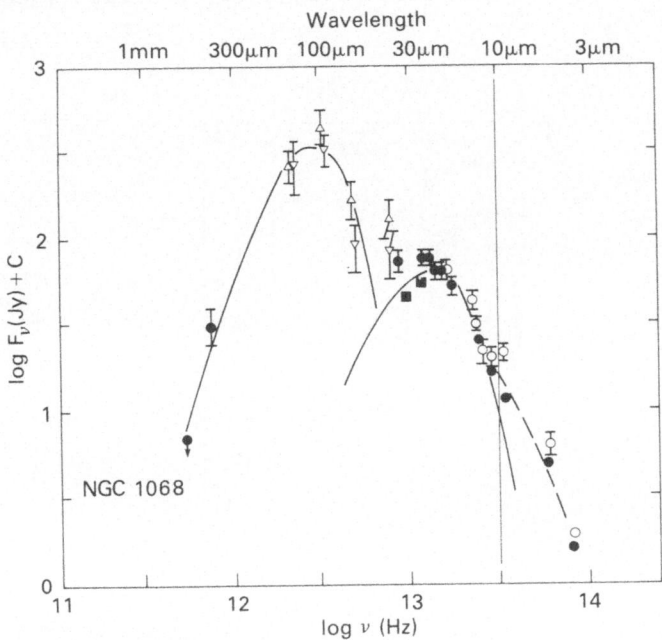

Figure 8. Infrared spectrum of NGC1068 showing the two-component structure (adopted from Telesco and Harper, 1980). The wavelength corresponding to the 0.9" speckle measurement of Becklin et al. (10 μm)is indicated by the arrow.

We know little about the properties of dust in AGNs. However, following the perrenial seeker for the lost key in the dark street under the solitary lamppost, we assume that the characteristics of the emitting grains are similar to those in the interstellar medium of our galaxy at 150K with $\epsilon \sim 0.03$ and $\kappa \sim 25m^2 \, kg^{-1}$ (e.g. Hildebrand 1983; Draine and Lee, 1984). This gives R $\sim$ 15 to 150 pc (cf. 60pc for NGC1068) and $M_d \sim 10^3$ to $10^5 M_\odot$. With a "normal" mass to dust ratio of 200, the corresponding mass of gas would be $M_g \sim 10^{5.3}$ to $10^{7.3} \, M_\odot$. The mass would be greater if the gas is optically thick. These values of M_g appear slightly larger than might be expected from the mass of ionized gas in the narrow-line region (typically $\sim 10^3$ to $10^7 M_\odot$), although given the large uncertainties and the considerable assumptions the two quantities are not significantly different.

If the mid-IR components are indeed similar to the one in NGC 1068, the emission would be connected with dust close to the inner part of the narrow-line region. We have therefore conducted a preliminary examination between the properties of the IR emission and those of the narrow emission line components of the Seyferts (e.g. Steiner et al.

1981). So far this analysis has only met with limited success. There are
weak tendencies for the objects with flatter spectra to be associated
with more luminous narrow lines, larger values of Balmer decrement
larger reddening as measured by the forbidden [OII] and [SII] ratios
(e.g. Allen 1979), and high ionizations as measured by the most
asymetric [OIII]/Hβ ratio. Also, the more luminous IR sources tend to be
associated with the most asymmetric [OIII]λ5007 profiles (Heckman et al.

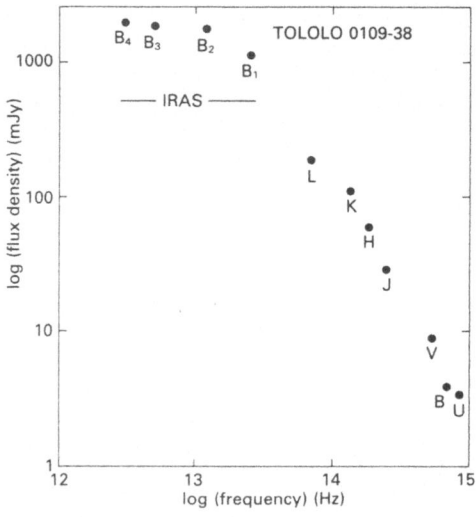

Figure 9. The optical and infrared spectrum of TOLOLO 0109-38. Data were
taken from Glass and Moorwood (1985), Veron-Cetty and Veron (1984) and
the IRAS Point Source Catalogue (1985).

1984). However, none of these trends are significant statistically. We
conclude that the mid IR components are not (or only weakly) associated
with the properties of the narrow-line regions. This result is still
consistent with the mid IR components originating from a region just
smaller than that which emits the bulk of the narrow emission lines. As
appears to be the case in NGC1068.
      A particularly interesting object in which there may well be a
connection between the mid-IR emission and the forbidden emission-line
spectrum is the Seyfert 2 galaxy TOLOLO 0109-38. Its continuum spectrum
shown in Figure 9 is distinguished not only by a large infrared excess,
but also by being remarkably flat from 12 μm to 100 μm. It is not a
strong radio source. Assuming the emission to arise from reradiating
dust with an emissivity of 0.03 powered by a single nuclear source the
constant spectrum from 12 to 100 μm implies a fairly continuous
distribution of dust from a distance of < 8 pc from the nucleus to > 175
pc. This galaxy has an exceptionally high-ionization spectrum (Fosbury
and Sansom 1983) and has been used to study the mechanisms responsible
for producing the high ionization lines in AGNs. The temperatures
derived by Fosbury and Sansom for the high and low ionization species

respectively were both consistent with the lines being emitted by means
of photoionization. However, they caution that differential reddening
between the high and low ionization zones would invalidate this
conclusion. The dust distribution suggested by the IRAS spectrum of
0109-38 would indeed cause significant reddening.

We consider one further consequence or the presence of the mid-IR
components. If the mid-infrared emission from Seyfert 1 galaxies has a
predominantly nonthermal origin while that from Seyfert 2 galaxies is
mainly thermal it seems strange that they just happen to produce similar
luminosities and spectral distributions in the IRAS bands. However, this
may not be a coincidence. The mid-IR component is presumably heated by
the nonthermal nuclear source and should have comparable energy output.
If Seyfert 2 galaxies are merely Seyfert 1s whose central source is
temporarily quiescent (e.g. Penston and Pérez 1984) a mid-IR thermal
component would remain. In that case the typical "off-time" should be
less than the light travel time from the nucleus to the observed
radiating dust (i.e. less than ~ 200 years for NGC1068).

5. HOST GALAXIES OF QSOs

Before leaving the optically selected AGNs, we should mention two
studies making use of the high-sensitivity directed mode observations
which throw some light on the host galaxies of QSOs. The first, by
Neugebauer et al. (1984) compared IRAS emission from 2 radio-quiet and 3
radio-loud quasars. There was no indication of any turnover in the
spectrum of the radio quiet QSOs for $\lambda < 100\mu m$. In fact there was an
indication of excess 100 μm emission in the radio quiet objects which
was interpreted within the above picture as possibly indicative of a
cold "spiral" component. If confirmed, this would be evidence that
spiral galaxies are associated with radio quiet QSOs.

The second result is from Neugebauer et al. (1985) who showed that
the quasar 3C48 has an IR luminosity $3 \times 10^{12} L_0$ and dominates the power
output of the quasar. These authors the IR emission to thermal
reradiation of heated dust. The IR spectral characteristics of 3C48 are
similar to that of Mkn 231 and indicate the presence of a thermal
component with T ~ 60K. The measured luminosity would then imply that
the grains are located about 2 kpc from the heating source, presumed to
be a single QSO nucleus. From the large implied mass, it is argued that
the IR emission comes from a galaxy surrounding the quasar.

Is there any characteristic of 3C48 which sets it apart from other
luminous quasars so far studied by IRAS? 3C48 is indeed distinguished by
its radio properties. Although most steep-spectrum radio sources have
dimensions $\geq$ 100 kpc, 3C48 is a member of a small class or radio
sources whose radio radiation is dominated by luminous steep-spectrum
emission on the scale of a kiloparsec (e.g. van Breugel et al. 1984). It
has been conjectured that these "steep-spectrum radio cores" (e.g. Miley
1980) represent radio sources where the synchrotron jet is prevented
form escaping from the parent galaxy by a dense dissipating medium
within or close to the nucleus. By indicating that the host galaxy of
3C48 is anomalously dust-rich compared with the elliptical-type hosts

known to be associated with most luminous radio sources, the IRAS data
lend support to such a viewpoint. Analysis of deep observations of other
objects in this class is needed before such a relationship can be
confirmed.

6. USING IRAS TO DETECT "NEW" AGNs

Now let us turn to the newly discovered AGNs. Bearing in mind that known
active galaxies have generally flatter IR spectra than galaxies without
AGNs, we decided to investigate the properties of sources in the IRAS
Catalogue which had similar spectral characteristics. Our hope was that
by using the "warm" IR spectra as an indicator of nuclear activity, we
might find hitherto unknown AGNs. This study had been successful and the
preliminary results have been published (De Grijp et al. 1985).

We have now compiled a catalogue of 563 sources in the "strong
source" survey all at high galactic latitude ($|b| > 20°$), having
spectral indices between 60 and 25 microns of $-1.5 < \alpha(25,60) < 0.0$ (De
Grijp and Miley 1985). These wavelengths were chosen as most efficient
for indicating flat-spectrum nuclear IR emission. A cold galactic disk
would dominate the 100 to 60 micron spectrum, possibly masking a warm
nuclear component. Our criterion using the 60 to 25 micron colours is
therefore more efficient than that proposed by Glass (1985) using the
100 to 60 micron colours. This can be seen by inspecting the colour-
colour diagram in Fig. 6. However, we caution that, although efficient,
our method does not provide a complete sample of Seyferts, as is the
case for most survey techniques (Veron 1985).

Of the 563 sources in our catalogue, 79(14%) are contained in known
AGN catalogues, 249(44%) are identified with objects in other catalogues
and not known to be AGNs, (109 galaxies, 76 stars and 64 galactic
nebulae). We have attempted to identify the remainder using the Palomar
and ESO/SRC Sky Survey plates. For 249 of these we now have optical
spectra and 110 have been studied by us in the radio using the VLA.

Our results are summarized in Table 3. Using the emission-line
classification scheme given by Baldwin et al (1981), 69% of a
representative sample have Seyfert spectra, compared with a Seyfert rate
of about 10% for Markarian galaxies. Hence, a flat IR spectrum is a
remarkably efficient predictor of the Seyfert phenomenon. Carter (1984)
independently reaches this conclusion from spectroscopic· observations of
counterparts of 13 "warm" IRAS sources.

Some of these IR Seyferts are remarkably luminous e.g.
IRAS1051-237P11 has a 25 micron luminosity of $10^{25.3}$ W/Hz which is the
most luminous Seyfert known at this wavelength.

How many of these IR Seyferts are contained in the IRAS Catalogue?
3% of strong (60 micron defined) IRAS sources have $\alpha(25,60)$ within our
flat-spectrum range. Assuming (i) that there is no change in spectral
index distribution at lower flux levels, and (ii) that the spectral
index distribution of the IR Seyferts is similar to that of the
optically selected Seyferts discussed by Miley et al. (1985), there
should be about 1400 infrared Seyferts present in the published strong
point source survey and a factor of ~4 more could be found if the

complete survey data were coadded. (Of course we could only use our
colour selection criterion to pick out those Seyferts with IR flux
densities sufficiently strong to be detectable at both 60 and 25
microns.) Comparing this number of several thousand IR Seyferts in the
IRAS survey data base with the 554 known active galaxies and 2251 known
QSOs tabulated by Veron-Cetty and Veron (1984), one can safely predict
that within the next few years IRAS will have led to the discovery of a
significant fraction of all known AGNs.

   Of particular interest is the question if and how these "infrared
Seyferts" differ from the Markarian Seyferts? Possibilities are that the
IR objects (i) are intrinsically similar in the optical but more
distant, (ii) are intrinsically fainter optically at same average
distance, (iii) are intrinsically redder.

   It is clear from Fig. 11 that the IR Seyferts are systematically
more distant than the Markarian Seyferts. About 5% have redshifts in
excess of 0.1. The infrared luminosity function of the infra-red
selected sample imply space densities, which are nominally about double
that of the Markarian Seyferts, but the difference is not statistically

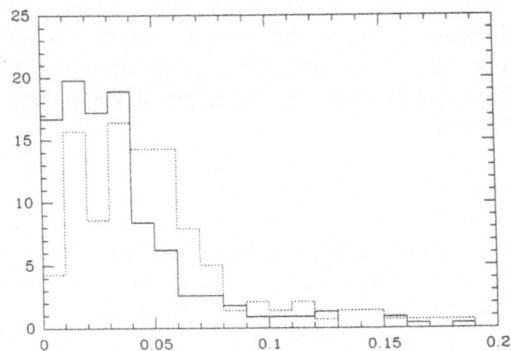

Figure 10. Redshift distributions for (a) the optically defined
Markarian Seyferts considered by Miley et al. (1985)/full line and (b)
the infrared-defined Seyferts selected from the IRAS catalogue (dashed
line).

significant. Previous indications that the space density of infrared
Seyferts is even larger was at least partially caused by the IRAS
candidates initially observed being biassed towards closer objects.

   Although as yet there is no accurate photometric data available for
the IR Seyferts, crude estimates indicate that many of them have
apparent magnitudes close to or less than 15.5, the magnitude below
which the Markarian Catalogue becomes complete. Also, regarding their
colours, the same dust that is presumably responsible for the IR
emission would significantly redden the nonthermal emission from the
nucleus, thereby reducing or eliminating the UV excess required for
inclusion in the Markarian survey. We conclude that the IR Seyferts are
probably similar but systematically more distant than the Markarian
Seyferts. However, we cannot rule out the existence of an intrinsic
difference between the two classes of objects.

   In this respect we point out that optically selected Seyferts are

associated predominantly with spirals and radio selected ones occur
mainly in elliptical-type galaxies. One IR Seyfert, 0421+040P06, which
has been studied in some detail provides a link between the optical and
radio Seyferts. A multispectral study of the associated spiral galaxy
has been carried out by Beichman et al. (1985). VLA measurements show a
30 kpc-sized radio source extending beyond the optical confines of the
galaxy. The radio source is several times larger than structures
previously seen in spiral galaxies and suggest that this may be a
transitional object between small kpc-sized sources often seen in the
nuclei of active spirals and the well studied large ($\gtrsim$ 100kpc) radio
sources commonly associated with elliptical galaxies.

8. CONCLUSIONS

To summarize, the IRAS data on Seyferts are most easily explained as due
to a mixture of three components:
(i)   a cold (c.30K) disk component having spectral indices similar to
        those in spiral galaxies with no active nuclei.
(ii)   a nonthermal nuclear component whose spectrum extrapolates into
        the optical and X-ray domain with an approximate power law index.
(iii)   a mid-infrared nuclear component which peaks between 20 and 60
        microns and by analogy with NGC1068 is probably thermal in origin
        (e.g. Jones et al. 1977) and is located within the central hundred
        parsecs of the nucleus.
The luminosity-colour relation can then be explained as a systematic
effect due to relative differences between the disk and nuclear
components, whereas the differences between the Seyferts is due to the
thermal and nonthermal nuclear components. Dominance of the flat-
spectrum nuclear component allows the Seyfert-AGNs to be efficiently
selected from the IRAS catalogue by their colours.
    From this vantage point, almost 2½ years after the launch of IRAS
(Anno 2.5 "AI") we have demonstrated two important contributions of IRAS
to the field of conventional AGNs. The first is the discovery that such
mid-IR components with powers $\sim 10^{24}$ W Hz^{-1} ($\sim 10^{10} L_o$) are common in
AGNs. The origin of such widespread nuclear dust and its relationship to
the other nuclear building blocks must be dealt with by any self
respecting model of AGNs. The second important contribution of IRAS (and
a corrollary of the first) is its use as an efficient tool for finding
hitherto unknown Seyferts, and the discovery that these appear in
surprising numbers.
    Analysis of the high-sensitivity deep fields and of the
intermediate-sensitivity coadded data should in the coming year provide
considerable additional information about both of these topics.
    The Seyfert studies wil be extended to carfully selected samples of
radio galaxies and QSOs with the aim of exploring and comparing their IR
luminosity functions and colours. In this work particular attention will
be paid to the place of the mid-IR component. In addition, further
correlative studies are carried out between the IR data and other
properties of the nucleus, such as optical line and X-ray continuum
emission. Here, data from the optically-defined and the infrared

selected AGNs will be combined. Since the mid-IR component may be an indicator of dust close to the narrow-line region, these studies are relevant to probing the nature of both the food needed to generate nuclear activity and the nuclear waste products.

Regarding the IR Seyferts, of especial interest are followup observations of objects in the deep fields which have IR colours which satisfy our Seyfert-indicating criterion. The median redshift of 0.03, and highest of 0.3 obtained for the "strong" IR Seyferts extrapolate to redshifts of 0.12 and 0.9 respectively for intrinsically similar objects which are observed to be a factor of 15 fainter. This would bring us into realms where cosmological evolution is known to be important.

Although IRAS is giving us a glimpse of active galaxies in the mid and far infrared, there are several limitations. One of these is the course coverage of the spectra provided by the four available bands. ISO and SIRTF should in the next decades provide enormous improvement in spectral resolution and coverage.

Perhaps the major limitation in the IRAS data is however the lack of spatial resolution. One picture is worth a thousand spectra. The situation is somewhat akin to that in radio astronomy in the mid-fifties before interferometers begun to chart the beautiful structures that eventually revealed that collimated outflow of energy is common in AGNs. Angular resolutions of at least 0.1 arcseconds will be needed before useful geometrical information for a significant number of AGNs can be obtained. At 30 microns this would imply an interferometer baseline with a length in excess of 60 m and points the direction to the next-but-one generation of infrared satellite missions.

ACKNOWLEDGEMENTS

It is a pleasure to acknowledge the professionalism of the engineers and managers from all three participating countries which helped make IRAS such a success. We are grateful for interaction with other members of the IRAS Science Team, and in particular G. Neugebauer, P. Clegg, S. Harris and M. Rowan Robinson. GM acknowledges a helpful discussion with Dr. B. Draine.

REFERENCES

Allen, D.A., 1979, Mon. Not. Roy. Astron. Soc. 186, 1P.
Baldwin, J.A., Phillips, M.M. and Terlevitch, R., 1981, Publ. Astron. Soc. Pac., 93, 5.
Becklin, E.E., Matthews, E., Neugebauer, G., and Wynn-Williams, C.G., 1973, Ap. J. (Letters), 186, L69.
Beichman, C., Wynn-Williams, C.G., Lonsdale, C.I., Persson, S.E., Heasley, J.N., Miley, G.K., Soifer, B.T., Neugebauer, G., Becklin, E.E., and Houck, J.R., 1985, Ap. J. (Letters), in press.
Carter, D., 1984, Astron. Express, 1, 62.
De Grijp, M.H.K., Miley, G.K., Lub,J. and de Jong, T., 1985, Nature, 314, 240.

Draine, B.T., and Lee, H.M., 1984, Ap. J., <u>285</u>, 89.

Fosbury, R.A.E., and Sanson, A.E., 1983, Mon. Not. R. Astron. Soc., <u>204</u>, 1231.

Glass, I.S., 1985, Present Symposium.

Glass, I.S. and Moorwood, A.F.M., 1985, Mon. Not. R. Astron Soc., in press.

Heckman, T.M., Miley, G.K. and Green, R.F., 1984, Ap. J., <u>281</u>, 525.

Hildebrand, 1983, Quart. J. R. A. S., <u>24</u>, 267.

The IRAS Point Source Catalog 1985, prep. by Joint IRAS Science Working Group (Washington D.C., U.S. Government Printing Office).

Jones, T.W., Leung, C.M., Gould, R.J. and Stein, W.A., 1977, Ap. J., <u>212</u>, 52.

Meurs, E.J.A., and Wilson, A.S., 1984, Astr. Ap., <u>136</u>, 206.

Miley, G.K., 1980, Ann. Rev. A. A., <u>18</u>, 165.

Miley, G.K., Neugebauer, G., Clegg, P.E.; Harris, S., Rowan Robinson, M., Soifer, B.T., and Young, E., 1984, Ap. J. (Letters), <u>278</u>, L79.

Miley, G.K., Neugebauer, G., and Soifer, B.T., 1985, Ap. J., in press.

Neugebauer, G., 1978, Physica Scripta, <u>17</u>, 149.

Neugebauer, G., Soifer, B.T., Miley, G., Young, E., Beichman, C.A., Clegg, P.E., Habin, H.J., Harris, S., Low, F.J., and Rowan Robinson, M., 1984, Ap. J. (Letters), <u>278</u>, L83.

Neugebauer, G., Rowan Robinson, M., 1985, Proc. on Structure and Evolution of Active Galactic Nuclei, Trieste, in press.

Penston, M. and Pérez, 1984, Mon. Not. R. Astron. Soc., <u>211</u>, 33P.

Rieke, G.H., 1978, Ap. J., <u>226</u>, 550.

Rieke, G.H., and Lebofsky, M.J., 1979, Ann. Rev. A. A., <u>17</u>, 447.

Smith, 1983, Proc. XVIth ESLAB Symposium, Toledo, ESA Publications, p. 365.

Smith, M.G., 1985, Active Galactic Nuclei, Ed. J.E. Dyson, Manchester University Press, in press.

Steiner, J.E., 1981, Ap. J., <u>250</u>, 469.

Telesco, C.M. and Harper, D.A., 1980, Ap. J., <u>235</u>, 397.

Van Breugel, W.S.M., Miley, G.K., Heckman, T.M., 1984, Astron. J., <u>89</u>, 5.

Veron, P., 1985, Proc. on Structure and Evolution of Active Galactic Nuclei, Trieste, in press.

Veron-Cetty, M.P. and Veron, P., 1984, A Catalogue of Quasars and Active Nuclei, Munich: European Southern Observatory, Sci. Rep. 1.

Ward, M., Allen, D.A., Wilson, A.S., Smith, M.G., and Wright, A.E., 1982, Mon. Not. R. Astron. Soc., <u>199</u>, 953.

AUTHOR INDEX

Abram, I. I.	447	Dingle, R.	307
Aeppli, G.	365	Dmitrienko, V. E.	377
Agranovich, V. M.	113	Dornhaus, R.	299
Akhmanov, S. A.	409, 517		
Apanasevich, P. A.	457	Efrima, S.	509
Aslanyan, L. S.	409	Endemann, M.	437
		Eremenko, V. V.	237
Basoon, S. A.	95		
Belyakov, V. A.	377	Farrow, R. L.	299
Benedek, G.	389	Feigenblatt, R.	189
Bergman, J. G.	167	Ferrell, R. A.	1
Bhattacharjee, J. K.	1	Fleury, P. A.	357
Birgeneau, R. J.	365	Friedman, J. M.	403
Birman, J. L.	131		
Bloembergen, N.	423	Gadzhiev, F. N.	409
Borovik-Romanov, A. S.	175	Geschwind, S.	189
Brewer, R. G.	159	Ginzburg, V. L.	331
Bruce, R. H.	229	Gossard, A. C.	307
Bruns, D. G.	347	Grimsditch, M. H.	249
Bunkin, A. F.	409		
Burstein, E.	479	Haller, K. E.	71
Byer, R. L.	437	Halperin, B. I.	47
		Heritage, J. P.	167
Callender, R.	391	Hizhnyakov, V.	269
Camley, R. E.	207	Hochberg, A.	29
Cardona, M.	249	Hochstrasser, R. M.	447
Carey, M. C.	389	Hohenberg, P. C.	23
Chang, R. K.	299		
Chen, C. Y.	479	Ipatova, I. P.	83
Choyke, W. J.	199		
Clark, N. A.	59	Jotikov, V. G.	175
Cohen, E.	293		
Cummins, H. Z.	229, 518	Kaplyanskii, A. A.	95
		Kardontchik, J. E.	293
Devlin, G.	189	Kazantsev, A. P.	471
DeVoe, R. G.	159	Kirtley, J. R.	499

Klein, M. V.   347
Klochikhin, A.   215
Klyshko, D. N.   283
Koroteev, N. I.   409, 437
Kreines, N. M.   175

Levanyuk, A. P.   331
Litster, J. D.   365
Lundquist, S.   479
Lyons, K. B.   357, 403

Maniv, T.   509
Martin, R. M.   299
Mazer, N.   389
Meadows, M. R.   39
Metiu, H.   509
Mills, D. L.   207
Missel, P.   389
Mockler, R. C.   39
Mollow, B. R.   467
Morozenko, Ya.   215

Nelson, D. R.   47
Nishio, I.   29

Olego, D.   249
O'Sullivan, W. J.   39

Pattanayak, D. N.   131
Permogorov, S.   215
Pershan, R. S.   365
Pinczuk, A.   307
Pitaevsky, L. P.   61
Pollak, F. H.   229
Popkov, Yu. A.   237

Rand, S. C.   159
Rebane, K. K.   257
Rebane, L. A.   71
Romestain, R.   189

Saari, P.   315
Scheibner, B. A.   39
Scott, J. F.   199
Sergienko, N. A.   237
Shekhtman, V. L.   95
Shumai, I. L.   409
Sigov, A. S.   331
Sobyanin, A. A.   331
Sooryakumar, R.   347
Störmer, H. L.   307
Subashiev, A. V.   83
Sun, S.-T.   29
Swinney, H. L.   15
Swislow, G.   29
Szabo, A.   159

Tanaka, T.   29
Toms, D. J.   199
Travnikov, V.   215
Tsang, J. C.   499

Varma, C. M.   81
Voitenko, V. A.   83

Waters, R. G.   229
Wicksted, J.   229
Wiegmann, W.   307
Worlock, J. M.   307

Young, C.   389
Yu, P. Y.   143

SEYFERT GALAXIES IN THE IRAS SURVEY AND JHKL PHOTOMETRY

I.S. Glass
South African Astronomical Observatory
P.O. Box 9
Observatory 7935 Cape
South Africa

SUMMARY.   The near- to far-infrared properties of an optically selected
sample of ordinary spiral galaxies, H2 (starburst) galaxies and Seyferts
types 1 and 2 are summarized.   The spectral indices between 1 and 100µm
are shown to be almost independent of galaxy type.   New JHKL photometry
for a sample of Seyferts found from IRAS data by de Grijp et al. (1985)
is given.   No essential differences are found in the near-infrared
behaviour of Seyferts selected by the two different methods.   The de
Grijp et al. criteria are shown to exclude many Seyfert galaxies known
from optical work.

1. OPTICALLY SELECTED GALAXIES

A search was made of the IRAS Point Source Catalogue for the "ordinary"
galaxies of Glass (1984) and the various types of emission-line objects
of Glass & Moorwood (1985), all of which have been measured previously
in the JHKL region.   Entries, where found, were used to calculate
spectral indices between the various wavelengths.   Those galaxies
having diameters in excess of 5 arcmin were rejected.   The 100µm IRAS
data were combined with J (1.25µm) measurements to obtain overall
spectral indices from the near- to mid-infrared (Table 1).
     The following trends may be discerned:
     (1) The spectral index between J (1.25µm) and 100µm is fairly
independent of whether a galaxy is of ordinary spiral or of Seyfert
type.   Only the H2 galaxies show significantly steeper slopes than the
rest.   This finding implies that the 1.25µm and 100µm emissions are
intrinsic to the underlying galaxies rather than to active nuclei.
     (2) The J and 100µm flux correlation is about as good as the B and
100µm one noted by de Jong et al. (1984).   This can be regarded as
surprising because the J flux is usually considered to arise from an old
population of red giants whereas the B and 100µm fluxes are thought to
be heavily dependent on star-formation activity.
     (3) The H2 and S2 galaxies show fairly similar IRAS colours,
further supporting the suggestion of Glass & Moorwood (1985) that they
are closely related.   The nuclear fluxes from both types are dominated

F. P. Israel (ed.), Light on Dark Matter, 487–488.

by radiation from dust which is at higher temperatures than in ordinary spirals.

(4) The low indices of the Seyferts between 60 and 100µm reflect the turnover in the spectra of their active nuclei at wavelengths shortward of 60µm.

2. COMPARISON OF IRAS AND OPTICALLY SELECTED SAMPLES

De Grijp et al. (1985) have succesfully used IRAS data to search for new Seyfert galaxies. 16 of the Seyferts from their list have been observed at JHKL.

Based on the de Grijp et al. criteria, the vast majority of the optical sample would not have been picked up as Seyferts from the IRAS data because their 25-60µm indices lie outside the acceptable range. In the case of S1's, both higher and lower values of α are encountered, whilst amongst the S2's, mostly lower ones (< -1.25) are found. In fact, there is a clear correlation for S1's between degree of dominance of nuclear flux in the near infrared and increasing 60-100µm spectral index, the range being from ~ -2 (ordinary galaxies) to ~ 0 in high-luminosity examples like IC 4329A.

The JHKL colours of the IRAS galaxies are basically similar to those reported by Glass & Moorwood (1985) for the optical sample. The J-H, H-K two-colour diagram, contains many points near the positions of ordinary galaxies. As in the optical sample, J-H is usually not much affected by moderate levels of nuclear activity while excesses become more noticeable in H-K. Also, as with the optically selected S2's, most of the IRAS ones show only moderately elevated values of H-K meaning that they are dominated at K by ordinary stellar radiation. With the S1's, the nuclei are stronger than the background galaxies at this wavelength, reflecting their generally higher luminosities. All the galaxies measurable at L (3.5µm) show strong excesses.

REFERENCES

de Grijp, M.H.K. et al., 1985. Nature **314**, 240.
de Jong, T. et al., 1984. Astrophys. J., **278**, L67.
Glass, I.S., 1984. M.N.R.A.S., **21**, 461.
Glass, I.S. & Moorwood, A.F.M., 1985. M.N.R.A.S., in press.

TABLE 1. Slope α Where $f_\nu \propto \nu^\alpha$

Wavelengths (µm)	n	12-25	n	25-60	n	60-100	n	1.25-100
Spirals	6	-0.78±.28	11	-2.46±.12	14	-2.01±.15	14	-1.08±.05
H2 galaxies	10	-1.86±.16	10	-1.86±.16	10	-0.42±.14	10	-1.34±.03
S2 galaxies	21	-1.54±.11	22	-1.70±.14	22	-0.77±.14	21	-1.12±.05
S1 galaxies	15	-1.01±.14	13	-0.83±.25	13	-0.91±.28	13	-0.94±.06
IRAS S-types	5	-1.12±.24	13	-0.87±.05	8	-0.63±.16	8	-1.10±.05

Note: n = number in sample

THE MOST LUMINOUS GALAXIES

R.P. Norris
CSIRO Division of Radiophysics
PO Box 76, Epping, NSW 2121, Australia

P.F. Roche and D.A. Allen
Anglo-Australian Observatory
PO Box 296, Epping, NSW 2121, Australia

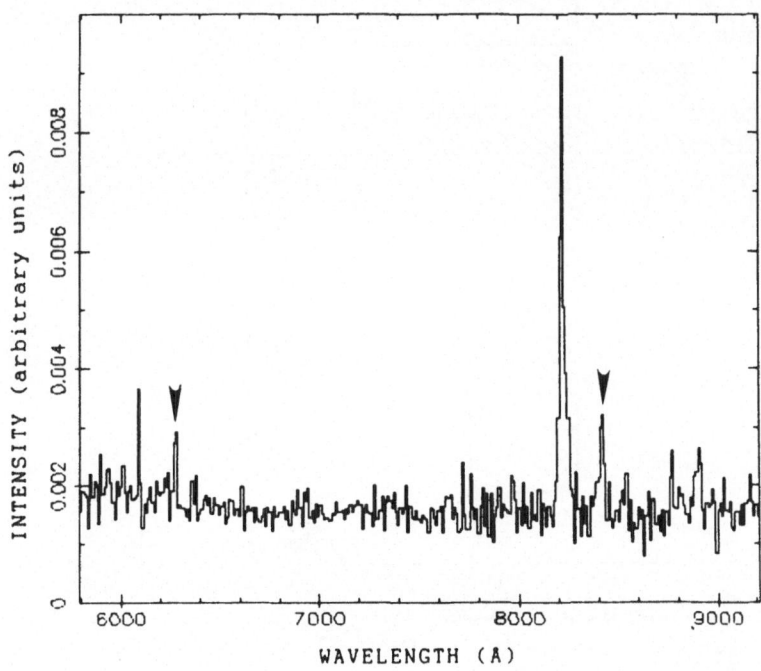

This spectrum of IRAS 14254-2655 B typifies the optically faint galaxies discovered by IRAS and which we are studying spectroscopically using the Anglo-Australian Telescope. Its redshift, 0.252, is the highest yet recorded for an IRAS galaxy. The strongest emission line is a blend of Hα and [N II]. [S II] lies to the red of this, and [O III] is seen in the blue (arrowed).

In contrast to the optically brighter galaxies in the IRAS catalogue, particularly those from Circular 11 surveyed by

489

F. P. Israel (ed.), Light on Dark Matter, 489–490.

Carter (1984) and de Grijp et al (1985), this and the other faint
galaxies in our sample of 40 do not appear to be Seyferts. Whilst
active nuclei may be obscured within some, the dominant energy source
is probably a massive burst of star formation. Many are interacting
or disturbed systems, leading to the speculation that galaxy mergers
trigger the activity.

The infrared luminosities of these galaxies are extreme. The small
wavelength coverage available does not permit an accurate measurement of
temperatures and hence luminosities. However, any reasonable fit using
even H_o = 100 shows many galaxies to exceed $10^{12}L_o$ in their infrared
output. These histograms show that the 100 μm/60 μm ratios of our sample

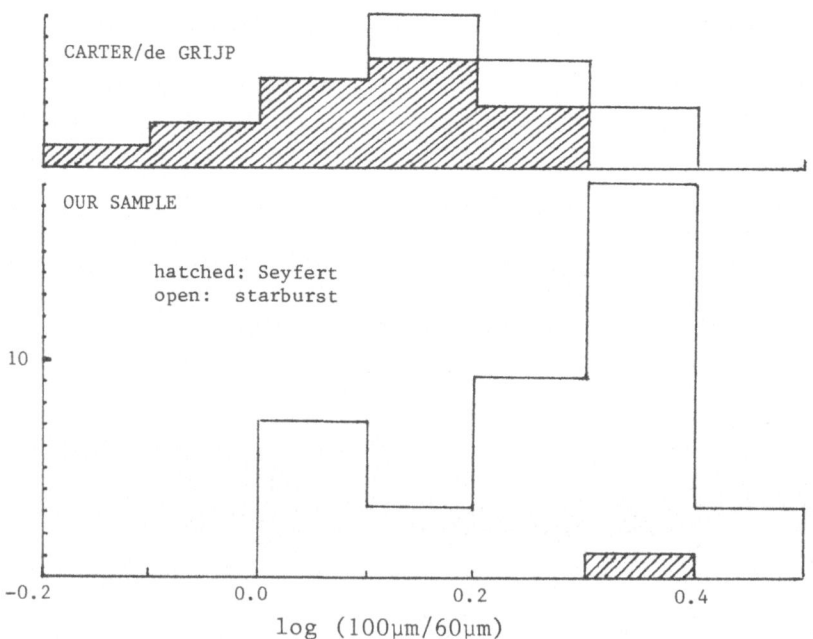

log (100μm/60μm)

are significantly higher than those of Carter/de Grijp. Although few of
our sample were detected at 25 μm, the upper limits show this redness to
be more extreme in the 60 μm/25 μm ratio: at least 62% are redder than
any Seyfert from Carter/de Grijp.

We estimate that 1000 of these galaxies at redshift >0.2 emit
strongly enough to be detected by IRAS. This number contrasts with
only a handful of Seyferts. Within the IRAS bandpass these
super-starburst galaxies are by far the most luminous objects.

Carter, D., 1984. *Astronomy Express*, 1, 61.
de Grijp, M.H.K., et al, 1985. *Nature*, 314, 240.

SUBMILLIMETRE TO INFRARED OBSERVATIONS OF ACTIVE GALAXIES

Lucinda M.J. Brown
School of Physics and Astronomy
Lancashire Polytechnic
Preston, PR1 2TQ
UK

ABSTRACT. We have investigated the submillimetre to infrared spectra
and variability of a sample of active galaxies comprising 13 blazars,
the quasar 3C273 and the peculiar Seyfert 2 galaxy NGC1275. We have
also measurements represent the first stages in a long-term
investigation into the links between various classes of active
galaxies.

INTRODUCTION.

We have been monitoring 13 blazars, the quasar 3C273 and the peculiar
Seyfert 2 galaxy NGC1275 since September 1982 in all available
atmospheric windows between 1 and 1100 μm. The aims of our monitoring
programme are as follows:

a) to discriminate between various models of the continuum emission
   of active galaxies.
b) to investigate the properties of the most compact regions of
   sources of non-thermal radiation.
c) to determine whether thermal radiation from dust contributes to
   the continuum emission.

FACILITIES.

All our submillimetre observations have been obtained using the
United Kingdom Infrared Telescope (UKIRT) with the QMC/Oregon
photometer (Ade et al., 1984). Infrared observations were made with
UKIRT and the IRTF, millimetre observations with UKIRT and the NRAO
12m telescope. Extended spectral coverage was achieved using VLA,
optical and EXOSAT observations. IRAS data are available for NGC1275,
3C273, and several blazars (from both catalogue and AO's): the
additional spectral information provided by these observations has
proved very valuable, particularly in the case of NGC1275 (see
below).

F. P. Israel (ed.), Light on Dark Matter, 491–492.

RESULTS

a. Spectra

The 1 μm to 2 mm spectra of 3C273 and the blazars are characterized
by power laws in the millimetre/submillimetre and infrared regions,
with several blazars showing signs of a high-frequency cut-off (Clegg
et al., 1983; Gear et al., 1985a). To date we have obtained no
convincing evidence of any contribution to the continuum emission of
these sources from thermal radiation by dust. We interpret the
emission as synchrotron radiation from an unresolved relativistic jet
aligned close to the line of sight (Gear et al., 1985a). The non-
stellar emission from NGC1275 is clearly the sum of two components: a
non-thermal power law and a separate 'thermal' component presumably
due to re-radiation from dust grains heated by star formation (Gear
et al., 1985c, submitted) possibly triggered by cooling flows from
the Perseus cluster (Fabian, 1981).

b. Variability

During the 1983 flare event observed in 3C273 (Robson et al., 1983)
the turnover frequency was observed to decrease while the peak flux
initially increased and subsequently remained fairly constant. This
behaviour was inconsistent with any current relativistic jet model
and with normal expansion models and led to the development of a new
model (Marschner and Gear, 1985, in press). In this model the flare
arises in a small shocked region in a relativistic jet. Gear et al.
(1985b, in press) show that this model is also consistent with the
observed variability of the blazars.

c. Thermal sources

We have mapped the central regions of the starburst galaxy NGC253 at
a wavelength of 350 μm. The submillimetre emission traces the dust
being heated by young stars. The observations indicate that the total
mass of interstellar material in the central 1.6 Kpc is ~1.6×10^8 M_O
and that conditions in the interstellar medium may be similar to
those in our own galaxy (Gear et al., 1985c, in preparation).
However, further observations are required to confirm this. We have
also observed and mapped other galaxies (some interacting) at
submillimetre wavelengths.
      The above results indicate the power of submillimetre-infrared
observations in elucidating the physical processes occurring in
active galaxies.

REFERENCES

Ade, P.A.R. et al., 1984, Infrared. Phys., 24(4), 403.
Clegg, P.E. et al., 1983, Ap.J., 273, 58.
Fabian, A.C. et al., 1981, Ap.J., 248, 47.
Gear, W.K. et al., 1985a, Ap.J., 291, 511.
Robson, E.I. et al., 1983, Nature, 305, 194.

ACTIVE GALACTIC NUCLEI IN THE IRAS PSC

M. DENNEFELD   and   M.P. VERON-CETTY
Institut d'Astrophysique   Observatoire de Haute-Provence
98 bis, bd Arago   F-04870 St Michel l'Observatoire
F-75014 Paris

We have investigated the far-infrared properties of previously known Active Galactic Nuclei (AGN) detected by IRAS. This is a necessary step in order to understand the various selection effects which could affect the use of the IRAS PSC for further studies like luminosity function or space density of AGN.

We have extracted from the PSC the 286 sources identified with one of the 2805 objects from the "Catalogue of quasars and active nuclei" (Véron-Cetty and Véron, 1984). The histogram of the V magnitudes of the detected objects is shown in Fig. 1a. Most of the objects fainter than V=16.0 are wrongly identified, due generally to a poorly known position of the optical source. From Fig. 1b, giving the percentage of objects in the quasar catalogue detected by IRAS as a function of the V magnitude, it appears that most objects brighter than V=15 have been detected. The non-detection of fainter objects is clearly due to the IRAS sensitivity limit as shown by the loose correlation found between V magnitudes and 60 μ flux density. We are left with 250 well identified objects, the identification being made on the basis of positionnal coïncidence only. The accuracy of IRAS positions is found to be better than 10" r.m.s. for those objects with accurate optical coordinates.

The possible differences between Sey 1, Sey 2 and HII-type galaxies has been investigated. We plot for example the 100 to 60 μ spectral index versus the 60 to 25 μ spectral index (fig. 2). Although there is no clear separation between the three groups, the Sey 1 appear to have on the average a larger (100,60) index for a given (60,25) one, while the Sey 2 are in between Sey 1 and the HII-type galaxies. The explanation of this has to be found in terms of different relative contributions of the non-thermal nuclear component and a thermal dust component heated by stars or by the central source. On the other hand, in the mid-infrared, Seyfert galaxies are believed to have a much flatter spectrum due to the nuclear component

F. P. Israel (ed.), Light on Dark Matter, 493–496.

as opposed to the cold dust emission in star-forming
galaxies. This has led de Grijp et al. (1985) to propose to
select Seyfert galaxies by choosing a (60,25) spectral index
between -0.5 and -1.25. Although they had a 70 % success rate,
Fig. 3 shows that they would have missed about 1/3 of the
known Sey 1 and 2/3 of the known Sey 2. This method can
therefore by no means provide a complete sample.

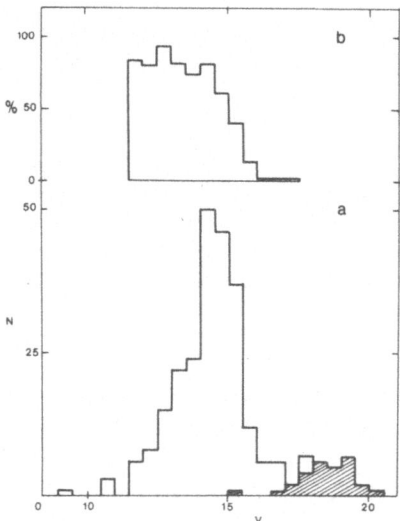
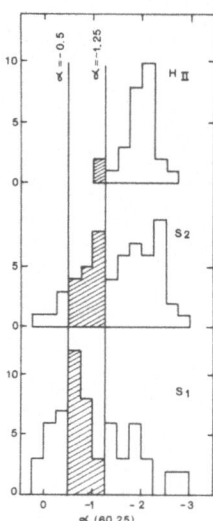

Fig. 1 a) Histogram of the V magnitudes of the AGN in the
Véron-Cetty and Véron Catalogue identified with IRAS sources.
The hatched area corresponds to objects which were shown to be
wrong identifications.
b) The fraction of objects in the same catalogue effectively
identified with an IRAS source.

Fig. 3 : Histograms of the α(60,25) spectral index for Sey 1,
Sey 2 and HII-type objects. The hatched areas represent the
limits of the search by de Grijp et al.

Fig. 4 gives a U-B versus the (60,25) index plot for the
known Sey 1 only. It shows that Sey 1 with the highest UV
excess have preferably a low (60,25) spectral index. This is
expected if a high (60,25) [or(100,60)] index reveals the
presence of a cold dust component which also increases the
reddening. The active galaxies with low spectral index could
probably be found all, whether or not they have a large UV
excess and provided there is enough dust, by extending the
limits of the search between $-1.25 < \alpha(60, 25) < 0$. We suspect
that the AGN found by de Grijp et al. might preferentially
have a positive U-B index. On the other hand, the Sey 1 with
high spectral index, which are excluded in the de Grijp et

al. search, are precisely those with a small UV excess, not found in other surveys and most of them are weak Sey 1 which were only found by detailed spectroscopic observations. They cannot be selected by lowering the spectral index limit alone because this would reintroduce the HII type objects (see fig. 2). We note an apparent anticorrelation between the (25,12) index and the (100,60) one for the 84 objects with four fluxes and well determined type. This possibly suggests the presence of an excess at 12 µ in the objects with the coldest dust component (among which we find the weak Sey 1), as seen in normal galaxies.

Finally, 9 QSO's only are found in the PSC, three of them having already been discussed by Neugebauer et al. (1985). As most quasars discovered after the first edition of the quasar catalogue are fainter than V=16.0, we do not expect this number to change significantly when we have completed the search for identifications of these new objects in the PSC.

A detailed discussion of all the topics presented here can be found in our paper submitted to Astronomy and Astrophysics.

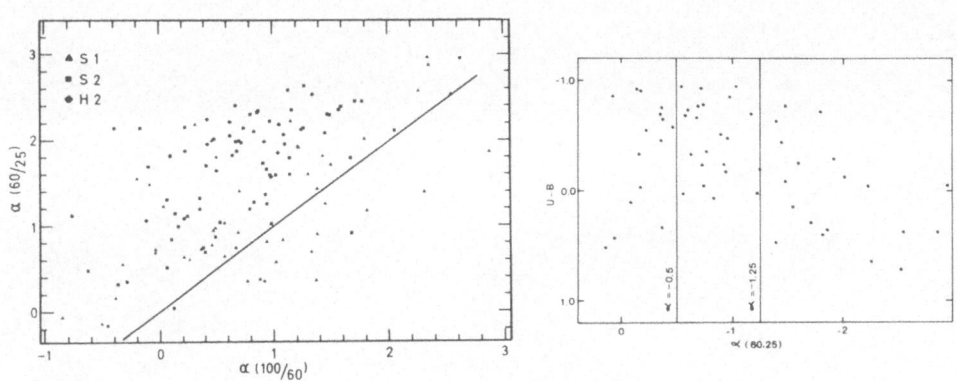

Fig. 2   A plot of α(60,25) spectral index versus the α(100,60) one. The solid line is the locus of constant spectral index. Note the unfortunate change of sign in the definition of the indexes in this figure.

Fig. 4 A plot of U-B versus the α(60,25) spectral index for the well known type 1 Seyfert. The vertical lines show the limits of the de Grijp et al.'s search.

References
- de Grijp, M.H.K., Miley, G.K., Lub, J. and de Jong, T. : 1985, Nature, 314, 240.
- Neugebauer, G., et al. : 1984, Ap.J. Letters, 278, L83.
- Véron-Cetty, M.P. and Véron, P. : 1984, ESO Scientific report n°1.

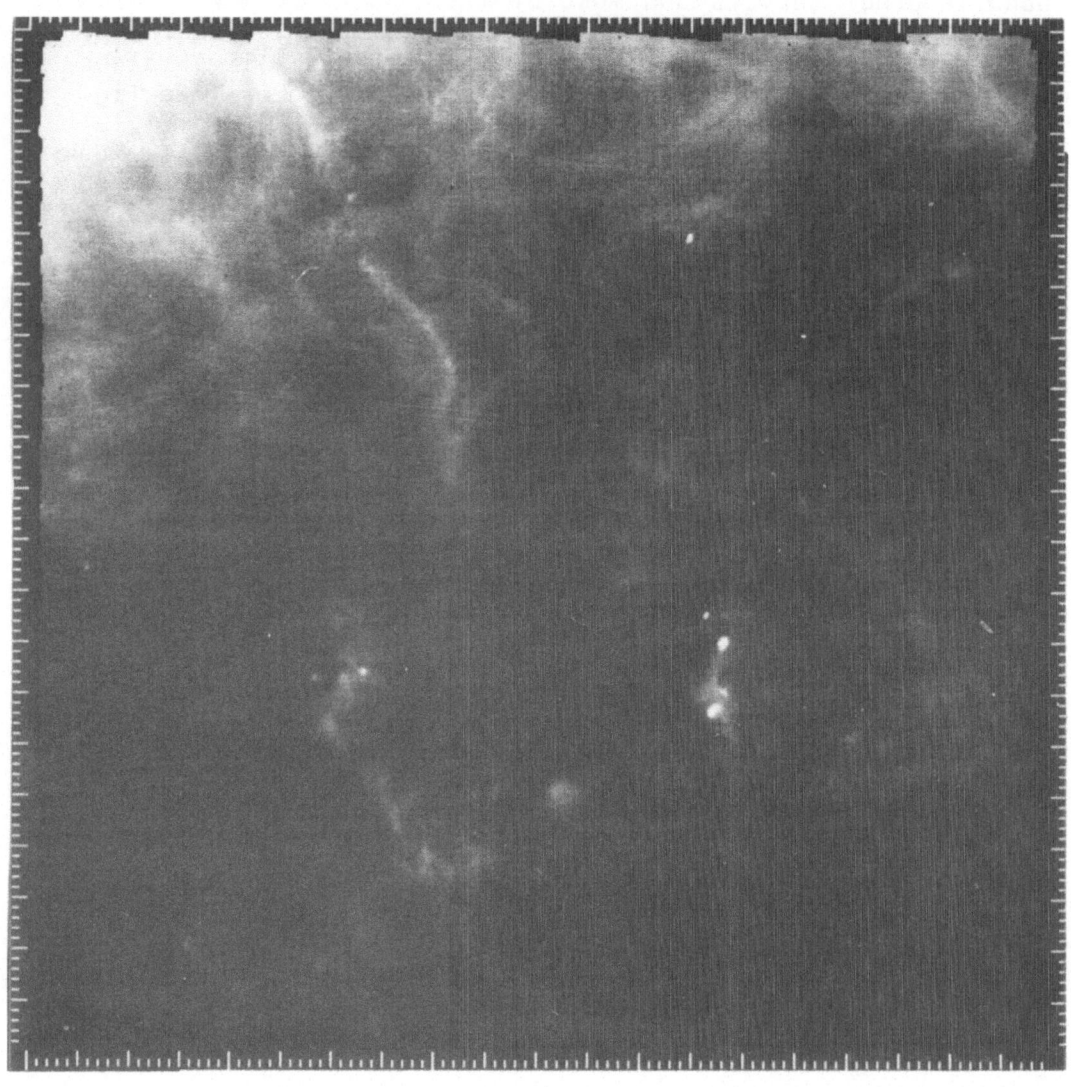

IRAS Field 207   α = 12^{h}00^m, δ = -75°, HCON-3, 100 μm. Field in
                  Chamaeleon, Musca and Carina; extensive cirrus in
                  Musca. The long filament left and above the center
                  corresponds to a dark filament seen near Y Muscae.

SECTION 7.

COSMOLOGY

COSMOLOGICAL RESULTS FROM IRAS

M. Rowan-Robinson
Theoretical Astronomy Unit
Queen Mary College
London E1

ABSTRACT. IRAS is shown to be a powerful cosmological tool. At high galactic latitudes the problem of emission from interstellar dust is greatly reduced and virtually all 60 μm sources are identifiable with galaxies. Source-counts at 60 and 100 μm show that there is a 20% higher source-density at b > 60° than at b < -60°. A redshift survey in the north galactic polar cap shows that the characteristic depth of the IRAS survey is about 200 Mpc and the inhomogeneity responsible for the source-count anisotropy must be of comparable dimensions. Studies of the 2-dimensional covariance function confirm the existence of significant structure on this scale.

The local gravitational field has been mapped using IRAS 60 μm sources and analyzed in spherical harmonics. The dipole component agrees well in direction with that of the cosmic microwave background anisotropy, suggesting that the latter is the result of motion induced by the combined gravitational attraction of galaxies and clusters within 200 Mpc.

The 100 μm background radiation has been deconvolved into contributions from interplanetary dust, interstellar dust and a residual, relatively isotropic component of intensity 5.7 ± 2.5 MJy/sr. The latter could be a calibration offset, cool dust in the outer solar system, or a cosmological background from high redshift starburst galaxies.

1. INTRODUCTION

I want to show how the IRAS survey data can be used to obtain some powerful cosmological results. Nick Gautier has shown (this volume p. 49) that there is extended emission from interstellar dust at 100 μm from a substantial fraction of the sky. A similar result is otained using the CIRR1 flag in the IRAS Point Source Catalog, which records the number of 100 μm only "point" sources found in a 1 sq deg area centred on each source. Figure 1 shows the distribution of those bins in which CIRR1 > 1, in galactic coordinates, and it can be seen that the cirrus extends up to 45° away from the plane at some longitudes. However in the polar caps (|b| > 60°) there are only a few small cirrus clouds that affect the IRAS point-source recognizer, the most prominent being cloud A of Low et al (1984) in the south. Many of these clouds can be associated with neutral hydrogen feaures.

F. P. Israel (ed.), Light on Dark Matter, 499–506.
© *1986 by D. Reidel Publishing Company.*

Outside these clouds the effects of the cirrus can be neglected and
we have a homogeneous and complete survey of the sky, unaffected by
interstellar extinction.

2. SOURCE COUNTS IN THE POLAR CAPS

We have compared the differential source-counts at 60 and
100 μm for b > 60° and b < -60° (Fig. 2) after exclusion of obvious
stars (defined by S(25) > 3 S(60)), areas affected by cirrus, and a
20° x 20° area centred on Virgo (Rowan-Robinson et al 1985a). The
straight line fits are noise-weighted least-square fits with slope -
1.5. Both at 60 and 100μ there is a 20% higher source-density in the
north polar cap than in the south.

To establish the cosmological significance of this anisotropy we
have undertaken a programme to identify and obtain spectra of all
the 60 μm > 60° and 180° < 1 < 290°, b < -60°). The work in the
northern sample is now complete and we find that virtually
all 60 μm sources in this area can be identified with galaxies on the
Palomar Sky Survey (Lawrence et al 1985). Spectra for those galaxies
with S(60) > 0.85 Jy for which redshifts are not available in the
literature have been obtained with the INT on La Palma and the
preliminary redshift distribution for the complete sample is shown in
Fig. 3. The 60 μm luminosity distribution for the sample is shown in
Fig. 4. It can be seen that the median value for νL_ν (60)
is ~ 10^{11} $L_\odot$. The 60 μm luminosity function for IRAS galaxies has

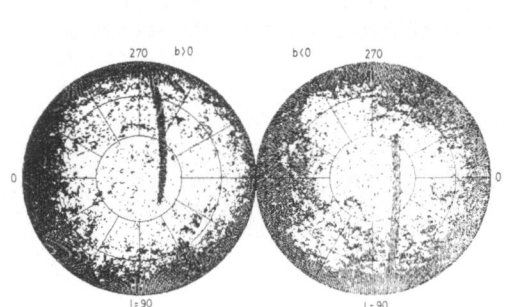

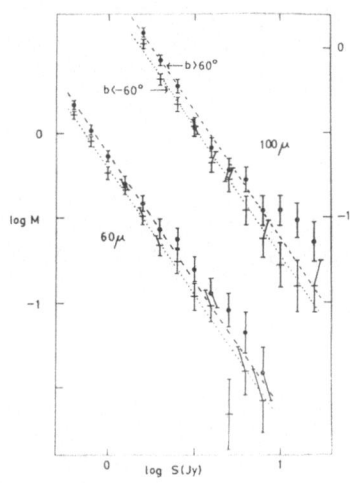

Fig 1: Distribution on sky in galactic
coordinates of 1 sq deg bins in which
CIRR1 > 1 (or in which there was no
IRAS coverage.

Fig 2: IRAS differential source
counts at 60 μm (LH scale)
and 100μ (RH scale) for b>60° (dots)
and b<-60° (crosses) (Rowan-Robinson
et al 1985a).

been calculated both for this sample and for a larger sample of
brighter galaxies consisting of IRAS galaxies with m_{pg} ≤ 14.5, b >
60°, S(60) > 0.5 Jy (Rowan-Robinson et al 1985b) and the result are
shown in Fig. 5. This luminosity function has been used to calculate

the 60 μm source-counts and the results are in excellent agreement
with the IRAS observations (Fig. 6). These calculations have been
carried out for an Ω = 1 cosmological model both for the cases of no
evolution and with strong luminosity evolution (L(z) α exp 5 (1 -
1/(1+z)), and either case is consistent with present observations.

From this analysis we can conclude that almost all 60 μm sources
at high galactic latitudes are galaxies and that the characteristic
depth surveyed by IRAS is 200 Mpc (for H = 50). The inhomogeneity
responsible for the source-count anisotropy illustrated in Fig. 2.
must also be of at least comparable dimensions. There is no
possibility that it can be caused by the Local Supercluster.

3. MAPPING THE "LOCAL" GRAVITATIONAL FIELD WITH IRAS

If we assume that IRAS galaxies are a good tracer of the matter
distribution within 200 Mpc, then the IRAS 60μ sources in regions
free of cirrus can be used to map the local gravitational field. We
have analyzed the graitational field traced in this way in terms of
spherical harmonics up to the quadrapole term (Yahil et al 1985). The
dramatic result is that the direction of the dipole component of the
local gravitational field, (l,b) = (248 ± 9, 40 ± 8), is in good
agreement with the direction of the dipole anisotropy in the cosmic
microwave background radiation, (l, b) = (277,29). The latter can
therefore be explained as due to motion induced by the net

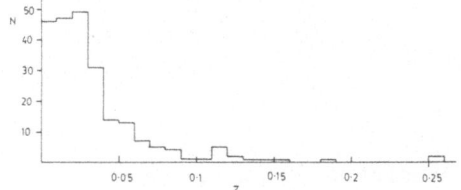

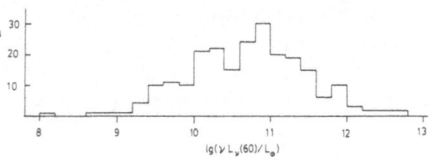

Fig 3: Redshift distribution for IRAS
Galaxies with S(60)>0.85 Jy, b>60°,
0° < l < 110° (Lawrence et al 1985).

Fig 4: 60 μm luminosity
distribution for the IRAS
galaxy sample of Fig 3.

gravitational attraction of galaxies and clusters within about 200
Mpc.

The actual distribution of IRAS galaxies is far from dipolar.
Fig. 7 shows a contour map of the surface-density of 60 μm sources
brighter than 0.6 Jy. The sky was divided into 720 equal areas 10° in
1 x 0.1 in sin b. 1 sq deg bins in which CIRR1 > 1 anywhere were
excluded from the source-density estimate. Areas in which more than
25% of the area was exluded in this way were assigned the mean
surface-density for |b| > 30°, which is 0.39 per sq deg, in the
contour plotting routine. The hatched area therefore delineates the
IRAS "zone of avoidance". Contours are shown at 20 and 40% above and
20% below the mean source-density. The strongest concentration of
sources is in the north polar cap and covers well over a steradian.

Although the Local Supercluster contributes to this concentration, it
is largely defined by much more distant clusters like Coma, A1367 and
the NGC5416 cluster. Several other concentrations of size 10° -60°
are seen, some of which can be associated with known superclusters.

      To investigate the degree of clustering of IRAS sources we have
studied the 2-dimensional covariance function for 60μ sources
brighter than 0,6 Jy, with $|b| > 30°$, outside the CIRR1 > 1 mask
(Rowan-Robinson and Needham 1985). The effective area in each angular
annulus has been calculated by Monte Carlo simulation. Figure 8 shows
the resulting w (θ) for IRAS galaxies. For θ < 10° the results are
consistent with the powerlaw form found in optical studies of the
galaxy population. The amplitude is broadly consistent with the
results of Peebles and Hauser (1973) for galaxies in the Zwicky and
Shane-Wirtanen catalogues, allowing for the intermediate depths
surveyed by IRAS. Thus there is no strong evidence that IRAS
galaxies, which are predominantly spirals, are clustered differently
from galaxies as a whole. Figure 8 shows significant excess power on

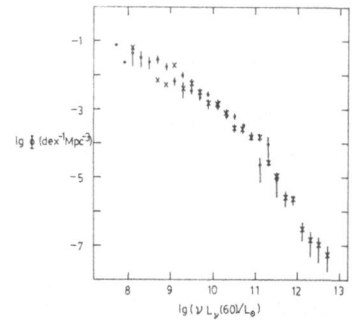

Fig 5: 60 μm luminosity function for the IRAS
galaxy sample of Fig 3 (x's) and for IRAS
galaxies with $m_{pg} \leq 14.5$, b>60°, S(60)> 0.5
Jy (dots, Rowan-Robinson et al 1985b).

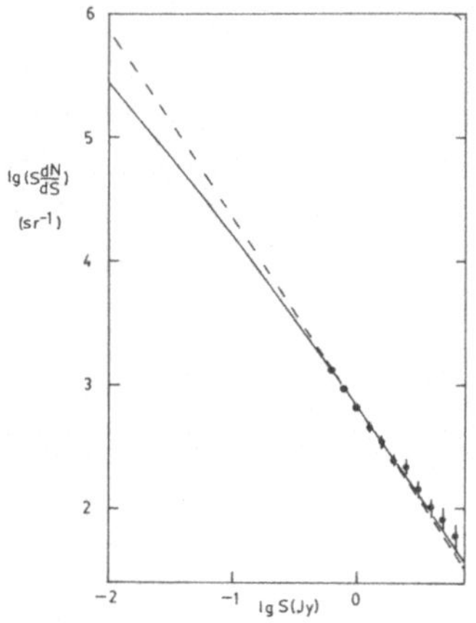

Fig 6: Predicted 60 μm differential
source counts for Ω = 1 with no
evolution (broken line, see text),
compared with IRAS counts for b>60°

the angular scale 20° - 60°, as might be expected from the source-
density maps (Fig. 6). Previous studies have either not been deep
enough or have not covered enough of the sky for this scale to be
studied effectively.

4. IS THERE A FAR INFRARED COSMOLOGICAL BACKGROUND?

      Finally we have been studying the IRAS 100 μm background
radiation. To do this it is necessary to remove the contribution of
the "cirrus". Table 1 summarizes the properties of a grain mixture
which gives an extremely good fit to the interstellar extinction
curve at visible and ultraviolet wavelengths (Fig. 9).
For 0.01μ ≤ a ≤ 0.03μ the grain properties are taken from Draine and
Lee (1984), but the properties of the larger (0.1 μm) grains are

taken from studies of circumstellar dust shells (Rowan-Robinson et al 1985c), which show that these grains have high absorption efficiency at 20-100 μm. Beyond 100 μm the grain properties are uncertain. Figure 10 shows the predicted for infrared spectrum of the cirrus for different extrapolations beyond 100 μm. If a part of the material I have assigned to 0.01 μm grains in Table 1 is in fact in much smaller grains, then there will be enhanced radiation at 10-60 μm (cf Draine and Anderson 1985). The predicted 100 μm intensity per unit column of dust can be expressed as

$$I(100) = 55\ E(B-V)\quad MJy/sr. \qquad\qquad (1)$$

Regions relatively free of interstellar dust can be used to model the emission from interplanetary dust. Using point sources with CIRR1 = 0 to define such regions, the 100 μm intensity is well fitted by a

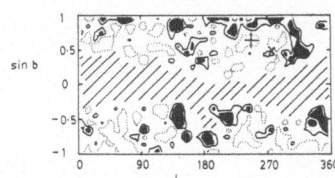

Fig 7: Contour map of surface density IRAS 60 μm sources brighter than 0.5 Jy. Dotted and solid contours corres-pond to 0.31 and 0.47/sq deg, heavy shaded areas to 0.55/sq deg. Cross denotes direction of dipole component of gravitational field, x denotes microwave background dipole direction.

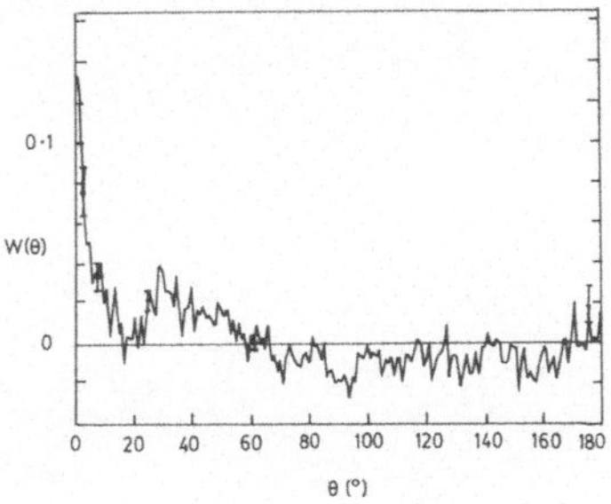

Fig 8. 2-D covariance of function for IRAS 60 μm sources with |b|> 30° out-side CIRR1> 1 mask (Rowan-Robinson and Needham 1985).

cosec |β| relation of the form

$$I(100) = 5.7 + 2.50\ cosec\ |\beta|\ MJy/sr, \qquad (2)$$

where β is the ecliptic latitude.
The statistical uncertainties in these numbers are very small. The main uncertainty lies in the absolute calibration.

How much of the 5.7 MJy/sr residual background should be attributed to a smoothly distributed interstellar dust component? Most studies have found the polar caps to be relatively free of reddening outside the known concentrations of gas and dust like cloud A. Hilditch et al (1983) quote E(B-V) < 0.005 for most of the sky at b > 80°. Knude (1986) has studied 4800 lines of sight to A and F stars at b > 70° and finds that about 6% of these have E(B-V) > 0.07, the dust having a tendency to be concentrated in filaments (few degrees x ten or more degrees). The median reddening in clear regions is equivalent to E(B-V) ~ 0.015, with photometric uncertainties being

of the same order. It would seem fair to take this as the upper limit
for any general reddening in the polar caps. Using eqn (1), the upper
limit for the contribution of interstellar dust to
the 100 μm intensity residual component (the constant in eqn (2) is
then

$$I(100)_{i.s.} \lesssim 0.8 \text{ MJy/sr} \qquad\qquad (3)$$

(the Hilditch et al (1983) upper limit would give $I(100)_{i.s.} < 0.3$
MJy/sr).

    To test the contribution of eqn (2) from the total
100 μm intensity and used eqn (1) to make maps of E(B-V). These are
found to be in good agreement with estimates derived from neutral
hydrogen, even down to galactiv latitudes of 25° (the limit we have

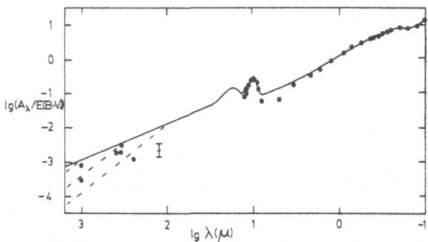

<u>Fig 9</u>: Fit of grain mixture of Table 1 to interstellar extinction
curve. The effect of taking $Q_\nu \propto \nu^2$ for $\lambda > \lambda_2$ for the 0.1 μm grains
has been indicated by the broken lines, for $\lg \lambda_2(\mu m) = 2$, 2.5, λ_2
(Rowan-Robinson 1985)

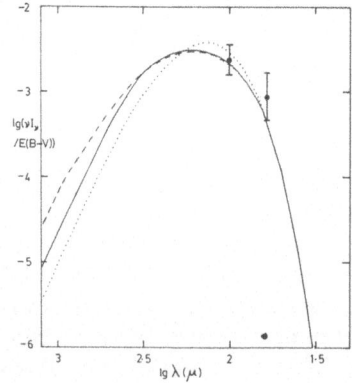

<u>Fig 10</u>: Predicted spectrum of cirrus emission in the far infrared for
$\log \lambda 2 = 2$ (dotted curve), 2.5 (solid curve), 3 (broken curve)
compared with average for clouds observed by Low et al (1984)
corrected to October 1984 calibration

studies to date). We infer that the 5.7 MJy/sr residual is relatively isotropic over the sky (±20%).

We conclude that there is an isotropic component in the 100 μm background radiation of intensity

$$I(100) = 5.7 \pm 2.5 \text{ MJy/sr} \qquad\qquad (4)$$

the uncertainty being entirely in the absolute calibration uncertainty (increased from 1.6 MJy/sr quoted in the IRAS Catalog Introductory Supplement to 2.5 MJy/sr by Nick Gautier at this meeting). Possible origins of a background of this magnitude are discussed by Rowan-Robinson and Walker (1985) and summarized in Table 2. A spherically symmetric distribution of dust in the outer solar system is a possibility, though the specific model proposed by Bailey (1983) for an inner extension to the Oort cometary cloud fails by an order of magnitude. Background radiation from starburst galaxies with strong evolution (cf Fig. 6), which could be associated with the generation of the bulk of the Population I metals in spiral galaxies during the first 10^9 years of their lives, would give a background of the right order.

REFERENCES

Bailey, M.E., 1983, M.N.R.A.S. 204, 603.
Draine B.T. Lee H.M., 1984, Astrophys.J. 286, 89.
Draine B.T. Anderson N., 1985, Astrophys.J. 292, 494.
Hilditch R.W., Hill G., Barnes J.V., 1983, M.N.R.A.S. 204, 241.
Knude J., 1986, this volume, page 55.
Lawrence A., Walker D., Rowan-Robinson M.,Penston M.V., Leech K., 1985, in preparation.
Low F.J., et al, 1984, Astrophys.J. 278, L19.
Peebles, P.J.E., Hauser M.G., 1973, Astrophys.J.Supp. 28,19.
Rowan-Robinson M., 1985, M.N.R.A.S.
Rowan-Robinson M. and Needham G., 1985, in preparation.
Rowan-Robinson M. and Walker D., 1985, in preparation.
Rowan-Robinson M., Chester T., Soifer T., Walker D. Fairclough 1985a, M.N.R.A.S.
Rowan-Robinson M., Lock A., Walker D. and Harris S., 1985c, M.N.R.A.S.
Yahil, A., Walker D. and Rowan-Robinson M., 1985, Astrophys.J.

TABLE 1: Model for interstellar grains

Grain type	evidence	$\dfrac{A_V}{E(B-V)}$	Abundance (x 10^5)	Temperature in interstellar radiation field
0.1μm amorphous silicate	M star c.d.s.*	1.32	2.5	12.3
0.03μm m silicate	i.s. extinction	0.12-0.19 μm 0 16	0.8 3.4 (cf 3.3 (cosmic)	18.4
≦ 0.01μm silicate	0.10-0.12μm i.s. extinction	0.03	0.1	19.7
0.1μm amorphous carbon	C star c.d.s.*	0.84	8.0	14.5
0.03μm graphite	0.22-0.24μm i.s. extinction	0.33	4.7 20 (cf 48 cosmic)	21.8
≦0.01 graphite	0.19-022μm i.s. extinction	0.30	6.9	22.8

* Rowan-Robinson et al 1985c.

TABLE 2: Possible explanations of isotropic 100 μm background.

explanation	Predicted magnitude (MJy/sr)
1. Calibration offset	± 2.5
2. Small interstellar grains with no associated larger grains	
3. Shell of 100 μm grains in solar system, beyond orbit of Pluto	
4. Enhanced Oort cloud (Bailey 1983)	< 0.5
5. $\Omega = 1$ due to earths	0.000008
Jupiters	0.00012
brown dwarfs	0.00009
minimum mass stars	0.00004
6. Quasars, no evolution	0.0002
luminosity evolution	0.003
7. Starburst and Seyfert galaxies, no evolution	0.25
luminosity evolution	4
8. Population III	?

EVIDENCE FROM IRAS DATA FOR LARGE-SCALE ANISOTROPY IN THE HUBBLE FLOW

C A Collins[1], R D Joseph[1], & N A Robertson[2]
[1]Blackett Laboratory, Imperial College, London SW7
[2]Dept. of Natural Philosophy, Glasgow University, Glasgow G12

1. INTRODUCTION

Galaxies are generally assumed to have evolved from small primordial density fluctuations which have grown into the structures we presently observe. One way to confront this idea observationally is to use galaxies as test particles, to map out the large-scale gravitational potential. In this case, one can search for deviations from isotropic Hubble flow induced by fluctuations in the large-scale mass distribution (including any "dark" matter present) which have arisen from the growth of the primordial fluctuation spectrum as the Universe has evolved.

We have been engaged in an investigation of the isotropy of the Hubble flow using near-IR photometry for an all-sky sample of spiral galaxies. Herein we describe the result of using IRAS 100 μm photometry for this same galaxy sample, and compare the solutions obtained with those using near-IR and optical photometry.

2. DATA ANALYSIS

The galaxy sample we are studying is the sample of 96 ScI-II spiral galaxies identified and classified by Rubin et al. (1976a). There are IRAS detections at 100 μm for 64 of these galaxies in the IRAS Point Source Catalogue. Since these galaxies are in a narrowly-defined luminosity class, we can take them to be "standard candles," and use their magnitudes to infer a velocity, assuming isotropic Hubble flow. We then compare this notional Hubble velocity, v_H, with the observed recessional velocity, v_{obs}, and look for a systematic difference across the sky by solving for a dipole velocity anisotropy, v_D, which minimises, in the least squares sense, the difference between the residual $v_{obs} - v_H$, and v_D. Since we have subtracted the solar motion in the Galaxy, v_D may be interpreted either as a Local Group motion with respect to the reference frame defined by this distant galaxy sample, or as a mean deviation from uniform Hubble flow for these galaxies.

The solution for the apparent Local Group motion, v_D, obtained using the IRAS 100 μm data, is shown in Table I. For comparison we have also included the solutions derived using near-IR photometry for the 43 galaxies in the sample we have so far observed, and the solution

F. P. Israel (ed.), Light on Dark Matter, 507–508.

reported by Rubin et al. (1976b) based on optical photometry <u>not</u>
corrected for galaxy diameter (cf. Schechter 1977). Despite the fact
that these solutions are obtained by photometry at very different
wavelengths, in apertures between 35 arcsec and several arcmin, and for
different subsets of galaxies in the sample, Table I shows that all the
solutions are identical within the errors. There appears to be a
significant dipole anisotropy in the peculiar velocities of these
galaxies with respect to the Hubble flow.

TABLE I. Magnitude and direction of dipole anisotropy.

Observations	v_D(km s^{-1})	l(deg)	b(deg)
IRAS 100 μm, 64 galaxies	660 ± 220	190	-6
Near-IR, 43 galaxies	750 ± 350	180	-9
Blue, 96 galaxies	580 ± 200	203	-11
3K CBR dipole anisotropy	620 ± 40	265	35

3. DISCUSSION

This dipole anisotropy in galaxy peculiar velocities has a direction
roughly orthogonal to that inferred from the dipole anisotropy in the
3K CBR, shown also in Table I (cf. Lubin et al. 1983). The latter is
likely to be the best indication of Local Group motion with respect to
the local co-moving frame. We therefore suggest that this velocity
indicates systematic departures from isotropic Hubble flow. These
galaxies have an average redshift of 5100 km s^{-1} and so this represents
a deviation from isotropic Hubble flow on a scale of ~100 h^{-1} Mpc.

This result provides strong evidence that the Universe is far from
smooth on scales of 100 h^{-1} Mpc. It contradicts the evidence offered
by de Vaucouleurs & Peters (1984) that the Hubble flow should become
isotropic outside the Local Supercluster. An anisotropy of this
magnitude on such a large spatial scale tends to favour a high density
Universe. Comparison of this result with the predictions of various
galaxy formation scenarios (cf. Schaeffer & Silk 1985), which should
provide constraints on some of the ingredients of these models, is in
preparation.

REFERENCES

Lubin, P.M., Epstein, G.C., & Smoot, G.F., 1983. Phys. Rev. Lett.,
    50, 616.
Rubin, V.C., Ford, W.K., Jr., Thonnard, N., Roberts, M.S., & Graham,
    J.A., 1976a. Astr. J., **81**, 687.
Rubin, V.C., Thonnard, N., Ford, W.K., Jr., & Roberts, M.S., 1976b.
    Astr. J., **81**, 719.
Schaeffer, R., & Silk, J., 1985. Astrophys. J., **292**, 319.
Schechter, P.L., 1977. Astr. J., **82**, 569.
Vaucouleurs, G. de, & Peters, W.L., 1984. Astrophys. J., **287**, 1.

SUMMING UP

ASTRONOMY AFTER IRAS

M.S. Longair
Royal Observatory
Blackford Hill
Edinburgh EH9 3HJ
Scotland

1. THE ACHIEVEMENT

This has been a wonderful week and it is a privilege to be invited to
review what has been achieved so far and to try to set the IRAS
project in the context of the major goals of contemporary research.
Clearly, this will be a personal view of the work of IRAS and I will
try to identify highlights and problems which I feel should be
specially mentioned.

I consider IRAS to have been one of the most courageous of space
projects. We have all been well aware of the technical problems
which the programme faced. The cryogenic telescope and the detector
systems were all at or beyond the state-of-the-art when the project
was conceived and yet it has more than achieved its objectives.

Already, the IRAS database has become an integral part of contemporary
astronomy. I found it striking how rapidly observers in other
wavebands have begun the scientific exploitation of the IRAS data.
What was apparent from the discussions was that IRAS has significance
for essentially all branches of astronomy. The great virtue of the
data is that it provides new and 'clean' samples of infrared objects
for astrophysical study. This means that it will now be possible to
undertake systematic studies of large, complete samples of different
classes of object, unaffected by the effects of obscuration. This is
of special importance for galactic studies of objects such as stars
with circumstellar shells and planetary nebulae and for extragalactic
sources such as Seyfert galaxies, some of which may have been missed
because of obscuration close to the centres of these galaxies. It is
my own view that, in the long term, this will be the aspect of the
project that will prove to be IRAS's most important contribution to
science. Time and again, many of the most important contributions to
astrophysical understanding come from systematic work on complete
samples of objects.

F. P. Israel (ed.), Light on Dark Matter, 511–521.
© *1986 by D. Reidel Publishing Company.*

2.   THE GREAT PROBLEMS OF CONTEMPORARY ASTRONOMY

On a number of occasions, I have recently been invited to review major
problems of contemporary astronomy.   In these, I have indicated what
I believe we know and what we need to know.   Let me give a list of
some of the most important of the problem areas:

- the formation of planetary systems
- the formation of stars
- mass-loss in all classes of star
- the structure and mass distribution of our Galaxy
- the nature and distribution of dark or hidden matter
- the origin and evolution of galaxies
- the astrophysics of active nuclei of all types
- cosmology

It is remarkable how close the correlation is between this list of
problems and the subject matter of the papers presented this week.
The first three topics in particular are among the least well
understood branches of astrophysics and much of the new data is
crucial for these as will be discussed in more detail later.
However, it is invidious to select from the list.   IRAS has immense
potential in all the areas listed above.

3.   THE DIAGNOSTIC TOOLS

The prime diagnostic tool of the far infrared waveband as observed by
IRAS is the emission of heated dust grains.   Unfortunately, dust
grains are much more difficult diagnostic tools than 'simple' tools
such as the emission and absorption spectra of atoms, ions and
molecules.   Current understanding of the nature of interstellar
grains was admirably surveyed by van de Hulst, Greenberg and Mathis.
The problems of understanding the nature of interstellar dust grains
have not diminished.   Careful reading of these papers will indicate
the nature of the uncertainties with respect to chemical constitution
and particle size distributions.   Another key point which they
emphasised is that there is definite evidence for variations in the
extinction curve in different environments within our own Galaxy and
also within the Magellanic Clouds.

How important are these problems for the interpretation of IRAS data?
I have reproduced a schematic version of the extinction curve for
interstellar dust grains in Figure 1.   It was very noticeable that
most of the theoretical discussion concerned optical and ultraviolet
wavelengths ($\lambda^{-1} > 1 \ \mu m^{-1}$).   This is of the greatest interest but, so
far as IRAS is concerned, these wavelengths are only important in so
far as they underline radiation.   What the IRAS scientists need to know
is the extinction, and hence emissivity, of interstellar grains in the
far infrared waveband.   The optical and ultraviolet properties of the
grains are essentially irrelevant since all the radiation from many of
the IRAS sources has been absorbed.   It was remarkable that the

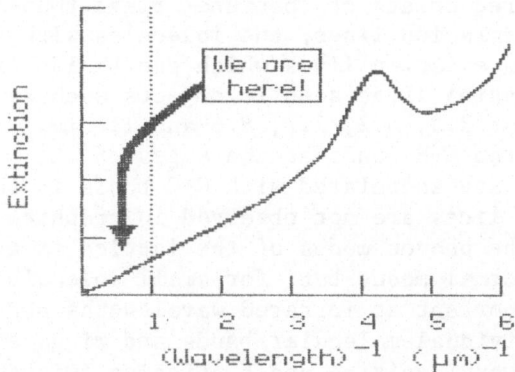

Figure 1.   A schematic diagram showing the extinction curve for
             interstellar dust grains (see the article by M. Greenberg)

theoretical question of the nature of the emissivity function of
grains in the infrared to millimetre waveband was not addressed.   We
heard of a range of different assumptions which were made – for
example, $\varepsilon_\lambda \propto \lambda^{-1}$ or λ^{-2}, the former corresponding to the simple
theory outlined by van de Hulst and the latter to the long wavelength
behaviour of real materials according to the Kramers–Kronig relations.
Which is applicable depends upon the nature of the grains and their
sizes.   I strongly recommend that, at the next IRAS symposium to be
held in Pasadena, the infrared and sub-millimetre properties of grains
should form the central concern of the papers on interstellar dust
grains.

An example of the riches to be gained from infrared studies of dust
grains is the work reported on small dust grains.   The story is in
two parts.   The first is Sellgren's beautiful analysis of the near-
infrared emission from gaseous nebulae.   The idea that this emission
is the thermal emission from very small dust grains
(a ≤ 10Å) heated to high temperatures by the arrival of single high
energy photons is, in my view, wholly convincing.   A number of
posters and presentations have shown that this is likely to be quite a
common process in a variety of ionised hydrogen and dusty regions.

The second part of the story concerns the nature of these small
grains.   Puget, Leger, Boulanger and their colleagues made a
beautiful case for associating at least some of these grains with
Polyaromatic Hydrocarbons (PAH).   The idea that small polyaromatic
molecules like coronene could be small dust grains follows naturally
from Sellgren's work.   The remarkable discovery of the French workers

is that, unlike the large ordered sheets of 'benzene' rings found in graphite which have very weak emission lines, the molecules with ~5 - 15 benzene rings possess strong emission lines which can be identified with a number of prominent emission lines seen in objects such as M82 in the near infrared waveband at 3.3, 6.2, 7.7, 8.6 and 11.3 μm.   The 3.3 μm line is probably due to the C-H bonds at the edges of the molecule while the other lines are associated with C=C bonds in the rings.   The reason that these lines are not observed in graphite is because in the large sheets, the phonon modes of the lattice as a whole are the most important normal modes but, for small molecules, the large scale modes are unimportant at infrared wavelengths and the resonances associated with individual molecular bands and rings are observed.   I find this story very exciting and convincing (without having read the papers!) and obviously, if it is confirmed by future work, a new area of astrophysics will be opened up.

This story is an excellent example of the interaction between ground-based observations, IRAS data and laboratory/theoretical studies.

4.   THE VEGA PHENOMENA

A great deal has already been written about what is unquestionably one of IRAS's great discoveries - dusty rings or discs around nearby stars.   Gillett gave a very clear and thought-provoking description of the observations and their interpretation.   I think everyone is convinced that the data are to be interpreted as dusty rings or discs. What is less clear is the nature of the grains and their size distributions.   The models discussed by Gillett are useful as bench-marks for the range of input parameters which may well be relevant. My own concerns relate to the time scales over which the phenomenon lasts.   Some of the time scales which came out of the analysis seemed to be very long ($\sim 10^9$ years) and surely something must happen to the dust over this time scale.   For example, the stars must have passed through a number of molecular clouds which will sweep out the gaseous and dust components.   The same interactions may, of course, help replenish the dust.   Follow-up observations and theoretical studies should help clarify the nature of these dust rings.

One very striking fact is how common these systems must be.   IRAS has detected 36 stars within about 25 pc and at least 12 of these show far infrared excesses similar to that observed in Vega.   Four of the 12 (including Vega) have been studied in detail.   Of these 4, three are associated with A-stars which are rather rare and the fourth is a K2 star at 3.3 pc.   As Gillett emphasised, the detection of dust rings is very sensitive to a number of astrophysical parameters, in particular the luminosities of the stars and the sizes and emissivities of the dust grains.   Granted these very strong selection effects against detecting anything at all, I find the detection statistics already very impressive.   My subjective impression is that there must be dust rings about a large fraction of all the stars in the Galaxy.

What is the relation of these rings to the zodiacal bands described by Houck? We have heard that the luminosity of the Vega dust emission is more than 100 times the luminosity of the zodiacal emission and it is not at all clear that we are talking about the same phenomenon.   I took away from Houck's talk the fact that we can account for the distribution of zodiacal emission in terms of what we might expect from optical observation of the zodiacal light but with a somewhat different albedo.   What I could not understand was the origin of the dust bands.   Houck suggested resonance phenomena associated with well-known dynamical resonances in the planetary system.   That the dust distribution is associated with these dynamical processes is suggested by the fact that the dust appears to be coplanar with the orbital planes of Venus and Mars at their radii and that ridges can be correlated with resonances within the asteroid belt.   I may well confess immediately to being no expert on the celestial mechanics of dust but it is clearly very important to understand dynamical processes in dusty discs as this must be an important part of the formation of planetary systems.

No one went so far as to claim that extra-solar system planetary systems have yet been detected but I get the strong feeling that most people believe that the dust observed in Vega is probably just the low mass end of a mass spectrum which may extend up to asteroids and if so, why not to planets as well.   However, this has yet to be proved. It is clear that we now have a prime finding list for systems to be studied intensively with the Space Telescope.

5.   STAR FORMATION

Star formation has always been regarded as the birthright of infrared astronomy.   Stars form deep inside optically obscured regions but far infrared dust emission from stars just before, during or just after the onset of nuclear fusion are observed readily in the far infrared waveband.   Elmegreen presented an outstanding review of how much can be learned from 'simple' analyses of IRAS data in conjunction with observations made in other wavebands.   His review is a treasure-trove of ideas which can be tested by careful inspection of the IRAS data. There is no question in my mind but that there will be complications in analysing the data and I fully expect controversy in the interpretation of some of the data.   However, the important point is that the basic data on where the youngest stars are located is now available.   This has not been available before.

Beichman and Myers showed in their careful presentations how the studies advocated by Elmegreen can be carried out using IRAS data. Their studies brought out a theme which I found running through much of the symposium - the interaction between IRAS data and other wavebands, particularly the near infrared, the sub-millimetre and millimetre wavebands will be the key to the full scientific exploitation of the IRAS database.

Another aspect of small grains came up in Elmegreen's discussion of mechanisms of star formation. He couched his discussion in terms of the small size cut-off in the particle distribution. Expressed in its simplest form, the question is how many small grains there are present in the collapsing gas cloud. The dust grains will be on average negatively charged because of the greater probability of collisions between grains with electrons rather than protons. In a simple way, it can be understood that the larger the number of grains, the greater the electrical conductivity and hence the longer the dust and gas will be frozen into the collapsing gas cloud. Elmegreen suggested that this would strongly influence the outcome of the star-formation process because the longer the dust and gas remain tied to the magnetic field during collapse, the more angular momentum can be transported away and single stars with planetary systems may form. If the conductivity decreases before much angular momentum is lost, binary or multiple systems may form containing most of the angular momentum of the cloud.

These ideas provoke many important questions. Are there sufficient small grains in regions of star formation? How can we find out if there are? Are these small grains Sellgren particles? Can they be polyaromatic hydrocarbons?

6. STELLAR EVOLUTION AND MASS LOSS

The outstanding feature of the statistics of stars catalogued by IRAS is that they are mainly associated with stars in the process of formation or in their last stages of evolution before final collapse to some form of dead star. Before tackling the latter aspect of the discussion, I should mention one area which was not discussed explicitly in the review papers but which deserves careful attention.

It will be very important to know the far infrared properties of all classes of star and not just the spectacular far infrared objects. We heard some discussion of the photospheric emission from stars but we must quantify this in more detail. It will be important to know how well the theory of stellar atmospheres can account for the observations of bright stars and also whether or not all stars, to some extent, possess infrared excesses (or even deficits!). The posters by Coté and Walters and by Leggett suggested that there are real discrepancies to be explained even in entirely 'normal' stars. The Einstein X-ray Astronomy Observatory detected X-ray emission from essentially all classes of star which, interpreted naively, suggests that they are all capable of possessing hot coronae and stellar winds. Do all stars also have a far infrared signature? It will be important to know.

The second dominant phase of stellar evolution observed by IRAS is mass loss. In my lectures to undergraduates, I emphasise that mass loss is a ubiquitous process in astronomy as illustrated in Figure 2. Again, one would very much like to know whether or not there is far-

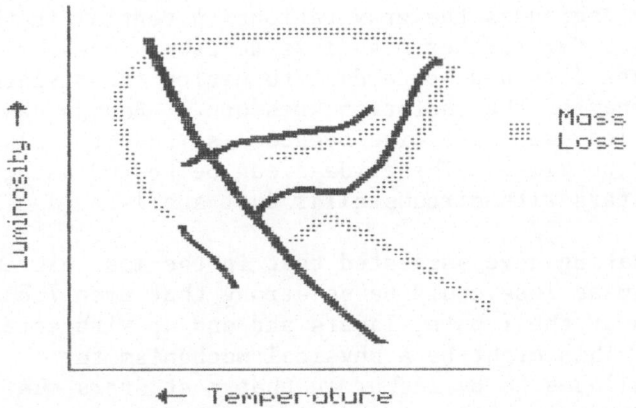

Figure 2.   A schematic H-R diagram showing the evolutionary track of
a $1M_\odot$ star.   The stippled areas indicate regions in
which there is evidence for mass-loss.

infrared evidence for mass loss in all these stages.   What was
clearly demonstrated by Zuckerman, Habing, Frogel and Feast was that
those stars on or approaching the Asymptotic Giant Branch are among
the most prominent far infrared emitters.   This is the region of the
H-R diagram populated by longer period variables and stars which are
pulsationally unstable.   Feast proposed that there is a continuous
sequence of long period variables extending from the classical Mira
variables with mass loss rates of $\sim 10^{-7}$ $M_\odot$ yr^{-1} and periods $\sim$ 100-200
days to the long period variables associated with OH/IR stars with
mass loss rates $\sim 10^{-4} M_\odot$ yr^{-1} and periods $\sim$ 800-1000 days.

We heard the beginnings of what must become a major IRAS industry.
There is an outstanding opportunity for making quantitative estimates
of mass loss rates for all sorts of stars as they move to the tip of
the asymptotic giant branch and then sweep across the H-R diagram
ejecting the shells we identify with planetary nebulae.   This work
can be beautifully complemented by radio observations of the OH
emission and infrared spectroscopy of emission and absorption lines
from the expanding shells.

One intriguing topic was raised by Habing and expanded upon by Bedijn.
Habing has made the case that the mass loss rate of stars on the
asymptotic giant branch may increase with time and he and Baud have
interpreted the luminosity function of OH/IR stars in terms of
accelerated mass loss.   Bedijn described some important results he
has obtained from physical models of long period variable stars such
as Mira variables.   He has studied the behaviour of the envelopes of
these stars as they become pulsationally unstable and found a

plausible physical mechanism for accelerated mass loss. In simple
terms, the loss of mass decreases the gravitational potential in the
envelope making it easier for further mass loss to take place.
Furthermore, as more mass loss and hence dust formation takes place,
the greater the efficiency of the radiation pressure. Bedijn has
described this as mass loss with positive feedback so that the mass
loss rate accelerates with time. These ideas can be tested using
IRAS data on the many stars with circumstellar dust shells.

Bedijn and Herman and Habing have suggested that in the most extreme
cases, the accelerated mass loss could be so strong that even $7-8M_\odot$
stars could be stripped of their outer layers and end up with core
masses of about $1 M_\odot$. This might be a physical mechanism for
resolving the problem alluded to by Zuckerman that most stars must die
as white dwarfs rather than in explosive events which form neutron
stars or black holes because of the very large proportion of white
dwarfs observed locally.

Proceeding further along the track to the death of stars, Pottasch
showed how effectively planetary nebulae can be studied by IRAS,
particularly by the low resolution spectrograph. He showed how
planetary nebulae, particularly the young ones require an additional
energy source to Ly-α heating and suggested that this may involve
heating by the stellar continuum. In addition, he came up with a
list of new candidates for planetary nebulae which will repay further
study.

7. GALACTIC STRUCTURE

I found Habing's map of the distribution of 'red' and 'very red' IRAS
sources one of the most beautiful pictures to be produced by the
mission. Although evidence for the bulge component had been seen in
a previous balloon flight by the Japanese astronomers, the IRAS map
shows very clearly the inner part of the galactic bulge delineated by
individual stars. Habing showed convincingly that the bulk of these
objects are long period variables which are related to the Mira and
OH/IR stars. This conclusion could only be drawn thanks to the
excellent long-term stability of the IRAS detectors. These studies
are of obvious importance for defining the structure of the inner
Galaxy and the galactic distance scale.

Burton's analysis of the large scale distribution of far infrared
emission in our Galaxy raised a number of problems to which there is
no immediate answer. These include the facts that there is much more
far infrared luminosity within a few kpc of the Galactic Centre than
would be expected from the gas distribution and the fact that the
scale height of the emission in galactic latitude is about 100 pc,
twice as large as the scale height of cold compressed gas and even
more than that of regions of active star formation. It may well be
significant that the scale height of the far infrared emission is
similar to that of cold diffuse hydrogen. It is established that the

central regions of M31 also display an excess of far infrared
emission.

8.   GALAXIES

The identification of high latitude IRAS sources has been remarkably
successful.   Wolstencroft, Helou, Rowan-Robinson and Becklin
described studies of galaxies associated with IRAS sources.
Wolstencroft and his colleagues showed that at most 8 out of more than
300 IRAS sources in the region of the South Galactic pole in the point
source catalogue cannot be reliably identified with stars and
galaxies.   Of particular interest is the fact that the maximum of the
apparent magnitude distribution of the galaxies is at about m_J = 16,
well above the limit to which reliable identifications can be made.
We can therefore claim that the nature of the bulk of the sources
appearing in the IRAS point source catalogue is understood, although
there is still a little room for the discovery of new classes of stars
and galaxies, for example brown dwarfs and highly obscured galaxies or
quasars.

Becklin derived the luminosity functions for galaxies in the far
infrared waveband.   The typical far infrared galaxy is a late or
irregular type galaxy in which there is already evidence for gas or
dust.   Detections of elliptical galaxies are very rare, the only
examples discussed by de Jong being objects in which there is
something 'peculiar' about the galaxy.   These include elliptical
galaxies with dust lanes or with active galactic nuclei.   Naively
interpreted, these results confirm the standard picture in which
normal elliptical galaxies are more-or-less gas free systems.

Some of the galaxies detected by IRAS are very strong infrared
emitters indeed but these are very rare.   Obviously these are of
great physical interest since they probably represent galaxies in
which the most extreme forms of star formation are taking place.   A
personal interest of mine has always been to know what primaeval
galaxies would look like.   My guess is that they may not appear too
different from some of the most extreme IR luminous galaxies detected
by IRAS.   In this context, it will be very important to make good
estimates of the amount of star formation going on in these extreme
cases.   From this, it will be possible to make more plausible
estimates of how detectable primaeval galaxies might be in the far
infrared waveband.

One interesting fact noted by de Jong and others is the strong
correlation between the far infrared luminosity and the non-thermal
continuum radio emission.   The details of this relation will repay
further study.   My own view is that it is very good evidence for the
relativistic electrons responsible for the radio emission originating
in supernovae which are associated with regions of star formation.
The most massive stars which become supernovae and sources of
relativistic electrons complete their evolution in a period of 10^6 to

10^7 years so that the regions in which they were formed are still
likely to be undergoing significant star formation activity.   If this
interpretation is correct, it is further evidence for the origin of
the bulk of cosmic ray electrons in supernova explosions.

An interesting methodological issue concerns the 'standard'
decomposition of the spectrum of a galaxy into a 'disc' component at
25-30K and the emission from HII regions at about 60K.   Although
there are plausible physical arguments for this hypothesis, I have my
suspicions that this decomposition is largely determined by the fact
that the spectra are normally only defined by two or at most three
spectral points.   My intuition tells me that there is much more
likely to be a continuum of temperatures and we must hope that more
detailed studies will help define the temperature structure of the
interstellar medium in more detail.

The ultimate astrophysical question must concern the origin of the
material responsible for the strong far infrared emission.   Hawarden
described some remarkable results on the types of galaxies which
become intense far infrared emitters.   He related the collection of
dust and gas in the central regions of galaxies to dynamical processes
involving both the stars and gas in the central bars of barred spiral
galaxies.   Joseph showed clearly how important interactions and
mergers of galaxies are in producing intense far infrared emission.
These are impressive results and provide further evidence for the
importance of interactions between galaxies in producing supplies of
gas and dust for star formation activity and possibly for fuelling
active nuclei.   There are, however, many other processes which must
be studied in more detail, in particular, the 'standard' mass-loss
processes which were described above.

9.   ACTIVE GALAXIES

De Grijp and Miley described how successfully the IRAS catalogue can
be used to identify new examples of active galactic nuclei.   Their
most spectacular result concerned the very high success rate of
finding new Seyfert galaxies from their (25-60 μm) colours.   Of 156 of
the galaxies with spectral indices in the range -1.5 to 0 (out of a
total of 467 candidate galaxies), about 68% are Seyfert galaxies.
From the point of view of UK and Netherlands astronomers, it is
particularly gratifying to note that much of this work has been
carried out with the 2.5m Isaac Newton Telescope on La Palma which
testifies to the high quality of that instrument and the site.   The
key question is how the selection criteria used in the choice of the
IRAS sources influences the proportions of active galaxies which are
detected.   De Grijp pointed out that in general the IRAS
identification technique selects more distant and dustier Seyferts
than the other techniques.   It will be of the greatest interest to
understand the relation of these Seyferts to Seyferts selected by
other techniques and to other classes of active galaxies like radio
galaxies and quasars.   The quasars scarcely appeared in the story

and I suspect this is because most of them are simply too distant to be detected by IRAS.

10. COSMOLOGY

Finally, we come to the remarkable paper by Rowan-Robinson which discussed how IRAS data can be used for cosmological studies. The catalogue of impacts of IRAS data upon cosmology was very broad, extending from the determination of the local gravitation potential and inhomogeneities in the large scale distribution of galaxies to counts of galaxies and the problem of measuring the far infrared background radiation. The possibility of measuring a value for the average density of the Universe is present in these data. His paper will repay much detailed study.

It is important to note the reason for this large potential impact of IRAS data upon cosmology. The very great advantage of the IRAS catalogue is that, provided all the calibrations have been carried out correctly, the selection of galaxies for study can be made from statistically complete samples of objects. Once the catalogues are known to be reliable and complete, one can begin to look for large scale inhomogeneities and the other effects which Rowan-Robinson described. The situation is unlike that in the optical waveband in which the selection criteria are very much more complex and account must be taken of internal and external extinction in the galaxies.

Even when these are all understood, cosmological studies of the type described by Rowan-Robinson are among the most difficult in all astronomy. But there are very exciting possibilities in these data and they may well prove to be of central importance for cosmology.

11. CONCLUSION

There is little to add to this catalogue of outstanding successes which have already been gained from the IRAS mission. It is important to remember that it is still very early days in the analysis of the IRAS data but already I believe the originators of the project can rest assured that the years of effort have been more than repayed by the results already obtained and that they have provided us all with a superb atlas of the far infrared sky which will become a part of our scientific culture. In my mind this is as great a scientific achievement as any in modern astronomy and everyone associated with the project deserves our highest appreciation and thanks for their magnificent efforts.

ACKNOWLEDGEMENTS

I am very grateful to Dr Matt Mountain for assistance in assembling the information contained in the papers presented at the meeting and in helping ensure that I have not totally misrepresented the authors. The responsibility for the above text and the opinions expressed are, of course, entirely my own responsibility.

INDICES

SUBJECT INDEX

AAT   489
Absorption Bands   233-236, 247-252, 323-324, 352, 435, 467-
                                470
Abundances   181-182, 506
Ammonia   295-296, 307-312
Anglo Australian Telescope   197
Anthracene   211
Asymptotic Giant Branch   46, 93-100, 103, 107-108, 113-118, 119-
                                126, 336-337, 517

Baade Windows   339-342, 345, 349-351, 416
Balloon Observations   305-306, 370
Barred Spiral Galaxies   455-462
Be Stars   83-85
Benzene   174, 514
Bipolar nebulae   315-317, 319-320
Blazars   491-492

Carbon   177-188, 194-196
Carbon Stars   101, 113-118, 173
Carina   373-374
Chamaeleon Dark cloud   197-199, 266
Circumstellar Shells/Disks   5, 7, 9, 12-17, 67, 93-100, 101-102, 109-
                                110, 111-112, 113-118, 119-126, 198, 284,
                                313-34, 329-337, 344, 416, 422, 514-515
Cirrus   5, 9 19-20, 49-54, 171-176, 239, 282,
                                288, 299, 370, 378, 384-35, 426
Classification of Sources   32-33
Clumping   277 ,307-312, 321-322
CO Emission   53, 321-322
CO Surveys   358, 373, 377-378, 437
Colour Diagrams   34-35
Comets   187, 225-228
Copernicus Satellite   79
Cosmological Background   499-506, 507-508
Cosmology   499-506, 507-508, 521
COSMOS Measuring Machine   23, 425
CPC Observations   155-158, 300
CTIO   351

Dark Clouds   29, 313-314, 375-376
Dust Albedo   166, 171-176
Dust Emission   136-142, 151-152, 155-158, 177-188, 194-
                                196, 201-202, 203-207, 213-26, 217-220,
                                239, 383-396, 401-404, 421-424, 512-514
Dust Grains   39-44, 47-48, 49-54, 55-57, 161-165, 171-
                                174, 177-188, 191-196, 213-216, 225-228,
                                229-232, 233-236, 269-273, 506, 512-514
Dust Grain Charge   270-272

525

Dust Grain Composition   171-176, 190, 248-250
Dust Heating   201-202
Dust Lanes   455-456
Dust Origin   171-172
Dust Grain Simulation   245-246, 247-252
Dust Grain Sizes   40, 52, 174-175, 214

Effelsberg Telescope   313
Elliptical Galaxies   329, 349-350, 415-417, 430
Emission Features   33, 132-136, 143-145, 174-175, 203, 209-
                              212, 218, 221-224, 229-232, 261-262, 99-
                              400, 435, 467-470, 451, 514
Emission-Line Stars   35-37
ESO   399
Extinction   52, 54, 55-57, 165-169, 172-173, 177-188,
                              191-196, 506, 512-514
Extinction Curve   161, 166-169, 189-190, 193-195

Fluorescence   209-212, 225-228

G type Stars   75-76
Galactic Anticenter   375
Galactic Bulge   5, 9, 22, 107-108, 329-337, 339-348, 349-
                              356, 518
Galactic Disk   330-332, 357-371
Galactic IR Emission   205
Galactic Pole, (N, S)   23-25, 55-57, 425-426, 499-506, 519
Galactic Structure   329-337, 339-348, 357-371, 373-374, 377-
                              378, 380, 75-371, 518-519,
Galaxies   5, 7, 9, 17-19, 24-27, 29, 405-413, 415-
                              419, 421-424, 425-434, 35-436, 437-438,
                              439-440, 441-442, 463-466, 443-444, 445-
                              446, 447-454, 455-42, 467-470, 471-486,
                              487-488, 489-490, 491-492, 493-495, 507-
                              508, 518-521

Gamma Ray Emission   365
Globular Clusters   329, 343, 355
Globules   29, 267
Gould's Belt   19
Grains: see Dust Grains
Graphite   161-170, 171-176, 177-188, 194-196, 506
Gravitational Field   501
Groningen Fields   345

Heptane   212
Herbig Haro objects   296, 322
Hercules Cluster   451
HI   357-371, 379-380, 384-385, 403-404, 445-
                              446
HI Surveys   439-440
High Velocity Clouds   379-380

HII Region Evolution   303-304
HII Regions   5, 29, 35, 163, 205, 303-304, 305-306,
                                313-314, 373, 399-400, 401-404, 468, 455
Houck Objects   406-407
Hubble Flow   507-508
Hydrocarbons   52-53, 174-175, 203, 209-210, 213-216,
                                218-219, 233-236, 37-240, 250, 261, 403,
                                513-514, 516
H_2 189-190, 358, 399-400, 451, 461, 464

Ice   174, 183, 221-224, 247-252
Identification of IRAS   23-25, 279-290,
    Sources   321-322, 405-408, 425-434
Initial Mass Function   277-278
Interacting Galaxies   430, 445, 447-454
Interplanetary Dust   39-44
Interstellar Dust
    Distribution   357-371
Interstellar Radiation
    Field   172
IRAS Maps   4, 13, 15, 16, 18, 20, 21, 47-48, 48, 49-
                                54, 50, 51, 135, 157, 18, 280, 283, 288,
                                289, 301-302, 302, 317, 331, 358, 360,
                                361, 363, 367, 373-34, 374, 385, 385-393,
                                387-390, 392-393, 402, 407
IRAS Plates   28, 58, 74, 86, 100, 106, 148, 154, 170,
                                200, 208, 260, 276, 292, 318, 338, 414,
                                420, 496
IRAS Source Types   3-22
IRTF   415
IUE   79-82, 173, 151, 191-195 225

Jeans Mass   273

KPNO   315

Late-Type Stars   93-100, 101-102, 111-112, 113-118, 119-
                                126, 221-224, 323-34, 329-337, 339-348,
                                349-356, 422

Lenticular Galaxies   430
Line Emission   132-136
Local Supercluster   408
LRS Catalog   31-38, 105, 115-118, 132-136, 143-145,
                                221-224
Luminescence   209-212

M Giants   102, 349-356
Magellanic Clouds   383-394, 397-398, 399-400, 409, 512
Magnetic Diffusion   272
Markarian Galaxies   441-442, 471-486

Masers (see also OH/IR
    Stars)   313-314
MERLIN   109-110, 443
Metal Abundances   351
Mie Theory   161
Mira Variables   103, 114, 116, 118, 119-126, 129, 324,
                                 329-337, 339-348,
Molecular Clouds   5, 189-190, 266-268, 291-294, 295-296,
                                 299-300, 297-298, 307-312, 319-320, 325-
                                 326, 357-371, 377-378
Molecular Cloud Models   229-232
Molecular ring   365

Naphtalene   211
NLR
Novae   146-147, 149-150, 151-152
NRAO   491

OH Survey   325-326
OH/IR Stars   103, 105, 107-108, 109-110, 119-126, 129-
                                 130, 138-139, 142, 01-202, 329-337, 339-
                                 348, 517
Oort Cloud   505-506
Open clusters   268-269
Organic Molecules   229-232, 238-239
Outflow Sources   315-317, 319-320, 321-322
O, B, A Stars   77-78

Period-Luminosity relation 125, 126
Perseus   311
Perylene   212
Photoprocessing   182-183, 248-250
Planetary Nebulae   5, 35-36, 119-126, 127-128, 129, 131-142,
                                 143-144
Planetary Systems   515
Pointing-Robertson Effect   63
Polarization   172-173, 177, 197-199, 218, 315-317,
Polycyclic Aromatic
    Hydrocarbons (PAH)   174-175, 203, 209-212, 213-216, 218-219,
                                 233-236, 237-240, 261, 403, 513-514, 516
Pre-Main Sequence Stars   268, 286-288, 322
Protostars   221-224

QSO   405-408, 422, 472-473, 481-482, 491-492,
                                                               506, 520-521

Radiation Pressure   63, 119-126
Red Rectangle   239
Redshift   501
Reflection Nebulae   175, 203, 209-212, 217-220, 261-262, 300

SAO Stars	12,17
Scattering	165-167
Seyfert Galaxies	220, 415, 418, 421-424, 469, 474-485, 487-488, 451, 487-48, 490, 491-492, 493-495, 506, 520,
Silicate	50, 161-170, 171-176, 177-188, 194-196, 214-215, 241-244, 323-324, 354, 469
SO Galaxies	416
Source counts	499-506
Source Statistics	3-22, 31-38, 279-291
Space Shuttle	246
Space Telescope	515
Spectral Decomposition	47-48
Spheroidal Galaxies	350
Spiral Galaxies	417-418, 430, 437-438, 485, 487-488,
Star Formation	265-275, 277-278, 296, 297-298, 301-302, 307-312, 417-48, 437, 515-516, 518
Star Formation Rate	119-126
Starburst (Galaxies)	220, 277, 415-419, 421-424, 438, 443-444, 445-446, 447-454, 455-462, 463-466, 467-470, 487-488, 506
Stars	3-5, 7, 9, 11-17, 37, 77-78, 91-92, 93-100, 279-290
Stars with IR Excess	61-69, 71-73, 73-74, 79-82, 84-85, 87-90, 284
Stellar Evolution	119-126, 268, 516-518
Stellar Magnetic Field	110
Stellar Masers	109-110
Stellar Mass Loss	79-82, 83-85, 93-100, 107-108, 109-110, 119-126, 516-518
Stellar Winds	79-82, 266
Submm Observations	29
Sulfur	225-228
Super Shells	389-391
Supernova Remnants	153, 155-158, 397-398, 443-444
Symbiotic Objects	323-324
S_2	225-228
T Tauri Stars	308-309
Tetracene	211
UK Schmidt	23-25
UKIRT	435, 464, 491
UKST	425
UV Extinction	191-195
Variable Stars	119-126, 329-337, 339-348
Virgo Cluster	415-419, 438

Water masers   313-314
White Dwarfs   120-126
Wolf Rayet Stars   97-82, 85-88

Yellow Stuff   173,174

Zodiacal Dust Albedo   45-46
Zodiacal Dust Origin   41-42
Zodiacal Emission   39-44, 45-46, 47, 360-361, 515

OBJECT INDEX

Alpha Orionis	106
AFGL 618	96
AFGL 1992	100
AFGL 2088	100
AFGL 2143	100
AFGL 2298	100
AFGL 2333	100
AFGL 2350	100
AFGL 2477	100
AFGL 2968	100
AFGL 5146	100
AFGL 5359	100
AFGL 5369	100
AFGL 5379	100
AFGL 5384	100
Alpha Eridani	83-85
Alpha Lyrae	61-69
Alpha Piscium Austr.	61-69
Ara	338
Arp 220	238, 406, 412, 450
AS 353	92
Auriga	74, 308
Barnard's loop	205
BD 30 03639	132-133 140-141
Beta Pictoris	61-69, 75-76
Betelgeuse	106
3C48	473, 481
3C120	477
3C273	473, 422, 491-492
3C345	473, 477
3C371	473
3C390	477
3C446	473
Canis Maioris OB 1	154, 208, 325
Carina Nebula	305-306, 414, 496
Cas A	155-158
Centaurus	414, 420 426
Cep OB3	313-314
Chameleon	496
Cha F11	198
Cha F16	198
Cha F30	198
Cha T21	198
Chi Ophiuchi	83-85
Circinus	420
CK Vul	150
CW Tau	310
Cygnus	104
Cygnus Loop	47-48
Delta Centauri	83-85
DF Tau	92
DM Tau	92
DO Tau	92
30 Doradus	388
Elias 16	183
Epsilon Eridani	61-69
ESO103-635	478
ESO141-G55	473
ESO350-IG58	478
FH Ser	146
Fomalhaut	75-76
G 333.6-0.2	338
G 351.6-1.3	305-306
Galactic Center IRS 7	183, 231, 292
Gamma Muscre	496
Gamma Velorum	87-90
Gemini	74, 282
GGD 12-15	458-459
GGD 4	322
GL 215	102
GQ Mus	145-146, 149
GW Ori	91-92
HD 101584	75-76
HD 155603	76
HD 97300	198-199
He 2-131	140
Heiles cloud 2	308
Helix Nebula	134-136, 143-145
HH 1	313
HH 6	300
HH 12	300
HH 19-27	313-314
HH 100	297

HH 101	297	IRAS00453+4427	100
HK Tau	310	IRAS00537-441	473
HL Tau	183	IRAS00537+238	322
Horsehead Nebula	295-296	IRAS0710+457	473
HP Tau	92	IRAS01204-3029	26, 437
HR Del	145, 146,	IRAS01209-3306	26, 437
	147	IRAS01252-3055	26, 437
		IRAS01256-3236	26, 437
IC 348	28	IRAS01257-3157	26, 437
IC 405	74	IRAS01281-2702	26, 437
IC 410	74	IRAS01284-2737	26, 437
IC 418	132-133,	IRAS01300-3203	26, 437
	140	IRAS01319-2940	26, 437
IC 443	74, 106	IRAS01358-3300	26
IC 446	301	IRAS01367-2856	26, 437
IC 694	450	IRAS01367-3010	26, 437
IC 1318	104	IRAS01377-2817	26, 437
IC 2162	106	IRAS01379-2942	26, 437
IC 2169	301	IRAS01380-2909	26
IC 2177	154	IRAS01380-3203	26
IC 4329	488,	IRAS01396-2847	26, 437
	478	IRAS01405-2804	26, 437
IC 4553	406, 412,	IRAS01458-2828	26, 437
	423, 450	IRAS01470-3259	26, 437
IC 4628	338	IRAS01472-2719	26
IC 4997	140	IRAS01472-2756	26, 437
IK Tau	123, 125,	IRAS05170+0535	76
	342	IRAS05383-0228	296
IQ Tau	92	IRAS05384-0229	296
IRAS00013-2903	27	IRAS05386-0229	296
IRAS00016-3056	24, 27	IRAS05388-0224	296
IRAS00019-3226	27	IRAS05413-0104	314
IRAS00023-3216	27	IRAS07134+1005	76
IRAS00024-2759	27	IRAS0713+1005	100
IRAS00029-2909	27	IRAS0737-4021	100
IRAS00040-3252	27	IRAS0743-3750	100
IRAS0050+124	478	IRAS0801-3627	100
IRAS00108-2932	27	IRAS0807-3615	100
IRAS00117-3156	27	IRAS09079-1942	76
IRAS0134+329	473	IRAS10282-5231	76
IRAS00147-2719	26, 437	IRAS1051-237	482
IRAS00148-3153	27	IRAS11059-7721	76
IRAS00157+001	473	IRAS11108-7627	76
IRAS00205-2756	27	IRAS11294-5909	76
IRAS00214-3248	27	IRAS11385-5517	76
IRAS00215+2822	100	IRAS12067-4508	76
IRAS00247-3308	27	IRAS1226+023	473
IRAS00248+030	473	IRAS1254+571	473
IRAS00256-2851	27	IRAS13395-6153	222-224
IRAS00420-014	473	IRAS1351+640	473
IRAS00421+040	463-466	IRAS1353+186	473
IRAS00423+5336	100		

IRAS14254-2655	489	IRAS23515-2917	24, 27
IRAS1440+357	473	IRAS23540-3138	27
IRAS15420-3408	76	IRAS23543-3141	27
IRAS15532-4210	76	IRAS23547-2742	27
IRAS15556-2248	76	IRAS23548-3156	27
IRAS1610-4205	100	IRAS23552-3252	27,
IRAS16105-4205	222-224	IRC 10011	123, 125
IRAS1613+658	473	IRC 10216	96, 115
IRAS1634-3814	100		
IRAS1634+706	473	Keplers SNR	155-158
IRAS1641+399	473	Kuk-05099	76
IRAS1646-4022	100	Kuk-05138	76
IRAS1700-4119	100	Kuk-07226	76
IRAS1700+518	473	K3-62	127-128
IRAS1705-022	129		
IRAS17109-3942	76	Lamda Orionis	106
IRAS17411-3154	222-224	Lk Hα 120	92
IRAS1744-4048	100	LMC	341,
IRAS1803-2201	100		383-384,
IRAS1807+698	473		388-396,
IRAS1809+2704	100		397-398,
IRAS1812-051	129		399-400
IRAS1818-1623	100	LMC-N11	389
IRAS1823+218,	129	LMC-N44	389
IRAS1824-0839	100	LMC-N48	389
IRAS18384-2800	76	LMC-N49	398
IRAS1905-750	129	LMC-N51	389
IRAS1912+172	129-130	LMC-N55	389-391
IRAS1916-587,	473	LMC-N59	389
IRAS1945+293	128	LMC-N63	398
IRAS1952+279	129-130	LMC-N79	389
IRAS1956+3423	100	LMC-N159	391,394
IRAS2002+3910	100	LMC-N186	398
IRAS2005+185	129-130	Lyra	86
IRAS20117+1634	76	L810	29
IRAS2041-109	473	L1551	266,
IRAS2043+3825	100		315-317
IRAS2130+099	473	L1605	301-302
IRAS2131+5631	100	L1624	301-302
IRAS2134+4508	100	L1630	313-314
IRAS2137+4540	100	L1641	313-314
IRAS2148+5301	100		
IRAS2155+5907	100	Mkn 3	475, 478
IRAS2155+6204	100	Mkn 6	475
IRAS2214+5206	100	Mkn 231	473, 481
IRAS2223-052	473	Mkn 266	442
IRAS2227+5435	100	Mkn 273	442
IRAS2250-596,	322	Mkn 273	442
IRAS23452-3048	27	Mkn 296	442
IRAS23499-2837	27	Mkn 306	442
IRAS23512-2811	27	Mkn 314	442

Mkn 335	478	NGC 253	277, 416,
Mkn 348	475		448, 452,
Mkn 376	473		458-460,
Mkn 463	442, 473,		467,
	475, 478		491-492
Mkn 478	473	NGC 520	445
Mkn 480	442	NGC 838	416
Mkn 509	473	NGC 1052	416
Mkn 544	442	NGC 1068	277, 471,
Mkn 673	442		477-481,
Mkn 739	442		464, 484
Mkn 788	442	NGC 1097	455-458,
Mkn 789	442		460
Mkn 876	473	NGC 1275	477,
Mkn 930	442		491-492
Mkn 1014	453, 473	NGC 1333	28, 300
Mkn 1027	442	NGC 1499	28, 74
Mkn 1210	478	NGC 1512	455
Mon R2	293-294,	NGC 1579	28, 74
	325-326	NGC 1808	467-468
Monoceros	106,	NGC 1907	74
	203-207,	NCG 1912	74
	268,	NGC 1931	74
	279-290,	NGC 2023	175, 206,
	291-294,		218, 238,
	325-326		262-261
Musca	496	NGC 2146	445-446
Mu Ser	150	NGC 2175	106
MWC 1080	92	NGC 2245	301
M1-67	88	NGC 2247	301
M8	292	NGC 2259	301
M16	200	NGC 2264	106,
M17	200		301-302
M20	292	NGC 2327	154
M31	365,	NGC 2335	154
	391-394,	NGC 2359	154
	401-404	NGC 2438	140
M32	414	NGC 2440	140
M33	58,	NGC 2467	260
	391-394	NGC 2568	260
M38	74	NGC 2867	140
M45	28	NGC 2903	417
M81	50	NGC 3227	419
M82	50, 238,	NGC 3242	140
	277, 448	NGC 3310	458
	443-444,	NGC 3690	450
	465, 467	NGC 3783	468-469
M 87	416	NGC 3918	133, 140
		NGC 4151	468-469
NGC 40	133	NGC 4388	419
NGC 246	140	NGC 4418	435-436,
			470

NGC 5033	457-458	Omicron Cen	76
NGC 5128	450	Ophiuchus	49, 50,
NGC 5253	468		54, 170,
NGC 6072	140		268,
NGC 6153	133-134		299-300,
NGC 6240	450, 461,		309-311,
	463-466		409
NGC 6303	133	Orion	49, 50,
NGC 6357	133, 292		54, 106,
NGC 6369	133		189-190,
NGC 6522	340, 342,		203-207,
	345-346		268,
NGC 6543	140		279-290,
NGC 6572	132-133,		291-294,
	140		313-314,
NGC 6578	133		325-326,
NGC 6781	140		426, 464
NGC 6790	132-133,		
	140	P Cygni	81-82
NGC 6820	104	Perseus	28, 268,
NGC 6823	86		299-300
NGC 6857	104	Perseus Cluster	492
NGC 6870	86	Phi Persei	83-85
NGC 6888	104	PKS	473
NGC 7009	133-134	PKS 2048-573	478
NGC 7023	175, 206,	Pleiades	28
	218, 238,	Puppis	208, 260,
	261-262		318
NGC 7027	96, 238	Pyxis	276
NGC 7293	134-136,		
	141,	R Coronae Borealis	173
	143-144	R CrA	206,
NGC 7354	134		297-298
NGC 7479	435-436	R Sge	75, 76
NGC 7662	134, 140	RAFG 5502	100
Norma	338	RAFGL 2343	100
Nova Aql 1982	150	RAFGL 5102	100
Nova Del 1967	146	RAFGL 5206	100
Nova Mus 1983	146	RAFGL 5250	100
	149	RAFGL 5254	100
Nova Per 1901	163	RAFGL 5416	100
Nova Pic 1925	146-147	RAFGL 5497	100
Nova Ser 1970	146	RAFGL 6815	100
Nova Ser 1983	150	RCW 57	305-306,
Nova Sgr 1982	146		414
	149	RCW 106	338
Nova Vul 1672	150	RCW 108	305-306,
			338
OH 25.6+0.6	103	RCW 122	305-306
OH 26.5+0.6	123, 124	Red Rectangle	210, 212

Rosette Nebula	106	Tololo 0109-38	480
RR Lyrae Variables	342, 345,	Triangulum	58
	352	Tr 3	301-302
RR Pic	146	TU Pyx	76
RS Oph	151-152	TY CrA	297-298
RU Cen	75-76	Tycho's SNR	153,
			155-158
S 146	313		
S 225	458	UX Tau	92
S 237	74		
S 255	106	VB 80	326
SAO105871	76	Vega	61-69,
SAO112630	76		71-73,
SAO154972	76		75-76,
SAO183986	76		514-515
SAO187317	76	Vela	276
SAO208569	76	Virgo	426
SAO223245	76	Vulpecula	86, 104
SAO226389	76	VX Sagittarii	109-110
SAO238126	76	VX Sgr	344
SAO239145	76	VY Mon	92
SAO239822	76	Vy 2-2	132-134
SAO96709	76	V1370 Aql	150
Sagittarius	200, 292	V4077 Sgr	144-146,
Saturn	305-306		149
Sco-Cen-Oph Complex	268	V615 Ori	296
Scorpius	338, 426		
Scutum	200	W4	267
Serpeus	170	W31	305-306
Sgr I	340,	W33	183
	342-346	W33A	225, 250
Sgr II	340	WR72	88
SMC	386-388,	WR104	88
	392-394,	WR112	88
	399-400	WR118	88
SMC-N66	394	WR124	88
SMC-N76	396	WX Ser	342
SMC-N81	396-397,		
	399	Xi Serpentis	170
SMC-N83	396		
SMC-N88	396-397,	Zeta Puppis.	81-82
	400	Zw958+16	435
S Mon	301	IZw1	473
SS Vir	116	IIZw40	468
		IIZw136	473
T Tau	91-92		
Taurus	28, 49,		
	58, 74,		
	148, 308,		
	311		
TMC 1	301-302		

AUTHOR INDEX

Aitken, D.K.   241-244, 435-436
Albinson, J.S.   149-150, 151-152, 297-298, 397-398
Allamandola, L.J.   233-236, 261-262
Allen, D.A.   489-490
Appleton, P.N.   443-444
Armstrong, J.T.   439-440
Axon, D.J.   443-444

Bailey, J.A.   197-199
Becklin, E.E.   415-419, 463-466,
Bedijn, P.J.   119-126
Beichman, C.A.   279-291, 463-466,
Blaauw, A.   325-326
Blitz, L.   377-378
Bode, M.F.   149-150, 297-298, 397-398
Boreiko, R.T.   127-128
Boulanger, F.   203-207, 293-294, 321-322
Braun, R.   47- 48, 155-158
Bregman, J.D.   261-262
Breukers, R.   105
Brown, L.M.J.   491-492
Burger, J.H.   103
Burton, W.B.   357-371, 373-374
Butner, H.M.   373-374

Callus, C.M.   149-150, 151-152
Cameron, D.H.M.   299-300
Casoli, F.   321-322
Caspers, H.C.M.   445-446
Chapman, J.M.   109-110
Chester, T.   3- 22
Chini, R.   29- 30
Chlewicki, G.   191-196
Clowes, R.G.   23- 27, 425-434
Cohen, R.J.   109-110
Collins, C.A.   507-508
Combes, F.   321-322
Cote, J.   77- 78
Coulson, I.M.   129-130
Cox, P.   201-202
Crawford, J.   303-304, 421-424
Cudlip, W.   299-300

Daniel, R.R.   305-306
Davis, D.S.   189-190
De Grijp, R.   471-486
De Muizon, M.   221-224

Dennefeld, M.   493-495
DePoy, D.L.   463-466
Desert, F.X.   213-216
Deul, E.R.   357-371, 373-374
Dinerstein, H.L.   145-147
Dumont, R.   45- 46
Dupraz, Ch.   321-322
D'Hendecourt, L.B.   221-224, 237-240, 247-252

Elmegreen, B.G.   265-275
Evans, A.   149-150, 151-152, 297-298, 397-398

Fairclough, J.H.   455-462
Feast, M.W.   339-348
Feitzinger, J.V.   375-376
Forrest, W.J.   315-317
Fricke, K.J.   441-442
Frogel, J.A.   349-356

Gautier, T.N.   49- 54
Gemund, H.-P.   29- 30
Gerin, M.   321-322
Ghosh, S.K.   305-306
Gillett, F.C.   61- 69
Glass, I.S.   487-488
Graham, J.R.   397-398
Greenberg, J.M.   177-188, 191-196, 225-228, 229-232
Greidanus, H.   47- 48, 155-158,
Grim, R.J.A.   225-228

Habing, H.J.   105, 325-326, 329-337
Harris, S.   101-102
Hauser, M.G.   39- 44
Hawarden, T.G.   455-462
Heasley, J.N.   463-466
Helou, G.   405-413
Herman, J.   103
Hill, G.J.   463-466
Hirst, C.J.   299-300
Hofmann, R.   189-190
Houck, J.R.   39- 44
Hough, J.H.   197-199
Hrivnak   127-128

Israel, F.P.   383-396, 399-400
Iyengar, K.V.K.   305-306, 437-438

Jennings, R.E.   299-300
Jenniskens, P.M.M.   325-326
Jongeneelen, A.A.W.   357-371
Joseph, R.D.   447-454, 455-462, 507-508

Jurriens, T.A.   87- 89

Kalafi, M.   23- 27, 425-434
Kirrane, T.M.   197-199
Knude, J.   55- 57
Kollatschny, W.   441-442
Koornneef, J.   399-400
Kreysa,, E.   29- 30
Krugel, E.   29- 30, 201-202,
Kunkle, T.D.   245-246
Kwok, S.   127-128

Lamers, H.J.G.L.M.   79- 82
Larson, H.P.   189-190
Leene, A.   142-143
Leger, A.   237-240
Leggett, S.K.   23- 27,   71- 73, 425-434, 455-462
Levasseur-Regourd, A.C.   45- 46
Lock, A.   101-102
Longair, M.S.   511-521

Maddalena, R.J.   293-294
Magnani, L.   377-378
Marsden, P.L.   91- 92, 153
Mathis, J.S.   171-176
McGillivray, H.T.   23- 27, 425-434
McKenty, J.W.   463-466
Meikle W.P.S.   397-398
Menten, C.   295-296
Menzies, J.W.   129-130
Mezger, P.G.   29- 30, 201-202
Miley, G.K.   471-486
Milone, E.F.   127-128
Moneti, A.   315-317
Mountain, C.M.   455-462
Myers, P.C.   307-312

Norris, R.P.   489-490

Odenwald, S.F.   75- 76
Olnon, F.M.   31- 38,   87- 89
Olofsson, G.   209-212

Papoular, R.   111-112
Pedlar, A.   443-444
Pegourie, B.   111-112
Perault, M.   203-207
Perrier, C.   221-224
Pipher, J.L.   315-317
Pottasch, S.R.   131-142, 143-144

Reipurth, B.   295-296
Puget, J.L.   203-207
Rengarajan, T.N.   305-306, 437-438
Robertson, N.A.   507-508
Robinson, E.L.   145-147
Roche, P.F.   435-436, 467-470, 489-490
Rouse, M.F.   197-199
Rowan-Robinson, M.   101-102, 303-304, 421-424, 499-506

Sandell, G.   295-296
Savage, A.   23- 27, 425-434
Schutte, W.   229-232
Schwartz, P.R.   301-302
Schwering, P.B.W.   383-396,401-404
Sellgren, K.   217-220, 261-262
Shane,, W.W.   445-446
Smith, C.   435-436
Smith, M.D.   319-320
Stephens, J.R.   245-246
Strazulla, G.   253-260
Strom, R.G.   47- 48, 155-158
Strong, I.B.   245-246
Stuwe, J.A.   375-376

Tandon, S.N.   305-306
Te Lintel, P.   105, 325-326,
Thaddeus, P.   293-294
The, P.S.   87- 89, 90

Ungerechts, H.   295-296
Unger, S.W.   443-444

Van de Hulst, H.C.   161-169
Van der Hucht, K.A.   87- 89, 90
Van der Laan, H.   47- 48, 155-158
Van der Veen, W.E.C.J.   105, 107-108
Van der Zwet, G.P.   233-236
Verma, R.P.   305-306, 437-438
Veron-Cetty, M.P.   493-495

Wakker, B.P.   379-380
Walker, D.W.   101-102
Walker, H.J.   91- 92, 357-371, 373-374
Walmsley, C.M.   295-296, 313-314
Walterbos, R.A.M.   401-404
Waters, L.B.F.M.   77- 78, 83- 85
Werner, M.W.   261-262
Whitelock, P.A.   129-130, 323-324
Whittet, D.C.B.   197-199, 297-298
Wiertz, M.   105
Wilkinson, P.N.   443-444